T0142345

Lecture Notes in Networks and Systems

Volume 62

Series editor

Janusz Kacprzyk, Polish Academy of Sciences, Warsaw, Poland
e-mail: kacprzyk@ibspan.waw.pl

The series "Lecture Notes in Networks and Systems" publishes the latest developments in Networks and Systems—quickly, informally and with high quality. Original research reported in proceedings and post-proceedings represents the core of LNNS.

Volumes published in LNNS embrace all aspects and subfields of, as well as new challenges in, Networks and Systems.

The series contains proceedings and edited volumes in systems and networks, spanning the areas of Cyber-Physical Systems, Autonomous Systems, Sensor Networks, Control Systems, Energy Systems, Automotive Systems, Biological Systems, Vehicular Networking and Connected Vehicles, Aerospace Systems, Automation, Manufacturing, Smart Grids, Nonlinear Systems, Power Systems, Robotics, Social Systems, Economic Systems and other. Of particular value to both the contributors and the readership are the short publication timeframe and the world-wide distribution and exposure which enable both a wide and rapid dissemination of research output.

The series covers the theory, applications, and perspectives on the state of the art and future developments relevant to systems and networks, decision making, control, complex processes and related areas, as embedded in the fields of interdisciplinary and applied sciences, engineering, computer science, physics, economics, social, and life sciences, as well as the paradigms and methodologies behind them.

Advisory Board

More information about this series at http://www.springer.com/series/15179

Mustapha Hatti

Editor

Renewable Energy for Smart and Sustainable Cities

Artificial Intelligence in Renewable
Energetic Systems

 Springer

Editor
Mustapha Hatti
EPST-CDER
Unité de Développement des Equipements
Solaires
Bou-Ismail, Algeria

ISSN 2367-3370   ISSN 2367-3389 (electronic)
Lecture Notes in Networks and Systems
ISBN 978-3-030-04788-7   ISBN 978-3-030-04789-4   (eBook)
https://doi.org/10.1007/978-3-030-04789-4

Library of Congress Control Number: 2018962148

This Springer imprint is published by the registered company Springer Nature Switzerland AG
The registered company address is: Gewerbestrasse 11, 6330 Cham, Switzerland

Short Preface

The energy transition so coveted by the entire world would be effective only through the increased integration of renewable energies in the advent of the concept of the smart city. This book aims to showcase trends in deployed technologies and research niches in the context of the considerable development of information and communication technologies at both the domestic and urban levels. The smart city makes it possible to efficiently and sustainably use all the renewable energy resources available to it in order to generate an added value in the proposed service and/or a reduction of the costs for the citizens while adopting a decentralized and open dimension to all citizens. The end goal of a smart city promises to improve the quality of life of all citizens in the city and in the countryside, in a sustainable way and respectful of the environment. Renewable energies are an indispensable answer in the supply of a smart city. From an environmental and technological point of view, renewable energies allow an optimal supply of the electricity network while emitting little or no pollution. Renewable energy innovations subject to the use of artificial intelligence technologies are more than likely to bring multiple benefits to the deployment and growth of smart, sustainable cities. Artificial intelligence is as useful for energy suppliers as it is for consumers. The latter can know in real time the price of electricity and adapt according to their consumption. The use of artificial intelligence today makes it possible to improve decentralized energy management by optimizing flows. The authors of this book have sought to clarify the issue of renewable energy related to the development of information and communication technologies, especially to the Industrial Internet of Things (IIoT) which is becoming increasingly important.

Short Biography

Dr. Mustapha Hatti was born in El Asnam (Chlef), Algeria. He studied at El Khaldounia school, then at El Wancharissi high school and obtained his electronics engineering diplomat from USTHB Algiers, and his post-graduation studies at USTO, Oran. He worked as research engineer, at CDSE, Ain oussera, Djelfa, CRD, Sonatrach, Hassi messaoud, CRNB, Birine, Djelfa, and senior scientist at UDES/EPST-CDER, Bou Ismail, Tipasa. He leads the "Tipasa Smart City" initiative and is an IEEE senior member, he is the author of several scientific papers and chapter books, and his areas of interest are smart sustainable energy systems, innovative, fuel cell, photovoltaic, optimization, and intelligent embedded systems.

Contents

Innovative Renewables

Inverters-Converters and Power Quality

Artificial Intelligence

The Approach Used for Comfortable and High Efficient Buildings in Sustainable Cities by Different Actors

Ouahiba Tizouiar(✉) and Aicha Boussoualim

Architecture and Environment Laboratory (LAE),
Polytechnique School of Architecture and Urbanisme (EPAU), Algiers, Algeria
{o.tiziouar, a.boussoualim}@epau-alger.edu.dz,
otizouiar@yahoo.com, a_boussoualim@hotmail.com

Abstract. In the aim of developing new methods to ensure construction quality and improve the living conditions of inhabitants. Awareness of extent for innovative techniques, tools and building materials is needed. This in view of their integration into new projects with an eco-construction spirit that leads to buildings and neighbourhoods that are less polluting and that can serve sustainable development. Relationship between the sustainable and comfortable built environment with intelligent lifestyle, can be ensured through the adopted technologies, but also the passive architectural devices and materials used.

In this sense, we verify in this paper if conceptors ensure comfort and energetic efficiency for their projects, this is by using two surveys. The first survey is intended for researchers to get their opinions about the question of environmental approach in architecture. These opinions are grouped and synthesized, then completed by another survey conducted with some actors of building such as designers, engineers, promoters and owners of projects, on the performance of their buildings and the environmental approach followed by them.

It is then the impact of a good mastery in this field by doing trainings and providing materials resources, this about intelligence on the building's energy consumption that has to do with sustainable building and also to ensure both comfortable and sustainable living without negative consequences on the energy performance and the environmental quality of the area.

Keywords: High performance building · Comfort · Energy efficiency
Environmental quality · Renewable energies

1 Introduction

In Horizon 2040, the Mediterranean region will face a threefold challenge: demographics with a minimum doubling of the urban population. Energetic due to an increase in energy demand of nearly 60%. And climate change due to an increase in greenhouse gas emissions in the order of 50% [1]. Thus, and like the countries of this region, Algeria should have a showcase in the field of green building technology, environmental preservation, sustainable development and scientific research applied to the construction and energy saving.

© Springer Nature Switzerland AG 2019
M. Hatti (Ed.): ICAIRES 2018, LNNS 62, pp. 3–13, 2019.
https://doi.org/10.1007/978-3-030-04789-4_1

In this context, we are helping to change the energy-saving uses in building sector, which is considered to be the most energy-consuming sector. This is through a process of design and environmental-conscious realization, by optimally exploiting natural resources, modifying building materials and replacing gas and electricity with solar energy. We are also interested about concepts related to bio-climate, environmental quality, intelligent building, with their management. More precisely also the approach, the culmination and success of efficient buildings. A project that deserves a search for control and approach to design, realization and exploitation.

In addition, efficient and intelligent buildings are both a key to sustainable development, and also a challenge for the architect, the engineer and the planner; who must bring to users the conditions of hygiene and comfort required, while saving energy and preserving the environment [2–4].

2   Problematic

The main problems that arise about the subject studied are the following:

«How do the different building actors work to create an energy efficient and environmental quality projects in sustainable cities? What is their level of master about environmental process and what are the constraints encountered.»

3   Methodological Approach

We studied the subject by adopting the method of surveys and interviews [5]. The first survey is launched on an online discussion platform (researchGate.net), and intended for researchers to get their opinions about the question of environmental approach in architecture. These opinions are grouped, synthesized, then completed by another survey conducted with some actors of building, such as designers, engineers, promoters and owners of projects, on the performance of their buildings and environmental approach followed by them.

From a set of answers recovered from these interviewers; they have been grouped, structured and analyzed in order to estimate the current state of involvement of these actors in taking into account the environmental and energy aspects in development and implementation of their projects. Questions asked revolve around three main themes; first the environmental and bioclimatic approach, second the energy aspect of the buildings, and third the tools for realization of such objectives. Namely the definition of environmental approach and its consideration in the project implementation process, and this in design, implementation and operation of indoor and outdoor spaces. The method used and by which projects are bioclimatic and take advantage of natural factors such as sun, light, wind, humidity… etc. The level of staffing buildings by intelligent systems for the comfort control, safety, management and energy consumption… etc. as well as the attempts and constraints encountered in integrating renewable energies into their buildings. Also on the use of BIM (Building Information Modeling) including the adoption of its sixth dimension related to energy analysis and management of pluridisciplinarity through this tool. In addition, an open question was asked about the prospects and expectations for uninsured performance that are usually expected to materialize in future projects. The choice of questions asked do not concern a single typology of buildings. This in order to define a global approach and not specific or contextualized recommendations (Fig. 1).

Fig. 1. First survey launched on (researchGate.net) and intended for researchers.

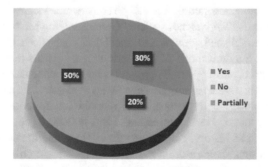

Fig. 2. Appreciation about answers related to the environmental approach

Second survey conducted with some building actors about performance of their projects and environmental approach.

4   Results and Discussion

4.1   Advice of Certain Researchers on Questions Launched on (researchGate.net)

The questions revolve around the following issues:

"What is the environmental approach? How is it taken into account in the process of realization of architectural projects?"

"Do you think that current projects are performing? If not, what are the performances that are lacking and which must be ensured in the future?"

"How is pluridisciplinarity managed through BIM? And how is the 6th dimension related to energy analysis addressed?"

Answers to these questions asked to researchers through the researchGate.net platform can be summarized as follows:

Some researchers define environmental approach as minimizing negative environmental impact and maximizing functionality to respond to people's needs, they are two major principle that sustainable and environment friendly architecture should follow. It is crucial to take into account resource efficiency (using less to produce more), minimizing GHG emissions through careful life-cycle assessment in all stages of architectural projects (from initial design to demolishing a structure). Also It depends to the integration of sustainable elements into the building design. So, they recommended to refer to those passive design strategies stipulated in the green building rating tools, such as LEEDS, BREEAM, GBI, Green Mark etc. Passive design strategies such as thermal mass, external shading, building orientation, cross ventilation, and better insulation in buildings, are the key element of sustainable building. they find that It is not easy to answer because environmental has its potency to be a hazardous for architecture. So that, environmental approach is limited by current conditions. we cannot understand how environmental change in the future.

Others researchers consider that It is very much depends on the context - there are many integrated approaches/models that can be considered. Meanwhile and in addition to established "scientific" approaches such as "Green Building Methods", "LEED", etc., they has supported established, proven, traditional design practices. One of the criticisms of Modernism as a "modern mode of conception" has been its dissociation and non-adaptation with nature, and beyond its principles and components. Humans have considered most natural resources as infinite and always available. The sense of "depletion" of resources reintroduces the retrospection of our old and common-sensual practices. Thus, they consider that one can begin to look at the holistic aspects of sustainability in the environment such as history, culture, people, ecology and the economy.

According to others, Green Building Rating Systems are a good place to start. But they think that these need modification, especially when applied to architectural projects in developing countries. They do not evaluate the social, economic and environmental impacts on existing communities. In such contexts, they believe that being "green" is not enough and that environmental considerations should be integrated with socio-economic and cultural considerations. For example, can a "net positive energy building" be used as a mini power plant, so that it can share energy with neighboring communities? Can the design use local materials and involve the local community in its design, construction and ongoing management, thus creating more jobs. Can passive heating and cooling be used without the need for expensive HVAC equipment and the problem of its maintenance and replacement?. So, they prefers to approach the issue holistically. The areas of sustainability are important in any environmental approach to architecture. The importance may depend on how we view and theorize the built environment. The point of view of the researcher plays an important role here. He would like to stress the rhetoric of Sitte (1965): "Is it enough then to place this mechanical project, designed to adapt to any situation, in the middle of an empty place without organic relationship with its environment or with the dimensions of a particular building?".

In the complexity thinking the main goal is to take consciousness about everything that surrounds the object of research, in this case an architectural project. Therefore, as has been said in the previous answers, it is kind of a speciality in modern architecture, but they would recommend to look into the history and theory of architecture because all the mayor parts of those designs have two main goals: (1) respond to the characteristics of the site (weather, soil, topography, etc.) and (2) be a symbolic reference for the people who live around. Nevertheless, at the 19th century the mechanical and dynamic development grave to the world the technic and technological progress we all know for the construction of contemporary architecture, but the sensibility of caring about the environment as part of the design was getting lost in the way until the 70's when it was taken back, but it was seen more as an obstacle. Now, these days are more aware of it, but in some cases sustainability it's presented more like a shopping list of machines and certain materials.

4.2　Interpretation of the Answers to Interviews Conducted with Some Actors of Buildings on the Performance of Their Projects and Environmental Approach Followed

First, for the questions related to environmental and bioclimatic approach, answers are summarized as follows:

"*Question 01*: What is the environmental approach for you? Can you define how is it taken into account in realization process of your projects, during design, realization and exploitation (interior and exterior spaces are concerned)?"

(30%) of interviewers give a positive answer (yes), they consider that the environmental approach is taken into account in realization process of their projects, whether it is outside by the control of their impacts on environment and by offering green spaces, or inside by creation of a healthy, comfortable and energy efficiency interior environment.

(50%) of them say that they cannot speak of a 100% environmental approach (partially), but they try every time to reduce the environmental impact, especially during the implementation phase. For example, the management of site's waste and the reduction of nuisances is of major importance described and respected in specifications of load notebook's projects. For exploitation, the environmental approach is practically absent. However, it has been noted that some building owners require instructions regarding energy management, especially in the home, (choice of plaster, types of joinery, etc.). Others complete by saying that it is respected only in the design, for realization the environmental approach remains theoretical. Their environmental approach is essentially based on integration into the immediate environment, by introducing spaces developed from private to collective, such as relaxation areas, sports, parking, green spaces setting.

The rest of interviewers (20%) give a negative answer (no). The environmental approach for them is almost absent in the process of realization of their projects, and this is due mainly to the requests of customer who gives importance to quantity and not to spaces environmental quality (as is the case for several promoters). On one hand, it means that we do not take into account the location of rooms as well as the dimensions of windows that were never made in relation to the dimensions of rooms but the most important it is that each room has a window for lighting and ventilation. On the other hand, even if the environmental dimension is taken into account in various stages of studies, it will be neglected during the realization by the entrepreneurs, this for the purpose of economy and sometimes lack of knowledge and expertise. And so, very little importance is given to the question of energy and its consumption, especially with the arrival of the economic crisis and particularly that of housing (Fig. 2).

≪Question 02: How your projects are bioclimatic and take advantage of natural factors (sun, light, wind, humidity…)?≫

Regarding to the bioclimatic aspect [6], (62.5%) of the interviewers give a positive answer (yes), they consider that the architecture of their projects is bioclimatic, HQE high environmental quality and that outdoor landscaping and location of their projects allows that. The consideration of natural factors is very important to ensure comfort and well-being of the customers, this is achieved by the orientation of buildings to have the most of the sun, also by the type, position and dimensions openings, overhangs, as well as the treatment of the envelope to ensure thermal and acoustic comfort. They take then into account these factors during design and during realization. So, the choice of openings, their size and layout are managed in relation to the Sun's trajectory and urban easements. Also, architectural features are set up to protect against prevailing winds, moisture management especially indoors is well thought out by separations in double bulkheads and waterproof coatings. Therefore their approaches offer bioclimatic projects, this by the establishment and the orientation of buildings in relation to the site and its specific bioclimatic and creation of intermediate spaces, this to ensure comfort and economy.

(25%) of them take into account the approach partially, their projects are not studied in relation to the factors (wind and humidity), on the other hand the sun and light are considered only for lighting and aeration of the parts. They think that for bioclimatic architecture most architects keep a definition of school without really grasping it.

(12.5%) of the interviewers found that nowadays we do not talk about architecture, we talk about construction and therefore this field of bioclimatic architecture is neglected in most cases by public authorities. The big worries is to house as many people as soon as possible (Fig. 3).

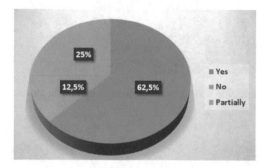

Fig. 3. Appreciation about answers related to bioclimatic architecture

Second, for the questions related to energy aspect of buildings, answers are summarized as follows:

≪Question 03: Are your buildings equipped with intelligent systems for the control of comfort, safety, management and energy consumption… etc.? If so which ones?≫

(12.5%) of the interviewers consider that intelligent systems [7] concern the decrease of energy consumption by the establishment of photovoltaic solar panels on the roof of buildings, the setting up of heat pumps which exploits in large part of the ground energy, then the heating or air conditioning equipment will only have to provide a small Delta difference to ensure a better energy comfort. As well as by insulation of the buildings preferably by the outside to reduce the losses by thermal bridges, adequate insulation according to the thermal regulations.

(62.5%) have projects whose intelligence is only limited to smoke detectors and CO, to certain systems for the control of comfort and safety such as remote controlled gates and gates, remote monitoring system in all common areas, digicode and videophone, car parks with smoke and fire system safety, fire doors and semi-armoured entrance door. The interior comfort is ensured by the air conditioning and central heating, thus staffing of the buildings by these systems remains partial and to be completed by studies of energy saving and above all the impact on environment.

In some replies, it is specified that only equipments such as hotels, administrative buildings and hospitals that are equipped with intelligent systems, for control of access, management of fire networks, air conditioning… etc.

(25%) find that these systems require an interesting financial envelope, and that despite they offer them to the customers, they refuse to install them because they do not manage to cover all these Charges. Architecture of intelligent buildings has not yet been born in Algeria and this is due to several parameters such as the need of a certain knowledge bases for manoeuvers on the techniques of implementation as well as the proceeds of understanding and functioning (Fig. 4).

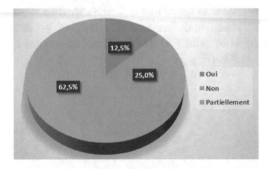

Fig. 4. Appreciation about answers related to intelligent systems in building

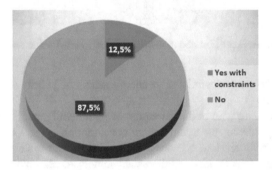

Fig. 5. Appreciation about answers related to the use of renewable energies into building

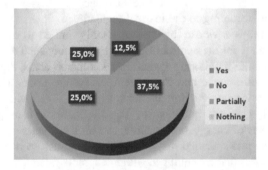

Fig. 6. Appreciation about answers related to the use of tools for energy analysis

≪Question 04: are there any attempts to integrate renewable energies into your buildings? If (no) what are the constraints for this use?≫

(12.5%) integrate renewable energies, except that the costs of the facilities remain high and the depreciation periods relating to this investments are lengthened.

(87.5%) made no attempt to integrate renewable energies into their buildings, because the integration of renewable energy is initially very costly, in addition it requires and a very developed techniques and knowledge.

So generally their integration is not used very much in projects, because even if they advise customers of the importance of these systems of energy saving, customers do not have culture and awareness in the matter. On the other hand they are expensive on the market. Also, the construction remains governed by the need in terms of number of dwellings and equipment.

They consider that renewable energies is phantom sector in our country, how can we integrate it without control and lack of appropriations allocated by manoeuvers of projects (Fig. 5).

Third, for the questions related to environmental integration tools, answers are summarized as follows:

≪Question 05: are your projects being done using BIM (Building Information modeling)? If so, how do you manage the pluridisciplinarity through this tool? And how do you approach its 6th dimension related to energy analysis?≫

(12.5%) are interested in the energetic analysis of buildings during the development of the studies, using computer tools. (37.5%) use them only for certain types of buildings, especially sanitary equipment. On the other hand (50%) operate as traditional multidisciplinary consultancies reinforced by a network of consultants.

Fourth, for the questions about the prospects and expectations for the performance to be ensured in their future projects, answers are summarized as follows.

≪Question 06: Do you think your projects are performing? If so, which and how else are the performances lacking that you intend to ensure in your future projects?≫

(12.5%) of the interlocutors confirm that their projects are energy efficient, bioclimatic (passive buildings) and that they are artificially intelligent buildings and that the thermal aspect is always taken into account (calculation Automatic temperature differences between the interior and the outside.

(25%) try to design with environment in accordance with the request of the owner, and propose projects that ensure good acoustic, visual and thermal comfort, thanks to the insulation (walls, floors, roof, windows in double glazing… etc.). Good ventilation which allows to regulate the humidity and the energy performance thanks to the solar inputs in heat and light.

However, They find that there is a lack in relation to the improvement of energy performance to minimize the energy used for heating and air conditioning (for example by the use of local materials that suit the climatic conditions of each region, or the improvement of insulation).

(37.5%) say that projects are studied functional, structural, performance and aesthetic but lacks about thermal, environmental and energy performance above all renewable energies and sustainable development that are not respected and supported when designing and carrying out projects. They think that it is unfortunate to learn during the study course the importance of these systems of energy saving and the environmental dimension in the building but in the professional field we are faced with several constraints whether it is the demands of the customers, the regulations and the requirements of the various organizations that require us to neglect this dimension.

The question for (25%) of the interviewers is not understood, they speak of another thing which concern rather the optimization of spaces, quality, cost and delays and the incoherent development policy (Fig. 6).

5 General Conclusion

Energy efficiency of buildings is an important lever in the fight against global warming; because this concept has become an indispensable step in the efficiency of construction worldwide. Developing countries, particularly Algeria, need to put a special emphasis on this concept and integrate various passive and technological techniques at the level of all these sectors, particularly housing.

Indeed, and according to the researchers interviewed in our study, being "green" is not enough and environmental considerations should be integrated with socio-economic and cultural considerations. Such as the sharing of energy, the use of local materials and involvement of the local community…

However, how can these considerations and these holistic aspects be taken in the case of realization of durable, efficient and environmental quality buildings on a virgin, isolated site, without any existing fabric nor history aside from its landscape, mening a new city who wants to be smart and sustainable?

And that through the environmental approach that must take into account all these considerations, how can the projects be efficient and what are the priority performances to ensure, during the conception, realization and exploitation that it is for the interior and external spaces?

For Algeria, the initial causes of the flagrant lack about efficient and intelligent buildings is that we remain in the initial stage; and few studies on these architectural devices and implementation techniques exist. This is certainly due to the various deficiencies and difficulties encountered in the construction of this type of building;

The major problem cited by the architects interviewed was the question of need in terms of housing and equipment (construction is governed by need). Nowadays we do not talk about architecture, we talk about construction and so this field of home automation, performance and renewable energy seems to be neglected. In addition, it requires a certain knowledge base in terms of manoeuvers on the techniques of implementation as well as on the understanding and operation procedures. Adding to this, these buildings are subject to enormous costs and conditioned by the will of owners, which until now does not give it a great importance.

So this concept remains in the initial stage of application, and our country is far from completely applying this kind of concept on all types of construction including the residential sector; and especially with the arrival of the economic crisis and particularly that of housing.

As prospects and in the future, we will contribute to propose a new approach for a project case that wants to be sustainable, efficient and environmental quality at the level of an existing site.

References

1. Algiers Energy Transition Congress, 4th MEDENER, International Conference (25 May 2016), Hotel El djazair, Algeria
2. Dutreix, A.: Bioclimatisme et performances énergétiques des bâtiments. Edition Eyrolles, Bd Saint Germain, Paris (2010)
3. Gonzalo, R., Habermann, K.: Architecture et efficacité énergétique: principes de conception et de construction. BIRKHAUSER Edition, Boston, Berlin (2006)
4. Liébard, L., De Herde, A.: Traité d'architecture et d'urbanisme bioclimatiques: concevoir, édifier et aménager avec le developpement durable. Observ'ER Edition (2005)
5. Evola, R.: Manuel d'enquête par questionnaire en Sciences sociales expérimentales. Publibook Edition (1999)
6. Tizouiar, O.: Disponibilité de l'éclairage naturel en milieu urbain dense: Investigation sur les performances de puits de lumière naturelle, publication in ligne, Editions universitaires européennes (2018). http://www.editions-ue.com/
7. Jeuland, F.-X.: Réussir son installation domotique et multimédia. 2nd Edition Eyrolles, Saint Germain, Paris (2008)

Multi-agent Approach to Analysis Data from Social Media for Building Smart Cities

Brahim Lejdel[✉]

El-Oued University, El Oued, Algeria
Brahim-lejdel@univ-eloued.dz

Abstract. With the objective of reducing costs and resource consumption in addition to more effectively and actively engaging with their citizens, smart cities can use multiple technologies to improve the performance of the business, health, transportation, energy, education, and electric and water services leading to higher levels of comfort of their citizens. One of the recent technologies that have a huge potential to enhance smart city services is social big data analytics. Also, one of the current main challenges in data mining related to big data problems is to find adequate approaches to analyzing huge amounts of data. As we know, a social media become an important part of everyday life of people and collected data has resulted in the accumulation of huge amounts of data that can be used in various beneficial application domains. Effective analysis and utilization of social big data is a key factor for success in many service domains of the smart cities. Thus, we think that for building smart cities, we have to hear what citizens talk in social media about parking, lighting, incidents, waste and many others. In this paper, we will review the utilization of social big data to build of smart cities. And, we will propose a smart model which permits to analyze huge data collected from the social networks to predict future events, as crime, incidents and public opinions for politic or business purposes. In addition, we will identify the requirements that support the implementation of big data applications for smart city services. Finally, we implement our smart model.

Keywords: Smart city applications · Data mining · Social big data
Social media

1 Introduction

The social network has created a special regional feeling of citizens and is helping to boost the local economy of any country. Thus, we think that social media represent the main tool to see the advantage and disadvantage of the application of such strategies in such city. Cities can use twitter, google+, and YouTube and Facebook platforms to improve the communications between the citizens and decision-makers, and also improve services, in the objective of creating smart city.

The social media can represent a convivial environment to listen to the insufficiencies which can found in such city. Much social media as Facebook or twitter use geo-tagged posts to locate the urban areas which represent valuable information for decision makers to define the location of such event. This information can be used by

© Springer Nature Switzerland AG 2019
M. Hatti (Ed.): ICAIRES 2018, LNNS 62, pp. 14–23, 2019.
https://doi.org/10.1007/978-3-030-04789-4_2

predictive models that enable decision support and should be a vital part of any smart city application.

Actually, smart cities use different technologies to gather data as sensors, Wi-Fi, Bluetooth, etc. Nevertheless, the price of these technologies is very expensive and their configuration is delicate. In other hand, we can confirm that smart cities are not the infrastructures or architectures it offers, but the ways in which its citizens interact with these systems as well as each other. Thus, we think that the big data generated from the social media can be used to build smart cities. We have to listen in real time, what are people saying for marketing, tourism, healthcare, incidents, education and waste management, in Facebook, Twitter, Google+, YouTube, etc.

The decision-makers can use the social networks to notify users in the different domain, business, health, water and electric service, energy, transportation, and education. For example, to facilitate the management of congestion on the road, we can notify drivers to change bus routes and we can modify traffic light sequences. We can also deliver information to drivers via a mobile application which indicates approximate driving times and giving alternative routes. also, decision makers can analyze social media in real-time using big data technologies, computational intelligence and data mining, which able us to measure public opinion on important issues and services such as business, transportation, waste management or security allowing us to define the politic strategies, clean the city and discover crimes in real-time.

Figure 1 shows the employment of social big data to build smart cities. Social big data generate huge amounts of date while big data systems use this data to provide information to enhance smart cities applications, as transportation, healthcare, energy, education, and business. The big data systems will store, process, and mine smart cities applications information in an efficient manner to produce information to enhance different smart city applications. In addition, the social big data will help decision-makers to plan for any expansion in smart city services, resources, or areas.

Citizens and their social relation must be integrated into smart cities applications because the citizens who live there represent an important part that must be served by the smart cities applications to achieve their maximum work. We think that a city cannot be smart without taking into account citizens factors and their social relations; the interaction between person and computer, the study of the user's experience. For example, we can measure the user's opinion about such product. We can also add the psychological, social, cultural variables and everything related to the human live. The intersection between social domain and knowledge engineering is now essential to understand the world in which we live. Thus, a conceptual model that combines big data application, behavioral sciences and social sciences with knowledge engineering will be the suitable area, which can support smart city application.

Today, society lives in a connected world in which communication networks are intertwined with daily life. For example, social networks are one of the most important sources of social big data; for example, Twitter generates over 400 million tweets every day [1]. In social networks, individuals interact with one another and provide information on their preferences and relationships, and these networks have become important tools for collective intelligence extraction. These connected networks can be represented using graphs, and network analytic methods [2] and can be applied to them different techniques for extracting useful knowledge.

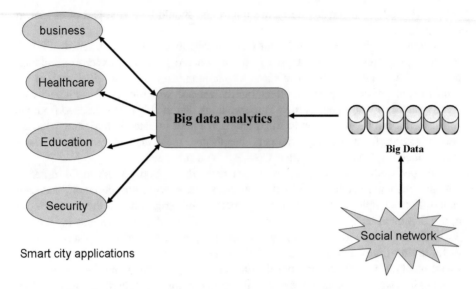

Fig. 1. The relation between social big data and smart city applications.

Figure 2 shows the Conceptual Representation of the Three Basic smart city applications: social media as a natural source for smart city applications; big data as a parallel and a massive processing paradigm; and city services, as transportation, education, tourism, healthcare, and marketing.

In this paper, we explore the different works related to social big data analytics. In Sect. 3, we propose a conceptual model for social big data and smart cities. Then, we discuss the different applications of social big data to support smart cities. In Sect. 4, we explore the challenges and benefits of incorporating social big data applications for smart cities. Finally, we present a conclusion.

2   Related Works

Social Big data is certainly enriching our experiences of how to build smart cities, and it is offering many new opportunities for social interaction between citizens and more informed decision-making with respect to our knowledge of how best to interact in cities. The methods of big data analytics can be applied to social big data for discovering relevant knowledge that can be used to improve the decision making in the smart city. Smart city applications include methodologies that can be applied in different areas such as shopping, healthcare, security, and many others; more recent methodologies have been applied to treat social big data. This section provides short descriptions of some works of these methodologies in domains that intensively use social big data.

In [2], the authors propose to use graphs and network analytic methods for extracting useful knowledge from big data of social network, because in social

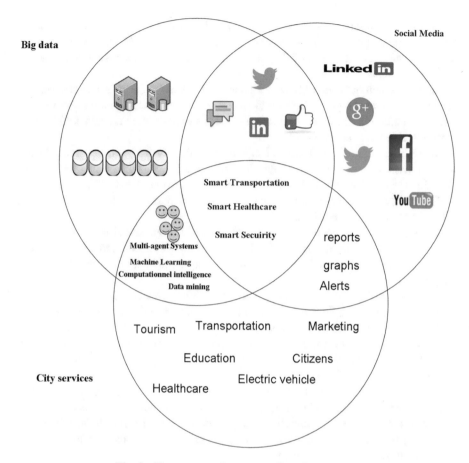

Fig. 2. The conceptual representation of our model.

networks, individuals interact with one another and provide information on their preferences and relationships. They present these connected networks using graph.

In [3], the authors present three models, along with three algorithms for selecting the best individuals to receive marketing samples. These models can diffuse both positive and negative comments on products or brands in order to simulate the real opinions within social networks. Moreover, the author's complexity analysis shows that the model is also scalable to large social networks.

In [4], the authors propose to use a graph-based dataset representation that allows extracting patterns from heterogeneous a real aggregated dataset and visualizing the resulting patterns efficiently. They present a crime data analysis technique that allows for discovering co-distribution patterns between large, aggregated and heterogeneous datasets. In this approach, aggregated datasets are modeled as graphs that store the geospatial distribution of crime within given regions, and then these graphs are used to discover datasets that show similar geospatial distribution characteristics.

In [5], the authors present an approach which using spatiotemporal tagged tweets for crime prediction. He uses Twitter as social networks. He applies a linguistic analysis and statistical topic modeling to automatically identify discussion topics across a city in the United States.

In [6], the authors explore how "Facebook ads" placed on users social streams that have been generated by the Facebook tools and applications can increase the number of visitors. In addition, the authors present an analysis of real-time measures to detect the most valuable users on Facebook.

In [7], the authors use document classifier to identify relevant messages. In this work, Twitter messages related to the flu were gathered, and then a number of classification systems based on different regression models to correlate these messages with CDC statistics were compared; the study found that the best model had a correlation of 0.78 (simple model regression).

In [8], the authors apply mining techniques for user-generated content, generally relies on a descriptive analysis derived by analyzing keywords and link structures. They track the mood of the blogosphere by tracking frequencies of terms like tired or happy on the web.

In [9], the authors propose to use the online spherical k-means algorithm, which is a segment-wise approach that was proposed for streaming data clustering. This technique splits the incoming text stream into small segments that can be processed effectively in memory. Then, a set of k-means iterations is applied to each segment in order to cluster them. This means that it is necessary to perform data mining tasks online and only one time as the data come in.

In [10], the authors use Cloud computing to process a large amount of data. It uses data mining techniques, as clustering to analysis social big data, which generated from the social network. Cloud computing can provide the virtual infrastructure for utility computing that integrates monitoring devices, storage devices, analytics tools, visualization platforms and client delivery.

3   Proposed Model for Social Big Data and Smart Cities

The huge amount of data which collected from social networks can play an important role in developing a smart model for cities. Uncovering hidden patterns, correlations, and other insights from large amounts of social data can enable decision-makers to improve citizens' life. The analytics of very large of data collected from the social networks can assist in acquiring knowledge to predict future events, as crime, incidents and public opinions for politic or business purposes, analytic methods can help in strategically placing an advertisement, enabling people to make a valuable decision in terms of understanding customers and products, and can help in identifying the potential risks and opportunities for a company. Moreover, analytic methods can help enterprises make smart strategies after analyzing employee data from social networks. Analyzing the products that people search for and buy can help business owners increase their income by satisfying the demands of the customers based on their needs. After analyzing the datasets of customers in social networks, companies can analyze the products that lead to revenue loss. Also, companies can analyze the huge amounts of data collected from social networks to get out hypotheses which can be proposed to be used later when experimentally verified.

We know that social media contain millions of comments about products, politic, education, incidents and many others. Data can be gathered from multiple sources and are stored in multiple databases. Such data can be used by the smart city applications and big data analytic methods to predict future behaviour in transportation, healthcare, education, crime, security and many others. The outcomes of the analysis can be shown in a form of reports, graphs or alerts. The data mining techniques can play a great role in extracting knowledge from the unstructured big data. The ideal conceptual model requires a security model throughout the processes and examines security issues from a system perspective to provide secure value to an organization.

Figure 3 represents the conceptual model for social big data and smart cities.

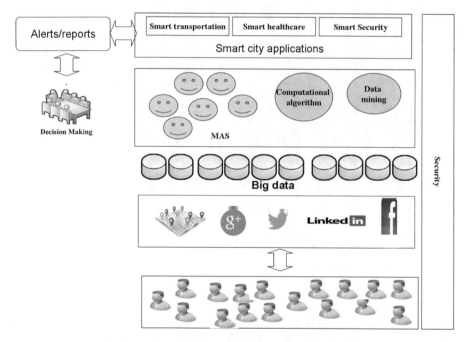

Fig. 3. The conceptual model for social big data and smart cities.

4   Experimentation and Results

In this experiment, we use principally open public data inspired by Facebook. Such data can be accessed by two types of users, customers, and decision-makers. We use, more than 100 databases were made available that cover a wide range of El-Oued regions, such as transport, business, security, news, health, and accommodation. To implement our agent we use Jade (http://jade.tilab.com) platform and we use Java (https://www.java.com/) to implement the different algorithms and functions.

In the implementation, we use two experiments. These experiments are

Experiment 01: In this experiment, one agent was used to control and optimize the different extractions, analysis, and visualization of data.

Experiment 02: In this experiment, we use Multi-Agents, which assigned to the business, security, and disaster management, news and accommodation. Each agent is used to control and manage the extraction, the analysis and the visualization of its specific domain. These experiments permit us to evaluate the performance of our proposed approach MAS-K-means.

In Fig. 4, we show the distribution of document data according to the topic, in El-Oued region. We use a Facebook data of El-Oued region for 4 weeks.

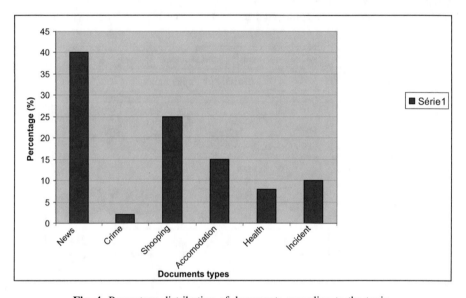

Fig. 4 Percentage distribution of documents according to the topic.

In Fig. 5, when we use the approach with MAS-K-means, each agent can be assigned to specific domains, and each agent can perform the k-means algorithm to carry out the classification of information from the big date. We observe that the time of treatment decreases according to the document size.

In Fig. 6, we present the classification ratio of documents of big data when we use multiple agents and when we use one agent to manage and control the extraction, analysis and visualization operations of knowledge from big data. Thus, when we multi-agent, when each agent can manage and control the extraction, the analysis and the visualization operations and each agent can perform the K-means algorithm to discover the optimal classification, we observe that the classification ratio increases according to document size. But, when we use one agent, we observe that the classi-fication ratio decreases according to document size.

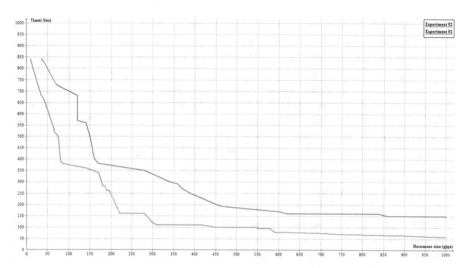

Fig. 5. Treatment time according to the size of the documents.

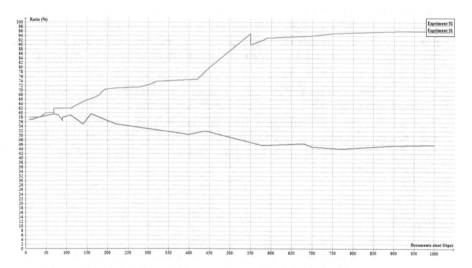

Fig. 6. Classification ratio according to documents size.

5   Conclusion and Future Works

In the last decade, social big data has become an important research topic because of the large number and rapid growth of social media systems and applications. The aim of this study has been to propose a conceptual model helping decision makers and customers to find the most relevant solutions that are currently available for extracting, managing, controlling, analysis and visualize knowledge in social media for better user experiences and services. In this paper, we present a framework for treating large volume social media content. This paper aims to offer a comprehensive view of the role of data mining and MAS (Muti-Agent Systems) for extracting meaningful knowledge from social big data. In this context, the conceptual model and architecture of the system with the aim of managing social big data were proposed, and the applications of data mining and MAS in which social big data analytics can play an important role were discussed. Different case studies were also outlined. Then, several open research challenges were explained. Finally, we conclude that social big data can play an important role in terms of gaining valuable information and for decision-making purposes. However, big data research in a social network is in its infancy and solution of the discussed challenges can make it a practical field of research.

This work can open multiple and varied topics of research, we can cite

- It appeared that the main advantage of K-means is that is easy to implement, but it has two big problems. Firstly, it can be very slow since, in each step, the Euclidean distance between each point to each cluster has to be calculated, which can be really expensive in term of time, especially that we use the large volume of datasets. Secondly, this method is really sensitive to the initial clusters, however, in this paper, we will improve the efficiently of this algorithm using the genetic algorithm or swarm optimization.

- The movement of people and objects at a large scale need to record such data about with the geo-location of these data, especially in the crime and tourism domain. Also, we need to integrate GIS application with our proposed model to analyse spatiotemporal data and solve spatiotemporal problems in social media. Thus, an Efficient GIS is critical to analysis and visualization application because it can provide interactive and easy-to-use platforms for the users; customers and decision makers. These platforms, however, call for the integration of 3D and touchscreen technologies to flow for example a criminal person. Such integration can enable policymakers to convert data into knowledge, which is critical in fast decision making [11]. The information extracted from the modeled data will be represented based on the user's need.

References

1. Bennett, S.: Twitter now seeing 400 million tweets per day, increased mobile ad revenue, says ceo (2012). URL http://www.adweek.com/socialtimes/twitter-400-million-tweets
2. Ott, L., Longnecker, M., Ott, R.L.: An Introduction to Statistical Methods and Data Analysis, vol. 511, Duxbury Pacific Grove, CA (2001)
3. Ma, H., Yang, H., Lyu, M.R., King, I.: Mining social networks using heat diffusion processes for marketing candidates selection. In: Proceedings of the 17th ACM Conference on Information and Knowledge Management, ACM, pp. 233–242 (2008)
4. Phillips, P., Lee, I.: Mining co-distribution patterns for large crime datasets. Exp. Syst. Appl. **39**(14), 11556–11563 (2012)
5. Gerber, M.S.: Predicting crime using Twitter and Kernel density estimation. Decis. Support Syst. **61**, 115–125 (2014)
6. Trattner, C., Kappe, F.: Social stream marketing on Facebook: a case study. Int. J. Soc. Hum. Comput. **2**(1–2), 86–103 (2013)

7. Culotta, A.: Towards detecting influenza epidemics by analyzing Twitter messages. In: Proceedings of the First Workshop on Social Media Analytics, ACM, pp. 115–122 (2010)
8. Mishne, G., Balog, K., de Rijke, M., Ernsting, B.: MoodViews: tracking and searching mood-annotated blog posts. In: International Conference on Weblogs and Social Media, Boulder, CO (2007)
9. Zhong, S.: Efficient online spherical k-means clustering. In: Proceedings of the 2005 IEEE International Joint Conference on Neural Networks, IJCNN'05, 5, IEEE, pp. 3180–3185 (2005)
10. Mell, P., Grance, T.: The NIST definition of cloud computing (2011). http://dx.doi.org/10.6028/NIST.SP.800-145
11. Ju, G., Cheng, M., Xiao, M., Xu, J., Pan, K., Wang, X., Shi, F.: Smart transportation between three phases through a stimulus-responsive functionally cooperating device. Adv. Mater. **25**(21), 2915–2919 (2013)

Parametrically Generated Building Designed in an Artificially Intelligent Environment

Abd el djalil Hamdaoui$^{(\boxtimes)}$ and Ismahane Haridi

Department of Architecture, Larbi Ben Mhidi University,
Oum El Bouaghi, Algeria
hamdaouihere@gmail.com, ismahane_haridi@hotmail.com

Abstract. Since the invention of computers and machines, Architects have continually pursued new and innovative processes to design intelligent buildings. Recently, many of these processes depend on artificial intelligence to produce smart designs. These processes are based on mathematical and algorithmic systems that implicitly follow a certain logic to come out with the final rendering production of the model. Moreover, the escalated evolution of digital technology in the field of parametric modelling and artificial intelligence is a massive advantage that could be integrated in architecture as a new way of optimizing the design of buildings, where the relationships between parametric elements are used in an artificially intelligent environment to manipulate a sustainable design. Instead of applying minor and repetitive changes, a predetermined algorithm will do the job saving a lot of time and effort.

Keywords: Parametric systems · Artificial intelligent · Sustainable design
Intelligent buildings

1 Introduction

Artificial intelligence is an interdisciplinary science and technology founded on many disciplines such as Computer Science, Biology, Psychology, Linguistics, Mathematics, and Engineering. According to the leader of Artificial Intelligence, John McCarthy, it is "The science and engineering of making intelligent machines, especially intelligent computer programs". Artificial Intelligence is a way that we can through it; make a computer, a computer-controlled machine, or a software that thinks intelligently, in a similar manner to how human intelligence works. AI is accomplished by studying how the human brain thinks and how humans learn, decide, and work in order to solve a problem, and then using the generated outcomes as a basis to develop intelligent solutions and systems to bypass a certain obstacle.

This approach can be assimilated in architectural design of intelligent buildings to give them a perception of modernity and to create interactive models susceptible to modification according to a set of provided conditions.

© Springer Nature Switzerland AG 2019
M. Hatti (Ed.): ICAIRES 2018, LNNS 62, pp. 24–30, 2019.
https://doi.org/10.1007/978-3-030-04789-4_3

2   The Philosophy of AI and Architecture

In the field of architecture, there are many designs and concepts that are created by imitating and modeling forms based on algorithms and equations (Fig. 1). Most of these processes are characterized by their innovative geometry, structure, and construction techniques, and have resulted in developments in many fields through groundbreaking, new and successful designs (Marble 2013). The implementations of the concept of parametrization as a method and artificial intelligence as a work environment in the field of architecture are mostly observed in the High-tech architectural design (Jabi 2013).

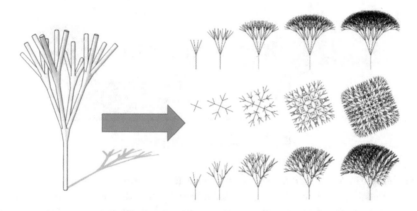

Fig. 1. A script that generates different iterations from a single form.

In the proposed study, besides those forms, structural behavior and the optimized response to internal and external agents, together with their geometrical configurations, have been studied to provide a methodology to understand logical and mathematical relationships to optimize structures and the design system in general.

3   Method

The modeling approach of the research (Fig. 2) is built around the idea of mathematical system that has been examined according to a set of parameters and developed to generate a mathematical model (Hahn 2012). A program has been written to generate the computational model of the selected system. Through a series of abstractions and assumptions, first mathematical then computational, a model of the actual concept has been obtained to explore the behavioral properties of its rotating structure.

Through this perception/thinking/designing/manufacturing method, an intelligent platform is formed to contain the new architectural concept.

As described by Howard Gardner, an American developmental psychologist, the Intelligence manifests in different forms: Linguistic intelligence, Musical intelligence…, this concept aims to make use of the logical-mathematical intelligence, which

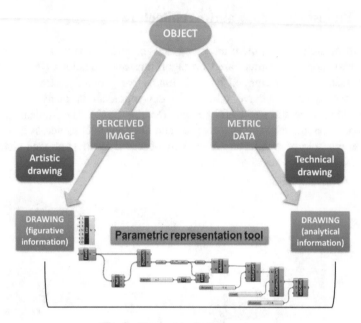

Fig. 2. Parametric design process.

is the ability to use and understand relationships in the absence of action, or objects. In addition to the spatial intelligence characterized by the ability to perceive visual or spatial information, change it, and re-create visual images without reference to the objects, construct 3D images, and to move and rotate them. This contribution consists of three essential steps (Fig. 3).

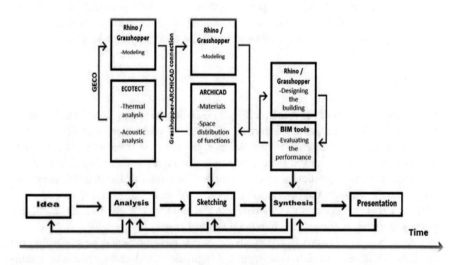

Fig. 3. The phases where different platforms are integrated into the virtual rendering process.

3.1 The Analysis Phase

The analysis phase of Rhino/Grasshopper and Ecotect, which can provide an analysis of the overall site conditions and the study parameters such as sunlight, shadows and weather condition. These entries can be integrated in the analysis phase because they focus mainly on identifying the elements of the site used in the sketching phase.

3.2 The Sketching Phase

In the sketching phase, Rhino/Grasshopper are integrated with Archicad and the study focuses on both the treatment of facades and internal parts and all the main aspects of the exoskeleton of the construction. The modeling with Rhino/Grasshopper and ARCHICAD in the sketching phase becomes more detailed and concerns the interior of the model, not only the form can be calculated and evaluated, but also and the different model functions and parts (Brian 2016).

3.3 The Synthesis Phase

Rhino/Grasshopper can be used with other BIM tools because the level of detail becomes so high that the calculation methods in ordinary tools are not precise enough, so we need to translate it into a percolated algorithm to bypass the intrication of form.

4 Results

The use of a parametric method (Fig. 4) of modeling and customization allows for an efficient design process to be realized (Woodbury 2010). A complex set of coupled parameters interact in a predetermined manner to generate a shape with different variations and outcomes that is the "best" possible in relation to the chosen parameters and the laws of the coupling (Benjamis 2001). The process of mass customization and modulation could be used in a way to design intelligent buildings paradigms by allowing a set of configurations to be simultaneously analyzed and solved according to the multitude of conditions determined previously by the designer like: the shape of the building, number and area of floors, structural system, etc. The nature of the architectural model provides a good opportunity to design systems that allow for diverse and interactive ideas that modify specific zones of the design, a complementary system allowing for a vast diversity in forms and functions. In this method, parametric tools can come out with systems where, a little bit of effort and time are invested into creating a much better model, a more sophisticated product of this way of design (Scheurer 2012). An algorithm is defined by connecting a parameter to a component input node via connecting wires. Then the component processes this data and it transfers it to its output node. In a fully connected parametric system, a line parameter input connects to a rotation component. A slider parameter is used to define the rotation angle of the model. Finally, a point parameter is used to define the center of rotation.

The new form is therefore the result of the parametric relationship between the two concepts (parametrization and intelligent design) for the purpose of creating a

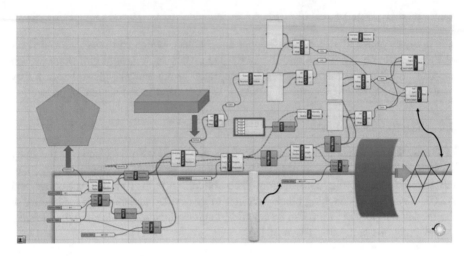

Fig. 4. Parametric input flow chart.

cooperative process where the two are mutually beneficial in both process of design and final result of this design method (Tuba et Benachir. 2011). The process of generative modeling is an effective technique for form exploration that consists of developing geometries based on parameters derived algorithms (Fig. 5) to create a prototype geometric profile for constructions (Kostas 2016). Moreover, to analyze the impact of the methods of parametric and generative modeling in the development of sustainable architecture.

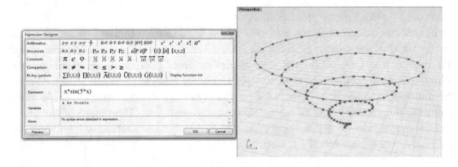

Fig. 5. Mathematical expression integrated in the parametric system.

The parameterization allows for a flexible design dotted with private garden terraces to provide oxygenation and sweeping views, balconies allow for expansion and induce wind turbulence to aid ventilation and curtain wall glazing offers clear and panoramic view. Concepts like organic tessellation (Fig. 6) and triangulation patterns are used to shape the envelope and to generate automatically openings and windows for each floor without the need to repeat the operation every time with the rest of the building floors (Daniel 2013).

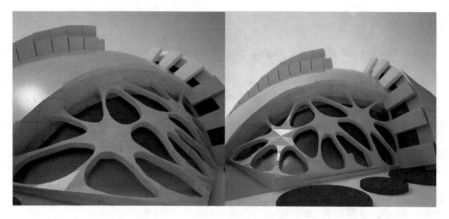

Fig. 6. Model with organic curtain wall realized parametrically.

The use of double skin facade system to give that transparency effect to the envelope and as buffer zone to optimize the heat distribution of the building. The double skin system is designed to possess an energy efficient benefit as well as to offer to designers a free space for aesthetic articulation (Yeang 2002). In most cases, the double skin also features shading systems and air exchange systems to help prevent glare and heat gain/loss. It has also improved sound insulation performance, and inner air quality (Yeang 1996). The use of custom curtain wall systems is due to the glass material and its detailing of connections that provide a high degree of flexibility and versatility when shaping it (Murray 2009). Heavy glass and other materials such as stone may require additional means of attachment to support their weight.

5   Conclusion

Parametric modeling tools and methods may offer some interesting potentials and opportunities in the field of architecture because there are often elements and features that are repeated frequently and always must adhere to certain pattern that can be translated parametrically (steps, ramps, handrails, etc.) (Jillian et Heike. 2016). Therefore, by utilizing this approach we convert architectural elements and a variety of building functions into controllable parameters, configured easily and efficiently throughout the design process. In this case, the model can create diverse and dynamic environment either internally or externally. This sustainable approach, in the form of a an intelligent building utilizing the parametric method with the artificial intelligent principles, will test, analyze, and modify the elements of the research to create durable solutions programmatically, functionally, and formally and to classify the complex propositions in multifold layers susceptible to modification.

References

Benjamis, D.: Beyond Efficiency. Digital Workflows in Architecture. MIT Press, Cambridge (2001)

Brian, R.: Design Computing: An overview of an Emergent Field. Routledge, Abingdon (2016)

Hahn, A.: Mathematical Excursions to the World's Great Buildings. Princeton University Press, New Jersey (2012)

Daniel, D.: Modelled on Software Engineering: Flexible Parametric Models in the Practice of Architecture. Routledge, Abingdon (2013)

Jabi, W.: Parametric Design for Architecture. Laurence King Publishing, London (2013)

Jillian, W., Heike, R.: Landscape Architecture and Digital Technology: Reconceptualising Design and Making. Routledge, Abingdon (2016)

Kostas, T.: Algorithmic Architecture. Routledge, Abingdon (2016)

Marble, S.: Digital Workflows in Architecture. Walter de gruyter, Berlin (2013)

Murray, S.: Contemporary Curtain Wall Architecture. Princeton Architectural Press, New York (2009)

Scheurer, F.: Digital Craftsmanship: From Thinking to Modeling to Building. Birhauser, Basel (2012)

Tuba, K., Benachir, M.: Distributed Intelligence in Design. Wiley, Cambridge (2011)

Woodbury, R.: Elements of Parametric Design. Routledge, Abingdon (2010)

Yeang, K.: Reinventing the Skyscraper: A Vertical Theory of Urban Design. Wiley, Chichester (2002)

Yeang, K.: The Skyscraper Bioclimatically Considered. Wiley, London (1996)

ESMRsc: Energy Aware and Stable Multipath Routing Protocol for Ad Hoc Networks in Smart City

Salah Eddine Benatia[1]([⊠]), Omar Smail[1], Meftah Boudjelal[1], and Bernard Cousin[2]

[1] Computer Science Department, Faculty of Exact Sciences, University Mustapha Stambouli, Mascara, Algeria
{se.benatia, o.smail, boudjelal.meftah}@univ-mascara.dz
[2] IRISA/University of Rennes 1, Rennes, France
bcousin@irisa.fr

Abstract. The mobile ad hoc wireless networks are characterized by the absence of central administration and any network element may be very mobile. There is no fixe element within an ad hoc network. In fact, within these networks, all elements must cooperate so as to create a temporary topology which facilitates communication. To create this topology and carry data, ad hoc networks must use efficient routing protocols.

In this paper, we propose a multipath routing protocol with low energy consumption in order to improve the performance of mobile ad hoc wireless networks. Our protocol ESMRsc (Energy aware and Stable Multipath Routing Protocol in smart city) uses a path selection strategy which is based on energy constraint and link stability. Our protocol will be designed on a realistic mobility model, contrary to most existing protocols which are based on random mobility models with some unrealistic behaviors such as sudden stop, abrupt acceleration. Simulation results demonstrate that our protocol has better performance in terms of energy consumption and network reliability.

Keywords: Mobile ad hoc network · Multipath routing
Mobility model in smart city · Energy efficiency · Link stability
Network reliability

1 Introduction

Wireless technology offers a high degree of flexibility in use and networking, enabling users to access information regardless of time and location factors. This technology continues to grow, providing humanity with several benefits in various fields. It enables them to save several constraints like hardware, duration of networking and equipment uninstalling, as well as great flexibility in terms of communication between nodes. Besides, it offers mechanisms to adapt to the various changes related to the mobility of mobile elements within a network. One of these mobile networks is called ad hoc network. In fact, this refers to a set of mobile entities interconnected by wireless technology forming a temporary network without the aid of any administration or fixed medium. During the last decade, a lot of research has been elaborated about ad hoc

© Springer Nature Switzerland AG 2019
M. Hatti (Ed.): ICAIRES 2018, LNNS 62, pp. 31–42, 2019.
https://doi.org/10.1007/978-3-030-04789-4_4

network. Indeed, the speed of implementation of this type of network is considered as an advantage during interventions in the context of wars or natural disasters. Recently, civil implementations of ad hoc networks have also been considered, they are used as soon as a network, which has to do with infrastructure, seems to be technically difficult or not economically effective. In an ad hoc network, all nodes must cooperate in order to create a temporary topology that makes communication easy. Thus, to create this topology and to route data, ad hoc networks must use efficient routing protocols.

Several routing protocols for ad hoc networks (Rivano et al. 2010; Radwan et al. 2011; Sherjung and Sharma 2017; Al-Karaki et al. 2017), which can be proactive, reactive (on-demand) or hybrid (Deshpande et al. 2013), have been useful and developed. However, among the drawbacks of these protocols are the problems of energy consumption and connectivity loss when the mobility of nodes is high. While some of these nodes are not much involved in routing, others are heavily congested and route much data in the network. Due to this inhomogeneous traffic load repartition, the loaded nodes consume quickly their limited physical resources (for instance battery poswer) and exhibit a high level of congestion. Yet, these effects hinder the good transmission of the packets and consequently make any network less efficient.

All nodes in an ad hoc network can be connected dynamically In this sense, it is possible to establish more than one path between a source and its destination at one time. This feature is used specifically in the routing process, a multipath routing. The purpose of this multipath technique is to find alternatives paths through a path discovery process, in order to find a better adaptation to frequent topology changes, to link breaks and to network overload. This technique also tries to improve the quality of service by reducing the data transfer delay.

In order to ameliorate the performance of ad hoc mobile wireless networks, we propose a stable, low energy multipath routing protocol. Our ESMRsc routing protocol (Energy aware and Stable Multipath Routing Protocol in smart city) is designed primarily for mobile nodes with battery-limited, where link failures and path breaks may occur frequently, the path selection strategy is based on energy constraints and link stability.

To design or study a new protocol for networks, it is important to simulate this protocol and to evaluate its performance. Hence, there exist several key factors and one of them is referred to as the mobility model. The mobility model is designed to describe the movement pattern of mobile users and how their location, speed and acceleration change over time. Since mobility models represent a crucial role in determining protocol performance, it is desirable that mobility models simulate the movement pattern of real-life targeted applications. Otherwise, the observations made and the conclusions drawn from these kinds of studies may be misleading. Then, while evaluating the ad hoc network protocols, it is necessary to choose the right model of underlying mobility.

The model of mobility which is frequently used in simulation of routing protocols for ad hoc networks is Waypoint random model (Broch et al. 1998), where nodes move independently to a random destination with a randomly chosen speed. The simplicity of the Waypoint model may be the reason for its widespread usage in this field. However, ad hoc networks can be used in applications designed for smart cities where mobility patterns are constrained by the city streets. Our ESMRsc protocol will be based on a not-random (realistic) mobility model, namely the Manhattan mobility model (Bai et al. 2003).

Indeed, it is designed for a city, contrary to most protocols which are based on random mobility models, that focuses on unrealistic behaviors such as sudden stop, abrupt acceleration and so on.

The paper is organized as follows. Section 2 provides a review of related works for multipath energy-aware and stable routing protocols in mobile ad hoc networks. Section 3 gives the design details of our ESMRsc protocol. Section 4 provides the simulation results of its performance evaluation. Section 5 concludes this paper.

2 Multipath Energy-Aware and Stable Routing Protocols

Ad hoc network routing protocols must efficiently transmit data packets using network nodes, depending on the node's lifetime. The node lifetime is related to its residual energy. Routing algorithms have an important role in saving node energy during the communication, smart routing may extend the service life of the nodes and thus of the entire network. Many works which have addressed routing based on energy saving exist in the literature (Bao and Garcia-Luna-Aceves 2010; Smail et al. 2014; Yadav et al. 2015; Arya and Gandhi 2015). These energy-aware routing protocols can be single path or multipath. A multipath routing can be an alternative mean to conserve energy in mobile ad hoc networks. Several on-demand routing multipath protocols (Aguilar and Carrascal 2010; Gole and Mallapur 2011; Nasehi et al. 2013; Bheemalingaiah et al. 2016), that preserve energy, have been proposed in order to avoid network failures as long as possible. All of these studies solve the problem of energy conservation, but the majority of power-saving mechanisms are based on the remaining power only; they cannot be used to establish the best route between source and destination nodes. On one hand, when a node is willing to accept any route requests because it currently has enough residual battery capacity, too much traffic load may be routed through that node. On the other hand, many energy saving mechanisms/algorithms neglect the power consumption induced by each node own message sending, which may cause network partitioning due to node battery exhaustion. Indeed, it reduces network performance. Hence, shared and balanced energy consumption is a remedy for these kinds of problems. Finally, another problem of these routing protocols is that they do not consider the stability of paths in their path establishment process. One important characteristic of mobile ad hoc networks is the mobility of the nodes in the network. Indeed, this characteristic makes the network topology very unstable and causes path disconnections. Recently, several protocols (De Rango et al. 2012; Noureddine et al. 2014; Moussaoui and Boukeream 2015) have been proposed to contribute to solve this problem. In the majority of these routing schemes a new path-discovery step is launched once a path failure is detected, and this process causes delay and node resources wastage which may reduce individual node lifetime and hence the network entire lifetime. In our solution, a path-discovery process is not systematically launched.

3   Energy Aware and Stable Multipath Routing Protocol in Smart City

The aim of this article is to develop an improved routing protocol, named Energy aware and Stable Multipath Routing protocol adapted to smart city (ESMRsc), for mobile ad hoc networks. ESMRsc is a reactive and multipath routing protocol which selects the energy-efficient path with stable links; it is designed to operate in city. ESMRsc uses the same types of messages as the multipath protocol ZD-AOMDV (Nasehi et al. 2013).

3.1   Multipath Routing Selection

This section describes path selection routing. When the node source receives the first *RREP (Route Reply)* packet, it waits for a given amount of time (*RREP_Wait_Time*) to receive more *RREPs (Route Request)* before selecting the best path. The choice of the best path between a source node *s* and destination node *d*, is done according to energy consumption and path stability. In this approach, two functions are defined: the cost function $fep_j(t)$ which characterizes a path *j* at time *t* from an energetic point of view and the cost coefficient $fsp_j(t)$ which represents the stability of path *j* at time *t*.

3.1.1   Energy Aware Cost Function

Let $fep_j(t)$, the minimum residual energy of nodes constituting the path *j* for a source node *s* to destination node *d* at time *t*, be expressed as

$$fep_j(t) = \min_{i=1}^{n-1}(fen_{i,j}(t)) \tag{1}$$

Where $feni,j(t)$ represents the energy cost function of node *i* belonging to the path *j*, formally:

$$fen_{i,j}(t) = \frac{Elev_{i,j}(t)}{DR_{i,j}(t)} \tag{2}$$

$Elev_{i,j}(t)$ denotes the energy level of node *i* belonging to the path *j* at time *t* during a discovery process between a source node *s* and a destination node *d*, given by:

$$Elev_{i,j}(t) = \frac{E_{i,j}(t)}{E_{average}} \tag{3}$$

Where $E_{i,j}(t)$ represents the node *i* residual energy belonging to the path *j* at time *t* and $E_{average}$ is the average residual energy of nodes that participated in the multipath discovery process between the source node *s* and the destination node *d*. $DR_{ij}(t)$ is the drain rate of the node *i* belonging to the path *j* at time *t*, which is defined as the rate at which energy is consumed at a given node, usually when a node is used by paths different than source *s* and destination *d*.

3.1.2   Link Stability Aware Cost Function

To calculate the link stability in mobile ad hoc networks, we consider the mobility of nodes as the main metric. The protocols based on nodes' mobility use some criteria inherent of the nodes mobility, such as their coordinates, their directions of movement or their speeds. Since the directions of nodes movement and the nodes speeds are calculated with specific devices and require time to be computed, we use the node coordinates for our protocol which is instantaneous way. Our protocol mainly utilizes message delay between the sending and receiving time to measure the link stability. We exploit discoveries messages and hello messages for collecting coordinate information of neighbors nodes, to avoid the overhead which will be generated by specific messages.

Each node adds its destination, hop count to the destination and sequence number into the original hello message. A new field $Distance_{i,j}$ is added to the Hello message in order to collect distances that separate nodes, is noted by d_{ij}, see Fig. 1.

Destination	Destination sequence number	Hop count	Distance $_{i,j}$	Expiration timeout

Fig. 1. Structure a destination node entry of a HELLO packet for ESMRsc.

Figure 2 shows the structure of an entry of the routing table of a node i. For each destination known by the node i there is an entry. *Route_list* contains all known neighbor nodes of node i which leads to that destination. Each neighbor for that destination is identified its *nexthop* address, and the *hopcount* field is the number of hops required to reach that destination using this neighbor. We add two new fields to the basic routing table, E_i and *distances_list*. The field E_j denotes the residual energy of a neighbor node and the field *distance_list* indicates registered distances between the node i and its neighbors over different time.

Destination
Sequence_number
Advertised_hopcount
Route_list {(nexthop1, hopcount1, E$_1$, distance_list {(d$_{i1}$,t$_1$),(d$_{i1}$,t$_2$),...}), (nexthop2, hopcount2, E$_2$, distance_list {(d$_{i2}$,t$_1$),(d$_{i2}$,t$_2$),...}),...}
Expiration timeout

Fig. 2. Structure of a routing table entry for ESMRsc.

Assuming a link between two nodes i and j, to calculate the stability of this link. The node i periodically sends message to the node j, when the node j receives the message, it compute its coordinates then sends back it to the node i. Based on this coordinates, the node i calculates the distance that separates node i from node j noted by $d_{i,j}$. The stability cost function of the link ij at time t, is denoted by: $fsl_{i,j}(t)$, given by:

$$fsl_{i,j}(t) = \frac{SDl_{i,j}(t)}{Ml_{i,j}(t)} \tag{4}$$

Where $Ml_{i,j}(t)$ represents the mean of the n distances recorded between the node i and node j, defined as follows:

$$Ml_{i,j}(t) = \frac{\sum_{t=t1}^{tn} d_{i,j}(t)}{n} \tag{5}$$

$SDl_{i,j}(t)$ denoted the mean absolute deviation of the distances recorded between the node i and node j, $SDl_{i,j}(t)$ is given by:

$$SDl_{i,j}(t) = \frac{1}{n} \sum_{t=t1}^{tn} \left| d_{i,j}(t) - Ml_{i,j}(t) \right| \tag{6}$$

The function $fsl_{i,j}(t)$ represents the coefficient of variation, also known as relative standard deviation. It is a standardized measure of dispersion of a probability distribution. The coefficient of variation formulated by $fsl_{i,j}(t)$ is used to quantify the measurement accuracy. In our case, it is the measurement of distances between two neighbor nodes. If the $fsl_{i,j}(t)$ tend toward 0, then we have a good distribution of distances, which means that the link is stable. If it tends toward 1 or to infinity, this corresponds to a poor distribution, which means that the link is instable.

Finally, we define $fsp_j(t)$ the path cost function stability of path j at time t, given by:

$$fsp_j(t) = \max_{i=1}^{n-1}(fsl_{i,j}(t)) \tag{7}$$

$fsp_j(t)$ is defined as the maximum of the stability costs of links constituting the path j.

3.1.3   Objective Problem Formulation

We design our multipath selection according to the energy consumption of their path nodes and the link stability of their path links. The corresponding objective function $fp_j(t)$ of path j at time t is defined by combining the energy aware cost function defined in formula (1) and the path cost function stability, formula (7):

$$fp_j(t) = fep_j(t) + \frac{1}{fsp_j(t)} \tag{8}$$

Our idea is based on sorting all paths between a source node s and a destination node d by the descending value of $fp_j(t)$. The path with the maximum $fp_j(t)$ is chosen to forward the data packets.

4   Performance Evaluation of ESMRsc

In this section, we present simulation results to demonstrate the efficiency of our proposed routing protocol. First we present the metrics used for performance evaluation and then we evaluate our protocol by comparing it with the protocol ZD-AOMDV (Nasehi et al. 2013), applying two types of mobility models: Random Waypoint (Broch et al. 1998) and Manhattan Mobility (Bai et al. 2003). We will use ZD-AOMDV (Nasehi et al. 2013) as a reference for our performance evaluation because it aims to improve the performance of mobile ad hoc network and has the same characteristics as our protocol, namely its reactivity, energy aware and multipath character. The evaluation is accompanied with an analysis and discussion of results.

4.1   Mobility Model

The performance of an ad hoc network protocol can change significantly when it is tested with different mobility models, but also when the same mobility model is used with different parameters. The performance should be evaluated with the mobility model which is near to the real scenario; this can facilitate a fair evaluation of the ad hoc network protocol.

One frequently used mobility model in MANET simulations is the Random Waypoint model (Broch et al. 1998). The random waypoint model assumes that each node is initially randomly placed in the network area. The node selects, uniformly and randomly, a target location of the network to move. The speed of movement to this location is also chosen uniformly and randomly, see Fig. 3. Once the node has moved to the chosen location, it waits at this location for a random time called pause time. This process is repeated at the end of simulation.

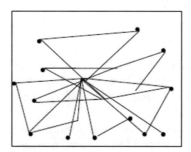

Fig. 3. Random waypoint model

Since cities are made up of streets, the movements of mobiles are very constrained by the geometry of the streets of the city. The Random Way Point model does not take into account these constraints, the Manhattan model is more suitable for such configuration.

Unlike the Random Waypoint model where the nodes can move freely, the mobile nodes in the Manhattan Mobility model (Bai et al. 2003) are only allowed to travel on the pathways. Manhattan Mobility model, it's also called the Urban Area Model. It forms a number of horizontal and vertical streets, see Fig. 4. Each mobile node can move along the grid of horizontal and vertical streets on the map. At the intersection of the horizontal and vertical street, the mobile node can turn left, right or straight by a probabilistic approach. The Manhattan mobility model is commonly used to model the movement of cars or people in a city.

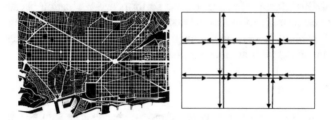

Fig. 4. Manhattan mobility model

The Manhattan Mobility model creates realistic movements for a section of a city, since it severely limits the displacements of mobile node. These nodes do not have the ability to roam freely without worrying about city streets, obstacles and other traffic regulations.

4.2 Performance Parameters

We evaluate two key performance metrics. Energy consumption is the average of the energy consumed by nodes participating in packet transfer from the source node to the destination node during the whole simulation. And the number of data packets successfully received by the destination node.

4.3 Performance Evaluation

We carried our simulations to determine the effectiveness of our protocol. The principal goal of these simulations is to analyze our protocol by comparing it with another protocol, mainly ZD-AOMDV (Nasehi et al. 2013). The values of simulation parameters are summarized in Table 1.

To evaluate ESMRsc, we use the network simulator ns-2. Two mobility models are used: RandomWaypoint model and "Manhattan" mobility model. For the Manhattan mobility model, the probability of direct move is 0.5, the probability of turning to the left is 0.5, and the probability of turning to the right is 0.5. Each node has a selected speed from 0 to 20 m/s (in discrete increments of 4 m/s). The shared radio model used is similar to the IEEE 802.11 standard, based on Lucent's Wave LAN (Feeney and Nilsson 2001). The number of simulation run is 20 per value, the source and destination nodes are chosen at random. It is assumed that a receiver or transmission node consumes 281.8 mW (Jung and Vaidya 2002), the consumption in idle mode is not considered (Kim et al. 2003), as it is the same for the two mobility model it has no impact on the results.

Table 1. Simulation parameters

Mobility model	Parameter	Value
Random waypoint	Simulator	NS2.35, NAM 1.13
	Channel type	Channel/wireless channel
	Routing protocol	ZD-AOMDV/ ESMRc
	Simulation time	500 s
	Number of mobile nodes	50/100/150
	Radio propagation model	Propagation/two ray ground
	Network interface type	Phy/wirelessphy
	MAC type	IEEE 802.11
	Interface queue type	Queue/DropTail/PriQueue
	Max packet In ifq	70
	Terrain range	800 m × 800 m
	Transmission range	250 m
	RREP_Wait_Time	1.0 s
Manhattan	Circular	false
	Dimension of movement output J	2D
	Xblocks,Yblocks	10, 10
	UpdateDist	5.0
	TurnProb	0.5
	SpeedChangeProb	0.2
	MinSpeed	12
	MaxSpeed	12
	PauseProb	0.02
	MaxPause	0.02

Figure 5 shows the energy consumed by ESMRsc using Random waypoint model (WM), or Manhattan model (MM), and ZD-AOMDV protocols. ESMRsc has limited performance when the nodes have a low moving speed or when there is a small number of network nodes (see Fig. 5 (a) for moving speed between 0 and 4), but as the number of nodes increases and as the speed increases (see Fig. 5a, b), links become unstable, path becomes longer. Nodes consume more energy which favour links fails and then the impact of our solution takes act. We can see that the energy consumed in ESMRsc is less than those consumed by ZD-AOMDV, for instance when the number of nodes of a network equal 150 nodes see Fig. 5 (b), the energy consumed of the ESMRsc protocol is nearly (on average) 19% for WM and 12% for MM, lower than the energy consumed of the ZD-AOMDV protocol. ESMRsc consumes less energy because ESMRsc is able to avoid nodes with low energy and it selects the path with more stability during the construction of the multipath.

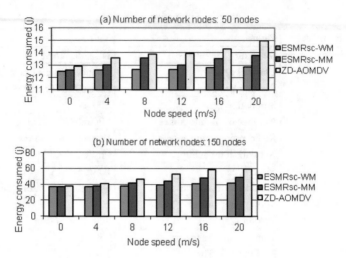

Fig. 5. Energy consumed versus node speed.

The number of successfully received packets is shown in Fig. 6 with different number of network nodes for various moving speeds. The number of successfully received packets is longer than ZD-AOMDV for two mobility models WM and MM. Our protocol receives more packets compared to ZD-AOMDV protocol mainly for a medium or high number of network nodes, which ensures a good reliability of our protocol for random and not-random models of mobility. ESMRsc selects the best nodes in term energy and stability, this prolongs the individual node lifetime and hence the entire network lifetime.

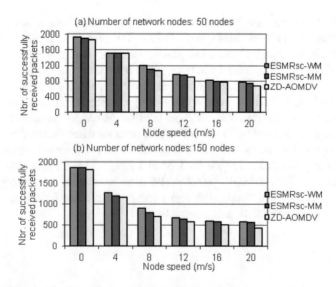

Fig. 6. Number of successfully recieved packets versus node speed.

5 Conclusion

A new multipath routing protocol, ESMRsc (Energy-aware and Stable Multipath Routing protocol in smart city) has been proposed in this paper. It ensures the network reliability and reduces energy consumption in ad hoc mobile networks. Alternative paths are precalculated then used whenever there is a path link failure. The choice of path depends on a weighted function. Nodes energy and links stability are used to calculate this weight. When a packet is to be sent, the best available path is selected.

In the literature, several mobility models exist, belonging mainly to two classes: random and non-random classes. The modeling of our protocol is done while these two classes are taken into account. Indeed, we chose two models: the well-known Random Waypoint model and a more realistic and non-random model based on a Manhattan grid. Through performance evaluation, we noticed that our ESMRsc multipath protocol improves the stability of the ad hoc mobile network and considerably ameliorate network performance compared to other routing protocols in literature.

References

Aguilar Igartua M., Carrascal Frías V.: Self-configured multipath routing using path lifetime for video-streaming services over ad hoc networks. Comput. Commun. **33**, 1879–1891 (2010). https://doi.org/10.1016/j.comcom.2010.06.019

Al-Karak, J.N., Al-Mashaqbeh, G.A., Bataineh, S.: Routing protocols in wireless mesh networks: a survey. Int. J. Inf. Commun. Technol. **11**, 445–495 (2017). https://doi.org/10.1504/IJICT.2017.087454

Arya, V., Gandhi, C.: Energy aware routing protocols for mobile ad hoc networks—A survey. Int. J. Inf. Commun. Technol. **7**, 662–675 (2015). https://doi.org/10.1504/IJICT.2015.072045

Bai, F., Sadagopan, N., Helmy, A.: Important: a framework to systematically analyze the impact of mobility on performance of routing protocols for ad hoc networks. In: Proceedings of IEEE Information Communications Conference (INFOCOM 2003), San Francisco. (2003). https://doi.org/10.1109/infcom.2003.1208920

Bao, L., Garcia-Luna-Aceves, J.J.: Stable energy-aware topology management in ad hoc networks. Proc. Ad Hoc Netw. **8**, 313–327 (2010). https://doi.org/10.1016/j.adhoc.2009.09.002

Bheemalingaiah, M., Venkataiah, C., Vinay, Kumar K., Naidu, M.M., Sreenivasa Rao, D.: Survey of energy aware on-demand multipath routing protocols in mobile Ad Hoc networks. Int. J. Adv. Res. Comput. Sci. Softw. Eng. **6**, 212–222 (2016)

Broch, J., Maltz, D.A., Johnson, D.B., Hu, Y.-C., Jetcheva, J.: A performance comparison of multi-hop wireless ad hoc network routing protocols. In: Proceedings of the Fourth Annual ACM/IEEE International Conference on Mobile Computing and Networking (Mobicom98), ACM. (1998). https://doi.org/10.1145/288235.288256

De Rango, F., Guerriero, F., Fazio, P.: Link-stability and energy aware routing protocol in distributed wireless networks. IEEE Trans. Parallel Distrib. Syst. **23**, 713–726 (2012). https://doi.org/10.1109/TPDS.2010.160

Deshpande, S.S., Asare, K.V., Deshpande, S.: An overview of Mobile Ad Hoc networks for the proactive, reactive and hybrid routing protocol. Int. J. Eng. Comput. Sci. **2**, 1134–1143 (2013)

Feeney, L.M., Nilsson, M.: Investigating the energy consumption of a wireless network interface in an ad hoc networking environment. IEEE INFOCOM (2001). https://doi.org/10.1109/INFCOM.2001.916651

Gole, S.V., Mallapur, S.V.: Multipath energy efficient routing protocol. Int. J. Res. Rev. Comput. Sci. (IJRRCS) **2**, 954–958 (2011)

Jung, E.S., Vaidya, N.H.: A power control MAC protocol for ad hoc networks. In: Proceedings of the ACM International Conference on Mobile Computing and Networking (MOBICOM), pp. 36–47 (2002). https://doi.org/10.1145/570645.570651

Kim, D., Garcia-Luna-Aceves, J.J., Obraczka, K., Cano, J.C., Manzoni, P.: Routing mechanisms for mobile ad hoc networks based on the energy drain rate. IEEE Trans. Mob. Comput. **2**, 161–173 (2003). https://doi.org/10.1109/TMC.2003.1217236

Moussaoui, A., Boukeream, A.: A survey of routing protocols based on link-stability in mobile ad hoc networks. J. Netw. Comput. Appl. **47**, 1–10 (2015). https://doi.org/10.1016/j.jnca.2014.09.007

Nasehi, H., Javan, N.T., Aghababa, A.B., Birgani, Y.G.: Improving energy efficiency in manets by multi-path routing. Int. J. Wireless Mob. Netw. (IJWMN) **5**, 163–176 (2013). https://doi.org/10.5121/ijwmn.2013.5113

Noureddine, H., Ni, Q., Min, G., Al-Raweshidy, H.: A new link lifetime estimation method for greedy and contention-based routing in mobile ad hoc networks. Telecommun. Syst. **55**, 421–433 (2014). https://doi.org/10.1007/s11235-013-9796-9

Radwan, A., Mahmoud, T.M., Houssein, E.H.: Performance measurement of some mobile ad hoc network routing protocols. Int. J. Comput. Sci. Issues **8**, 107–112 (2011)

Rivano, H., Theoleyre, F., Valois, F.: A framework for the capacity evaluation of multihops wireless networks. Ad Hoc Sens. Wireless Netw. (AHSWN) **9**, 139–162 (2010)

Sherjung, Sharma, R.K.: A survey on routing protocols in mobile Ad-hoc networks. Int. J. Innov. Res. Comput. Commun. Eng. (2017). https://doi.org/10.15680/IJIRCCE.2017.0505189

Smail, O., Cousin, B., Mekki, R., Mekkakia, Z.: A multipath energy-conserving routing protocol for wireless ad hoc networks lifetime improvement. EURASIP J. Wireless Commun. Netw. **139**, 1–12 (2014). https://doi.org/10.1186/1687-1499-2014-139

The Network Simulator ns-2. http://www.isi.edu/nsnam/ns. Accessed 05 May 2018

Yadav, R.K., Gupta, D., Singh, R.: Survey on energy efficient routing protocols for mobile Ad-hoc networks. Int. J. Innov. Adv. Comput. Sci. (IJIACS) **4**, 644–652 (2015)

Development of a Connected Bracelet Managed by an Android Application

El-Hadi Khoumeri[(✉)], Rabea Cheggou, and Kamila Ferhah

Ecole Nationale Supérieure de Technologie, Dergana, Algeria
{elhadi.khoumeri, rabea.cheggou, kamila.ferhah}@enst.dz

Abstract. Connected objects and the Internet of Things (IoT) are spreading gradually in all areas of daily life: transport, industry and the field of health and well-being. IoT has completely revolutionized the health sector through a continuous chain of diagnostics, treatments and omnipresent follow-ups. The watches and bracelets connected are one of its applications in this area that dress more and more our wrists. Connected wristbands are gradually entering the field of health thanks to their interest. Different models are designed by several companies and intended for different categories of users. The goal of this project is to design a connected bracelet that helps the user to improve their well-being and preserve their health while ensuring ease of use. From the technical point of view, the system measures and tracks physical activity (number of steps, number of calories burned and distance travelled), vital constants (heart rate and body temperature) and ambient temperature in real time. It allows sending alerts by SMS with the location of the person, in case of a fall as well as in case of anomaly in the pulse or the body temperature. If the option of alerts is activated and it displays the necessary information in case of emergency via the Android application, after scanning the QR code on the wristband. The connected bracelet is built around the microcontroller of the development board ESP8266 that provides system management, acquisition and processing of data retrieved by the various sensors as well as sends them to the Android application via Wi-Fi. Android application receives data sent by the ESP8266 using the TCP/IP protocol between a server and a client, the latter provides several functions such as processing and data backup sends SMS alerts and the display.

Keywords: Connected bracelet · Sport · Android application · Mobile

1 Introduction

Connected equipment and the Internet of Things (IoT) are spreading gradually in all areas of daily life: transport, industry and the field of health and well-being. The IoT has completely revolutionized the healthcare industry with a continuous chain of diagnostics, treatments and ubiquitous follow-ups. The watches and bracelets connected are one of its applications in this area that dress more and more our wrists. The knowledge of certain concepts is necessary for the realization of our project, in what follows we will define: IoT, connected bracelet, and the Android operating system.

© Springer Nature Switzerland AG 2019
M. Hatti (Ed.): ICAIRES 2018, LNNS 62, pp. 43–48, 2019.
https://doi.org/10.1007/978-3-030-04789-4_5

1.1   Internet of Things

The Internet of Things, or IOT (Internet Of Things) is considered the third evolution of the internet, referred to as Web 3.0. It is represented by the different devices of the real world and objects (vehicles, houses and other embedded elements with electronics, software, sensors and actuators) integrated into a network that ensures the exchange of information and between the elements that compose it. The IOT is the result of the convergence of wireless technologies, micro-electromechanical systems (MEMS) and the internet. It describes a vision that says everything must be connected to the internet, so that connected objects, communicating with each other (machine-to-machine communication without any interaction of the human being) and which are often designated by the term "Intelligent" or "Smart", will be fundamental, the forecasts say that in ten years (2015–2025) 150 billion of objects should connect with each other and with billions of people [1].

1.2   Android

It is a mobile operating system developed by Google, based on the Linux kernel and open whose source code is freely accessible. It equips mainly smartphones but also touch tablets, televisions, smart-watches… etc. Android designed by a startup specializing in the development of mobile applications in 2003 and acquired by Google in 2005. It has had several versions where each new version used to fix bugs and add new features [2]. The main components of this operating system are: the Linux kernel that manages the system services (security, memory and process management), libraries, the Android runtime engine, and the framework that offers developers the ability to create extremely rich and innovative applications as well as applications developed using the Java programming language and intended for the end-user [3].

1.3   Connected Bracelet

It can be defined as one of the connected objects that belong to the so-called Internet of Things; it allows communication generally with the Smartphone using a communication protocol such as Wi-Fi, Bluetooth or others. This project is particularly interested in activity bracelets or activity trackers, which are clothing and technological accessories that follow the physical activities (the number of steps taken, the distance traveled, the number of calories burned, heart rate…etc.) thanks to the measurements taken by a set of specific sensors. The display of the measured data is generally done via an Android application on the Smartphone [4].

1.4   Objectives of Connected Medical Bracelets

Connected objects and especially Wearables (bracelets, watches…) strongly help their users to achieve sport and health goals [5]. They allow you to monitor your activities, ensure your safety and improve your quality of life.

2   Why Connected Bracelet

The use of sensors for the visualization and monitoring of sports performance has been demonstrated in recent years [6–10]. As stated in [11], one of the main design systems for detection in sport is that sensors on the body (and sensor nodes) must be small and lightweight to avoid genes to the athletes during their activities. On the other hand, as noted in [8, 10], sensor nodes must support a rich set of functions and - to support large-scale deployment - the price of sensors must be accessible to the maximum number of people.

The goal is to design a connected bracelet that helps the user to improve their well-being and preserve their health while ensuring ease of use. From the technical point of view, the system allows:

- Measure and monitor physical activities (number of steps, number of calories burned and distance traveled), vital constants (heart rate and body temperature) and the ambient temperature in real time.
- Send alerts by SMS with the location of the person in case of a fall as well as in case of anomaly.
- Displays the necessary information in case of emergency via the Android application after scanning the QR code shown on the bracelet.

3   System Components

The system consists of several modules as illustrated in Fig. 1, each module performs one or more functions. Detection modules measure and send data to the central module that processes and transmits data via Wi-Fi for display on the graphical user interface.

Fig. 1. Structure of the connected bracelet system.

3.1   Central Module

It is the module, which manages the operation of the system, it retrieves the data from the various sensors to process them and send them to the Smartphone via Wi-Fi. We used the ESP 8266 that is a programmable microcontroller integrated circuit that supports Wi-Fi (802.11 b/g/n) developed in 2014 by the Chinese company Espressif [12]. Figure 2 shows version 12 of the ESP8266 module [13].

3.2   Detection Modules

The detection modules consist of the various MPU6050 accelerometer sensors [14], the SEN0203 [15] heart rate sensor and the MLX90614 [16] temperature sensor, which measure acceleration, pulse, ambient temperature and body.

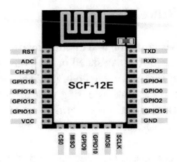

Fig. 2. The module: ESP8266-12E.

4 System Architecture

The connected bracelet built around the microcontroller of the ESP8266 development board, which manages the system, acquires and processes data collected by the various sensors and sends them to the Android application via Wi-Fi. The latest receives the data sent by the bracelet using the TCP/IP protocol between a server and a client; it provides several functions among which we quote: processing and data backup, sending alerts by SMS and display on graphical user interfaces. Figure 3 represents an explanatory diagram of the general operating principle of the system.

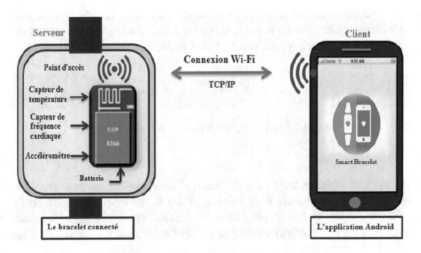

Fig. 3. General operating principle of the system.

5 Test and Result

The Android application receives the frame sent by the ESP8266 and displays the information read in the home interface. Figure 4 shows the result obtained.

Fig. 4. Display information received from the bracelet.

The Android application can graph the evaluation of the various data, after saving them in the database, as a function of time. Figure 5 shows an example, which represents the evaluation of the pulse as a function of time.

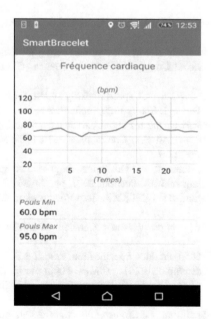

Fig. 5. Representative graph of the evaluation of heart rate as a function of time.

6 Conclusion

In this paper, we used three small, lightweight, portable and cost-effective sensors, for the realization of our Bracelet, were presented. The custom design system is cost-effective and highly flexible: its main unit includes an ESP 8266 processor board and a range of sensors, all of which could be attached to the system through a flexible band around the wrist. The wireless module supports data transmission to the application for playback, which installed on the phone. The applicability of the system demonstrated in this article, through radio information about an athlete in the race; the results of the experiment show that the system would be able to provide accurate information. Future work will allow us to add other needs such as the analysis of sleep phases, determination of stress levels or the addition of a food database.

References

1. Rouse, M.: Internet des objets (IOT). https://www.lemagit.fr/definition/Internet-des-objets-IoT
2. A. e. F. E. (DakuTenshi), Créez des applications pour Android. www.siteduzero.com
3. Platform Architecture. https://developer.android.com/guide/platform/
4. Suard, C.: Quel avenir pour le bracelet connecté? http://www.frandroid.com/produits-android/accessoires-objetsconnectes/bracelets-connectes/412741_le-bracelet-connecte-a-t-il-de-lavenir
5. Bastien, É.: les objets connectés peuvent aider à atteindre les objectifs fitness. https://www.objeko.com/etude-objetsconnectes-peuvent-aider-a-atteindre-objectifs-fitness-13424/
6. The SEnsing for Sports And Managed Exercise (SESAME) project. http://www.sesame.ucl.ac.uk
7. Chi, E.: Introducing wearable force sensors in martial arts. Pervasive Comput. Mag. **04**(3), 47–53 (2005)
8. Cheng, L., Hailes, S: Analysis of wireless inertial sensing for athlete coaching support. In: Proceedings of IEEE Global Communications Conference (GLOBECOM), New Orleans, USA, December 2008
9. Cheng, L., et al.: A low-cost accurate speed tracking system for supporting sprint coaching. Proc. Inst. Mech. Eng. Part P J. Sports Eng. Technol. **224**, 167–179 (2010)
10. Cheng, L., et al.: Practical sensing for sprint parameter monitoring. In: Proceedings of the 7th IEEE Sensor, Mesh and Ad Hoc Communications and Networks (SECON), Boston, Massachusetts, USA, June 2010
11. Kranz, M., Spiessl, W., Schmidt, A.: Designing ubiquitous computing systems for sports equipment. In: Proceedings of IEEE PerCom 2007, pp. 79–86
12. Boudou, A.: Présentation de l'ESP8266. https://www.ekito.fr/people/presentation-de-l-esp8266/
13. ESP8266-module-family. https://www.esp8266.com/wiki/doku.php?id=esp8266-module-family
14. MPU-6000 and MPU-6050 Product Specification Revision 3.4. Available: https://www.invensense.com/wp-content/uploads/2015/02/MPU-6000Datasheet1.pdf
15. DFRobot (2017). https://media.digikey.com/pdf/Data%20Sheets/DFRobot%20PDFs/SEN0203_Web.pdf
16. MLX90614 family—SparkFun Electronics, 30 Mars 2009. https://www.sparkfun.com/datasheets/Sensors/Temperature/SEN-09570-datasheet-3901090614M005.pdf

Security Mechanisms for 6LoWPAN Network in Context of Internet of Things: A Survey

Yamina Benslimane[1(✉)], Khelifa Benahmed[2],
and Hassane Benslimane[1]

[1] Laboratory of Physics and Semiconductor Devices, Tahri Mohamed
University, Béchar, Algeria
yami.benslimane@gmail.com, hassane_ben@yahoo.fr
[2] Exact Sciences Department, Tahri Mohamed University, Bechar, Algeria
benahmed_khelifa@yahoo.fr

Abstract. With the emergence of internet of things (IoT), the physical object belonging to our daily activity and to different domain as: home automation, industrial automation, monitoring environment and health care may be inter-acted and benefited from the world of internet. Thus, this communication provides several data that are circulate in the different networks as IPv6 network or the 6LoWPAN network. Since the 6LoWPAN network is the fundamental part of IoT, its security is challenge domain whether for the end-to-end security when the data are sent to the server outside the network or for the internal security with the intrusion detection system. In this paper, we present a survey about the proposed researches for the 6LoWPAN network security whether for inside or outside communication of network. The analysis of these proposed security mechanisms in the literature is discussed based on a taxonomy focusing on the following attributes: the selected internet security protocols as DTLS, HIP and IKE for the end-to-end security (out-side the 6LoWPAN network) and the attack detected as routing attack, DDoS attack,…etc. for the intrusion detection system (inside the 6LoWPAN network). We also give the Evaluation of these security mechanisms for 6LoWPAN network in term of different metrics. The aim of this work is to identify leading trends, open issues, and future research possibilities.

Keywords: Internet of things · 6LoWPAN network
Internet security protocols · End-to-end security · Intrusion detection system

1 Introduction

The idea to connect the smart objects with the internet is carried out with the internet of things. This paradigm allows the interconnection between the intelligent physical objects and the various entities of communication like the servers, the smart phone through the Internet. It gives the possibility of integrating into our daily life. Several fields as: health, environment monitoring, industry, private life, and ubiquitous computing are profited from IoT system. But the integration of the real world objects in the Internet opens the door of attackers to infiltrate on our daily activities. To protect our intelligent life, it is necessary to develop effective and suitable mechanisms for IoT security. However, many factors like high number of the connected devices, the

© Springer Nature Switzerland AG 2019
M. Hatti (Ed.): ICAIRES 2018, LNNS 62, pp. 49–69, 2019.
https://doi.org/10.1007/978-3-030-04789-4_6

limitation of the resources and the various protocols become the classic prevention and detection methods not respond to all the needs. For this, the several works are appeared in the literature whether to adapt the exist mechanisms of security for the requirements of IoT system or propose new mechanisms. These works concentrate on the 6LoW-PAN network which is the fundamental part in the IoT system. We can be distinguishing two lines of defenses. The first line is the end-to-end security mechanism that enables to secure channel when the 6LoWPAN node exchange of the data with a server located out-side its network. The second line is the intrusion detection system for the 6LoWPAN network that protects the network against internal attacks. IN our survey, we depended upon this principle to classify these works. this classification is based on the security mechanisms towards the outside of the 6LoWPAN network with the security protocols as HIP, DTLS and IKE and the security mechanisms towards the inside of the 6LoWPAN network represented in form of IDS with various suggested methods. In addition, we give an evaluation of these works according to different metrics. Also we argue the future research direction in this domain.

The rest of this paper is organized as follows. Section 2 gives the overview of 6LoWPAN network. Section 3 discusses related surveys and our positioning. Section 4 explains the security requirements for 6LoWPAN network. Challenges and limits in the usage of classic security mechanisms are discussed in Sect. 5. Section 6 presents our Taxonomy of proposed security mechanisms for 6LoWPAN network, details the proposed security protocols as DTLS, HIP and IKE for end-to-end communication between the 6LoWPAN node and the internet server and shows an analysis of the literature of IDSs for 6LoWPAN network respectively. One of the most relevant contributions of this work, an evaluation of proposed security mechanisms for 6LoWPAN network in IoT context according to different attributes, is presented at Sect. 7. Section 8 discusses the open issue and future research direction. Finally, the paper is concluded in Sect. 9.

2   6LoWPAN Network Overview

The IPv6 over the Low-Power Wireless Personal Area Network (6LoWPAN) is introduced and standardized by The Internet Engineering Task Force (IETF) workgroup. The 6LoWPAN standard puts mechanisms that enable the efficient use of IPV6 packet over low power wireless networks on simple embedded devices. These mechanisms consist of adding an adaptation layer between the link layer and network layer which allows IPv6 packet transmission over IEEE 802.15.4 links and that is through compression, fragmentation and reassembly mechanisms. The IEEE 802.15.4 device is characterized by short range, low bit rate (from 20 Kbits/s (868 MHz) to 250 Kbits/s (2.45 GHz)), small packet size (the maximum transmission unit or MTU on IEEE 802.15.4 links is 127 bytes). There are two types of devices: the full function device and the reduce function devices. The 6LoWPAN network is connected with internet through the edge routers (the 6LBR) that considered as the stub of the network. Their functions are:

- The data exchange between 6LoWPAN devices and the Internet (or other IPv6 network) through different media as Ethernet, Wi-Fi or 3G/4G.
- Local data exchange between devices inside the 6LoWPAN.
- The generation and maintenance of the radio subnet (the 6LoWPAN network).

The typical 6LoWPAN network as illustrate in Fig. 1 contains two device types: routers and hosts, the router is FFD device type; it can be routes the data toward another node in the network using RPL protocol defined from ROLL Working Group. The host is the RFD device type, it is not able to route data to other devices in the network. The host can also be a sleepy Device, waking up periodically to check its parent (a router) for data.

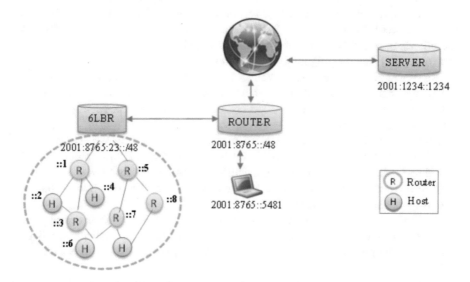

Fig. 1. 6LoWPAN network connected with Internet

The basic concept of 6LoWPAN stack is illustrated in Fig. 2.

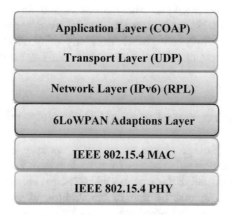

Fig. 2. The 6LoWPAN protocol stack

3   Related Surveys and Our Positioning

The internet of things security especially 6LoWPAN network security is a critical research domain.

There are several studies and surveys that are appeared in this domain. David Airehrour et al. focus on analyze of routing protocol for internet of things. Moreover, they present some existing mechanisms to secure routing protocols and discuss threats, open challenges and future research direction for improve the level of security in IoT routing protocol. Miorandi et al. (2012) gave an detailed survey about security, trust and privacy in IoT, they present the proposed researches in the domain of IoT security. Moreover, they have highlighting on future researches.

Nguyen et al. (2015) are concentrated in their survey on internet security protocols and its applicability in IoT domain where the major inconvenience is the resources limited. Further, the authors have done taxonomy of these protocols in term of key management mechanisms.

Zarpelão et al. (2017) are presented a study about the proposed researches in intrusion detection systems for IoT. They gave taxonomy of these proposed IDSs according to many factors as detection methods, IDS placement strategy, security threat and validation strategy. They also discuss the necessary needs to be a reliable IDS and able to meet the requirement of IoT.

The existing surveys take two ways different. In the first way, the surveys are presented the IoT security in general; they are concentrated on challenges, vulnerabilities, and future researches like researches (Airehrour et al. 2016; Miorandi et al. 2012).

In The second way, they are specified certain security aspects as internet security protocols and its adaptations for IoT, or the intrusion detection systems like researches (Nguyen et al. 2015; Zarpelão et al. 2017).

Our proposed survey focus especially on 6LoWPAN network security whether for outside or inside communication of the network where we look in depth into proposed end-to-end security mechanisms. These mechanisms enable the constrained devices establishes a secure channel with an other entities located outside its network and also we study the proposed IDSs that assure the internal security for 6LoWPAN network. Moreover, our survey proposes a global taxonomy of 6LoWPAN network security mechanisms in the context of IoT and an evaluation of these mechanisms in term the different attributes.

4   Security Requirements for 6LoWPAN Network

For real realization of the IoT system in the world, security is one the main requirements. Security should be needed whether for per hop basis between two neighboring devices in 6LoWPANs, and/or for end to end (E2E) between source and destination nodes. Generally, for Internet of things, the end-to-end communication between 6LoWPAN sensing devices and other external or Internet entities will require appropriate security mechanisms that guarantee confidentiality, integrity, authentication and non-repudiation of the transmitted data (Garcia-Morchon et al. 2013; Roman et al.

2011). These security properties should be realized in the context of communication protocol itself, or on the other end by external mechanisms. Another class of security requirement concerns the secure internal communication in 6LoWPAN network. This latter is exposed to several wireless attacks that inherited from the classical WSN. These mechanisms should assure the availability and resilience against such attacks. Finally, privacy and trust are also essential security requirements that should be realized for the social acceptance of the future IoT applications. Today's security researches are directed towards realizing these needs.

5 Challenges and Limits in the Usage of Classic Security Mechanisms

The use of existing security research proposals to protect IP based WSN or 6LoWPAN network faces several challenges as following

- Due to the constraint of WSN devices in term of the limitations of memory, processing power and energy, the usage of existing Internet security's will require the design and adoption of highly optimized mechanisms to support communications and security on such environments.
- Even considering that the improvement of today's devices sensing platforms will give more resources in the future, security operations are remain heavy to applique in these devices. Its integration in internet may also aide to appear new threats and attacks not encountered in classic WSN applications that utilize devices isolated from the Internet.
- Most security research existing base on the homogeneity of devices, unlike the IoT environments that characterize by the heterogeneity of devices and applications. For this, these researches must be adapted.
- The integration of IP based WSN with the Internet will require appropriate security mechanisms, which are able to provide fundamental security assurances to IoT applications, devices and communications.

6 Taxonomy of Security Mechanisms for 6LoWPAN Network

Our study focuses on proposed recent security solutions to secure the 6LoWPAN network traffic. We split these solutions into two classes: the security solutions directed to inside communication of 6LoWPAN network and the security solutions for outside communication of 6LoWPAN network. The Fig. 3 illustrates our proposed classification.

In this section, we describe the reference model that illustrates the scenarios in which the proposed security protocols and IDS systems can be deployed. Then, we present our taxonomy of the security mechanisms for 6LoWPAN network.

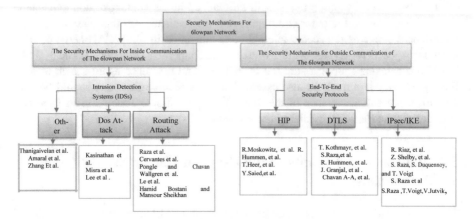

Fig. 3. The proposed taxonomy

6.1   Scenarios Under Consideration

The security mechanisms analyzed in this paper, are illustrated in Fig. 4, in which two segment are considered.

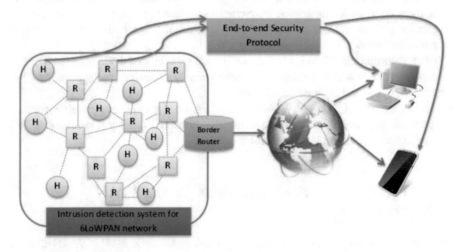

Fig. 4. The security mechanisms for inside and outside communications of the 6LoWPAN network

The first segment contains two entities. The One of them is a device with high resource constraints that is located in the 6LoWPAN network, whereas the second entity can be seen as another constrained resource device or an external Internet server (i.e., with rich resources) that is located outside the network. This communication is needed to the end-to-end security protocols. The second segment represents the

6LoWPAN network in which its nodes that are able to communicate with each other. The border router (6LBR) is the bridge that is connects the sensor node with the outside. The 6LBR may take part in the communication between two entities in a passive (transparent to the communicating parties) or active (as a mediator in the communication process manners. Our study concentrates mainly on securing unicast communications between two entities and the group communications inside the 6LoWPAN network.

6.2   Classification

In this paper, as mentioned early, the proposed security solutions for IoT are categorized into two main classes: the security solutions for outside communication the 6LoWPAN network (end-to-end security protocols) and the security solutions for inside the 6LoWPAN network (intrusion detection systems). This section describes the two first levels of the proposed taxonomy.

6.2.1   The Security Mechanisms for Outside Communication in the 6LoWPAN Network

When the 6LoWPAN sensor node may be connect and transmit data to an external entity in internet, the authentication methods and key establishment process are necessary. To achieve this goal, the security protocols like DTLS, HIP and IKE are exploited to create a secure channel to transmit data. But doe to the nature of this communication when the two entities (the 6LoWPAN sensor node and the server) are heterogeneous in terms of energy power and computation capability, it is impossible to applique these security protocols as they are. Several researches propose the techniques that enable to adapt the internet security protocols like DTLS, IKE and HIP for the IoT challenges. In our taxonomy, we are classified the proposed techniques for the end-to-end security mechanisms according to security protocols to the researches for HIP protocol, the researches for DTLS protocol, the researches for IKE protocol.

6.2.2   The Security Mechanisms for Inside Communication in the 6LoWPAN Network

The 6LoWPAN network as any wireless network is vulnerable to several kinds of attacks as sinkhole, wormhole, Sybil attack, etc. These attacks have a danger effect on the network. For the reduction of this effect, it is necessary to put the security mechanisms in inside the network. These mechanisms are the intrusion detection systems. In our taxonomy, we are classified the proposed IDS for 6LoWPAN network according to the attacks detected.

- *Intrusion detection system*

  An Intrusion Detection System (IDS) is a tool or mechanism that enables to detect an abnormal activity that carries out by an intruder to infect a system or a network through analyzing the activity in the network or in the system itself. Intruders may be external or internal. Internal intruders infect the network from inside. They have some degree of legitimate access. They attempt to raise their access privileges by misuse nonauthorized privileges. External intruders are outside users that target the network by trying to gain unauthorized access to system information.

IDS for the 6LoWPAN network must be ensured the internal security of network which is vulnerable from more attacks and at the same time minimize as much as possible the consumption of the resources, that is to say must be light weight.

6.3   End-to-End Security Protocols

6.3.1   The Proposed Techniques for DTLS Protocol

DTLS (Datagram Transport Layer Security) Rescorla and Modadugu (2012) protocol is based on the Transport Layer Security (TLS) protocol using UDP protocol for transporting service, and it provides equivalent security guarantees for end-to-end communication.

Figure 5 shows a fully authenticated DTLS handshake. These handshake messages are bundled into 6 message flights. Flight 1 and 2 are an optional feature to protect the server against Denial-of-Service (DoS) attacks. The client resends the clientHello message with the cookie sent in the ClientHello Verify message by the server to prove its capability to communicate with the server. That means it is able to send and receive data with the server, this message contains also the protocol version and the cipher suites supported by the client. Thereafter, the server sends Server Hello message that contains the appropriate cipher suite supported at client and a X.509 certificate to authenticate itself followed by a Certificate Request message if the server expects the client to authenticate.

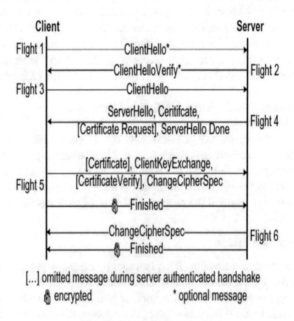

[...] omitted message during server authenticated handshake

🔒 encrypted   * optional message

Fig. 5. The DTLS protocol

The ServerHelloDone message terminates the flight 4. The Finished messages in flights 5 and 6 conclude the handshake. The half of the pre-master secret is encrypted with the server's public RSA key from the server's certificate and sends it in ClientKey Exchange message. The other half of the pre-master secret was transmitted unprotected in the ServerHello message. After that, the keying material is derived from the pre-master secret. The changeCipherSpec message shows that all following messages sends by client are encrypted with the keying material. The Finished message indicates that both client and server are functioned with the same, unaltered, handshake data. Finally, the server terminates the handshake by sends its own ChangeCiperSpec and Finished message DTLS protocol is used by COAP to assure the secure transport of data. However, for WSN, the applicability of DTLS is heavy. Therefore, numerous solutions are proposed to mitigate the overhead of this protocol.

In Kothmayr et al. (2012), they introduced the first fully implemented two way authentication security scheme for IoT, based on existing Internet standards, the Datagram Transport Layer Security (DTLS) protocol. Kothmayr et al. proposed the implementation of DTLS on sensor using hardware assistance. This solution assumes that each sensor is equipped with a TPM (Trusted Platform Module). A TPM is embedded chip that give a secured cryptographic keys and sealed storage as well as hardware support for cryptographic algorithms. The DTLS handshake is realized between a sensor (equipped with TPM) and a subscriber (external entity). Both peers exchange their X.509 certificate contained RSA keys signed by CA to initiate the authentication phase. This solution assures a high security level and provides message integrity, confidentiality and authenticity but necessitate additional hardware.

In Raza et al. (2012a, b, 2013a, b) the authors proposed a 6LoWPAN compression (6LoWPAN_NHC (Hui and Thubert 2011)) for DTLS headers. The Compression covers the following sub-protocols: record, basic, Client Hello and Server Hello messages in handshake protocol. This solution assures a secure communication using DTLS compressed between the internet host and sensor nodes in the 6LoWPAN network. In addition, it reduces the consumption of energy.

Hummen et al. (2013) proposed a solution that reduces the overhead of the DTLS handshake on node. The DTLS handshake procedure requires 15 message exchanges for establish a session key; necessitate also a high dynamic storage capability (RAM) during the communication and more processing time for realizing crypto-graphic tasks. In order to mitigate overhead of DTLS handshake, the authors proposed to offloading all handshake procedures including All certificate related tasks to a highly-resource entity, e.g. the gateway and only the session-state message is sent to the constrained device. With this message, the session can be established between them. This solution reduces effectively the communication overhead. However; the rich-resource entity must be trusted.

Authors in Granjal et al. (2013) proposed an architecture supporting low-power end-to-end transport-layer secure communications with mutual authentication using ECC public-key cryptography for Internet integrated sensing applications through the modification of DTLS. The main role in this architecture is played by the 6LoWPAN Border Router (6LBR) which mediates the exchange message of DTLS handshake protocol but it is transparent to peers of communication (6LoWPAN devices and the Internet host). The border router intercepts and forwards packets at the transport-layer.

From the side of internet, the traditional DTLS operates between the border router and the internet host using ECC based certificate for authentication. From the side of 6LoWPAN network, the 6LBR uses the pre-shared key security mode for protect their communication to the constrained sensing devices. In addition, the authors proposed an authentication protocol inside the 6LoWPAN network inspired from Kerberos protocol to protect internal communication against inside attack (Neuman and Ts'o 1994). When, the authentication is successful, the 6LBR encrypts the pre-master key sent from internet host to 6LBR in the Client Key Exchange message with secret session key shared between the sensing devices and the 6LBR and sends it to sensing device. Then, the sensing device calculates the materiel key. Actually, the end-to-end security is enabled between the 6LoWPAN device and internet host.

Authors in Chavan and Nighot (2016) implemented a compressed DTLS protocol with a raw public key mode in the constrained environment to establish secure communication among devices through COAP, this work minimizes the energy consumption.

6.3.2   The Proposed Techniques for HIP Protocol

HIP (Host Identity Protocol) (Henderson et al. 2015; Moskowitz et al. 2008) protocol is an alternative way to secure agreeing hosts; HIP separates the identity of a host from its location. HIP offers the end-to-end encryption and protection against certain DoS attacks, restores the end-to-end host identification in the presence of several addressing domains separated by Network Address Translation (NAT) devices, allows host mobility and ensures anonymous locations for end users, which is generally highly recommended in the majority of the applications of the Internet of things (IoT).

HIP defines a security handshake mechanism called Base Exchange (HIP-BEX). The HIP Base Exchange is a two-party cryptographic protocol used to establish communications context between hosts. The first party is called the Initiator and the second party the Responder. Only four messages are needed for realize the key negotiation (I1, R1, I2, R2). The message I1 initiates the exchange message. It contains the initiator and responder identities (HITI, HITR). Upon reception of I1, the responder sends a (possibly pre-computed) message R1 composed of a puzzle, its Diffie-Hellman public value, its public key (or Host Identifier) and a signature. Then, the initiator responses with an I2 message that included the puzzle solution (so as to prevent DoS attacks), its own Diffie-Hellman public value, its own (possibly encrypted) public key and a signature. At this stage, the initiator and the responder are able to compute the Diffie-Hellman shared key and derive the master key. Finally, the responder sends R2 message that finalizes the exchange message, contained a HMAC computed using the DH shared key, and a signature. HIP-BEX mechanism involves heavy asymmetric cryptographic operations and for this reason, it cannot be supported as it is by constrained sensor nodes. Therefore, several solutions have been proposed to lighten HIP and, make it more adapted (see Fig. 6).

In order to mitigate the complexity of cryptographic computations, the authors in Moskowitz (2011) introduced an HIP-DEX (Host Identity Protocol Diet Exchange) protocol, it is based on DH protocol to generate a session key between two entities but it removing all digital signatures and implements a static ECDH (Elliptic Curve Diffie-Hellman) to encrypt the session key, this protocol keeps the minimum level of security

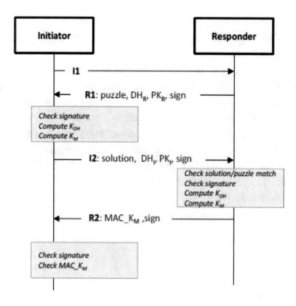

Fig. 6. HIP BEX protocol

by defined only a smallest set of cryptographic primitives (e.g. AES-CBC instead of cryptographic hash functions).

Hummen et al. (2013) proposed a variant of HIP-DEX protocol that used a session resumption mechanism as in TLS (Badra 2009). With this mechanism, the constrained node performs the expensive operations once and maintains session-state for reauthentication and reestablishment of a secure channel.

The authors in Saied and Olivereau (2012) introduced a Lightweight HIP (LHIP) is a variant of HIP, that is characterized by the minimal degree of security. That means, no Diffie-Hellman key is computed, and no secure IPsec tunnel is set up after the exchange. But only a hash chains are used to cryptographically attach successive messages with each other. LHIP protocol is poor in terms of security mechanisms as the mutual authentication and key exchange.

Saied and Olivereau in Heer (2007) presented a Distributed HIP Exchange (D-HIP) protocol inspired from HIP-BEX (Moskowitz et al. 2008). This approach based on distributed the two modular exponentiation operations of DH protocol to the proxy nodes. Thereby, the constrained node enabling to establish a session key with the server. For authorization and authentication of the involved proxies, a one-way hash functions are performed. Each proxy calculates the part Ki of session key K, in a parallel fashion, and sends it to external entity. When the external entity receives all the part Ki, it recalculates the final secret key. A major advantage of this approach is the delegation of all expensive computation tasks to the proxy nodes; this is permitted to keeps the life of constrained nodes. However, the communication cost is considerable and packet lost during communication can happen at any time.

6.3.3   The Proposed Techniques for IPsec/IKE Protocol

The Internet Key Exchange (IKE) protocol, described in RFC 2409 (Kaufman 2005), is a key management protocol standard which is used in conjunction with the IPs standard to establish a secure channel between two parties and enable them to mutually authenticate each other. IPsec can be configured without IKE, but IKE enhances IPsec by providing additional features, flexibility, and ease of configuration for the IPsec standard. All IKE messages are in the form of request-response pairs. An IKE transaction consists of two required request/response exchanges. The first exchange is IKE_SA_INIT that negotiates the security policies as cryptographic algorithms (SAi1, SAr1), exchanges nonces (Ni, Nr) and performs the Diffie-Hellman exchange to establish a master key Km between the two entities communicates. The second request/response exchange (IKE_AUTH) authenticates the previous messages.

The identities of both sides are authenticated, and a simple IPsec SA, called a child SA, is established (see Fig. 7).

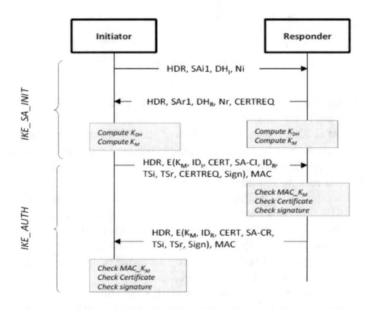

Fig. 7. Basic Internet Key Exchange (Establishment of a simple SA)

The IP-based-WSN or 6LoWPAN is characterized by the resources constraints and limited energy source. for this reason, the adoption of network-layer security approaches as IPSec and IKE in 6LoWPAN environments is the major problem which is discussed in previous contributions like (Riaz et al. 2009; Shelby and Bormann 2011) where the authors explain that the appropriate design for security mechanism must be compatible with the 6LoWPAN adaptation layer and is satisfied from the point of view of security properties such as confidentiality, integrity, authentication and non-repudiation.

The authors in (Shelby and Bormann 2011) proposed the 6LoWPAN compression for IPsec's sub-protocols AH (Authentication Header) (Kent 2005b) and SP (Encapsulating Security Payload) (Kent 2005a) for IoT security. The experimental evaluation of this proposal and its comparison against IEEE 802.15.4 link-layer security is described in Raza et al. (2014).

The Authors in Raza et al. (2012a, b) gave the 6LoWPAN compression for IKE headers in order to reduce the communication and computational cost resulting from the complex asymmetric cryptographic operations of IKE protocol. However, the authors are not evaluated the proposed scheme with others scheme like IP-sec for optimal result.

6.4   Intrusion Detection Systems

As mentioned earlier, the sensor nodes in the IoT are exposed to wireless attacks from the internet and WSN. So, Several IDSs have been proposed for IP-based WSNs in the literature. In our taxonomy, we group the proposed IDS according to the attack detected:

6.4.1   Routing Attacks

In addition to attack originating from the Internet, 6LoWPAN networks are exposed to most of the available attacks against WSNs (Karlof and Wagner 2003). The routing protocol standardized for the 6LoWPAN environment is RPL. This protocol as any vector protocol, is vulnerable to several attacks like sinkhole, selective-forwarding, wormhole attack...etc.

Regarding to select-forward attack which disrupts the path routing, malicious node forwards messages selectively. That means, it may refuse certain messages and simply drop them. This attack has damage effect when coupled with other attacks, for example, sinkhole attacks.

- In sinkhole attack, malicious node attracts its neighbors to forward its messages through it. It advertises itself to surrounding nodes as an optimal routing path. That means it presents a better rank to choosing it as a preferred parent.
- A wormhole is an out of band connection or high transmission between two nodes using wireless links. The packet transmissions between them are performed with high throughput and lower latency than those which is sent between the same pair of nodes over normal multi-hop routing. This attack has more dangerous effect, if it couple with other attack like sinkhole attack. Figure 8 illustrates the routing attacks; a) the selective-forwarding attack, b) Sinkhole attack, c) Wormhole attack.

In this section, we present the sub-category of proposed IDSs given in the literature that are protected the network from effect of these routings attacks.

Raza et al. (2013a, b) proposed a novel host based Intrusion detection system for IoT running on RPL that cope with different attacks, including sinkhole, selective forwarding and Sybil. Two centralized modules are defined in SVELTE, the first is

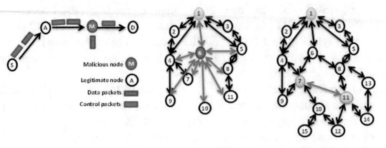

(a) A screenshot of select-forward attack   (b) A screenshot of sinkhole at-   (c) The screenshot of wormhole attack
                                              tack launched by node 6 in DODAG   launched by nodes 7, 11.

Fig. 8. The most dangerous Routing attacks for 6LoWPAN network

6Mapper which reconstruct the RPL topology with IDS parameters, and the second is a mini-firewall which protects the network against outside attacks. They implemented the proposed model in the Contiki operating system. Its simulations showed that its IDS has a suitable for the constrained nodes and is capable to detect most of malicious nodes that launch sinkhole and/or selective-forwarding attacks.

Cervantes et al. (2015) proposed an INTI system (Intrusion detection of Sink-hole attacks on 6LoWPAN for Internet of Things); this system identified sinkhole attack in IoT and reduces adverse effects resulting from IDS systems generally. It combines many strategies like watchdog, reputation and trust in analysis the node behavior. The INTI classifies the nodes as leader, associated or member nodes, composing a hierarchical structure. The role of each node does not fixated but can be changed over time due to the network reconfiguration or an attack event. Then, each node monitors a superior node by estimating it's inbound and outbound traffic. When a node detects an attack, an alarm message is broadcasted by it to the other nodes. So, the attacker is isolated. The simulation results showed the INTI performances and its capability in terms of attack detection rate. However, the authors did not discuss the effect of this solution on low capacity nodes.

Pongle and Chavan (2015) presented an IDS enabled to detect wormhole attacks in IoT devices. The authors depend on set of indicators to detect the presence of wormhole attack on the system, for example, a high number of control packets are exchanged between the two ends of the tunnel, or a high number of neighbors get formed after a successful attack. Using this logic, the authors proposed IDS based on three algorithms to detect such anomalies in the network. According to their experimentation, the system achieved a true positive rate of 94% for wormhole detection and 87% for detecting both the attacker and the attack. More than that, their IDS is suitable to IoT in term of power and memory consumption. However, no comparison with other research in the literature.

Wallgren et al. (2013) proposed a centralized approach in which the IDS is located in the border router. The proposed solution detects an attack within the physical domain. At regular intervals, the border router monitors all nodes by sending ICMPv6 echo requests and expecting the responses to detect unauthorized action from an attacks or availability issues. The experimental result shows that the proposed solution does not required allocation additional memory to nodes. Therefore, the overhead of network energy is minimal.

Le et al. (2016) grouped the network into small group called clusters. Each cluster has a same number of nodes and has a cluster head, which is a node that is able to communicate directly with all the cluster members. The proposed IDS instance is placed in each cluster head which monitors the cluster members by sniffing their communication. Cluster members should be reporting related information about itself and other neighbors to the cluster head. According to authors, this IDS is lightweight because the cluster head might be a more powerful node.

Bostani and Sheikhan (2017) proposed a real-time Hybrid of anomaly-based and specification-based IDS for Internet of Things using unsupervised OPF based on MapReduce approach. Their solution based on hybridization between local detection of specification based on agent placed in router nodes and global detection of anomaly detection based agent located in border router. At each time slot, each router node monitors its parents to detect the sinkhole attack and monitors its children to detect the select-forwarding attack through specification model based on preferred parent for sinkhole attack and packet dropper rate for selective-forwarding attack, then, it sends the list of suspect nodes to border router. This latter selects the real malicious nodes from suspect nodes through Optimum-Path Forest algorithm and voting algorithm. Also, the authors investigate the extension of their model to wormhole attack. The simulation result shows that the proposed solution has a high level of detection.

6.4.2   DoS Attacks

The 6LoWPAN network also is infected by the DoS attack (Denial of Service), this attack consumes network resources, and prevents legitimate user to serve network service requests. It can be coupled with other attacks such as selective-forwarding attack in order to maximize its effect. In this section, we present the sub-category of the proposed IDS that are addressed this attack in 6LoWPAN network.

Kasinathan et al. (2013) presented Denial of Service (DoS) attacks detection architecture for 6LoWPAN network. The authors integrated a signature-based IDS into the network framework developed within ebbits project (European project 2016). The protection manager analyzes the information like channel interference rate and packet dropping rate to detect attack after receiving alert from IDS. The proposed architecture was designed to allow the IDS deployment on a dedicated Linux host, avoiding problems related to low capacity nodes.

Misra et al. (2011) presented an approach to protect IoT middleware from DDoS (Distributed Denial of Service) attacks. The solution suggests specifying the maximum capacity of each middleware layer, when the number of requests to a layer exceeds the specified threshold, the system generates an alert.

Lee et al. (2014) proposed an IDS based on energy consumption level of node. They proposed model of regular energy consumption for mesh-under routing scheme and route-over routing scheme. Then, for each rate of 0.5 s, the node calculates its energy value. If the consumption energy of node deviates from the threshold value, the IDS classifies the node as malicious and removes it from the route table in 6LoWPAN. However, no false positive rate, no negative rate presented from the authors.

6.4.3   Other Attacks

This sub-category is about the IDS that intended for other attacks as RC intrusion, etc., in this section, we present these IDSs:

Thanigaivelan et al. (2016) briefly introduced a distributed internal anomaly detection system for IoT. Their IDS based on discrepancies in the network, monitoring the characteristics of one-hop neighbor nodes such as packet size and data rate to look any deviation. According to this monitoring, the system builds the normal behavior. However, unclear how the detection algorithm would work on IoT low capacity nodes.

Amaral et al. (2014) proposed the specification-based IDS, which is hosted by selected nodes called watchdog that allows to use a set of rules created by the network administrator for attack detection. When one of these rules is violated, the IDS sends an alert to the Event Management System (EMS). This later runs on a node whit high resource to correlate the alerts for different nodes in the network.

Zhang et al. (2015) presented the vulnerability of internal intrusions on self-organizing of RPL mechanisms, showed a new RPL internal intrusion called RC intrusion and evaluated the threat of a type RC intrusion. They introduced a design IDS including IDS detection methodology, system architecture, detection data and intrusion response that allows the defense against a type RC intrusion.

7   Evaluation of Security Mechanisms for 6LoWPAN Network in IoT Context

In this section, we conduct a classification of IDS proposed in the literature and classification of end-to-end security mechanisms for IoT context. For end-to-end security mechanisms, every work was classified according to following attributes: basic protocol, operational layer, adaptation cost, distribution, authentication, overhead, resilience, scalability, extensibility, communication cost, computation cost (see Table 1). For IDS proposals, each work was classified regarding the following Attributes: number for attack detected, IDS modules position, methods detection, attack type. The Table 1 summarizes the proposed end-to-end mechanisms for IoT and the Table 2 recaps the IDS proposals for IoT (see Table 2) according to different metric.

Where +, − denote respectively: supported, not supported in Distribution, Authentication, Resilience, Scalability and Extensibility metrics and denote respectively: Important, Low in Overhead, communication cost, computation cost and adaptation cost metrics.

Table 1. The comparison of different proposed end-to-end security mechanisms according to different metric

E to E security solution	Basic protocol	Operational layer	Adaptation cost	Distribution	Authentication	Overhead	Resilience	Scalability	Extensibility	Communication cost	Computation cost
[8]	DTLS	Transport	+	+	+	+	+	+	+	–	–
[9, 10]	DTLS	Transport	–	+	+	–	+	+	+	–	–
[12]	DTLS	Transport	–	+	+	–	+	+	+	–	–
[13]	DTLS	Transport	–	+	+	+	+	+	+	–	–
[14]	DTLS	Transport	–	+	+	–	+	+	–	–	–
[17]	HIP	Network	–	+	+	–	–	+	+	–	–
[18]	HIP	Network	–	+	+	–	+	+	+	–	–
[20]	HIP	Network	–	+	–	–	+	+	+	–	–
[21]	HIP	Network	–	+	+	+	+	+	+	+	+
[25]	IPsec	Network	–	+	+	–	+	+	+	–	–
[29]	IKE	Network	–	+	+	–	+	+	+	–	–

Table 2. The comparison of different proposed IDSs

Methods	Numbers of detected attack	IDS modules position	Detection methods	Attack type
Raza et al. [31]	Multiple	Hybrid		Routing attack
Cervantes et al. [5]	Single	Distributed	Hybrid	Routing attack
Pongle and Chavan [27]	Single	Hybrid	Specification	Routing attack
Wallgren et al. [40]	Single	Centralized		Routing attack
Le et al. [19]	Multiple	Hybrid	Specification	Routing attack
Hamid Bostani and Mansour Sheikhan [4]	Multiple	Hybrid	Specification and anomaly	Routing attack
Kasinathan et al. [17]	Single	Centralized	Signature	Dos attack
Misra et al. [22]	Single		Specification	Dos attack
Lee et al. [20]	Single	Distributed	Anomaly	Dos attack
Thanigaivelan et al. [39]		Distributed	Anomaly	
Amaral et al. [2]		Hybrid	Specification	
Zhang et al. [42]	Single	Hybrid		

8   Discussion, Open Issue, Future Research Direction

After classifying articles between articles for end-to-end security mechanisms and articles for intrusion detection systems for 6LoWPAN networks in the IoT context, we observe that these works do not explain the weak and strong points for each proposition, do not give any idea around the possibility of investigating these approaches in very specific fields of application of IoT, except some works like [38]. In the end-to-end security mechanisms, the research is essentially based on the three protocols: IKE, HIP, DTLS, to be adapted to the needs of IoT. But they do not give an idea about the possibility of parallel security in the network layer by IKE or HIP and the transport layer by DTLS. Is that if possible or not for an environment like the 6LoWPAN network. If one concentrates on the DTLS protocol as a functional protocol in the IoT domain, we observe that no work explains how to use this protocol within the 6LoWPAN network or between two 6LoWPAN networks in a different environment. Also, what is the closest owner among the approaches proposed for this protocol, which really meets the requirements of this new concept? All these questions and others open the door for further work in the future. Returning to works on intrusion detection systems, these works propose different methods for detecting between specification-based detection, anomaly-based detection, and signature-based detection, but there are no a consensus on the most appropriate method for IoT in terms of energy consumption and the most strategic location for IDS. Therefore, researchers need to do other experiments and tests to clearly clarify the strengths and weaknesses in each detection method under real conditions in the IoT environment. These experiments will deal with the following IDS characteristics: Attack Detection Accuracy, Signaling Speed, and Power Consumption. On the other hand, it is the extensibility of the detection system modules. These proposed works focuse on certain types of attacks such as routing attacks, DOS attacks. And as a conclusion, the future security mechanisms for 6LoWPANs networks must adapt to the application domains requirements where will be applied.

9   Conclusion

When a physical object of our life transmits and receives data on the internet, it can be dangerous on our private life. To protect our lives and also to put our intelligent lives, we need security mechanisms for the IoT system. Several works have appeared in this area. In this paper, we try to present a Survey around the security works that are proposed for IoT. In our Survey, we propose a classification of these researches that based on security protocols like DTLS, HIP and IKE for the outward security of 6LoWPAN network and detected attacks like routing attacks, DoS attacks for intrusion detection systems inside 6LoWPAN. We observe that this area of research remains at the beginning, several points is not studied whether it is for end-to-end security mechanisms like the end-to-end security between two networks 6LoWPAN, or for the IDS as the proposal of the IDS specifically for each IoT application area.

Finally, as a search for the future, researchers must focus on IoT's application areas and their need for IDS, address other types of attacks especially new attacks, and try to augment the detection rate.

References

Airehrour, D., Gutierrez, J., Ray, S.K.: Secure routing for internet of things: a survey. J. Netw. Comput. Appl. **66**, 198–213 (2016)

Amaral, J.P., Oliveira, L.M., Rodrigues, J.J., Han, G., Shu, L.: Policy and network-based intrusion detection system for IPv6-enabled wireless sensor networks. In: 2014 IEEE International Conference on Communications (ICC), pp. 1796–1801. IEEE (2014)

Badra, M.: Pre-shared key cipher suites for transport layer security (TLS) with SHA-256/384 and AES Galois Counter Mode (2009)

Bostani, H., Sheikhan, M.: Hybrid of anomaly-based and specification-based IDS for Internet of Things using unsupervised OPF based on MapReduce approach. Comput. Commun. **98**, 52–71 (2017)

Cervantes, C., Poplade, D., Nogueira, M., Santos A.: Detection of sinkhole attacks for supporting secure routing on 6LoWPAN for Internet of Things. In: 2015 IFIP/IEEE International Symposium on Integrated Network Management (IM), pp. 606–611. IEEE (2015)

Chavan, A.A., Nighot, M.K.: Secure and cost-effective application layer protocol with authentication interoperability for IOT. Procedia Comput. Sci. **78**, 646–651 (2016)

Garcia-Morchon, O., Kumar, S., Struik, R., Keoh, S., Hummen, R.: Security considerations in the IP-based Internet of Things (2013)

Granjal, J., Monteiro, E., Silva, J.S.: End-to-end transport-layer security for Internet-integrated sensing applications with mutual and delegated ECC public-key authentication. In: IFIP Networking Conference, 2013, pp. 1–9. IEEE (2013)

Heer, T.: LHIP lightweight authentication extension for HIP, draft-heer-hip-lhip-00 (IETF work in progress), February (2007)

Henderson, T., Heer, T., Jokela, P., Moskowitz, R.: Host identity protocol version 2 (HIPv2) (2015)

Hui, J., Thubert, P.: Compression format for IPv6 datagrams over IEEE 802.15. 4-based networks (2011)

Hummen, R., Wirtz, H., Ziegeldorf, J.H., Hiller, J., Wehrle, K.: Tailoring end-to-end IP security protocols to the Internet of Things. In: 2013 21st IEEE International Conference on Network Protocols (ICNP), pp. 1–10. IEEE (2013)

Karlof, C., Wagner, D.: Secure routing in wireless sensor networks: attacks and countermeasures. In: Proceedings of the First IEEE. 2003 IEEE International Workshop on Sensor Network Protocols and Applications, 2003, pp. 113–127. IEEE (2003)

Kasinathan, P., Pastrone, C., Spirito, M.A., Vinkovits, M.: Denial-of-Service detection in 6LoWPAN based Internet of Things. In: 2013 IEEE 9th International Conference on Wireless and Mobile Computing, Networking and Communications (WiMob), pp. 600–607. IEEE (2013)

Kaufman, C.: Internet key exchange (IKEv2) protocol. Report no. 2070–1721 (2005)

Kent, S. IP encapsulating security payload (ESP). Report no. 2070–1721 (2005a)

Kent, S.: RFC 4302: IP Authentication Header (AH). In: Request for Comments, IETF (2005b)

Kothmayr, T., Schmitt, C., Hu, W., Brünig, M., Carle, G.: A DTLS based end-to-end security architecture for the Internet of Things with two-way authentication. In: 2012 IEEE 37th Conference on Local Computer Networks Workshops (LCN Workshops), pp. 956–963. IEEE (2012)

Le, A., Loo, J., Chai, K.K., Aiash, M.: A specification-based IDS for detecting attacks on RPL-based network topology. Information **7**(2), 25 (2016)

Lee, T.-H., Wen, C.-H., Chang, L.-H., Chiang, H.-S., Hsieh, M.-C.: A lightweight intrusion detection scheme based on energy consumption analysis in 6LowPAN. In: Advanced Technologies, Embedded and Multimedia for Human-centric Computing, pp. 1205–1213. Springer (2014)

Miorandi, D., Sicari, S., De Pellegrini, F., Chlamtac, I.: Internet of things: vision, applications and research challenges. Ad Hoc Netw. **10**(7), 1497–1516 (2012)

Misra, S., Krishna, P.V., Agarwal, H., Saxena, A., Obaidat, M.S.: A learning automata based solution for preventing distributed denial of service in Internet of things. In: Internet of Things (iThings/CPSCom), 2011 International Conference on and 4th International Conference on Cyber, Physical and Social Computing, pp. 114–122. IEEE (2011)

Moskowitz, R.: HIP Diet EXchange (DEX): draft-moskowitz-hip-rg-dex-05. Internet Engineering Task Force, Status: Work in progress. Technical report (2011)

Moskowitz, R., Nikander, P., Jokela, P., Henderson, T.: Host Identity Protocol (2008)

Neuman, B.C., Ts'o, T.: Kerberos: an authentication service for computer networks. IEEE Commun. Mag. **32**(9), 33–38 (1994)

Nguyen, K.T., Laurent, M., Oualha, N.: Survey on secure communication protocols for the Internet of Things. Ad Hoc Netw. **32**, 17–31 (2015)

Pongle, P., Chavan, G.: Real time intrusion and wormhole attack detection in internet of things. Int. J. Comput. Appl. **121**(9) (2015)

Raza, S., Duquennoy, S., Chung, T., Yazar, D., Voigt, T., Roedig, U.: Securing communication in 6LoWPAN with compressed IPsec. In: 2011 International Conference on Distributed Computing in Sensor Systems and Workshops (DCOSS), pp. 1–8. IEEE (2011)

Raza, S., Duquennoy, S., Höglund, J., Roedig, U., Voigt, T.: Secure communication for the Internet of Things—a comparison of link-layer security and IPsec for 6LoWPAN. Secur. Commun. Netw. **7**(12), 2654–2668 (2014)

Raza, S., Shafagh, H., Hewage, K., Hummen, R., Voigt, T.: Lithe: lightweight secure CoAP for the internet of things. IEEE Sens. J. **13**(10), 3711–3720 (2013a)

Raza, S., Trabalza, D., Voigt, T.: 6LoWPAN compressed DTLS for CoAP. In: 2012 IEEE 8th International Conference on Distributed Computing in Sensor Systems (DCOSS), pp. 287–289. IEEE (2012)

Raza, S., Voigt, T., Jutvik, V.: Lightweight IKEv2: a key management solution for both the compressed IPsec and the IEEE 802.15. 4 security. In: Proceedings of the IETF Workshop on Smart Object Security, vol. 23 (2012)

Raza, S., Wallgren, L., Voigt, T.: SVELTE: real-time intrusion detection in the Internet of Things. Ad Hoc Netw. **11**(8), 2661–2674 (2013b)

Rescorla, E., Modadugu, N.: Datagram transport layer security version 1.2 (2012)

Riaz, R., Kim, K.-H., Ahmed, H.F.: Security analysis survey and framework design for ip connected lowpans. In: International Symposium on Autonomous Decentralized Systems, 2009. ISADS'09, pp. 1–6. IEEE (2009)

Roman, R., Alcaraz, C., Lopez, J., Sklavos, N.: Key management systems for sensor networks in the context of the Internet of Things. Comput. Electr. Eng. **37**(2), 147–159 (2011)

Saied, Y.B., Olivereau, A.: D-HIP: a distributed key exchange scheme for HIP-based Internet of Things. In: 2012 IEEE International Symposium on a World of Wireless, Mobile and Multimedia Networks (WoWMoM), pp. 1–7. IEEE (2012)

Shelby, Z., Bormann, C.: 6LoWPAN: The Wireless Embedded Internet, vol. 43. Wiley, New York (2011)

Thanigaivelan, N.K., Nigussie, E., Kanth, R.K., Virtanen, S., Isoaho, J.: Distributed internal anomaly detection system for Internet-of-Things. In: 2016 13th IEEE Annual Consumer Communications and Networking Conference (CCNC), pp. 319–320. IEEE (2016)

Wallgren, L., Raza, S., Voigt, T.: Routing attacks and countermeasures in the RPL-based Internet of Things. Int. J. Distrib. Sens. Netw. **9**(8), 794326 (2013)

Zarpelão, B.B., Miani, R.S., Kawakani, C.T., de Alvarenga, S.C.: A survey of intrusion detection in Internet of Things. J. Netw. Comput. Appl. (2017)

Zhang, L., Feng, G., Qin, S.: Intrusion detection system for RPL from routing choice intrusion. In: 2015 IEEE International Conference on Communication Workshop (ICCW), pp. 2652–2658. IEEE (2015)

Multi-biometric Template Protection: An Overview

Fatima Bedad[✉] and Réda Adjoudj

Evolutionary Engineering and Distributed Information Systems Laboratory,
EEDIS–Computer Science Department, University of Sidi Bel-Abbès,
Sidi Bel-Abbès, Algeria
bedad_fatima2006@yahoo.fr, AdjReda@yahoo.fr

Abstract. Today, biometrics is a research field in full expansion, several identification and verification systems are now being developed, however their performance is inadequate to meet the growing needs of security. The use of a single biometric modality decreases, in most cases, the reliability of these systems, which prompted us to combine several modality. This article provides a complementary aspect to the improved performance of biometric systems namely the protection of privacy. Techniques have been proposed in the last decade on the biometric template protection. Include BioHashing that wants a generalization of biometrics a hash function to protect the biometric data. The main objective is the protection of privacy. This paper presents state of the art specific algorithms, which are illustrated with recent work by researchers in this field.

Keywords: Multi-biometrics · Protection of privacy · Biometric cryptosystems
Revocable biometrics · BioHashing

1 Introduction

Literally, biometrics is a science that measures the characteristics of living things. In few years, biometrics refers more specifically to the identification or identity verification of individuals based on morphological data (face, fingerprint, iris…) or behavioral (voice, dynamic signature, keystroke dynamics…). It represents a simple and ergonomic way of identifying with what we are, as opposed to what one knows (password, PIN…) that you have (USB key, smart card, we generally speak of token), or how we do certain things (dynamic signature, keystroke dynamics). Many biometric applications have been developed include the electronic passport identity check, access control to secure buildings, or the fingerprint sensors present on newer laptops and smartphones.

In general, any use of a biometric system requires two phases:

- An enrollment phase, where the biometric data (fingerprint, face, iris, …) are captured, modeled (to build the reference biometric of the individual) and stored;
- A verification phase of comparing the new capture with the reference of the alleged individual.

M. Hatti (Ed.): ICAIRES 2018, LNNS 62, pp. 70–79, 2019.
https://doi.org/10.1007/978-3-030-04789-4_7

Each biometric feature has its advantages and disadvantages. Using a single modality sometimes involves some inconveniences which must be faced-off as bad data quality, non-universality of the selected characteristic, the poor results and the possibility of identity theft. That is why more and more research focuses on the merger of several sources of biometric information: multimodal biometrics helps alleviate some disadvantages of single modality, but mostly to increase performance (Rathgeb et al. 2015).

Despite the advantages of biometric systems over traditional systems, they are vulnerable to specific attacks that can significantly degrade their functionality and the value of deploying such systems. Thus, the safety assessment of biometric systems has become essential to ensure the operability of these systems.

The performance of a biometric system is commonly described by its False Acceptance Rate (FAR) and False Rejection Rate (FRR). The two measurements can be controlled by adjusting a threshold, but it is not possible to exploit this threshold simultaneously reducing FAR and FRR. FAR and FRR must be traded-off, as reducing FAR increases FRR and vice versa. Another index of performance is Equal Error Rate (EER) defined as the point where FAR and FRR are equal. A perfect system would have zero EER.

The biometric templates protection program that are commonly classified as - biometric cryptosystems and the revocable biometrics are designed to meet the four following properties (Kanade et al. 2010): (Diversity): secure model should not allow cross-match on databases, ensuring user privacy; (Révocability): it should be simple to revoke a compromised model and restart a new model based on the same biometric data; (Irréversibility): it must be difficult to obtain the calculations of original biometric template from the secure model. This property prevents an opponent to create a physical parody of the biometric feature from a stolen model; (Performance): the biometric template protection program should not degrade the recognition performance (FAR and FRR) of the biometric system.

This paper is organized as follows. Section 2 describes the general of Template protection, we present general principle of biometric cryptosystems and its work variations and we present the general principle of revocable biometrics and the technical BioHashing. In Sect. 3, describes the multi-biometric template protection and their related work then present a conclusion and discuss our work perspectives.

2 Template Protection

Biometric template protection technologies offer significant advantages to enhance the privacy and security of biometric systems, providing reliable biometric authentication at a high security level.

Biometric template protection schemes are commonly categorized as (A) biometric cryptosystems (also referred to as helper data-based schemes) and (B) cancelable biometrics (also referred to as feature transformation).

2.1 Biometric Cryptosystems (BCS)

The first approach to protect a biometric template is using a biometric cryptosystem. In a typical biometric cryptosystem, an external key is associated with the biometric data to obtain the so called secure sketch that does not reveal any information about the biometric template. During authentication, the query is used to recover the original biometric template from the secure sketch and the exact recovery of the original biometric data is verified to authenticate a user. In addition to hiding the biometric data, a biometric cryptosystem also serves as a mechanism to secure a key used in another cryptosystem. Note that key management is one of the most crucial issues related to a typical cryptographic design.

Fuzzy commitment (Feng and Choong Wah 2002) and fuzzy vault (Chavan and Gawande 2015) are two of the more practical biometric cryptosystems that can be used to secure biometric templates represented in the form of a binary vector and a set of points, respectively. A fuzzy commitment works on the basis of using an error correcting code to account for the difference between the template and query represented in the form of a binary string. Here, the secure sketch consists of the difference between a codeword obtained from a key and the binary biometric template and, during authentication, the sketch is added to the query to obtain a corrupted version of the codeword which can be easily decoded to recover the key.

2.2 Cancelable Biometrics

In 2001, (Ratha et al. 2001) proposed to change views on the secure storage of biometric templates. Instead of trying to derive cryptographic elements from biometrics, they suggest to directly distort the templates in the biometric space.

The cancelable biometrics (CB) approach consists in doing the verification using a biometric matcher. The principle is to replace a biometric template by a revocable one, through a kind of one-way transformation.

A cancelable biometrics system is defined through a family of distortion functions $F = \{f_i\}_i$. The functions $f_i: B \rightarrow B$ transform a biometric template b into another biometric template $f_i (b)$.

Two main categories of revocable biometrics are distinguished:

(a) *Non-invertible Transforms:* In these approaches, biometric data are transformed applying a non-invertible function. In order to provide updatable templates, parameters of the applied transforms are modified. The advantage of applying non-invertible transforms is that potential impostors are not able to reconstruct the whole biometric data even if transforms are compromised. However, applying non-invertible transforms mostly implies a loss of accuracy. Performance decrease is caused by the fact that transformed biometric templates are difficult to align (like in BCS) in order to perform a proper comparison and, in addition, information is reduced.

First approaches to non-invertible transforms, which have been applied to face and fingerprints, include block-permutation and surface-folding. Diverse proposals (Hämmerle-Uhl et al. 2009) have shown that recognition performance decreases noticeably compared to original biometric systems. Additionally, it is doubtable if sample images of transformed biometric images are non-invertible.

(b) *BioHashing:* The pioneering article is Article Ratha (Rathgeb and Busch 2014), which presents the general principle of BioHashing. This method distorts the biometric signal with a selected transformation function. Revocability is guaranteed because when transformed data is compromised, simply change the transformation function. Diversity is also ensured by the choice of different functions for different applications. However, finding such functions is not easy. In fact, besides the non-invertibility, these functions must exhibit two essential properties: intra-class robustness (that is to say, a robustness vis-à-vis variations of biometric data of an individual) and an inter-class sensitivity (one must be able to distinguish between two different individuals).

Processing allows issuing different models of a person for different applications using different processing parameters. Thus, the models can not be matched in the database for the protection of privacy. In addition, if it is found that the model is compromised, it can be canceled and a new model can be initiated using the same biometric feature. This ability to change the model and issue a new one called revocability.

Rathgeb and Busch (2014) proposed three different transformations for fingerprints (Cartesian, polar and functional). These transformations are unidirectional transformations in a manner that is not possible (or practically possible) to obtain biometric data from the original transformed data.

However, the focus in this article on revocable biometric system belongs to this class of systems using transformation functions: it is the BioHashing. This is a dual mechanism authenticator (the biometric data and a random key), one of the strengths is to increase the entropy of biometrics with the secret key stored on a token. The biometric data after processing is called BioCode. This BioCode depends on biometric data and external data (key) and may not be reproduced without the simultaneous presentation of these data. Tokens avoid the problem of replay attack using a dynamic mechanism.

The first work on the BioHashing was presented in 2003 in Goh and Ngo (2003) from face recognition. This technique was later applied to other biometric data: fingerprints (Rathgeb et al. 2015). Further work on the principle applied to BioHashing fingerprints was presented in Ratha et al. (2007), where a representative vector of the fingerprint image is derived using a filter of Gabor.

3 Multi-biometric Template Protection

The major goal of research in the area of multi-biometric template protection is to generate an industrial projects are presented on generic framework. The system should be capable of incorporating n templates, without the necessity to follow specific fusion levels for their representation, (k representations could be involved). The process is continued with a common representation and then the generic system is applied for the protection of the template (Fig. 1). In a fusion module a common representation of feature vectors is established and feature vectors are combined in a sensible manner. Subsequently, an adequate template protection scheme is applied to protect the multi-biometric template. Focusing on a generic fusion of multiple biometric templates in a template protection system several issues evolve (Rathgeb and Busch 2012).

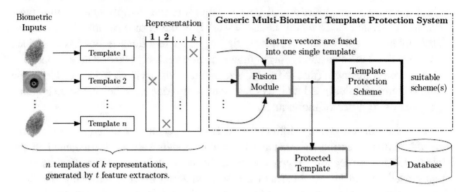

Fig. 1. A framework of a generic multi-biometric template protection at feature level (Rathgeb and Busch 2012)

Figure 1 shows the framework of a generic multi-biometric template protection at feature level (Rathgeb and Busch 2012).

Different techniques have been proposed, the paper wrote by Kanade et al. (2010) proposed an approach to combine multi-biometrics with cryptography for obtaining high entropy keys, and propose a novel method of Feature Level Fusion through Weighted Error Correction (FeaLingECc). With this method, different weights can be applied to different biometric data. The shooing scheme, which we applied earlier to the biometric data, is used in this system to randomize the error correcting codes data which helps make the system more secure. Two systems are proposed: (1) a multi-unit type system, and (2) a multimodal type system. Information from the left and right iris of a person is combined in the multi-unit type system to obtain long and high entropy crypto-bio keys. The second scheme is a multi-modal biometrics based system in which information from iris and face is combined. For the two-iris tests, we can obtain 147-bit keys having 147-bit entropy with 0% FAR and 0.18% FRR. And with multi-modal system (iris + face) we can obtain 210-bit keys having 183-bit entropy at 0.91% FRR and 0% FAR.

In 2010 (Argyropoulos et al. 2010) presented a framework for biometric template security in multimodal biometric authentication systems based on error correcting codes. This method binds the biometric template in a cryptographic key and SVM method where face and gait biometrics are employed un this method. The resultant operation point EER = 3.05% and EER = 2.54% for SVM method.

(Kelkboom et al. 2009) described the application of the protected template from two 3D recognition algorithms (multi-algorithm fusion) at feature-, score-, and decision-level. The authors show that fusion can be applied at the known fusion-levels with the template protection technique known as the Helper-Data System

In 2012 (Yang et al. 2012) presented an evaluation of decision level fusion results of fingerprint minutiae based pseudonymous identifiers generated by three biometric template protection algorithms developed in the European research project TURBINE. There are eight different fusion scenarios covering multiple samples, algorithms, sensors, instances, and their combinations in our tests. On the binary decision level, and evaluate their biometric performance and fusion efficiency on a multi-sensor fingerprint database with 71,994 samples. The resultant fusion operation point (FAR = 0.0012; FRR = 0.0703).

In the same year (Shanthini and Swamynathan 2012), proposed system the messages communicated between the users are encrypted by the cancellable cryptographic key generated from fingerprint features of the receiver by applying genetic operators and are embedded inside the scrambled face biometrics of the sender using steganography method. Fingerprint images and face images are chosen due to their unique physiological traits. Genetic algorithms and steganography are taken into our system. The method presented in this study remains as a preliminary approach to realize biometric security in applications which need high security and is designed for high security small group coalition operations and may not be suitable for enterprise usage.

In 2015, Rathgeb et al. (2015) presented a generic feature level fusion of protected templates, obtained from irreversible Bloom filter based transforms, and applied it to face and iris samples. A detailed privacy analysis of the presented approach is given and the experimental evaluation shows that the protected multi-biometric system maintains the biometric performance (EER _ 0:4%) compared to the score level fusion of the original unprotected systems. The privacy of the user is also increased: an eventual attacker would have to try up to _ 2268 sequences in order to recover the unprotected binary templates.

Kelkboom et al. (2009) presented a novel system for fingerprint privacy protection by combining two fingerprints into a new identity. In the enrollment, two fingerprints are captured from two different fingers. They extract the minutiae positions from one fingerprint, the orientation from the other fingerprint, and the reference points from both fingerprints. Based on this extracted information and their proposed coding strategies, a combined minutiae template is generated and stored in a database. In the authentication, the system requires two query fingerprints from the same two fingers which are used in the enrollment. The combined fingerprint issues a new virtual identity for two different fingerprints, which can be matched using minutiae based fingerprint matching algorithms. This system is able to achieve a very low error rate with FRR = 0.4% when FAR = 0.1%.

In the same year, (Wanjare and Tagalpallewar 2015) described from security point of view, fingerprints and biological data in general constitute sensitive information that has to be protected. The acquisition mode is responsible for acquiring fingerprint image, face image, Speech signal for user who intends to access a system. The verification module consist of four stages fingerprint verification, face recognition, speaker verification, decision fusion fingerprint verification. For the contribution to the above subject an algorithm is developed on banking security. For this consider a bank using biometric technology for its security purpose. The security is assured by using finger scan, voice scan and hand geometry scan and by requesting the password given by the bank for a particular user when necessary. The disadvantages of this approach is Reading always different: - Regardless of the method, extracted multibiometric data is typically different every time - even for the same person (environment, age ...).

Damasceno et al. (2015) proposed and evaluate a multi-privacy biometric protection scheme using set systems. Set systems, also known as multi-classifier systems or fusion of experts. The multi-privacy scheme combines the use of multi-algorithm and multiple protected biometric samples. Four cancellable transformations (Interpolation, BioHashing, BioConvolving and Double Sum) were used to protect a behavioral modality (TouchAnalytics). However, the authors using multiple matching algorithms or similarly. The EER of Biohashing, 28.6% in the best case (BioCBioH). This approach user inconvenience because the individual needs to present more than one key to encode his biometric sample during user verification.

In 2016, (Ramachandra et al. 2016) In this work a multi-biometric template protected system is proposed, based on Bloom filers and binarized statistical image features (BSIF). Features are extracted from face and both periocular regions and templates protected using Bloom filters. Score level fusion is applied to increase recognition accuracy. The system is tested on a database, consisting of 94 subjects, of images collected with smartphones. A comparison between unprotected and protected templates in the system shows the feasibility of the template protection method with observed Genuine-Match-Rate (GMR) of 95.95% for for unprotected templates and 91.61% at a False-Match-Rate (FMR) of 0.01%.

In 2017, (Yildiz et al. 2016) presented a biometric authentication framework that constructs a multi-biometric template by layering multiple biometrics of a user, such that it is difficult to separate the individual layers. Thus, the framework uses the biometrics of the user to conceal them among one another. The resulting biometric template is also cancelable if the system is implemented with cancelable biometrics, such as voice. We present a realization of this idea combining two or three different fingerprints of the user, using four different methods of template construction. Three of the methods use less and less information of the constituent biometrics, so as to lower the risk of leakage and cross-link rates. Results are evaluated on publicly available Finger Verification Championship (FVC) 2000, 2002 and NIST fingerprint databases. With the FVC databases, we obtain 2.1%, 3.9% and 3.4% Equal Error Rate on average using the three proposed methods, while the state-of-the art commercial system achieves 1.9%. Furthermore, we show low cross-link rates under 63% under different scenarios, while genuine identification rates are 100%, with such a small gallery of 55 templates.

Karthi and Azhilarasan (2013) developed an unlinkability and irreversibility analysis of the socalled Bloom filter-based iris biometric template protection introduced in (Ratha et al. 2007) Firstly we analyse unlinkability on protected templates built from two different iris codes coming from the same iris whereas Hermans et al. Analysed only protected templates from the same iriscode. Moreover we introduce an irreversibility analysis that exploits non-uniformity of the biometric data. Our experiments demonstrate new vulnerabilities of this scheme. Then we will discuss the security of other similar protected biometric templates based on Blooms filters that have been suggested in the literature since 2013. Finally we suggest a Secure Multiparty

Computation (SMC) protocol, that benefits of the alignment-free feature of this Bloom filter construction, in order to compute efficiently and securely the matching scores.

Recently, some works are being carried out to hybrid template protection technique which combines both Biometric CryptoSystem (BCS) and Cancelable Biometrics (CB) was proposed by Karthi and Azhilarasan (2013).

The proposed system uses key generating cryptosystem and feature transformation method. The limitation of the feature transform was overcome by this approach. We focus on these techniques using Iris, Fingerprint (i.e. multimodal Biometrics)as traits. The Experimental Results shows no degradation in results by combining both the Biometric CryptoSystem (BCS) and Cancelable Biometrics (CB).

Another Hybrid approach presented by Jain et al. (2008) for Securing Multibiometric Templates Based on Cancelable and Fuzzy Commitment Scheme Right and left irises of a single individual will be used as input templates. The experiment will be carried out using CASIA-v3 iris database to verify the soundness of the proposed system. This research is expected to show that the proposed hybrid template protection scheme can satisfy all template protection requirements without degrading the iris recognition performance.

4 Conclusion

The presented paper provides an overview of multi-biometric template protection. While technologies, multi-biometric recognition [21] and biometric template protection (Feng and Choong Wah 2002), suffer from serious drawbacks a sensible combination of these could eliminate individual disadvantages. Different template protection systems incorporating multiple biometric traits, which have been proposed in literature, are summarized. While, at first glance, multi-biometric template protection seems to solve several drawbacks, diverse issues arise. Based on a theoretical framework for multi-biometric template protection several issues, e.g. template alignment at feature level, are elaborated and discussed in detail. While generic approaches to the construction of multi-biometric template protection schemes have remained elusive we provide several suggestions for designing multi-biometric template protection systems.

The revocable biometrics including BioHashing is a major issue at present. Despite active research in recent years in the proposal for biometric data protection schemes, few studies have focused on the security and robustness of the protocols. This is, however, essential in an area such as biometrics for handling highly sensitive data.

In our work, we propose to address the multi-biometric data protection. It can be carried out by protecting each system involved but a joint protection could be interesting. We use the BioHashing technique, fusion biocodes (result of the protection of each biometric data), hybrid method.

References

Jain, A.K., Nandakumar, K., Nagar, A.: Biometric template security. EURASIP J. Adv. Signal Process. **2008**, 113 (2008)

Wanjare, M.P.A., Tagalpallewar, M.M.H.: Secure internet verification base on multibiometric system. Int. J. Res. Emerg. Sci. Technol. **2**(1) (2015)

Argyropoulos, S., Tzovaras, D., Ioannidis, D., Damousis, Y., Strintzis, M.G., Braun, M., Boverie, S.: Biometric template protection in multimodal authentication systems based on error correcting codes. J. Comput. Secur. **18**(1), 161–185 (2010)

Damasceno, M., Canuto, A.M., Poh, N.:Multi-privacy biometric protection scheme using ensemble systems. In: 2015 International Joint Conference on Neural Networks (IJCNN), pp. 1–8. IEEE (2015)

Feng, H., Choong Wah, C.: Private key generation from on-line handwritten signatures. Inf. Manag. Comput. Secur. **10**(4), 159–164 (2002)

Goh, A., Ngo, D.C.: Computation of cryptographic keys from face biometrics. In: IFIP International Conference on Communications and Multimedia Security pp. 1–13. Springer, Berlin (2003)

Hämmerle-Uhl, J., Pschernig, E., Uhl, A.: Cancelable iris biometrics using block re-mapping and image warping. In: International Conference on Information Security, pp. 135–142. Springer, Berlin (2009)

Kanade, S., Petrovska-Delacrétaz, D., Dorizzi, B.: Obtaining cryptographic keys using feature level fusion of iris and face biometrics for secure user authentication. In: 2010 IEEE Computer Society Conference on Computer Vision and Pattern Recognition Workshops (CVPRW), pp. 138–145. IEEE (2010)

Karthi, G., Azhilarasan, M.: Hybrid multimodal template protection technique using fuzzy extractor and random projection. IJRCCT **2**(7), 381–386 (2013)

Kelkboom, E.J.C., Zhou, X., Breebaart, J., Veldhuis, R.N., Busch, C.: Multi-algorithm fusion with template protection. In: IEEE 3rd International Conference on Biometrics: Theory, Applications, and Systems, BTAS 2009, pp. 1–8. IEEE (2009)

Rathgeb, C., Busch, C.: Multi-biometric template protection: Issues and challenges. In New trends and developments in biometrics. InTech, Rijeka (2012)

Chavan, K.P., Gawande, S.: Fingerprint Combination for Privacy Protection. ISSN 2321-8665 **03**(02), 0114–0123 (2015)

Ramachandra R., Sigaard, M.K., Raja K., Gomez-Barrero, M., Stokkenes, M., Busch, C.: Multi biometric template protection a security analysis of binarized statistical features for bloom filters on smartphones. In: 2016 6th International Conference on Image Processing Theory Tools and Applications (IPTA), pp. 1–6. IEEE (2016)

Ratha, N.K., Connell, J.H., Bolle, R.M.: Enhancing security and privacy in biometrics-based authentication systems. IBM Syst. J. **40**(3), 614–634 (2001)

Ratha, N.K., Chikkerur, S., Connell, J.H., Bolle, R.M.: Generating cancelable fingerprint templates. IEEE Trans. Pattern Anal. Mach. Intell. **29**(4), 561–572 (2007)

Rathgeb, C., Busch, C.: Cancelable multi-biometrics: mixing iris-codes based on adaptive bloom filters. Comput. Secur. **42**, 1–12 (2014)

Rathgeb, C., Gomez-Barreroy, M., Busch, C., Galballyy, J., Fierrezy, J.: «Towards cancelable multi-biometrics based on bloom filters: a case study on feature level fusion of face and iris»; (tec2012-34881) from Spanish Mineco, Fidelity (fp7-sec-284862) and beat (fp7-sec-284989) (2015)

Shanthini, B., Swamynathan, S.: Multimodal biometric-based secured authentication system using steganography. J. Comput. Sci. **8**(7), 1012 (2012)

Yildiz, M., Yanikoğlu, B., Kholmatov, A., Kanak, A., Uludağ, U., Erdoğan, H.: Biometric layering with fingerprints: template security and privacy through multi-biometric template fusion. Comput. J. **60**(4), 573–587 (2016)

Supply Chain Management a Help Tool Decision and These Impacts on the Transport and Safety of Merchandises in Algeria: Case of National Company of Industrial Vehicles

Billal Soulmana[✉] and Salim Boukebbab

Laboratoire Ingénierie des Transports Campus Universitaire Zarzara,
25017 Constantine, Algeria
{soulmana.bilal, boukebbabs}@umc.edu.dz

Abstract. Any company whatever their field of activity seeks to improve its situation, this is true when applying the right method, today forecasting method is a necessity for most companies. Forecasting is an essential element in the logistics chain because makes it possible to determine the quantity of stock needed, the material to be purchased, the quantity to be produced. To this end, we proposed the method of Box and Jenkins in order to predict the sales of the year 2017 for the company SNVI-CIR with appropriate forecasting models. Thus, the models we have chosen have improved the quality of forecasts compared to those used by the company.

Keywords: Supply chain · Forecast · Sales · Time series · Box and jenkins
ARIMA

1 Introduction

In the modern supply chain, predictions are needed for production companies (Fildes et al. 2006). Prediction is not an exact science (Despagne 2010) but it is a necessity for most companies. Rigoureau from Catalliances, indicates that predictions are a key point of quality of service. They are calculated from the sales histories, corrected by a certain number of parameters controlled by the companies (Armstrong 2001). In 1999, Armstrong and Brodie took back results of studies showing that in 1987, 99% of the companies prepare forecasts when they develop a marketing plan, 93% of companies indicate that forecasts are one of the most critical aspects, but very important to the success of the company (Armstrong and Brodie 1999).

In this framework, we offer to make a research on the importance of prediction and its consequences on the successful operation of the company and to apply an adequate predicting method on a production company in order to identify the best way he could be of foresee the future sales, for this we chose the Rouiba Industrial Unit specialized in the production of trucks and tractor trucks, buses and industrial bodywork equipment, We found that this company did not use a precise forecasting method to predict these future sales.

© Springer Nature Switzerland AG 2019
M. Hatti (Ed.): ICAIRES 2018, LNNS 62, pp. 80–86, 2019.
https://doi.org/10.1007/978-3-030-04789-4_8

The main objective of our work is to propose a forecast model to predict the sales of two vehicles: the k66 and the mini car.

2   Methods

2.1   Descriptive Study

We are interested in the descriptive side of chronological series, to know his graphic presentation, its essential elements and his models of decomposition.

2.2   Forecast Sales by the Box and Jenkins Method

We will start our method by the first step which is the study of the stationarity, then the identification of the model, the estimation and validation of the parameters of this model and end with the forecast (Box et al. 2015).

2.3   Data Analysis

All the results got in this part were obtained using software R. We thus used three packages: tseries, forecast, normtest.

We will first start with the Table 1 K66 vehicle and make the predictions then we go to the second vehicle which is mini car Table 2.

Table 1. Sales of product K66

K66	2012	2013	2014	2015	2016	Total
January	5	2	0	10	0	17
February	8	19	0	5	6	38
March	1	48	0	10	29	88
April	22	41	25	10	27	125
May	43	47	55	14	22	181
June	0	43	65	5	7	120
July	20	37	50	9	3	119
August	38	43	13	0	0	94
September	22	25	24	14	22	107
October	9	41	1	8	14	73
November	7	44	26	0	8	85
December	68	35	10	1	8	122
Total	243	425	269	86	146	1169

2.4   Presentation of the SNVI Group

The National Company of Industrial Vehicles (SNVI) is an Economic Public Company incorporated since May 1995, its mission is the design, manufacture, marketing and

Table 2. Sales of the mini car product

Mini Car	2012	2013	2014	2015	2016	Total
January	3	10	0	0	0	13
February	29	18	4	20	6	77
March	40	19	15	30	1	105
April	44	44	9	0	44	141
May	51	32	26	22	1	132
June	54	35	10	2	26	127
July	49	25	8	0	14	96
August	28	7	1	3	1	40
September	37	0	2	0	27	66
October	54	1	2	0	27	84
November	48	15	17	0	11	91
December	66	42	31	2	16	157
Total	503	248	125	79	174	1129

after-sales support of a large range of products. At the share capital of 2.2 billion DA, wholly owned by the Algerian state, the SNVI produces trucks and tractor trucks, buses and industrial body equipment. The SNVI group is constituted of a society mother and five subsidiaries.

3   Results and Discussion

3.1   Descriptive Study

According to Table 1 we find Average = 19.48, Min = 0, Max = 68, Standard deviation = 18.21, coefficient of variation CV = 0.93 for 5 years.

We find that the year 2013 is the year where the sales of the vehicle were the most important, that of 2015 where an average of only 7.17 vehicles was observed, which means there is a decline in sales. In addition, we also see a fairly high coefficient of variation CV for all years except for 2013 where sales did not vary significantly from one month to another Table 3.

Table 3. Descriptive statistics of the K66 vehicle

K66	Means	Min	Max	SD	CV
2012	20.25	0	68	20.41	1.01
2013	35.42	2	48	13.62	0.38
2014	22.42	0	65	23.09	1.03
2015	7.17	0	14	4.97	0.69
2016	12.17	0	29	10.36	0.85

3.1.1   Graphic Presentation Sales of the Vehicle K66

See Fig. 1.

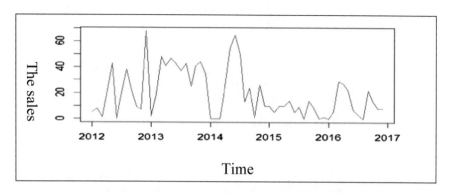

Fig. 1. Sales of K66 vehicle make by SNVI-CIR

3.1.2   Multiplicative Decomposition of the K66 Series

See Fig. 2.

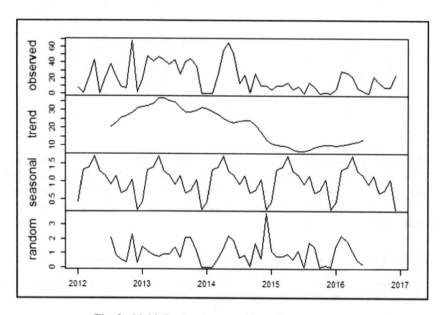

Fig. 2. Multiplicative decomposition of the K66 series

3.2   Prediction of Sales Vehicle K66 by the Box and Jenkins Method

Through AIC's Akaike criterion (Akaike 1998) we find that the ARIMA model (1, 1, 1) is the one that minimizes this criterion, so we will retain this model for the K66 series of the first six months of the year 2017 (Table 4).

Table 4. The forecast and the associated confidence interval

K66	Jan	Feb	Mar	Apr	May	Jun
Prevision	10	11	11	11	11	11
CI		[0,43]	[0,46]	[0,47]	[0,48]	[0,48]	[0,49]

To find out if our forecasts are reliable, we will use the ARIMA model (1, 1, 1) to forecast sales of 2016 (July to December), and we make a comparison between planned sales, forecasts of the company and sale real of the year 2016, we use the criterion of the mean squared error (MSE), the best prediction plan is the one that minimizes this criterion Table 5.

Table 5. Comparison between planned sales and company forecasts for 2016

K66	July	Aug.	Sept.	Oct.	Nov.	Dec.
Prevision (P)	11	12	13	13	12	13
Prevision of the l'Entreprise (PE)	76	20	80	100	95	88
Real (R)	3	0	22	14	8	8
MSE (P)	7.427					
MSE (PE)	71.248					

We find that there is a very large discrepancy between the forecasts of the company and the actual sales compared to our forecast. This translates into the value of the MSE which is 7.427 for our forecasts and a fairly high value of 71.248 for the company's forecasts. So the method we used to predict sales of the k66 product is much better than the one used by the company.

3.3   Prediction of Sales Vehicle Mini Car by the Box and Jenkins Method

We can retain the ARIMA model (4, 1, 2) because it is the model that minimizes the AIC criterion (490.86) Table 6.

Table 6. The forecast of 2017 and the confidence interval

Mini car	Jan	Feb	Mar	Apr	May	Jun
Prevision	9	10	16	19	19	16
CI		[0,35]	[0,36]	[0,43]	[0,48]	[0,50]	[0,50]

And for the comparison between the planned sales, the forecasts of the company and the real sales of the year 2016 from 6 months July to December we have (Table 7).

Table 7. Comparison of planned sales to company forecasts

Mini car	July	Aug.	Sept.	Oct.	Nov.	Dec.
Prevision (P)	12	18	22	24	23	19
Prevision of the l'Entreprise (PE)	46	86	45	44	38	37
Real (R)	14	1	27	27	11	16
MSE (P)	8.94					
MSE (PE)	40.89					

We find that there is a very large discrepancy between the forecasts of the company and the actual sales compared to our forecast. Where we find a value of the MSE which is 8.94 for our forecast and a fairly high value of 40.89 for the forecasts of the company. It is concluded that the method we used to predict sales of the mini car product is much better than the one used by the company.

4   Conclusion

We started our work with a descriptive study for both series. In this step, from the graphical representation as well as the descriptive statistics we was found that the sales of both vehicles recorded a sharp drop, especially in the last three years.

The application of the method of Box and Jenkins comes just after the descriptive study, we started with a study of the stationarity, then the identification and estimation of the parameters of the model. We found that the models ARIMA (1,1,1) and ARIMA (4,1,2) are the best models for both series (K66 and mini car respectively) in order to make predictions.

According to the MSE criterion, it was found that the forecasts given by the two models are better than the forecasts of the company, we can conclude that the Box and Jenkins method is the correct method to predict future sales for the CIR unit.

We recommend SNVI-CIR to use the Box and Jenkins method to forecast future sales.

References

Akaike, H.: Information theory and an extension of the maximum likelihood principle. In: Selected Papers of Hirotugu Akaike, Springer, pp. 199–213 (1998)

Armstrong, J.S.: Principles of Forecasting: A Handbook for Researchers and Practitioners. Springer, Berlin (2001)

Armstrong, J.S., Brodie, R.J.: Forecasting for marketing. In: Hooky, G.J., Hussey, M.K. (eds.) Quantitative Methods in Marketing, pp. 92-119. International Thompson Business Press, London (1999)

Box, G.E., Jenkins, G.M., Reinsel, G.C., Ljung, G.M.: Time Series Analysis: Forecasting and Control. Wiley, New York (2015)

Despagne, W.: Construction, analyse et implémentation d'un modèle de prévision. Déploiement sous forme d'un système de prévision chez un opérateur européen du transport et de la logistique. (phdthesis). Université Européenne de Bretagne (2010)

Fildes, R., Goodwin, P., Lawrence, M.: The design features of forecasting support systems and their effectiveness. Decis. Support Syst. **42**, 351–361 (2006)

A Human Behavior in Fire Building

H. Miloua[(✉)]

Department of Mechanical Engineering, Faculty of Technology,
University Djilali Liabes, 22000 Sidi Bel Abbes, Algeria
miloua_hadj@yahoo.fr, miloua_hadj@gmail.com

Abstract. An investigation by CFD code named Fire Dynamics Simulator (FDS) with sub-model of evacuation used to modelling a human behaviour during fire in smart building. Using the numerical simulation method to prevent against fire hazards by computing the heat release rate HRR of fire, gas temperature, smoke propagation in a building and evacuation situation for various cases to develop an emergency plan coherent for smart building in case of fire hazard, by optimisation of human displacements, reducing evacuation time, detection of black points in buildings, evacuate people faster and safer.

Keywords: Evacuation · Heat · Human · Building · Fire

1 Introduction

Fires impact people, property and the environment in all countries around the world. In some cases, the resulting losses are extraordinary, causing hundreds of deaths, widespread damage to property and contents and significant impacts on the environment. More often, fires may cause a single casualty or affect a single home, though the effects are still highly significant to those affected and collectively are substantial. In this paper we are interested in the fire Virtual Reality (VR). When fire occurs, the population is given a choice between: evacuating early, before fire reaches their area of residence, because "many people have died trying to leave at the last minute" (Country Fire Authority 2014); or stay and defend their house, only if very well physically and mentally prepared. In both cases, the decision a heavy death was registered (Teague et al. 2009; McLennan and Elliott 2011) and a plan must be prepared well in advance, because many residents had over-estimated their ability to face the fires, and were unaware and unprepared. the present work focused to the preventive simulation by description of people behavior in building during fire, we briefly outlines the factors that influence an occupant to take actions during his/her evacuation and identifies future areas of research that are needed to develop a predictive behavioral (action-based) model of an evacuation during a building fire.

© Springer Nature Switzerland AG 2019
M. Hatti (Ed.): ICAIRES 2018, LNNS 62, pp. 87–93, 2019.
https://doi.org/10.1007/978-3-030-04789-4_9

2 Results and Discussion

The simulation tool has been implemented to the Fire Dynamics Simulator (FDS) software, and called FDS + Evac. Only the documentation and user support of the FDS + Evac are administered by VTT (Korhonen and Hostikka 2009; McGrattan et al. 2013). The home page for FDS + Evac can be found at http://www.vtt.fi/fdsevac. Evacuation models, including engineering hand calculations and computational tools, are used to calculate the time it takes to evacuate a building, which can then be used in an engineering safety analysis. a human behavior in fire, "behavioral facts" have been obtained from a variety of incidents about what people do in fires. As shown by the three techniques, the current computer models attempt to simulate behavior during building fire scenarios. Evacuation models can simulate actions such as route choice, crawling, rerouting, moving at a slower speed, delay of response or stopping action, any kind of itinerary (spatial sequence of movement from one point to another), and even group sharing of information to make decisions on movement. These types of actions are likely to occur in building fire evacuations; however, the current models are essentially simulating separate "behavioral facts" rather than attempting to represent behavior based on a complete behavioral conceptual model. There are many benefits to the development of a comprehensive conceptual model for the field of human behavior in building fires. The inclusion of a conceptual model into computer evacuation tools will enable a comprehensive model that can actually predict occupant behavior in a building fire based only on initial input. There is a need in the field of computer evacuation modeling for a comprehensive conceptual model of human behavior in fire to more completely simulate behavior during building evacuation scenarios. Currently, separate, "behavioral facts" on human behavior in fire exist and are used to simulate behavior in computer evacuation models.

Each agent finds the exit from building, which exits may be used by various number and can be getting very hot with 600°C, the FDS flow solver is used to calculate an approximation to this potential flow and identify a dark point. Vertical air movement through a building caused by the temperature differential between the conditioned building air and the ambient outside air is known as the stack effect. During cold weather conditions, the stack effect causes air to move vertically upward in buildings. During very hot weather conditions, the stack effect causes air to move vertically downward in buildings. Stack effect can have a significant impact on the design of a stair pressurization system and must be considered in its design. Heat effect (Fig. 1e) can be the most serious form of all heat related injuries and should be consider a serious medical emergency. Common symptoms for heat stroke are but not limited to: nausea, seizures, confusion, disorientation and sometimes loss of consciousness or even possibly a coma. Other symptoms may include:

- Throbbing headaches
- Dizziness and light headedness
- Lack of sweating despite the heat
- Red, hot and dry skin
- Muscle weakness and/or cramps
- Nausea and vomiting

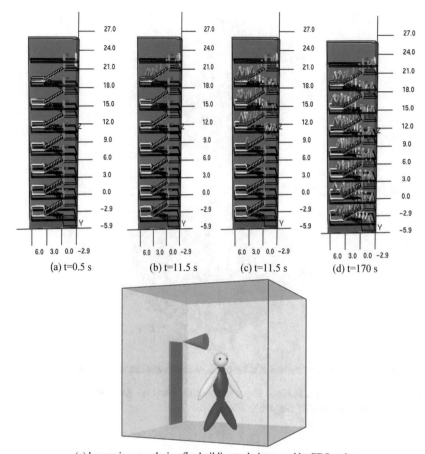

(a) t=0.5 s   (b) t=11.5 s   (c) t=11.5 s   (d) t=170 s

(e) human in room during fire building technique used by FDS code

Fig. 1. A example showing how the human evacuate hugely by stair during neighboring fire.

- Rapid heartbeat, which may be either strong or weak
- Rapid, shallow breathing
- Behavioral changes such as confusion, disorientation or staggering
- Seizures
- Unconsciousness

The houses collapse from fire damage when their resistance drops to 0. They then cease to offer protection, and the resident's motivation to defend them also disappears (Fig. 1e). Shelters are safe places. Residents agents can perform the following actions depending on their state e.g. Figure 1 in stairs and Fig. 3 move to top floor):

- Observe fires: action performed by all agents as a reflex (Fig. 2) at each cycle of the simulation. Their chance of actually detecting a (new) fire depends on their objective ability. Detected fires are added to the resident's list of known fires, and motivations are subsequently updated.

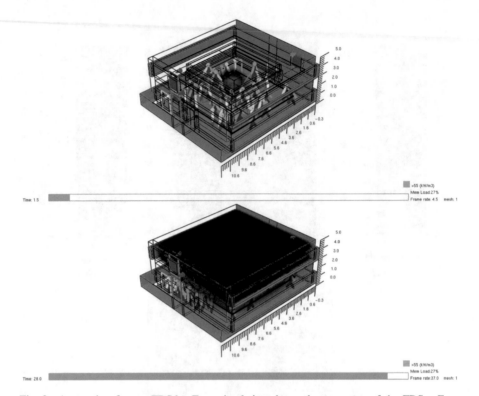

Fig. 2. A snapshot from a FDS6 + Evac simulation shows the geometry of the FDS + Evac model for the second floor of the public library.

Fig. 3. Gas temperature after 30 s. Red and blue colors indicate high and very low values, respectively.

- Prepare for fire: action performed while in preparing to defend preparing to defend or preparing to escape preparing to escape state. It consists in raising the resistance of the house (to simulate various actions such as watering, weeding, etc.) and the agent's health (to simulate the effect of wearing appropriate clothing, etc.). The value of the increment is computed based on the agent's objective ability and on a parameter of the resident agents. Success or failure of this action in monitored by the agent and influences its subjective ability.

- Fight fire: action performed while in defending state. Its effect is to decrease the intensity of nearby fires by a value based on objective ability and on another resident parameter. The agent monitors success (number of fires extinguish, total intensity decrement) to update its subjective ability, thus reconsidering its motivation over time;

- Escape: action performed while in the Escaping state. Its effect is to compute and follow a path towards the nearest shelter. The resident's evacuation (Fig. 2) speed and accuracy depend on its objective escape ability: an agent with a low ability has more chances to take a longer path (to simulate getting lost or not taking the best route), and it moves more slowly along that path (e.g. disabled or injured people). Agents might get injured while escaping if they travel too close to the fire.

About a fire modeling a very complex and detailed models of fire spreading already exist (Duff et al. 2013), but realistic fire behavior is not the focus here. Still with the goal of not adding unneeded complexity, we have designed a very simplistic model of fire that is sufficient to trigger and visualize the reactions of the population and calculation of Human density (Fig. 4).

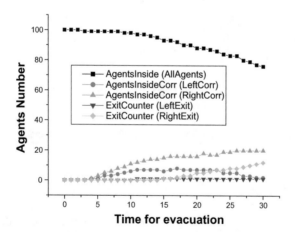

Fig. 4. Humans densities calculated in each exit during evacuation from library building under fire.

The fire is composed of fire agents (each with a location and an intensity representing its radius of action), having a reflex architecture, i.e. the following reflexes are triggered at each step of the simulation (Fig. 4):

- Increase or decrease intensity: probabilities are parameters; Propagate to a non-burning neighbor cell, creating a new fire agent. Probability of propagating, and starting intensity of new fires, are parameters;
- Deal damage to buildings in its radius of action (based on its intensity): the amount of damage is picked randomly between 0 and a maximum value, function of intensity and a "damage factor" parameter;
- Deal injuries to residents in its radius of action, also random amount between 0 and the maximum value based on its intensity and an "injury factor" parameter. If the person is in their house, the injury is moderated by its resistance weighed by a "protection factor" parameter;
- Disappear when its intensity is null.

The Heat Release Rate (Fig. 5) is most important parameter used to evaluate fire damage (Miloua et al. 2011; Miloua 2018) this amount released from fire is automatically converted to conduction, convection and radiation heat transfer, radiation is most dangerous because they propagate without a contact and they dehydrated a human in few second.

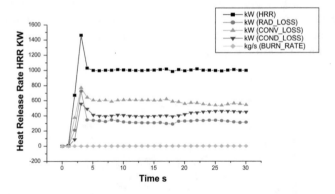

Fig. 5. Heat release rate generated from fire divided to three heat transfer modes conduction, convection and radiation.

3   Conclusion

A virtual reality by a CFD code was proposed to study human evacuation in many parts of building by general investigation presented to identify the factors that influence occupants to decide to take a specific action, and the factors that influence whether that action is ultimately performed. By participation to identify the factors that have been shown to influence each phase in the behavioral process, researchers can begin to develop a comprehensive, predictive, behavioral model for a building fire evacuation in order to limit the risks during a disaster.

Acknowledgements. This work was supported by the Algerian Research Organism DGRSDT, under the project [No. A11N01UN020120150001].

References

Country Fire Authority: Your guide to survival (2014). http://goo.gl/7UzH7g

Duff, T.J., Chong, D.M., Tolhurst, K.G.: Quantifying spatio-temporal differences between fire shapes: estimating fire travel paths for the improvement of dynamic spread models. Environ. Model Softw. **46**, 33–43 (2013). https://doi.org/10.1016/j.envsoft.2013.02.005

Korhonen, T., Hostikka, S. Fire Dynamics Simulator with evacuation: FDS + Evac. s.l., Finland: VTT Technical Research Centre of Finland, April 2009. VTT Working Papers 119 (2009)

Mclennan, J., Elliott, G.: Community members decision making under the stress of imminent bushfire threat—Murrindindi fire. Technical report, Bushfire CRC Extension, School of Psychol. Science, La Trobe University (2011). http://goo.gl/cyVgY3

McGrattan, K., et al.: Fire Dynamics Simulator User's Guide. Gaithersburg, Maryland, USA: s. n., November 2013. NIST Special Publication 1019. SFPE. 2003. Engineering Guide— Human Behavior in Fire. Society of Fire Protection Engineers, Bethesda, Maryland, USA, June 2003 (2013)

Miloua, H., Azzi, A., Wang, H.Y.: Evaluation of different numerical approaches for a ventilated tunnel fire. J. Fire Sci. **29**(5), 403–429 (2011)

Miloua, H.: Fire behavior characteristics in a pine needle fuel bed in northwest Africa. J. For. Res. (2018). https://doi.org/10.1007/s11676-018-0676-8

Oka, Y., Atkinson, G.T.: Control of smoke flow in tunnel fires. Fire Saf. J. **25**, 5–22 (1996)

Teague, B., Mcleod, R., Pascoe, S.: Final report. Technical report, Victorian Bushfires Royal Commission. Summary at http://goo.gl/ulCt4U (2009)

Adaptive Governance for Attractive Smart City

Slimane Zalene$^{(\boxtimes)}$

Consultant en Management, B.P. 52 B, 16063 Zéralda, Algeria
slim_zalene@yahoo.fr

Abstract. In his book entitled *"le social et le vivant"* J. FONTANET [1] explains that the multidimensional crisis of our societies is mainly due to a failure of our intelligence. This observation, dating back to the end of the 1970s, is still valid today because in the face of "complexity" we still continue to use mechanistic methods. From the *one best way* TAYLOR during the industrial revolution of the nineteenth century to the various *best practices* are required to fit into the post-industrial economy of the twentieth century, Enterprises have tried to adapt their management to the evolution of their environment. At the beginning of the twenty-first century, globalization and the digital economy required Enterprises to make intensive use of information and communication technologies and force them to constantly review their strategy to survive in a international market where performance is continually demanded. On one hand, to face a tough competition, and on the other hand, to respect multiple rules of QHSE (quality, hygiene, safety and environment), of sustainable development, of social responsibility … in a word of good governance in the most broad and not just limited to the company's traditional stakeholders. In this context of management under several constraints, the artificial intelligence (AI) brings an additional dimension of adaptation to a self-adaptive management for good governance. In this regard, this paper first discusses the concept of systemic adaptation, then its application to business strategy, then the possibilities of using artificial intelligence to improve the many expected results of managers who are inundated with sometimes contradictory information with requests for decisions that are often urgent, and finally the broadening of this adaptive approach to make a smart city more attractive.

Keywords: Identification · Adaptation · Flexibility and fractal methods

1 Introduction

This communication has no claim to expose a new management mode; the literature on this subject is already quite abundant. It only involves introducing, with the aid of artificial intelligence, an additional adaptation in the management of an enterprise or a city. This adaptability, vital for living organisms, has inspired the construction of self-adaptive systems that have proven themselves in various technical fields (ZALENE [2]).

© Springer Nature Switzerland AG 2019
M. Hatti (Ed.): ICAIRES 2018, LNNS 62, pp. 94–100, 2019.
https://doi.org/10.1007/978-3-030-04789-4_10

Their transposition in the managerial domain, on the other hand, encounters numerous constraints, notably the behavior of the human factor, which, because of its complexity and diversity, is difficult to model.

But Artificial Intelligence offers an opportunity to provide an enterprise or a city with a model that, explicitly or implicitly, is an essential reference for its adaptation to changes in its external or internal environment. This modeling difficulty is perhaps one of the reasons for the lack of applications of self-adaptive systems in management strategy. On the other hand for more than half a century, self - organization, agility, self - management, fractal organization, have been required by Enterprises to give their management more flexibility or even more "intelligence" (the meaning of the term in French "of understanding"). This need for intelligence is essential in management knowing that it tends to become complex in the face of a very changing or even chaotic environment.

In this context, a permanent questioning of management methods is imperative, the *one best way* has long been abandoned in favor of *best practices*, however the example of even famous Enterprises should not be a reference to follow, copy them blindly sometimes leads to serious difficulties because each case is unique and the critical elements that have helped managers to make strategic decisions have often been intuitive (BRUN [3]).

2 General on Systemic Adaptation

The word adaptation and derived terms (adaptive, adapted, adaptable, adaptability, ...) are very often used in books and articles on management without a precised definition or theoretical development. A brief reminder will clarify the concept of systemic adaptation.

1.1. Brief Reminders

These adaptive systems are among the latest developments in BERTALANFY's *General Systems Theory*. ASHBY borrows the term "adaptive" from biologists, who refer to the ability of living organisms to adapt to their environment.

If the statement of adaptation is clear in biology, for instance, its mechanism remains to be discovered; in the technical field, on the other hand, adaptation methods are known but there is still no single, precise definition of adaptive systems.

By looking at the various definitions given by various authors including BELL-MAN, FELDBAUM, TCHINAEV, and ZADEH we note a tendency to link the notion of *adaptation* to that of *learning*, whereas they are respectively distinguished as follows:

- the first (adaptation) tests the state of the system and uses this information to modify certain variables having the effect of improving the response of the system,
- the second (learning) tries different methods of correction to weight according to the results obtained.

1.2. Definition

Given the previous remark, the definition used is as follows:

"The self-adaptive system is a system designed to act in a slowly changing environment, so that the adjustment is made during the change of environment."

The restriction on the slow change of environment is introduced to allow the system to react in time, otherwise the adaptation is of no use because the environment has been subject to another change (this corresponds to the notion of responsiveness required in management).

1.3. Summary Description of Self-Adaptive Systems

An adaptive system must, according to Definition 1.2, perform the following functions:

- an "IDENTIFICATION" function, to capture the variations of the internal or external conditions of the system,
- an "ADAPTATION" function to develop the appropriate decision when changing these conditions.

The identification must be both internal and external:

- *external*, the purpose of which is to locate and specify the influences of the external environment on the system under consideration,
- *internal* to analyze information about the system itself in order to develop an adequate response to the change experienced or to be sustained.

This identification contributes to the construction of a model (or reference) which is at the base of the adaptation process. This makes a comparison between the evolution of the system in question (or its environment) with the reference or model chosen (see Fig. 1 below).

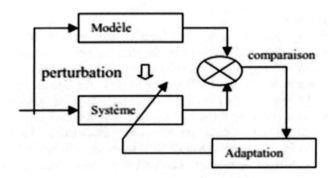

Fig. 1. .

Adaptation is achieved by providing the system with two (2) subsystems:

- one is fixed (or controlled) composed of elements having characteristics that cannot be modified,
- the other is variable (or adaptable) consisting of elements in which the characteristics can be modified.

1.4 Adaptive mechanisms are certainly numerous, but four (4) types are generally used by

- *direct action*, which acts directly on the parameters by adjusting them for a better response of the system;
- *indirect action*, in this case the modifications of the parameters can only be obtained by feedback actions (feed-back) or direct actions (feed forward);
- *additional signal injection,* this influences the behavior of the system and forces to filter to remove the undesirable effects;
- *disturbance of the entry,* this device raises less objection than the previous one.

3   Management and Environment

Several authors (in particular BURNS and STALKER) have studied the influence of the environment on the management also to approach simply this topic it was only retained the first studies made on this subject in particular those which show that an organization (enterprise or city) depends on its environment and, depending on the case, its management system is either mechanical or organic:

- the *mechanical system*, suitable for stable conditions, is characterized in particular by: a rigid hierarchical structure, a very advanced specialization, a precise definition of the tasks, vertical communications, a recovery of the problems towards the summit, relations based on the obedience and loyalty (*one best way*);
- the *organic system*, more adapted to changing conditions, with a network structure favoring participation, versatility, multidisciplinarity, a permanent revision of the definition of tasks, lateral communications, a solution of problems where they arise, relations based on trust and seeking competence (*best practices*).

Both systems can exist in the same organization. However there is no absolute structure but that everything is relative according to the theory of contingency (LAWRENCE and LORSCH), the organization will seek to adapt with its environment and its performances depend on the efforts of differentiation or integration that its managers give to their internal or external environment.

4   AI and Adaptation

To do this, it is necessary that the organization has variability, that is to say the possibilities of internal modifications, in order to adapt to changes affecting its environment. What are the elements of this variability? Four (4) elements are concerned:

3.1. Technology: A subsystem that combines, for production purposes, the material or immaterial resources that go into the goods or services to be provided to customers.

3.2. <u>Training</u>: A subsystem designed to increase the scientific and technical knowledge of trained and motivated staff who alone are able to push back the limits of equipment or processes of business transformation.

3.3. <u>Structures</u>: A subsystem regrouping the regulations concerning the management of the human, material, financial and informational resources which also evolve;

3.4. <u>Strategy</u>: A subsystem to optimize the use of the mentioned resources for the achievement of fixed objectives that can be countered by changes in its environment.

The restricted model of adaptive management is schematized Fig. 2 below.

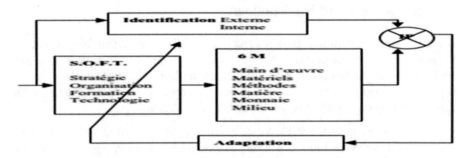

Fig. 2. .

5   Applications to Good Governance of Cities

The "CITY" is no longer just a geographical space gathering people working within it but has now become a true ENTERPRISE that must survive economically, culturally and ecologically to retain its inhabitants and attract investors so that its operation does not cost nor its financing: the recourse to the budget will no longer be automatic to ensure BALANCING its finances and solving its difficulties.

For this, the CITY needs INTELLIGENCE to develop a sustainable development strategy.

In this respect, the "smart city" is not a cluster of Information and Communication Technologies (ICT) for "modern" PARAMETER but is "efficient and effective" so that its inhabitants have an income sufficient and a pleasant living environment.

INTELLIGENCE is therefore vital for the CITY it must serve to adapt to the change in its environment and the mood of its inhabitants.

The CITY is not a simple administration but a living entity described by three (3) social systems (or 4 if the urban transport is considered see Fig. 3) in interaction positive and negative according to the case:

On the other hand the city must seek to be attractive, for that the city must balance its management even to release income to finance its investments or to decide on incentives to attract investors (PPP).

Fig. 3. .

Sustainable development requires reducing waste (see Fig. 4), especially treating waste that is also a source of pollution and its consequences on the health of inhabitants:

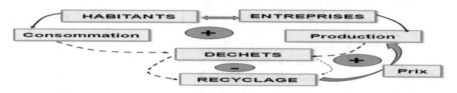

Fig. 4. .

The city must be attractive to develop by attracting Enterprises and offering a pleasant living environment) (see Fig. 5 below):

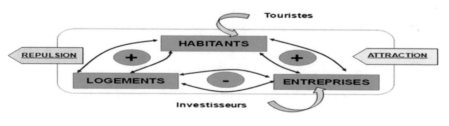

Fig. 5. .

It must also be repulsive for end-of-life businesses by helping them to convert and under qualified people by facilitating their development.

6 Retrospective and Outlook

Historically, in the 19th century, the concept of adaptation was the basis of the theory of evolution, of which one of the precursors DARWIN was influenced by the economist MALTHUS. It is therefore a fair return that the economy and management are inspired by natural phenomena to steer the City or the Enterprise in an increasingly turbulent environment. As early as the sixteenth century, one of the many advices of

MACHIAVEL to Prince Laurent of MEDECIS was that: *"Très peu d'hommes, quelque soit leur sagesse, savent s'adapter ... si tu savais changer de nature quand changent les circonstances ta fortune ne changerait point."* This advice is still relevant, HART wrote in his book on Systèmes Dynamiques: *"S'adapter ou mourir est certainement la devise qui convient le mieux à notre civilisation"*. TOFFER developed it in his book entitled *"S'adapter ou périr"*, It is therefore not surprising that, after the ideas of learning and intelligent Enterprises [4], the latter become adaptive ([5] and [6]) and even self-adaptive [7].

References

1. Fontanet, J.: "Le social et le vivant" Plon (1977)
2. Zalène, S.: Liaison Sensibilité - optimisation dans un système auto-adptatif. Université de Lyon (1975)
3. Brown, C.: l'irrational dans l'entreprise. Balland (1989)
4. Quinn, J.B.: Intelligent Enterprise. The Free Press, New York (1992)
5. Haeckel, S.H.: Adaptive Enterprise. Harvard Business School Press, Brighton (1999)
6. Robertson, B., Sribar, V.: Adaptive Enterprise. Addison Wesley Professional, Boston (2002)
7. Reeves, M., Venjara, A., Zeng, M.: The Self-Adaptive Company. Harvard Business Review (2016)

Control and Sliding Mode Control

Artificial Neural Networks Technique to Detect and Locate an Interturn Short-Circuit Fault in Induction Motor

S. Bensaoucha[1(✉)], A. Ameur[2], S. A. Bessedik[2], and Y. Moati[2]

[1] Laboratoire d'Etude et Développement des Matériaux Semi-Conducteurs et Diélectriques (LeDMaSD), Université Amar Telidji de Laghouat, BP 37G, 03000 Laghouat, Algeria
s.bensaoucha@lagh-univ.dz
[2] Laboratoire d'Analyse, de Commande des Systèmes d'Energie et Réseaux électriques (LACoSERE), Université Amar Telidji de Laghouat, BP 37G, 03000 Laghouat, Algeria
{a.ameur,s.bessedik,y.moati}@lagh-univ.dz

Abstract. Induction machines (IMs) faults diagnosis is of significance to enhance the reliability and security of the industrial productions. In recent years, the faults detection and diagnosis of IMs have moved from traditional techniques that using the signals analysis to artificial intelligence (AIs) techniques such as the neural networks (NNs). Unfortunately, stator faults related failures account for a large percentage of faults in IMs. In this context, this paper proposes a diagnostic technique based on NNs for detecting and locating the interturns short-circuit in one of three stator winding phases of IM. Often, the most important step in which NNs is the process of selecting the best inputs that enables higher performance and better diagnosis, in this study, the three-phase shift between the stator voltages and its currents are considered as inputs of the NNs in order to develop an automatic fault detection and classification system. The simulation results in MATLAB environment prove the efficiency of the suggested NN technique to detect and locate short-circuit between turns of the same coil when the SCIM is operating under various load levels and various fault severities.

Keywords: Induction machines · Faults diagnosis
Interturn short-circuit detection and location · Artificial neural networks

1 Introduction

The squirrel cage induction motors (SCIMs) plays an important role in the field of electromechanical energy conversion due to their: power density, robustness, ease of implementation and low cost and their reliability. Consequently, the diagnosis of electric machines is known for its great interest and much research to improve machine efficiency and protect it from the various faults [1–3]. Usually, the most traditional diagnostic techniques based on signal analysis used the different techniques of time domain, frequency domain and time–frequency domain, and high order spectra [4, 5],

© Springer Nature Switzerland AG 2019
M. Hatti (Ed.): ICAIRES 2018, LNNS 62, pp. 103–113, 2019.
https://doi.org/10.1007/978-3-030-04789-4_11

often, the results obtained from these techniques are complex, thus, the interpretation of the harmonics results is difficult and could not be done easily by the industrialists [6–8]. To avoid these constraints, frequently, artificial intelligence techniques (AIs) have been used as complementary tools, currently, a vast research projects is being carried out in the research centers to integrate the AIs such as a neural network (NNs), fuzzy logic, adaptive neural fuzzy, genetic algorithm, support vector machine, and deep learning in order to monitoring and diagnosis the different faults of IMs [2, 3, 6, 7, 9–13]. In this context, many studies based on AIs have been conducted to IMs faults detection using the different measurements signals data (inputs) such as the temperature of IM, the stator voltages and its currents, the vibration data taken from the IM, the motor speed and load torque, the characteristic's frequencies obtained from the spectral analysis of the different motor currents signals, the three-phase shifts between the line current and the phase voltage of the machine [5–7, 10–12]. In this paper, the authors are using the three-phase shifts $(\varphi_a^s, \varphi_b^s, \varphi_c^s)$ between the line currents and the voltages of IMs as three inputs of NN. In particular, NNs provided a suitable solution to solve the problems of fault diagnosis and automation of the monitoring procedure, generally, NNs represents a nonlinear system referring to its inputs-outputs behavior, this non-linear transformation results from the inner structure of a NN, in addition, NNs stand out from other AI tools by their ability to learn and generalize, it is not required the existence of a formal modeling of the equipment to be monitored, the use of NN has been the subject of recent studies related to IMs [5, 6, 10, 13].

On another hand, as mentioned earlier, numerous studies have shown that a large percentage of failures in IMs related to the stator windings (30–40% of full faults) defined as interturns short circuit or a short circuit between phases, short-circuit between coils of the same phase, short-circuit between coils of different phases [2, 3, 5, 14]. Particularly, this paper focuses on interturns short-circuit faults detection and location using NNs technique.

For the above-mentioned subject, this paper is structured as follows: Sect. 2 presents the modeling and simulation the SCIM in presence a stator fault to look for the best indicators $(\varphi_a^s, \varphi_b^s, \varphi_c^s)$ of each state (healthy state, faulty state) for training and testing the NN to identify the state of the machine, Sect. 3 is focused on the design a monitoring system based on NNs to perform a correct and robust diagnosis allowing to detection the SCIM fault, Sect. 4 presents a conclusion for this work.

2   Squirel Cage Induction Motor Model for Stator Faults Detection

A. *System Equations*

In the stator faulty case, the model of a SCIM in the reference frame d, q axis related to the rotor $(\theta = \theta_r)$ is defined as bellow [5, 12, 14]:

$$\begin{cases} \dot{x} = Ax(t) + Bu_{dq}^s(t) \\ y(t) = Cx(t) + Du_{dq}^s(t) \end{cases} \tag{1}$$

with $x = \begin{bmatrix} i^s_d i^s_q i^r_d i^r_q i_e \end{bmatrix}^T$, $u = \begin{bmatrix} u^s_d u^s_q \end{bmatrix}^T$ and $y = \begin{bmatrix} i^s_d i^s_q \end{bmatrix}^T$, where i^s_{dq} and u^s_{dq} are respectively the stator currents and its voltage, i^r_{dq} and i_e are the rotor bars and the end ring segment currents in axis d, q respectively.

$$A = \begin{bmatrix} R_s & -\omega_r L_{sc} & 0 & -N_r \omega_r M_{sr} & 0 \\ \omega_r L_{sc} & R_s & -N_r \omega_r M_{sr} & 0 & 0 \\ 0 & 0 & R_r & 0 & 0 \\ 0 & 0 & 0 & R_r & 0 \\ 0 & 0 & 0 & 0 & R_e \end{bmatrix}$$

$$B = \begin{bmatrix} L_{sc} & 0 & \frac{-N_r M_{sr}}{2} & 0 & 0 \\ 0 & L_{sc} & 0 & \frac{N_r M_{sr}}{2} & 0 \\ -\frac{3}{2} M_{sr} & 0 & L_{rc} & 0 & 0 \\ 0 & -\frac{3}{2} M_{sr} & 0 & L_{rc} & 0 \\ 0 & 0 & 0 & 0 & L_e \end{bmatrix}$$

$$C = \begin{bmatrix} 1 & 0 & 0 & 0 & 0 \\ 0 & 1 & 0 & 0 & 0 \end{bmatrix} \qquad D = \sum_{m=1}^{3} \frac{2.n^m_{cc}}{3R_s} P(-\theta)(\theta^m_{cc}) P(\theta)$$

Where $P(\theta)$ and (θ_{cck}) are given by:

$$P(\theta) = \begin{bmatrix} cos(\theta) & -sin(\theta) \\ sin(\theta) & cos(\theta) \end{bmatrix} \qquad (\theta_{cck}) = \begin{bmatrix} cos(\theta_{cc})^2 & cos(\theta_{cc})sin(\theta_{cc}) \\ cos(\theta_{cc})sin(\theta_{cc}) & sin(\theta_{cc})^2 \end{bmatrix}$$

where R_s are the stator windings resistances, M_{sr} are the stators to rotor mutual inductances and ω_r is the angle speed between rotor and stator, R_r are the rotor windings resistances and N_r is the total number of rotor bars. L_{sc}, L_{rc}, L_e are the cyclic inductances of the stator, rotor, and segment of end ring, respectively. $m = 1, 2, 3$ is positive integer to indicate one of the three stator phases, $P(\theta)$ is Park rotational matrix with θ is rotor angular position and n^m_{cc} is the parameter defined the interturns short circuit in stator winding phase defined as $n^m_{cc} = (N_{cc}/N_s)$, where N_{cc} is number of interturns short-circuit in stator windings phase, N_s is total number of interturns in healthy state, (θ^m_{cc}) is matrix allows to localize of the faulty winding in short circuit on the stator phases (a_s, b_s, c_s) with $\theta^m_{cc} = 0, 2\pi/3$ or $4\pi/3$ is the real angle between the short-circuit interturn stator winding and the first stator phase.

A. Simulation Results for Interturns Short Circuit Fault

In the stator faulty, the simulation result of SCIM is based on the stator fault mode given by Eq. 1, the simulated machine having four poles $(p = 4)$, a supply frequency $f_s = 50Hz$ and the nominal speed $\Omega_n = 1425$ rpm, the rated power $P = 1kW$, all SCIM parameters are shown in Table 1.

Figure 1 show the stator currents lines (i_{sa}, i_{sb}, i_{sc}) for a healthy motor and faulty motor (stator fault) at fixed load torque $(T_L = 7Nm)$ introduced at $t = 0.6s$.

Table 1. Induction machines parametres

Parameters			Values
V_s	220/380V	N_s	464
I_n	2.6/4.3A	N_r	28
R_s	9.81Ω	R_e	$5.6 \times 10^{-7}Ω$
M_{sr}	$1.7 \times 10^{-4}H$	L_e	$17 \times 10^{-7}H$
R_r	3.1Ω	f_d	0.00119Nm/rad.s
R_{bar}	$6.1 \times 10^{-7}Ω$	inertia(J)	$0.0125kg.m^2$

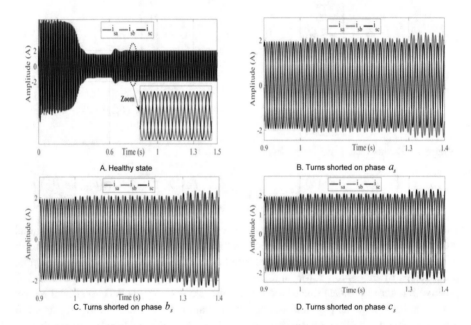

Fig. 1. Stator current's lines of induction motor under full load. (A) Healthy motor no shorted turns, (B) shorted turns phase a_s, (C) shorted turns phase b_s, (D) shorted turns phase c_s.

Figure 1(B–D) present respectively the profiles of the simulated three line's currents (i_{sa}, i_{sb}, i_{sc}) for six shorted turns $(N_{cc} = 6)$ at $t = 1s$ and fourteen shorted turns $(N_{cc} = 14)$ at $t = 1.3s$ on one of the three phases a, b or c.

As shown in simulation results (Fig. 1), it is clearly notice that the short circuit between the windings causes a significant increase in currents in the coil leading to a fast deterioration, in addition, the increase of this undulation depends directly on the number of short turns. Figure 2 is presents an example to describe the phase shift between two sinusoidal signals, and a schematic view of a short circuit between turns of same phase is shown in Fig. 3.

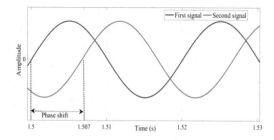

Fig. 2. A diagram of the phase shift between two signals wave

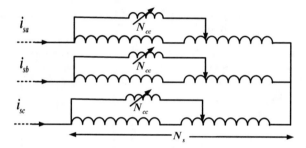

Fig. 3. Schema equivalent of interturns short circuit of stator winding phases

3 Artificial Neural Network for Stator Fault Detection and Location

A. *Artificial Neural Networks Basics*

Usually, the most useful neural networks in function approximation are multilayer layer perceptron (MLP) or radial basis function (RBF) networks, in this paper, the NN was used for a MLP classifier, a MLP consists of an inputs layers, several hidden layers, and an outputs layers with nodes (i) called a neuron connecting in parallel and series, it includes a summer and a nonlinear activation function, The activation function was originally chosen to be a relay function, but for mathematical convenience a hyperbolic tangent (tanh), or a sigmoid function are most commonly used [5, 6, 10, 12].

As mentioned above, this work depends on the shift phase $(\varphi_a^s, \varphi_b^s, \varphi_c^s)$ between the stator currents and its voltages that extracted from the motor where it operates under different conditions as no fault, short circuit in phase (a_s, b_s, c_s). In this paper, these inputs are calculated using MATLAB.

In MATLAB R2016 environment, the phase shift of each stator currents (i_{sa}, i_{sb}, i_{sc}) and its voltages (V_{sa}, V_{sb}, V_{sc}) can be calculated using the following code:

$$\begin{cases} \varphi(rad) = a\cos(dot(sig_1, sig_2)/(norm(sig_1) * norm(sig_2))) \\ \varphi\ (°) = \varphi(rad)\ * 360/(2 * pi) \end{cases}$$

where: $sig_1 \in (i_{sa}, i_{sb}, i_{sc})$ and $sig_2 \in (V_{sa}, V_{sb}, V_{sc})$.

Using this code all inputs were collected as shown in Fig. 4(A) and (C), under the different conditions of a SCIM (healthy state or interturns short circuit fault).

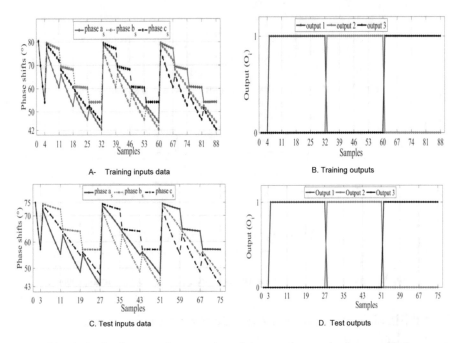

Fig. 4. Define inputs and outputs data of the neural networks for stator fault.

B. *Training and Test Database*

The selected NNs inputs are phase shifts $(\varphi_a^s, \varphi_b^s, \varphi_c^s)$ corresponding to fault on phases a_s, b_s, c_s at four different load conditions $(T_L = 1, 3, 5, 7Nm)$ with $N_{cc} = 3, 5, 7, 9, 11, 13, 15$. Thus, a total of 88 indicators $(88 = (4 + (4 \times 7) \times 3))$ are extracted for each phase shift input (Fig. 4A). The NN has three outputs to indicate the state of an SCIM as flows (Fig. 4B and D):

- If $O_i = [0\ 0\ 0]$ then SCIM is normal;
- If $O_i = [1\ 0\ 0]$ then the SCIM is abnormal (short circuit at phase a_s);
- If $O_i = [0\ 1\ 0]$ then the SCIM is abnormal (short circuit at phase b_s);
- If $O_i = [0\ 0\ 1]$ then the SCIM is abnormal (short circuit at phase c_s);

As shown in Fig. 4(C), all test data set is presented to the NNs for three load conditions $(2, 4, 6Nm)$ for $N_{cc} = 2, 4, 6, 8, 10, 12, 14, 16$, thus, a total of 75 samples $(3 + (3 \times 8) \times 3) = 75$.

C. *Neural Networks Architecture Using MATLAB*

In fact, the optimal number of neurons in the hidden layer is not known, the NN must be trained and then tested successively with separate data not previously introduced during the training in order to check both the training and test errors [2, 5, 10].

In this paper, the structure of the NNs is first defined (Fig. 5), the activation functions are chosen and the network parameters, weights and biases are automatically generated by the MATLAB command, the activation functions of one hidden with

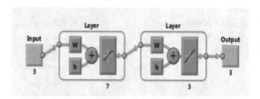

Fig. 5. Neural networks structure

seven neurons for "tansig" function and output layers with activation functions "purelin", this NNs has three inputs $(\varphi_a^s, \varphi_b^s, \varphi_c^s)$ and three outputs.

Figure 6 presented the parameters associated with the training algorithm like error goal, maximum numbers of epochs (iterations) are defined (5000 epochs), and also, the training Levenberg-Marquardt algorithm (trainlm) is used. Moreover, the mean square error (MSE) is defined as comparing the difference between the outputs produced by the ANN compared to the desired results (targets). From the Figs. 6 and 9, the MSE (performance) was reduced to an acceptably low value 3.07×10^{-11}, this value is suitable to classify the test set correctly. Through the simulation results of NNs (Figs. 7 and 8), NNs is very precise and effective technique to detect and locate the interturns short circuit fault in stator winding phases of SCIM. As shown in Figs. 7 and 8, the training and the test errors for each output are very low proving that the NN has well learned the input data and has correctly reproduced the desired output.

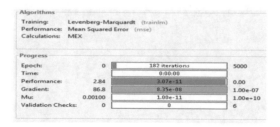

Fig. 6. Neural networks parameters

D. *Training Results of Neural Networks*

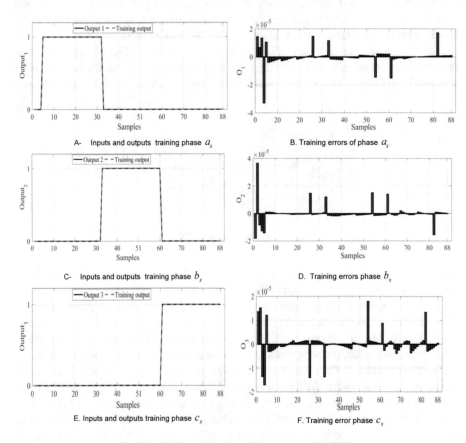

A- Inputs and outputs training phase a_s

B. Training errors of phase a_s

C- Inputs and outputs training phase b_s

D. Training errors phase b_s

E. Inputs and outputs training phase c_s

F. Training error phase c_s

Fig. 7. Training outputs and its errors of the NNs

E. *Test Results of Neural Networks*

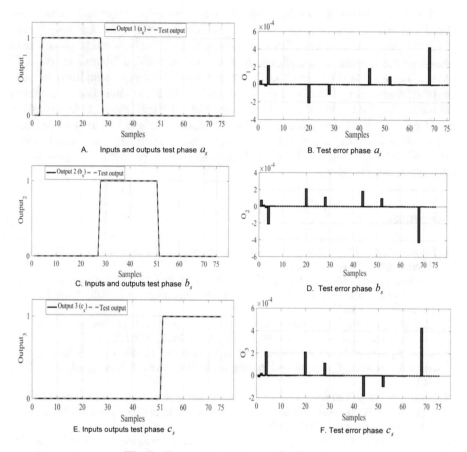

Fig. 8. Test outputs and its errors of the NNs

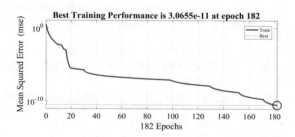

Fig. 9. Neural networks performance.

4 Conclusion

In this paper, a diagnostic method using NNs was proposed to detect and locate short-circuit between turns of the same coil in SCIM. From the training and test procedure of NNs, this study confirms the efficiency and accuracy of the NNs technique to detect and locate this type of fault. It should be noted that, in addition that this technique is able to diagnose the short circuit in SCIM, this technique can be simplified the procedure of diagnosis by identifying and locating the fault at the same time. In addition, the use of the phase shift between the voltage and the stator current as inputs of the neural networks does not require changing the signals analysis domain of this study (just the time domain), i.e., no need the use the spectral analysis of the signals in this case, which helps to reduce the cost of diagnosis, and kept additional space in the information storage devices.

References

1. Tavner, P.: Review of condition monitoring of rotating electrical machines. IET Electr. Power Appl. **2**, 215–247 (2008)
2. Palácios, R.H.C., da Silva, I.N., Goedtel, A., Godoy, W.F., Lopes, T.D.: Diagnosis of stator faults severity in induction motors using two intelligent approaches. IEEE Trans. Ind. Inf. **13**, 1681–1691 (2017)
3. Dias, C.G., Pereira, F.H.: Broken rotor bars detection in induction motors running at very low slip using a hall effect sensor. IEEE Sens. J. **18**, 4602–4613 (2018)
4. Saleh, S., Ozkop, E.: Phaselet-based method for detecting electric faults in 3ϕ induction motor drives—part I: analysis and development. IEEE Trans. Ind. Appl. **53**, 2976–2987 (2017)
5. Bouzid, M.B.K., Champenois, G., Bellaaj, N.M., Signac, L., Jelassi, K.: An effective neural approach for the automatic location of stator interturn faults in induction motor. IEEE Trans. Ind. Electron. **55**, 4277–4289 (2008)
6. Siddique, A., Yadava, G., Singh, B.: Applications of artificial intelligence techniques for induction machine stator fault diagnostics. In: Diagnostics for Electric Machines, Power Electronics and Drives, 2003. SDEMPED 2003. Symposium on 4th IEEE International, pp. 29–34 (2003)
7. Razik, H., Correa, M., Da Silva, E.: An application of genetic algorithm and fuzzy logic for the induction motor diagnosis. In: Industrial Electronics, 2008. IECON 2008. 34th Annual Conference of IEEE, pp. 3067–3072 (2008)
8. Devi, N.R., Sarma, D.S., Rao, P.R.: Diagnosis and classification of stator winding insulation faults on a three-phase induction motor using wavelet and MNN. IEEE Trans. Dielectr. Electr. Insul. **23**, 2543–2555 (2016)
9. Ballal, M.S., Khan, Z.J., Suryawanshi, H.M., Sonolikar, R.L.: Adaptive neural fuzzy inference system for the detection of inter-turn insulation and bearing wear faults in induction motor. IEEE Trans. Ind. Electron. **54**, 250–258 (2007)
10. Filippetti, F., Franceschini, G., Tassoni, C., Vas, P.: Recent developments of induction motor drives fault diagnosis using AI techniques. IEEE Trans. Ind. Electron. **47**, 994–1004 (2000)

11. Haroun, S., Seghir, A.N., Touati, S.: Multiple features extraction and selection for detection and classification of stator winding faults. IET Electr. Power Appl. **12**, 339–346 (2017)
12. Saddam, B., Aissa, A., Ahmed, B.S., Abdellatif, S.: Detection of rotor faults based on Hilbert Transform and neural network for an induction machine. In: 2017 5th International Conference on Electrical Engineering-Boumerdes (ICEE-B), pp. 1–6 (2017)
13. Sun, C., Ma, M., Zhao, Z., Chen, X.: Sparse deep stacking network for fault diagnosis of motor. IEEE Trans. Ind. Inform. (2018)
14. Bachir, S., Tnani, S., Trigeassou, J.-C., Champenois, G.: Diagnosis by parameter estimation of stator and rotor faults occurring in induction machines. IEEE Trans. Ind. Electron. **53**, 963–973 (2006)

Monitoring Tool for Stand-Alone Photovoltaic System Using Artificial Neural Network

Nassim Sabri[1]([⊠]), Abdelhalim Tlemçani[1], and Aissa Chouder[2]

[1] Laboratory of Electrical Engineering and Automatics LREA,
University of Médéa, 26000 Médéa, Algeria
sabri_nassim@hotmail.com, h_tlemcani@yahoo.fr
[2] Laboratoire de Géni Electrique (LGE),
University of Mohamed Boudiaf de M'sila,
BP 166 Ichbilia, 28000 M'sila, Algeria
aissachouder@gmail.com

Abstract. Fault classification using supervised machine learning Artificial Neural Network (ANN) is proposed to diagnose some defaults in Stand-alone photovoltaic (SAPV) system, where the data learning includes the voltage and current of PV panels, Battery and load are collected for different operation mode of the system (healthy and faulty). The proposed approach is applied to small SAPV system installed at LREA in the University of Médéa, Algeria in which the results of classification show a high accuracy up to 97%. In addition, a Graphical User Interface (GUI) Matlab is created in computer to display the results of classification by the developed ANN.

Keywords: SAPV system · ANN · Accuracy · GUI matlab

1 Introduction

SAPV system constitutes an alternative to bring electricity for many developing countries and remote areas (IEA 2016), the main issue with SAPV system is the safety problem, for PV panels, the current protection device (fuse) could be unreliable in low irradiance (Zhao et al. 2011), and for the Battery the risk is increased due to the lack of selecting the right sizing fuse (Nailen 1991). In addition a little published work that address the fault diagnosis in SAPV system including Battery. IEA (International energy agency) gives some energetic coefficients to asses the global performance of SAPV system (Mayer and Heidenreich 2003), where a translation of these coefficients into current parameters are given in Muñoz et al. (2006), and to improve the performance analysis, a new expression developed in Muñoz et al. (2009) and a monitoring is proposed using these formulas (Torres et al. 2012). To protect the Battery from reverse current in case of fault in PV panels, Wiles and King (1997) recommends the use of blocking diode or fuse. Otherwise, none effective method has been done to identify the faults in SAPV system. In this work, a fault classification of SAPV system is proposed using Artificial Neural Network (ANN), where this method has been widely used for fault detection and identification for grid-connected PV system (Chine et al. 2016; Yuchuan et al. 2009) and static PV system (Syafaruddin et al. 2011; Li et al. 2012;

© Springer Nature Switzerland AG 2019
M. Hatti (Ed.): ICAIRES 2018, LNNS 62, pp. 114–121, 2019.
https://doi.org/10.1007/978-3-030-04789-4_12

Xiao et al. 2014). The feedforward ANN proposed fault classification, is trained with Back-propagation Levenberg-Marquardt algorithm, where the data learning is acquired from small SAPV system installed at (LREA), University of Médéa, Algeria. The data learning includes the Normal operation and some faults realised, where the results of test classification show an accuracy of 97.7%, which indicate a high ability of ANN to correctly classify the faults. To facilitate access to the state of SAPV system using the ANN classifier, a Graphical User Interface (GUI) Matlab is created in computer to display the output of the ANN.

2   Description of SAPV System Data Acquisition

Data acquisition of experimental SAPV system installed at LREA, University of Médéa, Algeria, is performed, where the system as illustrated in Fig. 1 is composed of 2 PV panels connected with Battery and load as well as a Charge regulator to protect the Battery from overcharge and deep discharge. Table 1 give the electrical characteristics of the different components SAPV system.

The Current and Voltage of PV panels, Battery and load are acquired by mean of dspace DS 1104 controller board and computer and using the different sensors such as the current sensor Fluke i30s and the voltage sensor GW Instek DDP-100. Experimental facility of acquisition system is shown Fig. 2, the DS 1104 connect all the sensors with the computer, where the measurements can be monitored in real time via dspace ControlDesk software. The sampling time of data acquisition is chosen to 1 min in order to cover the operation of SAPV system.

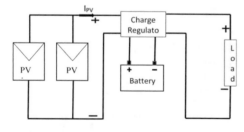

Fig. 1. Block diagram of SAPV system.

Table 1. SAPV   system   electrical characteristics

Element	Electrical specification
PV panels	Pmax = 180 Wp (2 × 90 Wp) Voc = 21.98 V, Isc = 11.08 A Vmpp = 18.25 V, Impp = 9.86 A
Battery	70 Ah (C20) VRLA Vs = 12 V
Charge regulator	30 A PWM charge regulator
Load	50 W Halogen lamp

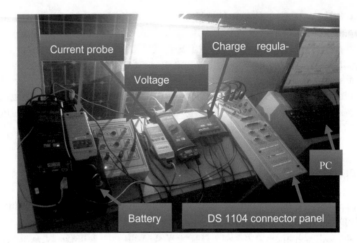

Fig. 2. Acquisition system at LREA, Médéa University, Algeria.

3 Proposed ANN Fault Classification

Three faults are established in SAPV system, where the Fault1 (F1) correspond to the short-circuit of PV module, the Fault2 (F2) is the open-circuit of PV module and the Fault3 (F3) is the external short-circuit of the Battery. Figure 3 shows these faults in schematic diagram.

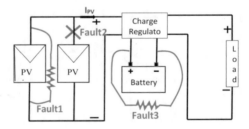

Fig. 3. Contrived faults in SAPV system.

3.1 Data Normalization

The data set issued from SAPV system are preprocessed before the training step, in which a normalization of the data is performed in order to make it in the same range [− 1, 1], the normalization of data allows the training algorithm to run quickly. The equation below is used to normalize the data:

$$y = \frac{(y_{max} - y_{min})\,(x - x_{min})}{(x_{max} - x_{min})} + y_{min} \tag{1}$$

where y is the normalized data limited between $y_{max} = 1$ and $y_{min} = -1$, x is the initial data which has a maximum value x_{max} and minimum value x_{min}.

3.2 ANN Approach

ANN has been widely utilised in fault detection and classification, pattern recognition, modeling and computer vision (Haykin 1994). In this work, a classification of faults in SAPV system is proposed using ANN, where an MLP (multi layer perception) feedforward neural network is used, in which it's characterised by one input layer, one output layer and one or more hidden layer (Lippmann 1987). The feedforward ANN adopted has one input layer, one hidden layer and one output layer, the input layer contains six neurons that correspond to the data acquired (voltage and Current of PV, Battery and load), the output layer contains four neurons that indicate one of classes correspond to the faults F1 to F3 as well as the Normal operation and the hidden layer contains 25 neurons that has been obtained by the manner trial & error, the architecture selected of ANN is shown in Fig. 4.

$$f(n) = \frac{e^n - e^{-n}}{e^n + e^{-n}} \tag{2}$$

$$f(n) = \frac{1}{1 + e^{-n}} \tag{3}$$

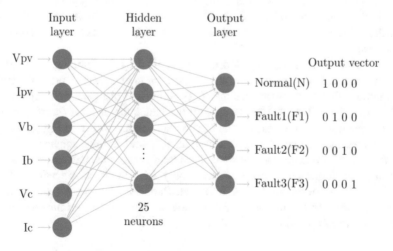

Fig. 4. Architecture of feedforward ANN faults classification.

The hidden layer uses the transfer function tangent sigmoid (tansig) defined in Eq. 2, while the output layer uses the transfer function log sigmoid (logsig) given in Eq. 3, and which has a binary output, as the ANN in our context is used for classification purpose (Hagan et al. 1996).

Back propagation Levenberg-Marquardt algorithm is used to train the ANN (Hagan et al. 1996; Hagan and Menhaj 1994), where an optimal values of weights and bias are found by minimizing the error between the predicted and desired (real) output, generally the error goal is the sum of square error defined in (Eq. 4), however in classification case the mean error is used instead.

$$\text{For } Q \text{ training data: } SSE = \frac{1}{2}\sum_{q=1}^{Q}(d_q - o_q)^2 = \frac{1}{2}\sum_{q=1}^{Q}e_q^2 \tag{4}$$

The Back propagation Levenberg-Marquardt algorithm use a modification to the Gauss-Newton's algorithm as demonstrated Eq. 5:

$$\Delta x = \left[J^T(x)\ J(x) + \mu\ I\right]^{-1}J^T(x)\ e(x) \tag{5}$$

where $J(x)$ is the Jacobean matrix containing the partial derivative of errors $e(x)$ with respect to the weights and biases for each training data q, and the parameter μ is increased by certain factor whenever the error E is increased, and decreased when the error E is reduced.

4   Results and Discussion

The training of ANN is carried out using 1236 samples which represent 80% of total data, while a number of 309 sample data is used for the test (equivalent of 20% of total data), the value of error goal is fixed to 0.01 which is sufficient to give good classification rate.

The convergence of training algorithm is verified by the learning curve given in Fig. 5, where it can be seen that the value of the error goal is reached after 66 iterations, which implies that the network parameters (weights and bias) are well determined.

To check the ability of ANN to classify the faults, the data set (training + test) is subject to the test, in which the results of classification are shown in Fig. 6, where the green squares indicate the correctly classified data and the red squares indicate the opposite. The confusion matrix shows that accuracy reaches the value of 97.7%, which reflect a good performance of predictive model ANN fault classification. The misclassified data recorded more, lies in 1st an 2nd class correspond respectively to Normal operation and Short-circuit PV module, in which the discrimination could be difficult in certain operation of SAPV system.

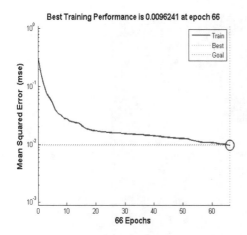

Fig. 5. Learning curve of trained ANN.

Fig. 6. Confusion matrix of data learning.

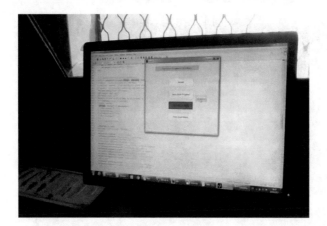

Fig. 7. Computer implementation of GUI matlab.

To monitor continuously the state of the system using the ANN classifier, an implementation of the developed ANN fault classification should be done, however the output of the network is a vector of 1 and 0 (Fig. 4), to clearly display the output of the network (the classes), a Graphical User Interface (GUI) Matlab is conceived to directly show the results of classification, where it's implementation is shown in Fig. 7 and the interface window is illustrated in Fig. 8.

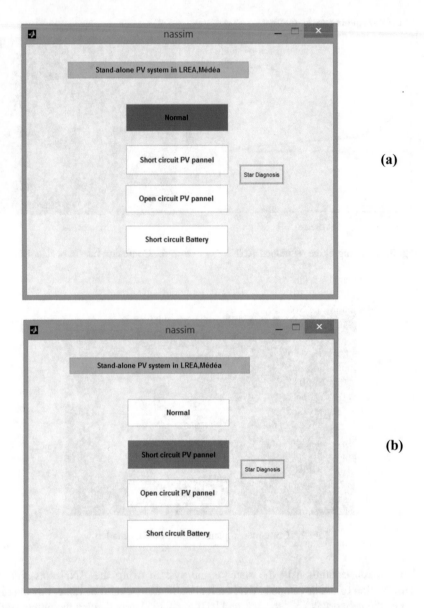

(a)

(b)

Fig. 8. Neural networks SAPV system faults diagnosis interface.

5 Conclusion

In this paper, a feedforward ANN is applied to classify the faults of small SAPV system installed at LREA, Algeria, where only the voltage and current of PV panels, Battery and load are used as features. The investigated faults includes short-circuit PV module, open circuit PV module and external short-circuit of Battery, in which the data of fault are acquired and preprocessed to use them to train the ANN classifier. A high classification rate is obtained by the test of ANN, where a computer GUI Matlab is created to display its results.

References

IEA. Trends 2016 in photovoltaic applications. Report IEA-PVPS T1-30:2016 (2016)

Zhao, Y., Lehman, B., Palma, J.-F., Mosesian, J., Lyons, R.: Fault analysis in solar PV arrays under: low irradiance conditions and reverse connections. In: 2011 37th IEEE Photovoltaic Specialists Conference (PVSC), pp. 002000–002005 (2011)

Nailen, R.L.: Battery protection-where do we stand? IEEE Trans. Ind. Appl. **27**, 658–667 (1991). https://doi.org/10.1109/28.85479

Mayer, D., Heidenreich, M.: Performance analysis of stand alone PV systems from a rational use of energy point of view. In: 2003 Proceedings of 3rd World Conference on Photovoltaic Energy Conversion, vol. 3, pp. 2155–2158 (2003)

Muñoz, F.J., Almonacid, G., Nofuentes, G., Almonacid, F.: A new method based on charge parameters to analyse the performance of stand-alone photovoltaic systems. Sol. Energy Mater. Sol. Cells **90**, 1750–1763 (2006)

Muñoz, F.J., Echbarthi, I., Nofuentes, G., Fuentes, M., Aguilera, J.: Estimation of the potential array output charge in the performance analysis of stand-alone photovoltaic systems without MPPT (Case study: Mediterranean climate). Sol. Energy **83**, 1985–1997 (2009)

Torres, M., Muñoz, F.J., Muñoz, J.V., Rus, C.: Online monitoring system for stand-alone photovoltaic applications analysis of system performance from monitored data. J. Sol. Energy Eng. **134**, 034502–034508 (2012)

Wiles, J.C., King, D.L.: Blocking diodes and fuses in low-voltage PV systems. In: Conference Record of the Twenty-Sixth IEEE Photovoltaic Specialists Conference, pp. 1105–1108 (1997)

Chine, W., Mellit, A., Lughi, V., Malek, A., Sulligoi, G., Massi Pavan, A.: A novel fault diagnosis technique for photovoltaic systems based on artificial neural networks. Renew. Energy **90**, 501–512 (2016)

Yuchuan, W., Qinli, L., Yaqin, S.: Application of BP neural network fault diagnosis in solar photovoltaic system. In: International Conference on Mechatronics and Automation (ICMA 2009), pp. 2581–2585 (2009)

Syafaruddin, S., Karatepe, E.S., Hiyama, T.: Controlling of artificial neural network for fault diagnosis of photovoltaic array. In: 2011 16th International Conference on Intelligent System Application to Power Systems (ISAP), pp. 1–6 (2011)

Li, Z., Wang, Y., Zhou, D., Wu, C.: An intelligent method for fault diagnosis in photovoltaic array. In: Xiao, T., Shang, L., Ma, S. (eds.) System Simulation and Scientific Computing, Part II, vol. 327, pp. 10–16 (2012)

Xiao, L., Pu, Y., Jiangfan, N., Jing, Z.: Fault diagnostic method for PV array based on improved wavelet neural network algorithm. In: 11th World Congress on Intelligent Control and Automation, pp, 1171–1175 (2014)

Haykin, S.: Neural Networks: A Comprehensive Foundation, 2nd edn. McMillan, Ottawa (1994)

Lippmann, R.: An introduction to computing with neural nets. IEEE ASSP Mag. **4**, 4–22 (1987)

Hagan, M.T., Demuth, H.B., Beale, M.H.: Neural Network Design. PWS Publishing, Boston (1996)

Hagan, M.T., Menhaj, M.B.: Training feedforward networks with the Marquardt algorithm. IEEE Trans. Neural Networks **5**, 989–993 (1994). https://doi.org/10.1109/72.329697

Prediction PV Power Based on Artificial Neural Networks

Lalia Miloudi[1(✉)], Dalila Acheli[1], and Saad Mekhilef[2]

[1] University of Boumerdès, Boumerdès, Algeria
lamiloudi@univ-boumerdes.dz
[2] University of Malaya, Kuala Lumpur, Malaysia

Abstract. The goal of this contribution is to estimate the power delivered by a multicrystals solar photovoltaic module based on artificial neural networks. Two structures of ANNs were tested: multiple-layer perceptron and radial basic function. The results obtained gave good coefficients of correlation, the statistical R^2-value obtained is about 0.96 to predict this important parameter.

Keywords: Artificial neural network (ANNs)
Multiple-layer perceptron (MLP) · Radial basic function (RBF)
Photovoltaic (PV) power

1 Introduction

Several studies have been conducted to predict parameters of PV systems, among these studies Almonacid et al. [1] presented the estimation output of a PV generator one hour ahead based on dynamic artificial neural networks. The results obtained of this study show that the proposed methodology could be used to forecast the power output of PV systems one hour ahead with an acceptable degree of accuracy. Ashhab [2] conducted the performance of an experimental photovoltaic solar system fixed panels, based on artificial neural networks. One structure of ANN was tested based on the Kaczmarz's algorithm, the model can be employed to estimate with precision the parameters of photovoltaic solar system which will produce maximum efficiency. Karatepe et al. [3] investigated a neural network approach for improving the accuracy of the electrical equivalent circuit of a photovoltaic module. ANN was applied to modeling a PV module, network input parameters are solar radiation and temperature. The current-voltage characteristics obtained by network training are closely similar to those measured and those of the conventional model of PV module proposed. The authors varied their investigations, indeed the extraction of PV power systems depend on many parameters such as weather conditions, geographical situation, materials of manufacture of the photovoltaic module and the electronic devices of tracking solar trajectory. The stabilization of the PV field is also depends on the performance of the energy storage system. Therefore, a large number of research studies have been performed to improve the efficiency and performance of the energy storage system as well as the overall PV system. Poullikkas [4] investigated a review on the battery large –scale for electricity storage. The analysis of this study showed that largest systems of energy

© Springer Nature Switzerland AG 2019
M. Hatti (Ed.): ICAIRES 2018, LNNS 62, pp. 122–128, 2019.
https://doi.org/10.1007/978-3-030-04789-4_13

storage are batteries containing sodium-sulfur for large-scale system. Whereas for the small systems the vanadium redox flow batteries are used. This paper is organized as follows: In Sect. 2 the concept of artificial neural network is presented. In Sect. 3 the photovoltaic module is modeling and the result of ANNs performance are discussed and finally the conclusions of the contribution are presented.

2 Artificial Neural Networks

Numerous research works demonstrated that, artificial neural networks (ANNs) may be used successfully in control and identification of nonlinear dynamical system [5, 6]. In the goal of prediction, feedforward or multi-layer perceptron (MLP) methods are well adapted. They can approximate virtually any measurable function up to an arbitrary degree of accuracy. It is usually constituted of two or three layers of neurons which are completely connected. The mathematical model is expressed as follows [7, 8]:

$$y = g\left(\sum_{i=1}^{m} v_j f\left(\sum_{i=1}^{n} w_{ij} x_i + b_i\right) + c\right) \tag{1}$$

Where y is the output of the network; $X = [x_1,...x_i]$ is the data inputs vector; (w_{ij}, v_j) and (b_i, c) are the weights and the biases of every hidden layer respectively. Activation functions (g) and (f) can be a simple threshold function, sigmoid, hyperbolic tangent or others.

The Radial Basic Function technique gives another alternative tool to learning in neural networks. The architecture of RBF-ANN as shown in the Fig. 1 [9], is simple and consists on one hidden layer and one output layer. The activation function is Gaussian as presented in Eq. (3). Each node (j) has a center value (c_j), where (c_j) is a vector whose dimension is equal to the number of input to the node. For each new input vector (X), the Euclidean norm of the difference between the input vector and the node center is calculated as in [10–12]:

$$v_j(x) = \left\| c_j - x \right\| = \sqrt{\sum_{i=1}^{N} (x_i - c_{j,i})^2} \tag{2}$$

The output of the network is given by

$$\hat{y} = \sum_{i=1}^{L} w_m \exp\left(-\frac{v^2}{\sigma^2}\right) \tag{3}$$

The most used statistical coefficients to estimate the efficiency of ANNs are: the mean bias error MBE, the root mean square error RMSE, and the coefficient of determination R^2. Where $Y_{i,est}$ and $Y_{i,mes}$ are respectively, the i_{th} estimated values and i_{th} measured values of the variables. Numerous works compared and presented their results about the following coefficients in [13–15].

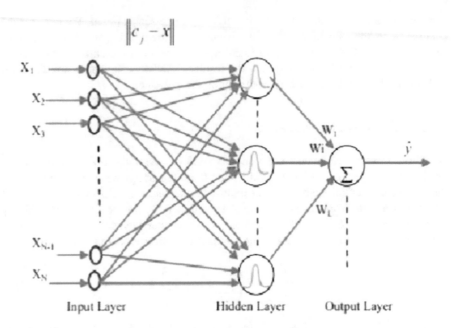

Fig. 1. Radial basic function artificial neural network [9]

The mean bias error:

$$MBE = \frac{1}{N} \sum_{i=1}^{N} \left(Y_{i,est} - Y_{i,mes} \right) \tag{4}$$

The root mean square error:

$$RMSE = \left[\frac{1}{N} \sum_{i=1}^{N} \left(Y_{i,est} - Y_{i,mes} \right)^2 \right]^{\frac{1}{2}} \tag{5}$$

Coefficient of determination:

$$R^2 = 1 - \frac{\sum\limits_{i=1}^{N} \left(Y_{i,mes} - Y_{i,est} \right)^2}{\sum\limits_{i=1}^{N} \left(Y_{i,mes} \right)^2} \tag{6}$$

3 Photovoltaic Module Representing

Any PV module exposed to the solar rays becomes generating of current (I_{ph}), and thus being able to extract an electrical power. The Fig. 2 presents the equivalent diagram of a PV cell under illumination. It corresponds to a generator of current in parallel on a diode. The serial resistance depends on the impedance of material, the resistance of shunt corresponds to a resistance of the junction. For an ideal cell, serial resistance tends towards zero, the resistance of shunt tends towards the infinite, (q) is the electron charge, (K) the Boltzmann constant, T is the cell temperature and A is diode ideal factor as calculated by the Eq. (8) [16–18]:

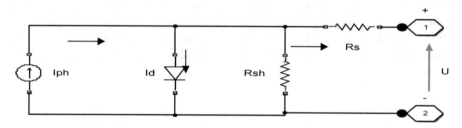

Fig. 2. General model of PV generator

The characteristic current-tension is put in the form as

$$I(U) = I_{ph} - I_d(U) \qquad (7)$$

The output current of PV module can be expressed by the following equation [17, 19, 20]

$$I = I_{ph} - I_d \left[\exp\left(\frac{q(U + R_S I)}{KTA} \right) - 1 \right] \qquad (8)$$

Mathematical model of a PV array which consists of N_S cells in series and N_P cells in parallel is given as

$$I = N_P I_{ph} - N_P I_d \left[\exp\left(\frac{q}{KTA} \left(\frac{U}{N_S} - I \frac{R_S}{N_P} \right) \right) - 1 \right] \qquad (9)$$

According the characteristic I-V is written as

$$I(G_c, T_c) = I_{sc} \left\{ 1 - \exp\left[\frac{(v_{oc} - v)}{\tau} \right] \right\} \qquad (10)$$

$$I_{cc2}(G_2, T_2) = I_{cc1}(G_1, T_1) \frac{G_2}{G_1} + \alpha(T_2 - T_1) \qquad (11)$$

$$V_{co2}(G_2, T_2) = V_{co1}(G_1, T_1) + A \ln\left(\frac{G_2}{G_1} \right) + \beta(T_2 - T_1) \qquad (12)$$

The estimated power extracted on the module is obtained by the product of the current and the output voltage. As presented in Eqs. (11), (12) and (13), the current and the tension depend directly on the temperature and the solar radiation

$$P = I * V \tag{13}$$

The various phases of training process to the RBF artificial neural network are presented in Fig. 3. Since a coefficient of correlation R equal to 0.89 in the first phase and equal to 0.927 in the second phase, to R equal 0.977 in the third phase until equal 0.9872 in the final phase. It is observed in Fig. 4, the photovoltaic power predicted by radial basic function ANN gives better estimation than the power predicted by MLP-ANN.

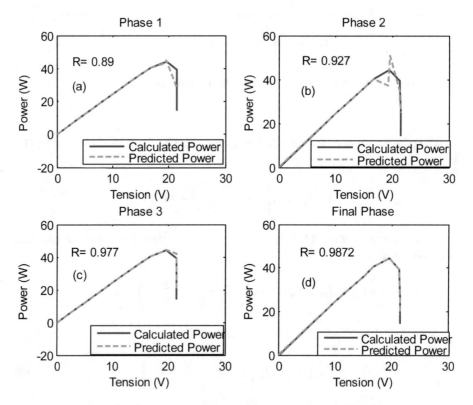

Fig. 3. Predicted PV power by radial basic function (RBF) ANN.

Different statistical coefficients for the studied cases are presented in Table 1. The mean bias error MBE, the root mean square error RMSE and the coefficient of determination R^2. These numbers are very close to 1, which indicates a good relation between the inputs or targets (X) and the outputs or response (Y) of neural networks. The relationship is given by: $Y = mX + b$, m corresponds to the slope and b the

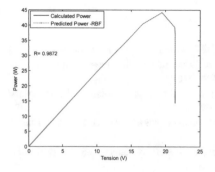 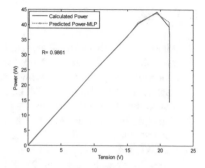

Fig. 4. Predicted PV power by radial basic function and multi layer perceptron (MLP) ANNs.

Table 1. Statistical performance for prediction photovoltaic power.

ANNs	MBE	RMSE	R^2	Best linear fit
MLP (a)	9.09e−006	0.8096	0.9724	Y = (0.972)X + (0.833)
RBF (b)	9.0909e−006	0.8028	0.9742	Y = (0.975)X + (0.765)

y-intercept of the best linear regression relating targets to network response. The coefficient R^2-value between the outputs and the targets if is equal to 1, then there is perfect correlation between targets and outputs.

4   Conclusion

The obtained results show that the artificial neural networks tested introduce a good estimation for of PV power can be extracted. This method was used successfully by training various ANNs with appropriate architecture MLP and RBF. The good choice of activation functions, the numbers of hidden layers and the numbers of nodes. Each neural network allowed to obtain predicted curves with a coefficient R^2-value better than 0.95. The ANNs operate like a "black box" model an advantage of using ANNs is their ability to treat a multidimensional system with many interrelated parameters. Therefore, the artificial neural networks are utilized to overcome these difficulties. This method can be a beneficial tool for the designers of PV systems, it can be applied before the implementation the PV installation, providing the power delivered by the system.

Acknowledgments. The first author would like to thank the System National of Documentation on Line (SNDL) for facilitate download articles as well as Pr S. Mekhilef from PEARL laboratory of University of Malaya, and Dr H. Sarimveis from National Technical University of Athens Greece.

References

1. Almonacid, F., Pérez-Higueras, P.J., Fernández, E.F., Hontoria, L.: A methodology based on dynamic artificial neural network for short-term forecasting of the power output of a PV generator. Energy Convers. Manag. **85**, 389–398 (2014)
2. Ashhab, M.S.: Adaptive prediction of the performance of a photovoltaic solar integrated system. Int. J. Therm. Environ. Eng. **3**, 43–46 (2010)
3. Karatepe, E., Boztepe, M., Colak, M.: Neural network based solar cell model. Energy Convers. Manag. **47**, 1159–1178 (2006)
4. Poullikkas, A.: A comparative overview of large-scale battery systems for electricity storage. Renew. Sustain. Energy Rev. **27**, 778–788 (2013)
5. Narendra, K.S., Parthasarathy, K.: Identification and control of dynamical systems using neural networks. IEEE Trans. Neural Netw. **1**, 4–27 (1990)
6. Miloudi, L., Acheli, D., Kesraoui, M.: Application of artificial neural networks for forecasting photovoltaic system parameters. Appl. Solar Energy **53**, 85–91 (2017)
7. Anders, U., Korn, O.: Model selection in neural networks. Neural Netw. **12**, 306–323 (1999)
8. Miloudi, L., Acheli, D., Kesraoui, M.: Sun trajectory and PV module I–V characteristics estimation using neural networks. In: 2017 8th International Renewable Energy Congress (IREC) IEEE Conference Publications, Amman, Jordann, pp. 1–6, 21–23 March 2017
9. Miloudi, L.: Méthodes heuristiques appliquées à l'optimisation du contrôle de l'orientation d'un panneau solaire photovoltaïque. Thesis, dlibrary.univ-boumerdes.dz (2018)
10. Musavi, W.A.M.T., Chan, K.H., Faris, K.B., Hummels, D.M.: On the training of radial basis function classifiers. Neural Netw. **5**, 587–598 (1992)
11. Sarimveis, H., Tsekouras, G., Bafas, G.: A fast and efficient algorithm for training radial basis function neural networks based on a fuzzy partition of the input space. Ind Eng Chem Res **41**, 751–759 (2002)
12. Miloudi, L., Acheli, D.: Prediction global solar radiation and modeling photovoltaic module based on artificial neural networks. In: Control, Engineering & Information Technology (CEIT), 2015 3rd International Conference on IEEE Conference Publications (2015)
13. Koussa, M., Malek, A., Haddadi, M.: Statistical comparison of monthly mean hourly and daily diffuse and global solar irradiation models and a Simulink program development for various Algerian climates. Energy Convers. Manag. **50**, 1227–1235 (2009)
14. Jiang, Y.: Computation of monthly mean daily global solar radiation in China using artificial neural networks and comparison with other empirical models. Energy **34**, 1276–1283 (2009)
15. Stone, R.J.: Improved statistical procedure for the evaluation of solar radiation estimation models. Solar Energy **51**, 289–291 (1993)
16. Miloudi, L., Acheli, D., Chaib, A.: Solar Tracking with Photovoltaic Panel. Energy Procedia **42**, 103–112 (2013)
17. Syed, I.M.: Comparative analysis of photovoltaic systems. Int. J., Electr. Comput. Energ. Electron. Commun. Eng. **09** (2015)
18. Equer, B.: Energie solaire photovoltaïque, vol. 1. Ellipses Editions Marketing and UNESCO 1993
19. Safari, A., Mekhilef, S.: Simulation and hardware implementation incremental conductance MPPT with direct control method using Cuk converter. IEEE Trans. Ind. Electron. **58**, 1154–1161 (2011)
20. Seyedmahmoudian, M., Mekhilef, S., Oo, A.M.T., Stojcevski, A., Soon, T.K., Ghandhari, A.S.: Simulation and hardware implementation of new maximum power point tracking technique for partially shaded PV system using hybrid DEPSO method. IEEE Trans. Sustain. Energy **6**, 850–862 (2015)

Application of Improved Artificial Neural Network Algorithm in Hydrocarbons' Reservoir Evaluation

M. Z. Doghmane[1,3(✉)], B. Belahcene[2], and M. Kidouche[3]

[1] Reservoir Evaluation Department, Exploration and Production Division,
Sonatrach, Hassi Messaoud, Algeria
MDoghmane@for.hmd.sonatrach.dz
[2] Gassi Touil Division Production, Sonatrach, Hassi Messaoud, Algeria
B.Belahcene@amt.sonatrach.dz
[3] Department of Automation, Faculty of Hydrocarbons and Chemistry
(Ex-INH), University M'hamed Bougara, Boumerdes, Boumerdes, Algeria
mkidouche@univ-boumerdes.dz

Abstract. The aim of this work is to develop an artificial neural network software tool in Matlab which allows the well logging interpreter to evaluate hydrocarbons reservoirs by classification of its existing facies into six types (clay, anhydrite, dolomite, limestone, sandstone and salt), the advantage of such classification is that it is automatic and gives more precision in comparison to manual recognition using industrial software. The developed algorithm is applied to eleven wells data of the Algerian Sahara where necessary curves (Gama Ray, density curve Rhob, Neutron porosity curve Nphi, Sonic curve dt, photoelectric factor curve PE) for realization of this technique are available. A graphical user interface is developed in order to simplify the use of the algorithm for interpreters.

Keywords: Artificial neural networks · Lithofacies classification
Industrial software · Algerian Sahara · Graphical user interface

1 Introduction

Well logging is the recording of data from a capture tool placed at the bottom of the hole, plotted against the depth of the well. The most common application of well logging in the oil industry is to look for areas of recoverable oil. It is useful to have a detailed account of the geological formations penetrated by a borehole. For the production of oil and gas, companies would have several kinds of information on a geological layer such as the oil content. To measure these properties, the sources and sensors in loaded enclosures known probes which can be lowered into an existing borehole (wireline logging) or can be mounted in a collar behind the drill bit to take measurements while the well is drilled. In wireline logging, electronic probes and cartridges are chained and reduced in a cased well to a cable which has an electronic signal wire. As the chain is lifted, the sensors measure a part or all of the following properties as a function of depth: the electrical resistivity, the electron density, the

© Springer Nature Switzerland AG 2019
M. Hatti (Ed.): ICAIRES 2018, LNNS 62, pp. 129–138, 2019.
https://doi.org/10.1007/978-3-030-04789-4_14

speed of sound, neutron moderation, absorption to thermal neutrons (induced) of the radioactivity, of natural and artificial spectral gamma rays, Compton scattering, size of the borehole, and the nuclear magnetic resonance occasionally. The data is transmitted through the wire to a computer at the surface where the data is recorded.

Neural networks have been applied in a variety of areas to solve problems such as classification, feature extraction, diagnosis, function approximation and optimization. Lithofacies classification for well logging is a complex and nonlinear geophysical problem due to several factors, such as shape of the pores, fluid saturation, grain size... etc. The complexity of the problem is characterized by non-linearity of the physical and statistical solutions; the latter is caused by the noise in geophysical measurements. Several researchers have worked in lithofacies classification using the conventional method as "cross-plot" and other statistical techniques. In Cross-plot technique (Pickett 1963 [1], Gassaway et al. 1989 [2]), two or more logs are cross plotted to get the lithology curve. The multivariate statistical methods have been used to study the drilling data. These techniques are, however, semi-automated and require a large amount of data, which are expensive and not readily available.

Existing methods for analyzing log data are very tedious and time consuming, especially when it comes to large number of complex and noisy drill measurements. The objective of this work is to develop an algorithm for the recognition of lithofacies and automatic classification based on the artificial neural network; we applied the algorithm to the data from eleven wells in the Algerian Sahara. The validation of the results was performed by comparing the automatic recognition using Matlab with manual recognition using Industrial software.

2   Overview of Well Logging Methods

2.1   Nuclear Logging Method

Measuring the radioactivity is of particular interest in the search for radioactive ores, in particular potassium and uranium salts, but also for the detection of clays benches or radioactive beds. It is a measure of the natural radioactivity of the formations traversed by a borehole. These measurements are useful for the detection of radioactive minerals existing in the formations namely: U, Th, K. These elements are particularly related to clay, and therefore, the gamma ray reflected especially shaliness.

In sedimentary formations, Gamma Ray is used for the delineation of layers depending on their clay content. The quantitative application will be in favorable cases the assessment of the clay percentage considering that the radioactivity of sedimentary rocks is mainly linked to K^{40} found in clay. We can write this condition using the following equation

$$V_{sh} = (GR_{lu} - GR_{min})/(GR_{max} - GR_{min})$$

$\qquad(1)$

2.2   Neutron Logging Method

Measuring the formations density in this method is similar to the previous method, but this time it is subjected to continuous training bombardment with energy equal to 662 kilo; the electron emitted by a source of 137Cs cesium. These gamma rays lose energy by collisions with electrons, it is a lithology and porosity tool; it measures the speed decrease of the number of thermal neutrons as a function of the distance from the source [3], this speed will depend mainly on the porosity. The application of neutron tools is the determination of porosity, where some corrections to the lithology and hole conditions are sometimes necessary [4]. For this technique, we can write the following equation.

$$\phi_{Ncor}(\%) = \phi_{Nlu} + 4 - V_{sh} \times \phi_{Nsh} \tag{2}$$

2.3   Sonic Logging Method

The sonic log is a continuous record, depending on the depth, of the speed of sound in the formations. It was found that the propagation of acoustic waves in training depends on the porosity. The transmitted wave is calibrated in amplitude and frequency, the signal recovered by the receivers, in comparison with that emitted, gives an idea on the speed of the acoustic wave in the medium which is related to its compaction, therefore, linked to porosity. The tool measures the transit time of the acoustic wave over a distance of 1 foot [5]. The transit time is the time it takes for a sound wave to cross one foot formation and it is expressed by (ms/ft). The BHC tool has two acoustic wave transmitters and four receivers. The relationship linking the measures with porosity is

$$\phi_{Scor} = \frac{(\Delta t_f - \Delta t_{lu})}{(\Delta t_f - \Delta t_{ma})} - V_{sh} \times \frac{(\Delta t_f - \Delta t_{sh})}{(\Delta t_f - \Delta t_{ma})} \tag{3}$$

3   Importance of ANN in Oil Industry

With the ability of classification and generalization, neural networks are commonly used in statistical problems, such as automatic classification and decision making. According to Dreyfus et al. 2004 [6], the benefit of ANN over other methods lies in their ability to perform precise equivalent models with less experimental data and their ability to create more accurate models from the same number of examples in addition to their ability to solve complex problems. This technique is implemented in several disciplines, the most well known [6]

- The seismic data processing [11], Veezhinatan 1990, McCormack 1991, Musuma 1992.
- The characterization of rocks (Derek 1990; Braunschweig 1990).
- Interpretation of logs (Baldwin 1989–1993; Avelino 1991; Riva 1992).
- Diagnostic of the drilling tool (Arehart 1989).

- Control Process (Lambert 1991).
- Recognition of lithological and facies analysis [12, 13].
- Lithological classification of seismic attributes (Mihoubi 2006) [14].

The propagation algorithm (PRA) is an example of supervised learning algorithm, it has been widely used for spectacular applications such as Rosenberg' demonstration (1987) [7] in which the PRA is used in a system that learns to read text. Another success was the prediction of the stock market (Refenes et al. 1994 [8]) (Fig. 1).

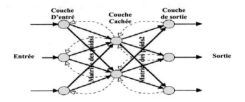

Fig. 1. An example of ANN architecture

4   Characterization of Reservoir Lithofacies

In this section, we focused on the algorithm used to achieve the objectives of this work; classification algorithms based on artificial neural networks consist of nodes in input layer, one or more hidden layers and an output layer. Each node in a layer has a corresponding node in the next layer, creating the stacking effect. Feed-forward is one of the popular structures from artificial neural networks; this structure is widely used to solve complex problems by modeling complex relationships. The developed algorithm procedure is described as follows

Step 1: Standardize data; The obtained data is mapped to the terminal [0, 1], so that to adjust the interval defined attributes and avoid saturation of neurons. Data normalization is calculated by

$$I_{new} = \frac{I_{old} - I_{min}}{I_{max} - I_{min}} \times (O_{max} - O_{min}) + O_{min} \tag{4}$$

I_{min} is the minimum value of the variables, I_{max} is the maximum value of the variables, O_{max} is the maximum value after normalization, O_{min} is the minimum value after normalization, I_{new} is the new value after normalization and I_{old} is the old value before normalization [15, 16].

Step 2: Set the network settings; Number of hidden layers: In this study, we took five hidden layers where the convergence and accuracy of the results can be guaranteed. We took the depths of 3300–3500 m where for each 0.1054 m we have a sample for the layer. Learning rate (η): In general, too fast or too slow a learning rate is detrimental to network convergence; in this study, we have selected values between 0.1 and 1.0 for testing networks [17]. Transfer function is the sigmoid function [18] for which its value is in the range [0, 1].

Step 3: Enter the data of training examples; the associated values of input and output training are entered using the GUI of the algorithm.

$$net_j = \sum_{i=1}^{m} w_{ji}x_i + b_j \tag{5}$$

The sigmoid function is used to convert signal for each neuron in the hidden layers. The signal from the output layer can be expressed by the target value of output neuron k and the error of each hidden layer

$$\delta_j^l = \sum_{i=1}^{l} \delta_i^{l+1} w_{ji}^l f' \left(net_j^l \right) \tag{6}$$

Progressive change for each weight of each learning interaction is computed using the first derivative of the sigmoid function.

Step 5: Calculate the error values: Steps 3–5 are repeated till the network converges [19].

5 Comparative Study ANN-Classical Techniques

5.1 Interpretation Using Conventional Techniques

In this section, the interpretation is done manually using the IP Log software, we used four curves; gamma ray curves (GR), the density curve, the neutron curve (NPHI) and the curve of the photoelectric factor (PE). In these curves, we have taken the depths area (3300–3500 m). The eleven wells are Well-1 (Pilot well), Well-2, Well-3, Well-4, Well-5, Well-6, Well-7, Well-8, Well-9, Well-10 and Well-11 for which all the necessary information in the curves of the selected area (3300–3500 m) are available. We applied the sequence for Well-1 after learning network; we repeated the same job for ten wells. The sequence is as follows

- We use the log gamma (GR) to determine the clay and non-clay benches in the logs for the selected area.
- From the graphs, we determine the matrix density and the fluid density which allow us to generate the lithology curve.
- Using conditions below, we have manually classified the facies for this well.

The conditions used for manual classification by IP Log are

If (GR > 70 API), we have clay else;
If ((NPHI-PHID) < 1%), we have limestone, else;
If (1% < (NPHI-PHID) < 7.5%), we have sandstone, else;
If (7.5% < (NPHI-PHID) < 13.5%), we have dolomite, else;
If (15% < (NPHI-PHID) < 45%), we have anhydrite, else
If ((NPHI-PHID) > 45%), we have salt,

The matrix parameters are calculated using Quick look method; the values used to plot the curve for each well are given in Table 1.

Table 1. The matrix parameters calculated for the different wells

Well	Dtmat	Rhomat
Well-1	50	2.85
Well-2	49	2.86
Well-3	50	2.85
Well-4	51	2.78
Well-5	54	2.81
Well-6	56	2.79
Well-7	51	2.84
Well-8	53	2.82
Well-9	51	2.84
Well-10	48	2.81
Well-11	49	2.86

5.2 Interpretation Using ANN

In this section, we have developed a code source for the automatic recognition of lithofacies, where all the targets achieved before with IP Log software are programmed using MATLAB, to simplify the use of the proposed algorithm; a graphical user interface (GUI) is developed. The results of the automatic lithofacies recognition can be plotted using Matlab plot tools as it is shown in the Fig. 2.

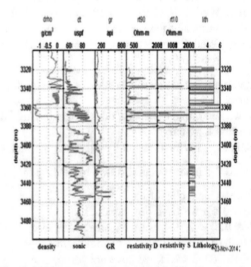

Fig. 2. The curves displayed by Matlab after learning

We note that the curve of the lithology is not well marked, for this reason and also to make an effective comparison, we export the results of Matlab to other visualization software, and we plot the curves in the same way of the manual recognition. To validate the use of ANN, we show in this part a brief comparison between the results of manual interpretation using conventional software and automatic recognition using ANN-Matlab, where we show the results of selected wells.

In order to facilitate the use of the developed algorithm without the need of having Matalb, we have developed the graphical interface, it has been compiled so that is can be run as standalone application (Fig. 3).

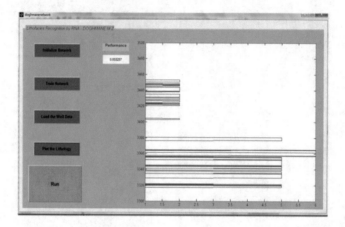

Fig. 3. Graphical user interface of the developed algorithm

- Well-2 (Fig. 4A) We have identified the presence of four areas that are
    Zone1 (3340 m–3376 m) we notice in this area, anhydrite and sandstone sequence, a presence of some clay.
    Zone2 (3376 m–3427 m) is the second area of clay layers separated by thin layers of sandstone.
    Zone3 (3427 m–3474 m): The third area is limestone layers separated by a considerable layer of clay; the presence of limestone is not taken into consideration.
    Zone4 (3474 m–3500 m) this area is a clay cover layer.
- Well-3 (Fig. 4C) We have identified the presence of six areas that are
    Zone1 (3316 m–3356 m): in this zone, the dominant facies is anhydrite, we can recognize a remarkable presence of the salt layer, the presence of limestone and sandstone is not considered.
    Zone2 (3356 m–3467 m) there is a layer of clay ended down by a thin layer of limestone.
    Zone3 (3467 m, 3477 m) a sandstone layer followed by a thin layer of limestone.
    Zone4 (3477 m–3392 m): a clay cover layer.
    Zone5 (3392 m–3408 m) another considerable layer of sandstone.

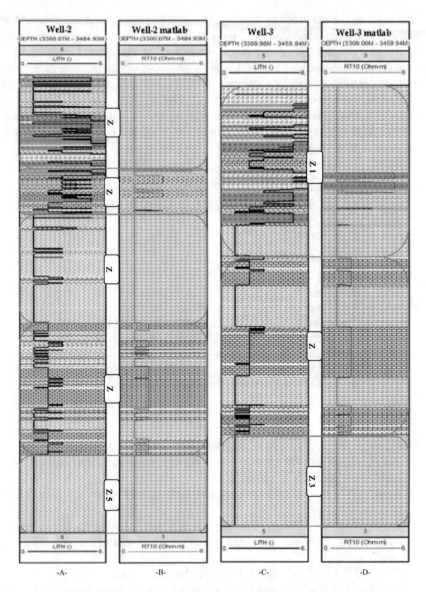

Fig. 4. Comparison of result for manual and automatic interpretation for wells Well-2 (A, B)/ Well-3 (C, D)

Zone6 (3408 m–3450 m): two clay layers opaque a layer of sandstone.

However, for the automatic lithofacies, we have found that

- Well-2 (Fig. 4B) We have identified the presence of four areas that are **Zone1** (3340 m–3376 m) we notice in this area, anhydrite and sandstone sequence, a presence of some clay.

Zone2 (3376 m–3427 m) is the second area of clay layers separated by thin layers of sandstone.

Zone3 (3427 m–3474 m): The third area is limestone layers separated by a considerable layer of clay; the presence of limestone is not taken into consideration.

Zone4 (3474 m–3500 m) this area is clay cover layer.

- Well-3 (Fig. 4D) We have identified the presence of three areas that are

  Zone1 (3316 m–3356 m) thick clay layers separated by a thin layer of anhydrite.

  Zone2 (3356 m–3467 m): a clays and stone sequence with the dominant facies is the sandstone.

  Zone3 (3467 m–3477 m): a thick clay layer that covers the upper layers.

As shown in Fig. 4, the recognition of the facies using ANN-MATLAB, has many advantages over the manual recognition using conventional software such as

- Simplify the work of interpretation by automatic procedures so that the time taken by the interpreter is minimized.
- The results found by the ANN are more precise; a neglected area by the Industrial software is considered by the algorithm.
- The implementation requirements for the classification and recognition by Matlab allow us to minimize the influence of the precision of the interval in the creation of virtual thin layers that do not exist in the geology of this region.

6 Conclusion

In this work, we have taken the goal of developing an algorithm for automatic lithofacies recognition based on artificial neural network, in addition to improving the accuracy of this type of procedure usually applied using software known as industrial software.

To achieve these goals, we have divided the work into stages, in the first stage, we used the Industrial software to do manual classification based on certain assumptions using available data, this procedure is applied for eleven wells in the Algerian Sahara, and the results were used as comparison samples to validate the results of the neural network classification. The next step is the training of the neural network using the lithology curve of pilot well, after that; we used five curves for automatic classification using the trained neural network in MATLAB. To simplify the use of this algorithm, we have also developed a graphical interface that allows users to simply use the necessary procedures. Finally, the last step of the project is the comparison between manual and automatic lithofacies recognition. Through this comparison, we concluded that

- The artificial neural network allows us to clarify the classification of lithofacies.
- Automatic recognition simplifies the work of interpreters.
- The use of Industrial software for manual recognition may cause some ignorance of important areas, but the neural network realizes a step by step interpretation.

The implementation of the condition in the Industrial software can cause non-existing layers especially if conditions intervals overlap. However, this algorithm will be ineffective in the absence of any geological information such as the number of lithofacies, lithology pilot curve. For this reason, it would be desirable to use as many carrot data as possible in order to increase the precision of the algorithm for far wells.

References

1. Schlumberger: Log Interpretation Charts (2013)
2. Gassaway et al.: The Graphical Cross-Plotting Technique (1989)
3. Mavko, G., Mukerji, T., Dvorkin, J.: The Rock Physics Handbook: Tools for Seismic Analysis of Porous Media. Cambridge University Press, New York (2009)
4. Serra, O.: Diagraphies Différés base de l'interprétation, Tome2. Etudes et productions Schlumberger, Montrouge (1985)
5. Halliberton: Logging and Perforating Products and services (2003)
6. Schlumberger: Log Interpretation Charts (2000)
7. Dreyfus, G. et al.: Réseaux de neurones, Méthodologie et applications, Editions Eyrolles (2004)
8. Rosenberg, C.R., Sejnowski, T.J.: The spacing effect on NETtalk, a massively-parallel network. In: Proceedings of the Eighth Annual Conference of the Cognitive Science Society, pp. 72–89. Lawrence Erlbaum Associates, Hillsdale, New Jersey (1987)
9. Zarpanis, A., Renese, A.P.: Principles of Neural Model Identification Selection and Adequacy, Centre of Neural Networks, Departement of mathematics. King's College, London (1999)
10. Sung, H., Lee, D.S.: Neuro-fuzzy recognition system for detecting wave patterns using wavelet coefficients. IEICE Trans. Inf. Syst. **84**-**D**(8) (2001)
11. Renders, J.M.: Algorithmes génétiques et réseaux de neurones. Editions Hermés, Paris (1995)
12. Johnston, D.H.: Seismic attribute calibration using neural networks. Soc. Expl. Geophys
13. Platon, E. et al.: Pattern matching in facies analysis from well log data - a hybrid neural network-based application. In: AAPG Conference and Exhibition, Barcelona, Spain. September 21–24 (2003)
14. Fournier, F.: Analyse automatique de faciès diagraphiques et sismiques. Réunion technique Société pour l'Avancement de l'Interprétation des Diagraphies SAID-Union Française des Géologues UFG du 25 Juin (2002)
15. Mihoubi, A.: Classification lithologique des attributs sismiques par les réseaux de neurone artificiels, Ph.D. thesis, Faculté des Hydrocarbures et de la chimie FHC – UMBB (2008)
16. Hassoun, M.H.: Fundamentals of Artificial Neural Networks. MIT Press, Cambridge, London (1995)
17. Masters, T.: Practical Neural Network Recipes in C++. Academic Press Professional Inc., San Diego (1993)
18. Fu, L.: Neural Networks in Computer Intelligence. McGraw-Hill, New York (1995)
19. Tabach, E.E., Lancelot, L., Shahrour, I., Najjar, Y.: Use of artificial network simulation metamodelling to assess groundwater contamination in a road project. Math. Comput. Model. **45**(7–8), 766–776 (2007)
20. Kumar, S.: Neural Networks. McGraw-Hill, New York (2005)

Fuzzy Logic Based MPPT for Grid-Connected PV System with Z-Source Inverter

Ali Teta[✉], Mohamed Mounir Rezaoui, Abdellah Kouzou, and Aicha Djalab

Applied Automation and Industrial Diagnostics Laboratory, University of Djelfa, Djelfa, Algeria
tetaali@hotmail.com, mm_rezaoui@yahoo.fr, kouzouabdellah@ieee.org, a.djalab@univ-djelfa.dz

Abstract. In recent years, renewable energy has become very important and has been given a lot of attention due to the fear of the exhaustion of non-renewable energy in addition to the fact that renewable energy is totally free-pollution and inexhaustible. One of the most well-known renewable sources is solar energy. This paper gives the detailed mathematical model and the characteristics of a photovoltaic array, then it illustrates an artificial intelligence based fuzzy logic FLC maximum power point tracking MPPT control method, next a single stage operating based on the new impedance Z-source inverter which has been used as an interface between the PV system and the grid is presented. Lastly the simulation is achieved using Matlab Simulink and the results are discussed.

Keywords: Renewable energy · Fuzzy logic · MPPT · Z-source inverter

1 Introduction

Over the last few years the non-renewable energy such as fossil fuel has become highly-priced energy as a result of its widespread use and aggressive consumption. As concern is growing over this situation, the renewable energy development such as photovoltaic energy has become significantly more necessary than ever (Reshmi and Nandakumar 2016; Liu et al. 2014). The dc power delivered by the PV systems can be exploited in two ways, the first is to use the generated dc power at the same place, astoring equipment is needed to avoid the interruptible operating. While the second one is to connect the PV system directly to the grid by considering the quality conditions (Bourguiba et al. 2016). In the two stages operating, the PV configuration is based on two converters, DC-DC converter which is responsible for extracting the maximum power from the PV generator and connecting it to the DC-AC converter which is considered as the main component in the system, Its importance stems from being the component that transform the power generated by the PV generator into a useful alternative power (Bourguiba et al. 2016; Selvaraj and Rahim 2009).

Due to its sensitive role in the PV systems, several developments having been applied to the inverter which aim to increase its efficiency and decrease its power losses. In order to overcome the disadvantages of the two stage operating systems such

© Springer Nature Switzerland AG 2019
M. Hatti (Ed.): ICAIRES 2018, LNNS 62, pp. 139–147, 2019.
https://doi.org/10.1007/978-3-030-04789-4_15

as power loss, high cost and control difficulties, Peng, F.Z had suggested a new promising topology in (Peng 2003) which is named impedance source or Z-source inverter topology (Ketabi and Tabatabei 2015; Rasin and Rahman 2012). Because of the unsteady environmental conditions and the nonlinearity in the V-I characteristics of the PV array, it was always challenging to extract the maximum power point from the PV cells. To achieve such purpose several MPPT methods have been developed as the perturbation and observation method and the incremental conductance (Ahmed et al. 2016), to overcome the drawbacks in the above-mentioned methods, this work proposes to use an artificial intelligence based fuzzy logic maximum power point tracking control strategy. This paper aims firstly to give the detailed mathematical model of each part of the studied PV system, secondly offer a satisfactory explanation for the adopted control strategy and finally provide and discuss the simulation results.

2   System Description and Modelling

The elaborated system comprises essentially a PV array connected to a Z-source inverter, based fuzzy logic MPPT control method is used to extract the maximum power from the PV array, and finally the grid network as shown in Fig. 1.

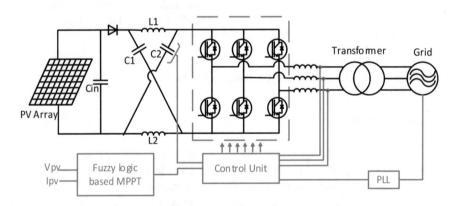

Fig. 1. PV based grid-connected ZS inverter system.

2.1   PV Cell Modelling

The PV array is composed of cells which transform solar energy to electrical power, these cells can be associated in series or in parallel depends on the desired value of current and voltage (Bourguiba et al. 2016). Figure 2 presents the equivalent circuit of the PV cell.

By applying Kirchhoff's Current Law

$$I_{pv} = I_{ph} - I_d - I_p \tag{1}$$

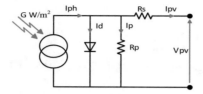

Fig. 2. Equivalent circuit for PV cell.

Where

I_{ph} is the photocurrent [A].
I_d is the diode current.
I_p is the current leak in parallel resistor.

According to the Eq. (1) the output current is given by

$$I_{pv} = I_{ph} - I_d - \frac{V_{pv} + R_s \cdot I_{pv}}{R_p} \tag{2}$$

Diode current equation

$$I_d = I_0 \left[exp\left(\frac{V_{pv} + R_s \cdot I_{pv}}{A \cdot N_s \cdot V_T}\right) - 1 \right] \tag{3}$$

That

I_0 is the reverse saturation or leakage current of the diode [A].
V_{pv} is the terminal voltage [V].
R_s is the series resistance [Ω].
R_p is the parallel resistance [Ω].
N_s is the number of PV cells connected in series.
A is ideality factor
V_T is the thermal voltage presented in Eq. (4) which is equal to 26 mV at 300 K for Silisium cell.

$$V_T = k \cdot V_{pv}\big/q \tag{4}$$

Where

K is Boltzmann constant $1.381 \cdot 10{-}23$ [J/K].
q is electron charge ($1.602 \cdot 10{-}19$ [C].
The reverse saturation current is given by

$$I_0 = DT_C^3 exp\left(\frac{-q\varepsilon_G}{A \cdot k}\right) \tag{5}$$

Such as

D is diode diffusion factor

T_C is the actual cell temperature [K]

ε_G is material band gap energy [eV], (1.12 eV for Si) (Belia et al. 2014).

2.2   PV Cell Specification

A 305 W SunpowerSPR-305-WHT PV has been used (Ahmd et al. 2016), all the parameters are shown in Table 1.

Table 1. Parameters of Sunpower SPR-305-WHT PV model

Parameter	Value
Number of cells per module	96
Referenced solar irradiance	1000 W/m^2
Referenced cell temperature	25 °C
V_{OC}	64.2 V
I_{SC}	5.96 A
V_{mp}	54.7 V
I_{mp}	5.58 A
P_{mp}	305 W

Figure 3 shows the power-voltage characteristics curves under constant temperature and different irradiances. As it is presented the PV power increases proportionally with the increasing of the irradiance.

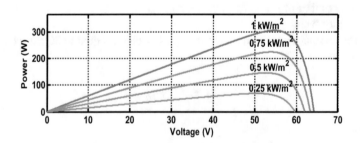

Fig. 3. P-V characteristics of PV cell.

2.3   Fuzzy Logic Controller Based MPPT

Recently, fuzzy logic based controllers have been broadly used especially in the field of the renewable energy because of its easy implantation, good performance and the ability to deal with the nonlinear characteristics of the PV system (Abdullah et al.

2012). To ensure that the PV system is operating in its maximum power point, a robust tracking strategy is needed. For such purpose a fuzzy logic tracker is proposed to handle the operating of the PV system under the unsteady environmental conditions. The main structure of the fuzzy logic process which consists of three steps: fuzzification, rules identification and defuzzification (Abdullah et al. 2012). In the traditional configurations the inputs of the FLC tracker are the error dp/dv and its derivative as in Eqs. (6) and (7), while the output is the duty cycle of the Dc-Dc converter (Youcef et al. 2014).

$$e(n) = \frac{P(n) - p(n - 1)}{V(n) - V(n - 1)} \tag{6}$$

$$\Delta e(n) = e(n) - e(n - 1) \tag{7}$$

Although in the proposed configuration the FLC tracker is modified to adjust the shoot-through duty ratio for the purpose of tracking the maximum power point by exploiting the boost feature in the z-source inverter according to the equation below.

$$V_{PV} = \frac{1 - D_0}{1 - 2D_0} \tag{8}$$

Where V_{pv} is the PV voltage and D_0 is the shoot-through duty ratio.

Figure 4 presents the chosen membership functions of the error for the linguistic variables for the inputs and the output which can be divided into: NL (negative large), NS (negative small), Z (zero), PS (positive small) and PL (positive large), the memberships of the error derivative and the output are the same but with interval of $(-0.03$ to $0.03)$ for the derivative and $(0$ to $0.4)$ for the output which is the shoot through duty ratio. While Table 2 shows the rule base of the selected fuzzy controller.

Table 2. Fuzzy rules base

	NL	NS	Z	PS	PL
NL	Z	Z	PL	PL	PL
NS	Z	Z	PS	PS	PS
Z	PS	Z	Z	Z	NS
PS	NS	NS	NS	Z	Z
PL	NL	NL	NL	Z	Z

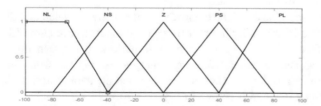

Fig. 4. Memberships of fuzzy logic controller

2.4   Z-Source Inverter

Due to the drawbacks in the conventional voltage and current source inverters, Peng, F. Z has proposed in (Peng 2003) a new topology that overcomes these drawbacks and provides a new feature of the ability to buck or boost the voltage with such a simple configuration which is missing in the conventional topologies. The ZS inverter equivalent circuit is presented in Fig. 5, in this topology a symmetrical impedance network is used as an interface between the source and the inverter, this network contains two capacitors and two inductors. Unlike the traditional V-source inverter which has two states, active state wherein the DC voltage is applied across the load, this state can be obtained by 6 switching states, and zero state which can be obtained by 2 switching states wherein the load terminals are shorted through either upper or lower three switches.

The Z-source inverter has one extra state called shoot-through zero state wherein both the upper and lower switches are turned on, this state is forbidden in the conventional topologies because a damage may occurs when the devises in the same leg gated on in the same time (Peng 2003). Z-source inverter is able to operate in two modes, non-shoot through mode and shoot through mode (Youcef et al. 2014).

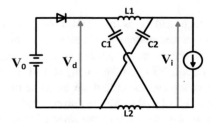

Fig. 5.   Equivalent circuit for Z-source inverter

3   System Configuration and Operating Principle

For the maximum power control Fuzzy logic based technique is used to extract the maximum power from the PV array by controlling the shoot through duty ratio. The shoot through duty ratio lines V_p & V_n will be compared with a carrier signal, the shoot through state will be generated according to this comparison to reach the desired PV voltage. For current control strategy, the output three phase voltage and current are transformed into d-q components.

The capacitor voltage will be compared with a reference value and the error signal is delivered to a PI controller in order to generate the reference current Idref, with the assumption that I_{qref} is zero a comparison is performed between the reference and actual currents. The generated references voltages is compared with carrier signal for the purpose of controlling the capacitor voltage and the current flow at the same time, the control block diagram is shown in Fig. 6.

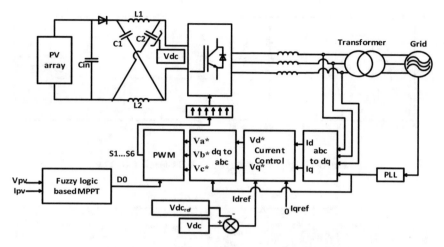

Fig. 6. The block diagram of the control system

4 Simulation Results

The simulation of the system is performed using Matlab Simulink and its parameters are listed in Table 1, Fig. 7a illustrates the irradiance, as it is shown there is a quick variation applied in 0.18 s–0.39 s to test the response speed and the accuracy of the fuzzy logic controller.

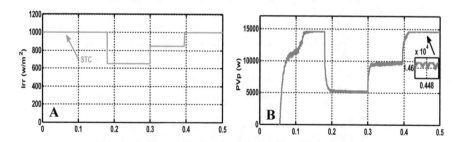

Fig. 7. Unsteady irradiance and the output PV power

Figure 7b shows the power obtained from the PV array which consists of 6 parallel and 8 series modules, this combination provides about 14.6 kW in the standard test conditions (1000 w/m², 25 °C), although under an unsteady irradiance the PV power is changing according to the variation of the irradiance, in Fig. 8a it is clearly obvious that the FLC MPP controller track the MPP voltage which equals in STC 328 V.

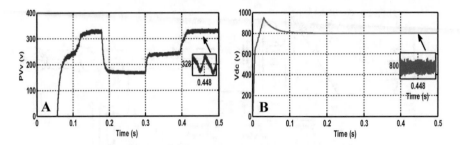

Fig. 8. Output PV voltage and DC capacitor voltage

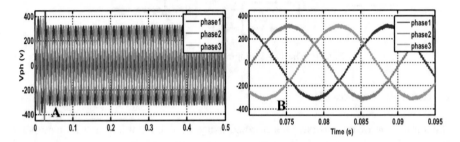

Fig. 9. Output voltage and its close up view

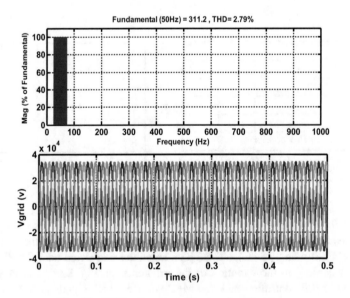

Fig. 10. THD of output voltage and grid voltage

A PI controller is used to regulate the DC capacitor voltage to keep it equal to the reference value (800 V) after a transient period of 0.1S as shown in Fig. 8b. Figure 9a and b show the output voltage and its close up view respectively, the system provides 220 V rms with frequency of 50 Hz and Thd below 5% to feed a domestic grid or to be connected with a grid of 25 kV as illustrated in Fig. 10.

5 Conclusion

This paper presented in the first part of the study a PV array connected to the grid through a z-source inverter, by using the topology presented in the paper the system will operate in a single stage operation therefore the dc-dc converter no longer used which increases the efficiency and decreases the complexity of the system. In the second part the fuzzy logic based MPPT algorithm had been used for the purpose of extracting the maximum power from the PV array, moreover an efficient strategy has been used firstly to control the inverter according to the reference voltage and to maintain secondly the capacitor voltage at a constant value. The obtained results were very convincing comparing to other research.

References

Reshmi, N., Nandakumar. M.: Grid-connected PV system with a seven-level inverter. In: International Conference on Next Generation Intelligent Systems, pp. 1–6 (2016)

Liu, Y., Lan, P., Lin, H.: Grid-connected PV inverter test system for solar photovoltaic power system certification. In: PES General Meeting Conference and Exposition, pp. 1–6 (2014)

Bourguiba, I., Houari, A., Belloumi, H., Kourda, F.: Control of single-phase grid connected photovoltaic inverter. In: 4th International Conference on Control Engineering and Information Technology, pp. 1–6 (2016)

Selvaraj, J., Rahim, N.A.: Multilevel inverter for grid-connected PV system employing digital PI controller. IEEE Trans. Ind. Electron. **56**, 149–158 (2009)

Peng, F.Z.: Z-source inverter. IEEE Trans. Ind. Appl. **39**, 504–510 (2003)

Ketabi, A., Tabatabaei, M.S.E.: Photovoltaic single-stage grid tied inverter with one-cycle control. In: 6th Power Electronics, Drives Systems and Technologies Conference, pp. 1–6 (2015)

Rasin, Z., Rahman, M.F.: Design and simulation of quasi-Z source grid-connected PV inverter with battery storage. In: IEEE International Conference on Power and Energy, pp. 1–6 (2012)

Ahmed, A.S., Abdullah, B.A., Abdelaal, W.G.A.: MPPT algorithms: performance and evaluation. In: 11th International Conference on Computer Engineering and Systems, pp. 461–467 (2016)

Bellia, H., Ramdani, Y., Moulay, F.: A detailed modeling of photovoltaic module using MATLAB. NRIAG J. Astron. Geophys. **3**, 53–61 (2014)

Abdullah, M.N., Khaled, E.A., Hussein, M.M. A fuzzy logic control method for MPPT of PV systems. In: 38th Annual Conference on IEEE Industrial Electronics Society, pp. 874–880 (2012)

Youcef, S., Mihcene, B., Sami, K., Kais, B.: Maximum power point tracking using fuzzy logic control for photovoltaic system. In: 3rd International Conference on Renewable Energy Research and Applications, pp. 902–906 (2014)

Fuzzy Logic Control Scheme of Voltage Source Converter in a Grid Connected Solar Photovoltaic System

Amel Abbadi[1(✉)], Fethia Hamidia[1], Abdelkader Morsli[1],
Djemel Boukhetala[2], and Lazhari Nezli[2]

[1] Electrical Engineering and Automatic LREA,
Electrical Engineering Department, University of Medea, Médéa, Algeria
amel.abbadi@yahoo.fr
[2] Laboratoire de Commande des Processus, Département d'Automatique,
École Nationale Polytechnique, El Harach, Algeria

Abstract. This paper focuses on control strategy of a voltage source converter based on fuzzy logic controller. A PV array is connected to alternative current (AC) grid via a boost converter and a three-phase three-level voltage source converter. The three phase VSC converts the DC link voltage to AC and keeps unity power factor. The VSC control system uses two controllers. The first one is an external fuzzy controller which regulates the DC link voltage to its reference value and the second fuzzy controller regulates active and reactive grid current components. The active current reference is the output of the DC voltage external controller. The reactive current reference is set to zero in order to maintain unity power factor. Vd and Vq voltage outputs of the current controller are converted to three modulating signals used by the PWM Generator. Result from the simulation shows that the fuzzy controller is capable to stabilize the DC link voltage, the active and the reactive components of the current VSC in the vicinity of references values in various temperature and irradiation condition.

Keywords: Fuzzy logic controller · Grid-connected photovoltaic system
Fuzzy MPPT · DC link voltage · Voltage source converter

1 Introduction

With the deterioration of the global environment, renewable energy, headed by photovoltaic power generation, has attracted worldwide attention because of its advantages of environmentally clean, pollution-free and simple installation (Dong-hui et al. 2010). In recent years, the distributed generation system of photovoltaic power generation is gradually developing from the island to the integrated grid, indicating that large-scale grid-connected PV will be the main mode of photovoltaic power generation (Jie and Qian 2009). The integration of PV systems with the distribution systems are increasing nowadays. The Voltage Source Converters (VSC) and DC-DC converters can be used to integrate PV arrays with the AC grid.

The controller is one of the main parts of any control system. The conventional PI controller requires precise linear mathematical model of the system, which is difficult to

© Springer Nature Switzerland AG 2019
M. Hatti (Ed.): ICAIRES 2018, LNNS 62, pp. 148–154, 2019.
https://doi.org/10.1007/978-3-030-04789-4_16

obtain under parameter variations, nonlinearity, and load disturbances (Saad and Zellouma 2009). Recently, the fuzzy logic system (FLS) has been widely applied to many control problems (Saad and Zellouma 2009; Mohanty and Parhi 2014; Abbadi et al. 2012; Attia et al. 2015), as they need no accurate mathematical models of the uncertain nonlinear systems under control. This paper deals with the control of currents and DC link voltage of 3-phase VSC based fuzzy control scheme for integration of the PV system into 3 phase system.

This paper is organized as follows. In Sect. 2, the system under study is described. The Fuzzy logic control strategy is presented in Sect. 3. In Sect. 4, the proposed control is validated by means of simulation and discussed. Finally, the conclusions are summarized in Sect. 5.

2 System Structure

Grid-connected PV system is combined with PV array, MPPT module, VSC module and distribution network. Figure 1 shows the system structure.

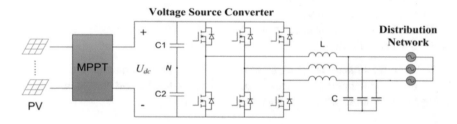

Fig. 1. System structure

The main function of grid-connected PV system is to transmit active power of PV array (100 kW at 1000 W/m² of sun irradiance) to utility grid (25 kV distribution feeder + 120 kV equivalent transmission system). For MPPT, the boost converter boosts DC voltage from approximate 273 V DC to 500 V DC. This operation is achieved by automatically varying the duty cycle. The 3-level 3-phase VSC converts the 500 V DC link voltage to 260 V AC and keeps unity power factor.

3 Proposed Fuzzy Controller Strategy

In this paper two mainly fuzzy control strategies are proposed. The first is the fuzzy MPPT controller and the second, the VSC fuzzy controller.

3.1 Fuzzy MPPT Controller

MPPT using Fuzzy Logic Control gains several advantages of better performance, robust and simple design. In the proposed system, the input variables of the FLC are slope of the PV cell's Power-Voltage (P-V) curve ($E = dP/dV$) and variation of slope (ΔE). The output of the FLC is dD (Abbadi et al. 2018) (Table 1).

Table 1. FLC rules base

Fuzzy rule		E				
		NB	**NS**	**ZE**	**PS**	**PB**
ΔE	**NB**	ZE	PB	PS	ZE	NB
	NS	PB	PS	ZE	ZE	NB
	ZE	PB	PS	ZE	NS	NB
	PS	PB	ZE	ZE	NS	NB
	PB	PB	ZE	NS	NB	ZE

3.2   VSC Fuzzy Controller

The VSC control strategy of VSC controller has the following tasks.

1. Control of active power supplied to the grid
2. Control of DC link voltage
3. Ensure high quality of injected power
4. Grid synchronization.

The control strategy adopted, as shown in Fig. 2, consists mainly of two cascaded loops, the fast inner current control loop, which regulates the grid current, and an outer voltage control loop, which controls the dc-link voltage.

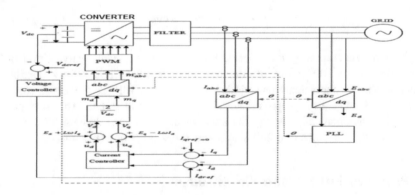

Fig. 2. Three phase grid connected converter control for PV system

- *Inner control loop*

The inner controller or current controller as input takes the error between the reference current and measured current. This error is carried through fuzzy regulator and the decoupling terms are compensated by feed-forward. As a result the desired converter voltage in dq reference frame is obtained. The feed-forward is used to minimize disadvantage of slow dynamic response of cascade control.

The structure of the inner current controller is presented in Fig. 2. The controller in Fig. 2 consists of two fuzzy regulators, for q and d axis respectively. The fuzzy control rules are illustrated in Table 2.

Table 2. FLC rules base for current controller

Fuzzy rule		e_Iq/e_Id				
		NG	NM	ZE	PM	PG
Δe_Iq/Δe_Id	NG	NG	NG	NM	NM	ZE
	NM	NG	NM	NM	ZE	PM
	ZE	NM	NM	ZE	PM	PM
	PM	NM	ZE	PM	PM	PG
	PG	ZE	PM	PM	PG	PG

- *Fuzzy DC Link Voltage Controller*

Linear control algorithm, such as PI, can stabilize DC link voltage in the vicinity of reference value. While, due to DC output voltage of PV is nonlinear, it does not improve output voltage vibration. Accordingly, Fuzzy control algorithm is more suitable to control DC link voltage of VSC. It will make system get a good dynamic performance and indirectly improve grid-connected performance. The DC link voltage control diagram with Fuzzy-PI algorithm is shown at Fig. 2.

In the proposed system, the input variables of the FLC are the error between the reference DC voltage and measured Dc link voltage (e_Udc) and variation of error (Δe_Udc). The output of the FLC is dI_{dref}. The fuzzy control rules are illustrated in Table 3.

Table 3. FLC rules base for DC link voltage controller

Fuzzy rule		e_Udc = Udc_ref - Udc						
		NB	NM	NS	ZE	PS	PM	PB
Δe_Udc	NB	NB	NB	NB	NM	NM	NS	ZE
	NM	NB	NM	NM	NM	NS	ZE	PS
	NS	NB	NM	NS	NS	ZE	PS	PM
	ZE	NM	NM	NS	ZE	PS	PM	PM
	PS	NM	NS	ZE	PS	PS	PM	PB
	PM	NS	ZE	PS	PM	PM	PM	PB
	PB	ZE	PS	PM	PM	PB	PB	PB

4 Simulation

Our application is 100 kW array connected to a 25 kV *grid* via a DC-DC boost converter and a three-phase three-level VSC. The simulation is performed under different temperature and sun irradiation conditions (Fig. 3).

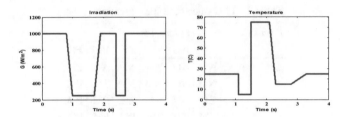

Fig. 3. The different temperature and sun irradiation conditions

4.1 PV Array

Figure 4 shows the voltage and PV current.

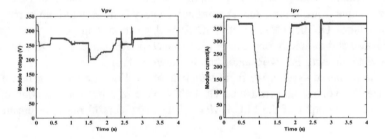

Fig. 4. PV voltage and current

4.2 DC-DC Converter

Boost converter is de-blocked at t = 0.05 s and MPPT is enabled. The MPPT regulator starts regulating PV voltage. The duty cycle is represented in Fig. 5(a). The steady state of the DC link voltage (Vdc = 500 V) is reached at about t = 0.2 s and remains hanging close to its reference value even after the changes imposed on test conditions (Fig. 5(b)).

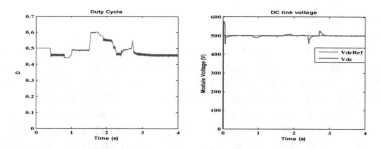

Fig. 5. The duty cycle and the DC link voltage

4.3 DC-AC Inverter

At t = 0.05 s, VSC converter is de-blocked too. Figure 6(a) depicts the voltage and current waveform of VSC, connected to utility grid. The harmonic spectra of the converter output current at solar irradiance levels of G = 1000 W/m² is shown in Fig. 6 (b). The phase voltage and current (Va and Ia) at 25 kV bus are in phase (Fig. 7(a)). Therefore, the system provides the power (Fig. 7(b)) to utility grid with unity power factor.

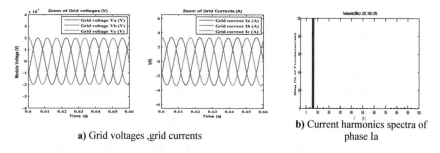

a) Grid voltages ,grid currents

b) Current harmonics spectra of phase Ia

Fig. 6. Utility grid voltages and currents

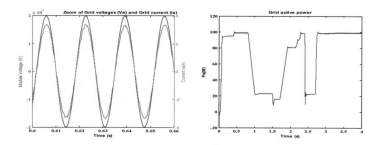

Fig. 7. Unity power factor of the utility grid and the grid active power

5 Conclusion

In this paper, a three phase grid connected VSC control structures for solar PV grid integration have been studied in detail. The proposed control strategy adopts an inner current control loop and an outer DC link voltage control loop. The proposed control strategies of the current and the DC link voltage of the VSC are based on fuzzy logic controllers. The developed control strategy is verified through simulation studies on 100 kW Grid connected photovoltaic system. The obtained results showed that the control system employing fuzzy logic controllers is very effective in the regulation of the DC link voltage, the active and the reactive components of the current VSC when irradiance and temperature levels change. The THD of the grid injected current is very satisfying. The simulated results, obtained for different operating conditions, have shown a good performances of the grid connected photovoltaic system with the fuzzy logic controllers.

References

Abbadi, A., Boukhetala, J., Nezli, L., Kouzou, A.: A nonlinear voltage controller using T–S fuzzy model for multimachine power systems. In: Proceedings of the 9th Annual IEEE International Multi-conference on Systems, Signals and Devices, Chemnitz, Germany, pp. 1–8 (2012)

Abbadi, A., Hamidia, F., Morsli, A., Boukhetala, J., Nezli, L.: Fuzzy-logic-based solar power MPPT algorithm connected to AC grid. Lect. Notes Netw. Syst. **35**, 206–214 (2018)

Attia, A.H., Rezeka, S.F., Saleh, A.M.: Fuzzy logic control of air-conditioning system in residential buildings. Alex. Eng. J. **54**(3), 395–403 (2015)

Dong-hui, L., He-xiong, W., Xiao-dan, Z., et al.: Research on several critical problems of photovoltaic grid-connected generation system. Power Syst. Prot. Control **38**(21), 208–214 (2010)

Jie, S., Qian, A.: Research on several key technical problems in realization of smart grid. Power Syst. Prot. Control **37**(19), 1–4 (2009)

Mohanty, P.K., Parhi, D.R.: Navigation of autonomous mobile robot using adaptive network based fuzzy inference system. J. Mech. Sci. Technol. **28**(7), 2861–2868 (2014)

Saad, S., Zellouma, L.: Fuzzy logic controller for three-level shunt active filter compensating harmonics and reactive power. Electr. Power Syst. Res. **79**, 1337–1341 (2009)

Fuzzy Logic Controller Based Perturb and Observe Algorithm for Photovoltaic Water Pump System

Amel Abbadi[1(✉)], Hamidia Fethia[1], Morsli Abdelkader[1],
Boukhetala Djemel[2], and Nezli Lazhari[2]

[1] Electrical Engineering and Automatic Research Laboratory LREA,
Electrical Engineering Department, University of Medea, Medea, Algeria
amel.abbadi@yahoo.fr
[2] Laboratoire de Commande Des Processus, Département D'Automatique,
École Nationale Polytechnique, El Harach, Algeria

Abstract. Photovoltaic (PV) panels are devices that convert sun light into electrical energy and are considered to be one of the major ways of producing clean and inexhaustible renewable energy. However, these devices do not always naturally operate at maximum efficiency due to the nonlinearity of their output current-voltage characteristic which is affected by the panel temperature and irradiance. Hence, the addition of a high performance maximum-power-point tracking, MPPT, power converter interface is the key to keeping the PV system operating at the optimum power point which then gives maximum efficiency. In this paper, new adaptive P&O method with variable step size is investigated by using fuzzy logic control. This control technique automatically adjusts the step size to track MPP. The proposed method can largely improve the MPPT response speed and accuracy at steady state simultaneously with less steady-state oscillation. The obtained results demonstrate the efficiency of the proposed MPPT algorithm in terms of speed in MPP tracking and accuracy.

Keywords: Fuzzy logic control · Maximum power point tracking
Photovoltaic (PV) · Perturb and observ algorithm (P&O) · Variable step size

1 Introduction

The energy generated from clean, efficient and environmentally-friendly sources has become one of the major challenges for engineers and scientists. Among all renewable energy sources, photovoltaic generation systems are one of these sources which attract more attention because they provide excellent opportunity to generate electricity. However, the outputs power of photovoltaic systems are depending on atmospheric conditions (Temperature and solar irradiation). The maximum power point tracking (MPPT) technology improves the effective use of solar energy and the efficiency of PV power generation system. A several maximum power point tracking (MPPT) algorithms have been proposed including Hill climbing (Koutroulis et al. 2001), perturb and observe (P&O) (Kwon et al.2008), incremental conductance (INC) (Hussein et al. 1995), and artificial-intelligence-based algorithms (Alajmi et al. 2013).

© Springer Nature Switzerland AG 2019
M. Hatti (Ed.): ICAIRES 2018, LNNS 62, pp. 155–161, 2019.
https://doi.org/10.1007/978-3-030-04789-4_17

In practice, the P&O method is the most commonly used technique, owing to its low cost, ease of implementation and its relatively good tracking performance. Nevertheless the P&O method fails to track the MPP when the atmospheric conditions change rapidly and oscillates around the MPP or near to it when the atmospheric conditions change slowly or constant. Consequently, part of available energy is wasted. In Aashoor and Robinson (2012), the fuzzy logic controller is used to generate the variable step-size of the P&O algorithm. This technique is proposed for further improvement in the tracking speed and steady state accuracy. 25 rules have been generated with the knowledge base of the system. In our work, more simplicity was brought to the calculation of the variable step size. Indeed, Inspired by Shiau et al. (2015) work, 5 rules were used to generate the variable step-size of the P&O algorithm with one input and one output.

2   The Proposed System

The water pumping system is a stand-alone 150 W system as shown in Fig. 1. The system consists of a single PV module, an MPPT controller, and a DC water pump.

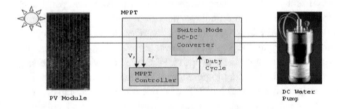

Fig. 1. Block diagram of Proposed System

In general, DC motors are preferred because they are highly efficient and can be directly coupled with a PV module. Brushed types are less expensive and more common although brushes need to be replaced periodically. The water pump selected was a submersible solar pump which is a diaphragm-type positive displacement pump equipped with a brushed permanent magnet DC motor. It operates with a low voltage (12 ~ 30 V DC), and its power requirement is as little as 35 W (Kyocera Solar Inc 2001).

3   The Solar Model

The PV module chosen is a 72 multi-crystalline silicon solar cells in series able to provide 150 W of maximum power. The electric model is shown in Fig. 2.

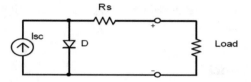

Fig. 2. Equivalent circuit used in the MATLAB simulations

The current-voltage relationship of the PV cell (Thompson 2003) is

$$I = I_{sc} - I_0 \left(e^{q((V + I.Rs)/nkT)} - 1 \right) \tag{1}$$

where: I is the cell current, V is the cell voltage, T is the cell temperature in Kelvin. The short-circuit current (I_{sc}) at a given cell temperature (T) is:

$$I_{sc}|_T = I_{sc}|_{T_{ref}} \cdot \left[1 + a(T - T_{ref}) \right] \tag{2}$$

where: I_{sc} at T_{ref} is given in the datasheet, T_{ref} is the reference temperature of PV cell in Kelvin (K), usually 298 K (25 °C), and a is the temperature coefficient of I_{sc} in percent change per degree temperature. The short-circuit current (I_{sc}) is proportional to the amount of irradiance. I_{sc} at a given irradiance (G) is:

$$I_{sc}|_G = (G/G_0)I_{sc}|_{G_0} \tag{3}$$

where: G_0 is the nominal value of irradiance. The reverse saturation current of diode (I_0) at the reference temperature (T_{ref}) is given below with the diode ideality factor added:

$$I_0 = I_{sc} / \left(e^{qVoc/nkT} - 1 \right) \tag{4}$$

The reverse saturation current (I_0) is temperature dependant and the I_0 at a given temperature (T) is calculated by the following equation

$$I_0|_T = I_0|_{T_{ref}} (T/T_{ref})^{\frac{3}{n}} \cdot e^{\frac{-qEg}{nk} \left(\frac{1}{T} - \frac{1}{T_{ref}} \right)} \tag{5}$$

Finally, it is possible to solve the equation of I-V characteristics (1) by simple iterations using the Newton's method

$$I_{n+1} = I_n - \left(\left(I_{sc} - I_n - I_0 \left[e^{q((V + I.Rs)/nkT)} - 1 \right] \right) \Big/ \left(-1 - I_0 \left(\frac{q\,Rs}{n\,k\,T} \right) e^{q((V + I.Rs)/nkT)} \right) \right) \tag{6}$$

The parameters chosen for modeling corresponds to the BP SX150 s module, as listed in Solar B P (2001).

4   DC Motor Model

Many PV water pumping systems employ DC motors because they could be directly coupled with PV arrays and make a system very simple. A permanent magnet DC motor (PMDC) is used here in this application as it can provide a higher starting torque. Figure 3a shows an electrical model of a PMDC motor.

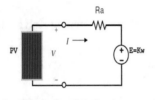
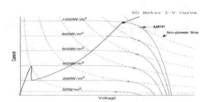

a) Electrical model of permanent magnet DC motor

b) PV I-V curves with constant power lines (dotted) and a DC motor I-V curve

Fig. 3. Electrical model and I-V curve of permanent magnet DC motor.

Figure 3b shows a major problem with a direct coupled PV-motor setup in efficiency because of mismatching of operating points of the motor with the maximum power point of the PV system. To overcome this problem, a MPPT algorithm should be used.

5   Conventional Perturb and Observe Algorithm (P&O)

The Perturb and Observe algorithm is considered to be the most commonly used MPPT algorithm of all the techniques because of its simple structure and ease of implementation.

It is based on the concept that, at the maximum power point, dP/dV goes to zero. The flowchart of P&O algorithm is shown in Fig. 4.

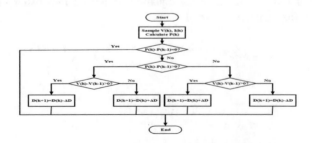

Fig. 4. Flowchart of the P&O algorithm

6 Fuzzy Logic Controller Based Perturb and Observe Algorithm (FP&O)

Generally the P&O MPPT algorithm is run with a fixed step size. If this step-size is set to be large, the algorithm will have a faster response dynamics to track the MPP. However, the algorithm with a large step-size results in excessive steady state oscillation. This performance situation is reversed when the P&O MPPT is running with a small step-size. Therefore, P&O MPPT with fixed step-size does not allow a good tradeoff between steady-state oscillation and dynamic response to changing operating conditions.

In this work a modified P&O MPPT algorithm with variable step-size is proposed. This controller is implemented using fuzzy logic control as shown in Fig. 5a, where the variable step-size (ΔD) of the P&O algorithm is the output of the FLC. The input of the FLC is defined as:

$$180° - \left[tan^{-1}\left(I_{pv}/V_{pv} \right) + tan^{-1}\left(dI_{pv}/dV_{pv} \right) \right] \qquad (7)$$

The membership functions of the input are expressed by triangular functions and the output fuzzy sets are defined as fuzzy singletons. The basic principle of the proposed variable step size is illustrated in Fig. 5b and c. The duty cycle has a variable step size (large step size in the far left and far right of MPP, however very small step size near the MPP and equal to zero at MPP).

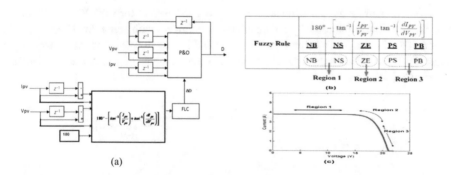

(a)   (b)   (c)

Fig. 5. Block diagram of the proposed FP&O algorithm with Fuzzy rules

7 Simulation Results

Our application is a water pumping installation destined to irrigation. The simulation is performed under the linearly increasing irradiance varying from 20 W/m² to 1000 W/m² with a moderate rate of 0.3 W/m² per sample and a buck-boost converter is used for water pumping system.

In Fig. 6(a) and (b) are given the P-V and I-V curves. It is clear that the trace of operating point is staying close to the MPPs during the simulation. Figure 6(c) shows the relationship between the output power of converter and its duty cycle. Figure 6(d) shows the current and voltage relationship of converter output which is equal to the DC motor load.

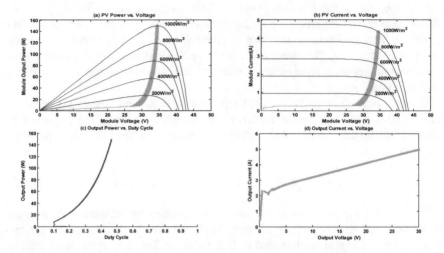

Fig. 6. Simulations results of the proposed MPP technique with the DC pump motor load (20 to 1000 W/m², 25 °C)

To demonstrate the performance of the proposed FP&O for such PV installations, the water volume pumped by an installation equipped with the FP&O technique is compared with the water volume pumped by a PV installation without MPPT technique (Fig. 7). The irradiance data used here are the measurements of a sunny day in January in Tindouf, Algeria.

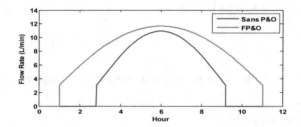

Fig. 7. Flow rates of PV water pumps

The results show (Fig. 7) that FP&O technique offers significant performance improvement. Indeed, the system with FP&O can utilize more than 99.9% of PV capacity and the system without MPPT has poor efficiency (54.63%). From the Fig. 8 (a) and (b), it is clear that the trace of operating point of the FP&O is staying more close to the MPPs during the simulation (less oscillation) then the classical P&O.

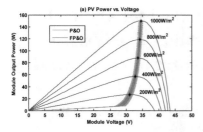
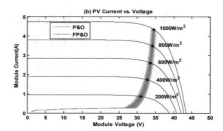

Fig. 8. Comparison of the simulation results: the classical P&O technique and the FP&O technique with the DC pump motor load (20 to 1000 W/m^2, 25 °C)

8 Conclusion

In this work, an adaptive P&O MPPT has been proposed and evaluated using fuzzy logic control to give variable step-size convergence to improve the efficiency of the photovoltaic water pumping system. The simulation results clearly show that, the FP&O has the ability to improve the dynamic performance of the photovoltaic power generator system. The results, also, validate the benefits of the FP&O technique which can significantly increase the efficiency of energy production from PV and the performance of the PV water pumping system compared to the system without MPPT technique.

References

Aashoor, F.A.O., Robinson, F.V.P.: A variable step size perturb and observe algorithm for photovoltaic maximum power point tracking. In: 47th International Conference Power Engineering, pp. 1–8 (2012)

Alajmi, B., Ahmed, K., Finney, S., Williams, B.: A maximum power point tracking technique for partially shaded photovoltaic systems in microgrids. IEEE Trans. Ind Electron. **60**(4), 1596–1606 (2013)

Hussein, K.H., Muta, I., Hoshino, T., Osakada, M.: Maximum photovoltaic power tracking: an algorithm for rapidly changing atmospheric conditions. IEE Proc. Gener. Trans. Distrib. **142**, 59–64 (1995)

Koutroulis, E., Kalaitzakis, K., Voulgaris, N.C.: Development of a microcontroller-based, photovotaic maximum power point tracking control system. IEEE Trans. Power Electron. **16** (I), 46–54 (2001)

Kwon, J.M., Kwon, B.H., Nam, K.H.: Three-phase photovoltaic system with three-level boosting MPPT control. IEEE Trans. Power Electron. **23**(5), 2319–2327 (2008)

Kyocera Solar Inc. Solar Water Pump Applications *Guide* 2001 (downloaded from www. kyocerasolar.com)

Shiau, J.K., Lee, M.Y., Wei, Y.C., Chen, B.C.: A study on the fuzzy-logic-based solar power MPPT algorithms using different fuzzy input variables. Algorithms **8**(2), 100–127 (2015)

Solar B.P.: SX 150–150 Watt multicrystalline photovol-taic module datasheet (2001)

Thompson M.A.: Reverse-Osmosis Desalination of Seawater Powered by Photovoltaics without Batteries, Doctoral Thesis, Loughborough University (2003)

A Genetic Algorithm Method for Optimal Distribution Reconfiguration Considering Photovoltaic Based DG Source in Smart Grid

Mustafa Mosbah[1]([⊠]), Salem Arif[1], Ridha Djamel Mohammedi[2],
and Samir Hamid Oudjana[3]

[1] LACoSERE Laboratory, Department of Electrical Engineering,
Amar Telidji University of Laghouat, Laghouat, Algeria
mosbah.mustapha@gmail.fr
[2] Department of Electrical Engineering, University of Djelfa, Djelfa, Algeria
[3] Unité de Recherche Appliquée en Energies Renouvelables (URAER),
Ghardaia, Algeria

Abstract. The distribution network have a very weakly meshed reconfiguration, with loops between different source stations, but the operation is carried out via a tree-based reconfiguration. This reconfiguration is determined by the opening and closing of switches in order to minimize the total power losses taking account the technical, security and topological distribution network constraints. In this paper, a Genetic Algorithm (GA) method based on graphs theory is proposed to design an optimal reconfiguration in presence of a photovoltaic based Distributed Generation source. The proposed method is tested on IEEE distribution network (69 bus) and validated on Algerian distribution network (116 bus). The proposed method was developed under MATLAB software. Certain results are better then others papers viewpoint active losses.

Keywords: Distribution network · Optimal configuration · Photovoltaic source

1 Introduction

An power system can be divided into deferent phases, generation phase, transmission phase and distribution phase. The distribution phase is the important part, that presents the liaison between transmission network and electric consumers. A particular attention has been done by researchers of distribution networks in those last years, that is more and more important, because of the introducing of renewable sources in those networks. Distribution networks operate in voltages lower than 50 kV and present looped structures, but exploited in open loop (radial configuration) [1]. This is translated by the existing of a one electrical way, between all points of the network and source stations [2]. The search of an optimal radial configuration is the process to change distribution

M. Mosbah—Engineer Operating in Company of Algerian Distribution of Electricity and Gas.

M. Hatti (Ed.): ICAIRES 2018, LNNS 62, pp. 162–170, 2019.
https://doi.org/10.1007/978-3-030-04789-4_18

network topology by changing the switchers devices stat (open or closed stat) so that we minimize the desired objectives [3, 4].

Many techniques have been published in literature on distribution network configuration optimization, as an example: Deterministic optimization methods have been presented in literature to determine the optimal distribution networks reconfiguration, as an example, Simplex Method [5, 6]. A Spanning Tree Method [7]. Mixed-Integer Convex Programming Method [8].

Many algorithms are based on Artificial intelligence and/or metaheuristic search algorithms have been used to solve distribution network reconfiguration for example, Genetic Algorithms [9], Antlion [10], Modified Taboo Search [11], Hybrid Big Bang Big Crunch, Non-dominated Sorting Genetic Algorithm (NSGA-II), Fireworks Algorithm, Memetic Algorithms, Ant Colony Algorithm, Simulated Annealing Algorithm, Fuzzy Logic Multiobjective, Harmony Search Algorithm, Honeybee Mating Optimization, Particle Swarm Optimization, Artificial Neural Networks Algorithm, Non-Dominated Sorting Particle Swarm Optimization, Hybrid Fuzzy Bees Algorithm, Cuckoo Search Algorithm, Bacterial Foraging Optimization Algorithm, Hybrid The Minimum Spanning Tree and Improved Heuristic Rules Algorithm, Refined Genetic Algorithm, Backtracking Search Optimization Algorithm, Artificial Immune System, Binary Gravitational Search Algorithm, Differential Search Algorithm, Differential Evolutionary Algorithm. Runner-Root Algorithm [12], Stochastic Dominance Concepts Algorithm. Reference examined some of the most recent methods for distribution network reconfiguration.

This paper presented application of Genetic Algorithm (GA) method based on graphs theory to design an optimal distribution network reconfiguration in presence photovoltaic based DG-source. This reconfiguration, determine the adjustment of switches state, in order to minimize the total active power loss. This study, proposed to adapt the principles of GA method to the strategy case of branches permutation, this opens and closes the switches devices. The proposed method is tested on IEEE distribution network (69 bus) and validated on Algerian distribution network (116 bus).

2 Problem Formulation

2.1 Objective Function

The objective of distribution network reconfiguration problem is to fine the best network configuration having minimal power losses by considering all exploitation constraints. Since many switching combinations in a distribution network exist, the search for an optimal configuration is a complex proses, non-linear, combinatory and a problem of non-differentiable constraints optimization. The objective function to minimize is represented in expression (1). Figure 1 shows an equivalent circuit model of a distribution network in the presence of looping switchers.

$$P_{loss}^T = I_{pq}^2 R_{pq} = \frac{S_{pq}^2}{V_p^2} R_{pq} = \frac{P_{pq}^2 + Q_{pq}^2}{V_p^2} R_{pq} \tag{1}$$

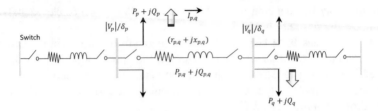

Fig. 1. One line diagram of a two-bus distribution network

$$F = Min \sum_{i=1}^{NB} P_{loss(i)} \qquad (2)$$

where $|V_p|/\delta_p$, is voltage and angle voltage at bus p, r_{pq} and x_{pq} are resistance and reactance of line connecting bus p and bus q, respectively, P_{pq} and Q_{pq} are active and reactive power through the branch between bus p and bus q, NB is number of lines and N is number of buses.

2.2 Equality and Inequality Constraints

In the optimization problem of distribution network configuration in presence of PV sources it is necessary to take in consideration all the following constraints:

Balancing constraints:

$$\{P_G + P_{DG} = P_D + P_L \qquad (3)$$

$$\{Q_G = Q_D + Q_L \qquad (4)$$

Voltage limit:

$$V_{imin} \leq V_i \leq V_{imax} \ for \ i = 1 \ldots \ldots N \qquad (5)$$

Line thermal limit:

$$S_k \leq S_{kmax} \ for \ k = 1 \ldots \ldots .NB \qquad (6)$$

Real power generation lim:

$$P_{Gimin} \leq P_{Gi} \leq P_{Gimax} \ for \ i = 1 \ldots .NG \qquad (7)$$

The DG source limit:

$$0 \leq \sum_{i=1}^{NDG} P_{DGi} \leq P_{DGi}^{max} \ for \ i = 1 \ldots \ldots NDG \qquad (8)$$

where (P_G, Q_G), are the total active and reactive power of conventional generator, respectively, (P_D, Q_D) the total active and reactive power of load, respectively,

(P_L, Q_L) is the total active and reactive power losses, respectively, P_{DG} is active power of DG (DG source modeled as photovoltaic power).

2.3 Radiality and Connectivity Constraints

The network topology should always be radial; the topology is radial if it satisfies the two following conditions [13]: No load out of service, and the network topology must be radial (no-loop).

2.4 Preserving Solution Feasibility

Note that the control variables are generated in their permissible limits using strategist preservation feasibility (perform a random value between the minimum and maximum value), while for the state variables, including the voltages of load bus, the power flowing in distribution lines, it appealed to penalties functions that penalize solutions that violate these constraints. The introduction of penalty in the objective function, transforms the optimization problem with constraints in an optimization problem without constraints [14–17], so it is easier to deal, in this case the Eq. 2 shall be replaced by:

$$F_p = Min \sum_{i=1}^{NB} P_{loss(i)} + k_v \cdot \left(V_{Li} - V_{Li}^{lim}\right)^2 + k_s \cdot \left(S_{li} - S_{li}^{lim}\right)^2 + k_m \cdot N_m + k_i \cdot N_i \quad (9)$$

where k_v, k_s, k_m and k_i, are penalty factors, N_m is the number of existing meshes, N_i is the number of isolated loads.

3 Applied Approach

Genetic Algorithms (GA) are stochastic optimization algorithms found on the natural selection mechanism of a generation. Their operation is extremely simple. We start by an initial population of potential solutions (chromosomes) chosen randomly. We evaluate their relative performance (fitness). In the base of their performance we create a new population of potential solutions by using simple evolutionary operations: The selection, crossover and mutation. We repeat this cycle until we find a satisfied solution. GA have been initially developed by *John Holland* [18].

4 Simulations and Results

In this section, the proposed algorithm is tested on 69 bus, then validated on Algerian distribution network (116 buses). IEEE 69 bus are generally known, but the Algerian network consists of 116 bus, 124 lines containing 09 loop lines, this load is spread over 09 feeders. The nominal voltage of 116 busses network is 10 kV. The substation is connected to MV network via a 30/10 kV transformer. The upper and lower of voltages limits considered in this paper are 0.95 pu and 1.05 pu, respectively. The DG size considered in this study is $P_{DG} = 0.3 * \left(\sum_{i=1}^{Nbus} P_{Di} \right)$ with $PF = 1$. This DG is plased in lowest bus voltage (see Table 1). This is to demonstrate the influence of photovoltaic based DG-source on the reconfiguration and the differents parameters of distribution network. Table 1 present the size and placement of photovoltaic for two distribution system.

Table 1. Size and placement of photovoltaic based DG-source

Distribution network	Distributed generation		
	Bus location	Size	
		P (kW)	PF
69 bus	65	1141	1
116 bus	81	7166	1

The optimal reconfiguration problem in presence of photovoltaic based-DG using GA method based graphs theory has been tackled with the objective of minimizing active loss. It has been recall that for each reconfiguration requires a calculation of a load flow, by backward forward method. In this study, where four simulation cases are considered, case1, presents initial reconfiguration (without reconfiguration and without DG installation), case2 reconfiguration before DG installation, case3 presents only DG installation and cas4 reconfiguration after DG installation. Following the different executions of the program under MATLAB software, the optimal parameters of GA method used in this simulation are, population size is 100, maximum iteration is 100, crossover probability is 1, mutation probability is 0.01 and one point crossover. Table 2, shows the switches state, active power losses and minimum bus voltage of each distribution network for different cases studied. In order to demonstrate the effectiveness of GA method comparisons were made with other works in literature (see Table 3). Figure 2 present the Algerian 116 node distribution network topology. Figures 3 and 4 shows voltage profile for same cases studied. From these figures it has been found that the improvement of the voltages is due to optimization of the reconfiguration and the presence of DG source. Figure 5 present the GA method convergence for 69 distribution network and 116 bus. The same resonance with the total loss values. Case4 (reconfiguration after DG installation) proved its effectiveness better than other cases by the minimization in addition to total losses with better voltage profile. From the results obtained, for the Algerian distribution network, the losses value in case3 is greater than case1, but after optimization of the reconfiguration (case4),the value of the demined losses decrease comparing to case1, which confirms the importance of reconfiguring the network in presence of DG.

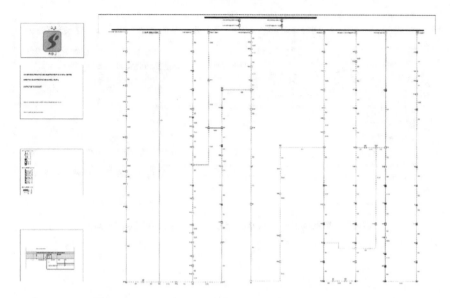

Fig. 2. Algerian 116 node distribution network topology

Table 2. Result determined by GA method

Distribution network	Before reconfiguration		After reconfiguration			Before reconfiguration		After reconfiguration		
	Before DG installation					After DG installation				
	Real power loss (kW)	Minimum bus voltage (pu)	Switches opened	Real power loss (kW)	Minimum bus voltage (pu)	Real power loss (kW)	Minimum bus voltage (pu)	Switches opened	Real power loss (kW)	Minimum bus voltage (pu)
69 bus	224.78	0.9092	14- 56- 61- 69- 70	99.58	0.942	116.37	0.955	12-57- 70- 69- 73	69.41	0.956
116 bus	402.02	0.9696	99- 75- 79- 105- 19- 121- 68- 60- 107	367.65	0.975	606.56	0.974	99- 75- 79- 89- 19- 121- 82- 13- 17	371.47	0.9825

Table 3. Comparisons with other works

Distribution network	Before reconfiguration			After reconfiguration			References
	Initial switches opened	Initial P_{loss} (kW)	V_{min} (pu)	Final P_{loss} (kW)	Final open switches	V_{min} (pu)	
69 bus	69- 70- 71- 72- 73	225.05	0.9092	99.61	Not reported	0.9428	Kashem et al. [19]
		225	0.9092	99.35	13- 18- 56- 61- 69	0.9428	Rao et al. [20]
		225	0.91	98.59	14- 58- 61- 69- 70	0.95	Ding and Laparo [21]
		225.07	0.9092	99.58	14- 58- 61- 69- 70	0.9428	Proposed GA

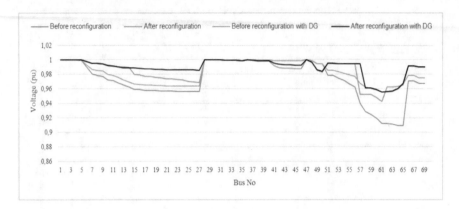

Fig. 3. Voltage profiles the different cases in 69 bus network

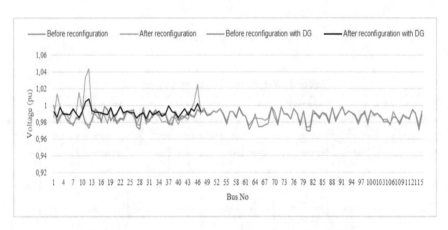

Fig. 4. Voltage profiles the different cases in 116 bus real network

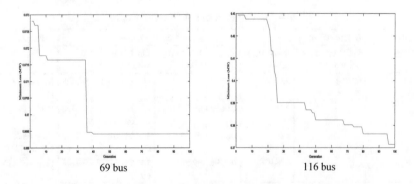

Fig. 5. GA method convergence for 69 bus distribution network and 116 bus

5 Conclusion

In this work, a hybrid genetic algorithm-graph theory is proposed in order to optimize distribution network reconfiguration considering photovoltaic based DG source. The objective function considered is minimization of real power losses under technical, security and topological constraints. The effectiveness of this method is shown in the quality of the results comparable to the few works of literature, by validating the algorithm proposed on IEEE distribution network (69 bus) and a real distribution network (116 bus).

References

1. Civanlar, S., et al.: Distribution feeder reconfiguration for loss reduction. IEEE Trans. Power Deliv. **3**, 1217–1223 (1988)
2. Shirmohammadi, D., Hong, H.W.: Reconfiguration of electric distribution networks for resistive line loss reduction. IEEE Trans. Power Deliv. **4**, 1492–1498 (1989)
3. Baran, M.E., Wu, F.F.: Network reconfiguration in distribution systems for loss reduction and load balancing. IEEE Trans. Power Deliv. **4**, 1401–1407 (1989)
4. Solo, A.M.G., et al.: A knowledge-based approach for network radiality in distribution system reconfiguration. IEEE Trans. Power Eng. Soc. Gener. Meet. (2006)
5. Aoki, K., et al.: New approximate optimization method for distribution system planning. IEEE Trans. Power Syst. **5**, 126–132 (1990)
6. Abnndams, R.N., Laughton, M.A.: Optimal planning of networks using mixed-integer programming. IEEE Proc. **121**, 139–148 (1974)
7. Mosbah, M., et al.: Optimum dynamic distribution network reconfiguration using minimum spanning tree algorithm. In: IEEE 5th International Conference on Electrical Engineering, Boumerdes (2017)
8. Jabr, R.A., et al.: Minimum loss network reconfiguration using mixed-integer convex programming. IEEE Trans. Power Syst. **27**, 1106–1115 (2012)
9. Tomoiaga, B., et al.: Optimal reconfiguration of power distribution systems using a genetics algorithm based on NSGA-II. Energies **6**, 1439–1455 (2013)
10. Mosbah, M., et al.: Optimal Algerian distribution network reconfiguration using ant lion algorithm for active power losses. In: IEEE 3rd International Conference on PAIS 2018, Tebessa, Algeria (2018)
11. Th Nguyen, T., et al.: Multi-objective electric distribution network reconfiguration solution using runner-root algorithm. Appl. Soft Comput. **52**, 93–108 (2017)
12. Chicco, G., Mazza, A.: Assessment of optimal distribution network reconfiguration results using stochastic dominance concepts. Susta Energy Grids Netw. **9**, 75–79 (2017)
13. Mosbah, M., et al.: Optimal reconfiguration of an Algerian distribution network in presence of a wind turbine using genetic algorithm. In: 1st International Conference on Artificial Intelligence in Renewable Energetic System, IC-AIRES 2017. Springer, Cham (2018)
14. Badran, O., et al.: Optimal reconfiguration of distribution system connected with distributed generations: a review of different methodologies. Renew. Sustain. Energy Rev. **73**, 854–867 (2017)
15. Mosbah, M., et al.: Optimal of shunt capacitor placement and size in Algerian distribution network using particle swarm optimization. In: IEEE Proceeding of ICMIC 2016, Algiers, Algeria (2016)

16. Mosbah, M., et al.: Genetic algorithms based optimal load shedding with transient stability constraints. In: Proceedings of the 2014 IEEE International Conference on Electrical Sciences and Technologies in Maghreb (2014)
17. Mosbah, M., et al.: Optimal sizing and placement of distributed generation in transmission systems. In: ICREGA 2016, Belfort, France, 8–10 February 2016
18. Holland, J.H.: Adaptation in Nature and Artificial Systems. The University of Michigan Press (1975)
19. Kashem, M.A., et al.: Loss reduction in distribution networks using network reconfiguration algorithm. Electr. Mach. Power Syst. **26**, 815–829 (1998)
20. Rao, R.S., et al.: Power loss minimization in distribution system using network reconfiguration in the presence of distributed generation. IEEE Trans. Power Syst. **28**, 317–325 (2013)
21. Ding, F., Laparo, K.A.: Hierarchical decentralized network reconfiguration for smart distribution systems Part II: applications to test systems. IEEE Trans. Power Syst. **30**, 744–752 (2015)

PV/Battery Water Pumping System Based on Firefly Optimizing Algorithm

Fethia Hamidia[1(✉)], Amel Abbadi[1], and Mohamed Seghir Boucherit[2]

[1] LREA Laboratory, Yahia Feres Medea University, Médéa, Algeria
fehamidia@gmail.com, amel.abbadi@yahoo.fr
[2] LCP Laboratory, Ecole National Polytechnique, Harrach, Algiers, Algeria
ms_boucherit@yahoo.fr

Abstract. Since the beginning of the century, global energy consumption has been growing strongly in all regions of the world. It seems likely that energy consumption will continue to increase, as a result of economic growth on the one hand, and of the increase in per capita electricity consumption on the other, whatever the scenarios considered.

For this reason, renewable energies appear today and in the long term as the appropriate solution that covers this energy need by reducing the major disadvantage emitted by fossil and fissionable energies. This paper proposes in one hand, an artificial neural network controller to track the maximum power point and to get better performance mainly on variation of load and weather condition. In second hand, we added to our system a battery and voltage PID controller based on Firefly Algorithm FA to tune their parameter.

Keywords: Induction motor · PVG · MPPT · Battery · Firefly algorithm

1 Introduction

Millions of peoples around the world live in rural villages with limited access to water where the deep groundwater is extracted via electric water pumps (Solar 2017). Since the last century, energy consumption has increased dramatically. However, our resources of oil and gas are not eternal and it is not also best to burn them more to avoid exacerbating pollution. Algeria has the effect of this significant renewable energy resource that can overcome particularly in the context of the production of electrical energy, the main vector of all economic and social development (Boukhalafa and Bouchafaa 2012).

Currently, as the electricity is often not available, we can find several forms of renewable energy, the most commonly used are: solar, wind and hydraulic.

Agricultural watering needs are usually greatest during sunnier periods when more water can be pumped with a solar system (Bouzeriaa et al. 2015). The solar energy is free, so by choosing solar, you have helped to reduce the cost of accessing water by using cheap and clean energy in your country for the next years. It's becoming increasingly evident that solar is simply better (Solar 2017).

© Springer Nature Switzerland AG 2019
M. Hatti (Ed.): ICAIRES 2018, LNNS 62, pp. 171–177, 2019.
https://doi.org/10.1007/978-3-030-04789-4_19

In recent years, heuristic optimization techniques have gained a lot of attention from researchers due to their better performance compared to mathematical optimization techniques in coping with large and complex optimization problems (Wong et al. 2014). The induction motor is used more and more for photovoltaic pumping systems. The low cost of the engine, the low maintenance requirements and the increased efficiency for solar pumping systems make it particularly major problem with the use of PV panels is their non-linear nature (Abouda et al. 2013).

In this paper, we present and discuss the application of direct torque control on the induction motor supplied with photovoltaic energy tracking their maximum power point using artificial neural network, this system is combined by the use of battery and controlled it by PI based on firefly algorithm. The panels are assembled in series and parallel to generate a 514 V voltage range in a MPP operation under different load changes.

2   MPPT Control Based on ANN

To optimize the power provided by the generator, a static converter, which operates as an adapter must be added. Many algorithms are used to Track the Maximum Power Point MPPT. In this work, we propose an intelligent technique called artificial neural network to control duty cycle of the switching transistor, as shown in Fig. 1.

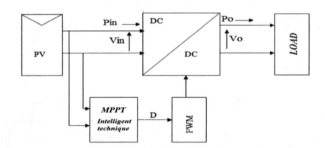

Fig. 1. Proposed intelligent MPPT technique

This technique presented in this work is proposed to resolve the problem of the classical controller (P&O), because neural network controller does not require a precise model, it is flexible and can provide the best results versus traditional solution.

The Fig. 2 show the structure of the proposed neural approach the most used training algorithm for supervised type is back-propagation algorithm that adapts this proposed structure in order to minimize the square of the error between desired and actual output. The two inputs are representing by the voltage and current, we select 15 neurons at the hidden layer, with the 'tansig' activation functions for all neuron layers, the sum squared error falls under 0.02 after 100 iterations.

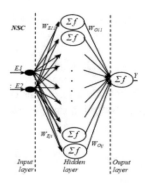

Fig. 2. Propposed neuro-MPPT

3 Simulation Results

Among most of the modern optimization algorithms are inspired by nature and such algorithms are collectively known as the meta-heuristic algorithms. Out of the 11 popular meta-heuristic algorithms used for the purpose of optimization, the Firefly algorithm is the best in terms of speed of execution when realized in a microcontroller. The Firefly algorithm was developed by Yang (Sundari et al. 2016; Nazarian and Hadidian-Moghaddam 2015; Hemalatha et al. 2016). This algorithm is based on the principle of attraction between fireflies and simulates the behavior of a swarm of fireflies in nature, which gives it many similarities with other meta-heuristics based on the collective intelligence of the group, such as the Particle Swarm Optimization (PSO) algorithm, Ant Colony optimization algorithm (ACO, bee colony optimization algorithm (ABC), and also the forage bacteria algorithm (BFA). According to recent works in literature, the performance of the Firefly algorithm in solving optimization problems exceeds those of other algorithms, such as genetic algorithms. This has been justified by recent research, where the performances of this algorithm have been compared with those of some known algorithms. The brightness of fireflies is determined according to an objective function.

The Firefly algorithm is formulated with two important things: The variation of the intensity of the light and the formulation of the attraction. For simplicity, the attraction of fireflies is determined according to the brightness, where the brightness is determined with the objective function. In the case of a minimization problem, the brightness I of a firefly at a position x can be defined as $I(x)$. However, the attraction β is relative to the position of the other fireflies. Consequently, it varies according to the distance r_{ij} between the firefly i and the firefly j. On the other hand, the intensity of light decreases with increasing distance from the source. This makes the attraction vary depending on the degree of absorption. For simplicity, the intensity of the light $I(r)$ changes according to the law

$$I(r) = I_s/r^2 \tag{1}$$

where I_s is the intensity at the source. For a constant value of γ, the intensity varies as a function of the distance r, which gives

$$I(r) = I_0 e^{-\gamma r} \tag{2}$$

where I_0 is the intensity of the light of the source. The combination of the two effects of the inverse square law and the absorption can be approximated with the following Gaussian formula.

$$I(r) = I_0 e^{-\gamma r^2} \tag{3}$$

Knowing that the attraction of a firefly is proportional to the intensity of the adjacent fireflies. The formula of this attractiveness β of a firefly can be defined as:

$$\beta(r) = \beta_0 e^{-\gamma r^2} \tag{4}$$

where β is the attraction *at* $r = 0$, To generalize, the calculation of $\beta(r)$ is defined as:

$$\beta(r) = \beta_0 e^{-\gamma r^m} \quad (m \geq 1) \tag{5}$$

On the other hand, the distance between two fireflies i and j at positions x_i and x_j is defined by the following Cartesian distance:

$$r_{ij} = \left\| x_i \quad x_j \right\| = \sqrt{\sum_{k=1}^{d} (x_{i,k} - x_{j,k})^2} \tag{6}$$

where $x_{i,k}$ represents spatial component k^{th} of the xi coordinate of the firefly i. The formula r_{ji} becomes:

$$r_{ij} = \sqrt{(x_i - x_j)^2 + (y_i - y_j)^2} \tag{7}$$

The movement of a firefly i attracted by another firefly j (brighter than i) is determined by

$$x_i = x_i + \beta_0 e^{-\gamma r_{ij}^2}(x_j - x_i) + \alpha\left(rand - \frac{1}{2}\right) \tag{8}$$

4   Simulations Results

In the first part, As is cited below, this paper proposes in one hand, a direct torque controlled IM supplied with photovoltaic panel to replace flux oriented control; and to track the maximum power point, this technique (MPPT) is based on Artificial Neural Network technique to get better performance specially on variation of load and weather condition. Some tests have been carried out to improve the performances of neural network MPPT method. To verify the effectiveness of the proposed technique, simulations are performed in this section by using MATLAB/SIMULINK.

As shown Fig. 3. It can be noticed that these results obtained by using P&O algorithm gives us high torque and flux ripples than the results obtained by using Neural network.

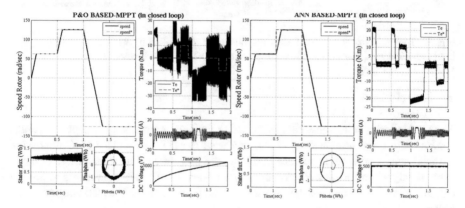

Fig. 3. Torque, flux, rotor speed and current responses in closed loop by using P&O and ANN (G = 1000 W/m², T = 298 K).

In the second part, we propose in the second part, a control system of the output voltage of the photovoltaic system. It consists of a photovoltaic panel, a DC-DC converter with its MPPT control, a storage battery, a boost converter with an output voltage control system.

In this paper and to control the voltage, we propose Firefly based algorithm proposed to tune the PID controller as shown Fig. 4.

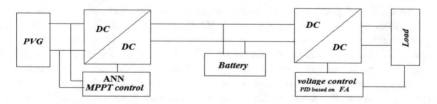

Fig. 4. Schematic diagram of proposed intelligent control

We noticed that estimated values flow their references and we have good performance by adding battery to our system, and also the proposed intelligent technique of PID-FA gives very acceptable responses (Fig. 5).

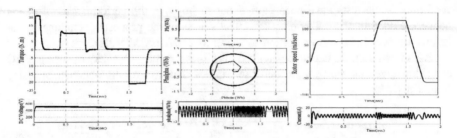

Fig. 5. Torque, flux and rotor speed responses (in closed loop) with load torque, speed, irradiation and temperature variation using Voltage PID-FA with battery

5   Conclusion

The simulation results show that proposed techniques give a very acceptable results compared to conventional one. Where, we used the DTC to replace FOC controller, artificial neural network to replace P&O and we used one of the most powerful meta-heuristic algorithm; named Firefly algorithm to tune the PI gain parameters to control the DC voltage, the input of the inverter and the power needed it to feed our pumping system.

References

Boukhalafa, S., Bouchafaa, F.: Analysis and control of a further maximum power point (MPPT) of PV array. In: Proceedings of the 2nd International Conference on Systems and Control, Marrakech, Morocco, 20–22 June 2012

Bouzeriaa, H., Fethaa, C., Bahib, T., Abadliab, I., Layateb, Z., Lekhchinec, S.: Fuzzy logic space vector direct torque control of PMSM for photovoltaic water pumping system. Energy Proc. **74**, 760–771 (2015)

Wong, L.A., Shareef, H., Mohamed, A., Ibrahim, A.A.: Optimal battery sizing in photovoltaic based distributed generation using enhanced opposition-based firefly algorithm for voltage rise mitigation. Sci. World J. **2014**, Article ID 752096, 11 p (2014). Hindawi Publishing Corporation

Abouda, S., Nollet, F., Chaari, A., Essounbouli, N., Koubaa, Y.: Direct torque control of induction motor pumping system fed by a photovoltaic generator. In: International Conference on Control, Decision and Information Technologies (CoDit), pp. 404–408 (2013)

Barazane, L., Kharzi, S., Malek, A., Larbès, C.: A sliding mode control associated to the field-oriented control of asynchronous motor supplied by photovoltaic solar energy. Rev. Energ. Renouv. **11**(2), 317–327 (2008)

Hamidia, F., Abbadi, A., Bocherit, M.S.: Maximum power point tracking of photovoltaic generation based on fuzzy logic. In: International Conference on Artificial Intelligence in Renewable Energetic Systems, IC-AIRES2017, Tipaza, Algeria (2017a)

Hamidia, F., Abbadi, A., Bocherit, M.S.: Neuro-fuzzy logic controlled induction motor supplied with PVG. In: CGE10, EMP (2017b)

Sundari, M.G., Rajaram, M., Balaraman, S.: Application of improved firefly algorithm for programmed PWM in multilevel inverter with adjustable DC sources. Appl. Soft Comput. **41**, 169–179 (2016)

Nazarian, P., Hadidian-Moghaddam, M.J.: Optimal sizing of a stand-alone hybrid, power system using firefly algorithm. In: Proceedings of Eleventh The IIER International Conference, Singapore, 15th February 2015, pp. 93–97 (2015)

Hemalatha, C., Rajkumar, M.V., Krishnan, G.V.: Simulation and analysis of MPPT control with modified firefly algorithm for photovoltaic system. Int. J. Innov. Stud. Sci. Eng. Technol. **2**(11), 48–52 (2016)

Solar Pumping (2017). www.worldbank.org/solarpumping

Innovative Renewables

Real Time Implementation of Sliding Mode Supervised Fractional Controller for Wind Energy Conversion System

Hamza Afghoul[1,2(✉)], Fateh Krim[2], Antar Beddar[3],
and Anouar Ounas[2]

[1] Ecole Supérieure de Technologies Industrielles, Annaba, Algeria
hamza.afghoul@gmail.com
[2] LEPCI Laboratory, Electronics Department, Faculty of Technology, Setif-1
University, Setif, Algeria
f_krim@ieee.org, anouar.ounas@gmail.com
[3] Electrical Engineering Department, University of Skikda, Skikda, Algeria
antar_tech@hotmail.com

Abstract. Wind energy conversion system is increasingly taking the place to be the most promised renewable source of energy, which obliges researchers to look for effective control with low cost. Thus, this paper proposes to apply a suitable controller for speed control loop to reach the maximum power point of the wind turbine under sever conditions. In literature, a major defect of the conventional PI controller is the slow response time and the high damping. Moreover, many solutions proposed the fractional order PI controller which presents also some weakness in steady state caused by the approximation methods. The main idea is to propose a Sliding Mode Supervised Fractional order controller which consists of PI controller, FO-PI controller and sliding mode supervisor that employs one of them. A prototype is built around real-time cards and evaluated to verify the validity of the developed SMSF. The results fulfil the requirements and demonstrate its effectiveness.

Keywords: MPPT · Wind energy conversion system · PI controller
FO-PI controller · Sliding mode control · DPC

1 Introduction

Recently, the high demand of electrical energy and the decrease in naturel resources lead to look for new, clean and inexhaustible sources of energy. Thus, Renewable energies have received much attention in the last decades [1, 2]. Therefore, wind energy is one of the most developed sources by looking to the installed capacity worldwide which is about 485 GW [3] and 12.63 GW for Europe [4]. Actually, permanent magnet synchronous generators (PMSGs) have been gaining much attention in modern wind energy conversation systems (WECSs) due to the variable speed operation, low converter cost, fast dynamical response [1, 5]. However, variable physical parameters, nonlinearity, and variable load torque are some of the important issues to be considered in the control process design of PMSG [6].

© Springer Nature Switzerland AG 2019
M. Hatti (Ed.): ICAIRES 2018, LNNS 62, pp. 181–191, 2019.
https://doi.org/10.1007/978-3-030-04789-4_20

Recent developments in control loops are responsible for optimum operation of the wind turbine [7]. Examples include torque control [8], power control loops [9], and pitch control [10]. Moreover, there is still a need for implementing each control loop separately using conventional proportional integral (PI) or proportional integral derivative (PID) controllers, and minimizing the couplings among loops by iterative adjustments [3]. However, the modern wind turbine structure, larger, more flexible and environmental conditions, make the conventional controllers not suitable in transient state even their good steady state. Therefore, the need for advanced control methods is increased. Previous works have focused on improving the dynamic of the PI controllers such as: applying fuzzy logic control [11], and sliding mode control [12]. The main limitation of these methods is the need for high speed power converters and powerful calculators.

Until now, PID controllers are still being used in the industry applications for their simple structure, ease of design, and inexpensive cost [13]. In addition, they have good performance, including acceptable overshoot with small settling time for slow industrial processes [13, 14]. This solution could not be a suitable alternative in high nonlinear systems. Hence, fractional-order (FO) controllers have been applied in several fields with better results in comparison with the traditional ones [1]. Appropriate FO integral (I)/derivate (D) can be utilized in order to improve the performance of the PID controller. In recent past, particular interest has been given to the fractional calculus theory and the approximation methods to build suitable controllers for complex systems. Firstly, A. Oustaloup was proposed Commande Robuste d'Ordre Non Entier (CRONE) controller in 1991. Afterwards, I. Podlubny started the FO PID (FO-PID) in the form of $PI^\lambda D^\mu$ in 1999 [15]. The powers λ, μ are real orders employed by researchers to give more freedom degrees to the FO-PID compared to the integer PID. But, the high order of the approximation method could be a major drawback for the FO-PID controller. Thus, compromising between the order of the approximation and the required performance in experimental application is the real challenge. Lately, more researches have been introduced to the structure of the FO-PI controllers employed in several domains. [14] tries to employ a fractional order integral plus proportional (FO-IP) controller in active power filtering to gain short response time and low overshoot but this kind of controllers presents some limitations in steady state with high distortions. Then, [16] proposed a combination between a conventional PI and FO-PI controllers and switch between them when external disturbances are detected but this approach still have to be improved in order to control the decision maker. In [17], authors introduced an intelligent solution by replacing conventional decision with a fuzzy logic supervisor but the cost of this approach could be a critical issue. After that, authors of [18] applied a new structure of hybrid FO-PI controller on the same system under study of this paper. The obtained results could be satisfying in steady and transient states but the complexity of the proposed controller presents some critical concerns in term of the powerful calculator units needed that lead to increase the implementation cost.

Nowadays, sliding mode control has been the choice of many researchers in control and could been a good alternative in supervising domain [1]. However, the natural chattering phenomenon of the sliding mode is a major problem for inverters control [1]. But, in our case, it could be the major advantage by up or down depending on sliding surface. Thus, this paper proposes a new controller structure named Sliding Mode Supervised Fractional order controller (SMSF) with simple design, suitable for WECS, easy to be implemented, earn the benefits from the efficiency of the conventional PI controller in steady state and accuracy of tracking the wind speed in sever conditions. The SMSF controller combines between conventional PI controller and FO-PI controller and set one of them based on a sliding mode supervisor (SMS). Moreover, the general structure of a PI controller is kept by employing two parallel paths which have tunable scalar gains in order to ensure that the SMSF controller can easily be implemented in control loops.

2 Wind Energy Conversion System

Figure 1 shows the general power circuit configuration of the WECS. The system is composed of three parts: an electrical part, a mechanical part, and a control part. Whereas, the mechanical part represented by a wind turbine, which is employed by DC motor with separate excitation used as an emulator. The electrical part includes a PMSG connected to the grid via two back-to-back converters. Indeed, the control part was implemented with two real-time dSPACE1104 cards. The first card contains the improved current vector control [1] with the wind turbine model and wind profile. The direct power control (DPC) [1] was implemented in the second one to control the GSC.

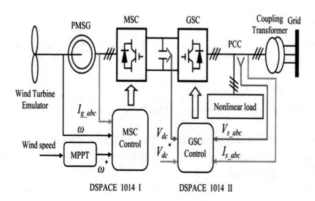

Fig. 1. Wind energy conversion system.

2.1   Wind Turbine Model

The aerodynamic power extracted from the wind is expressed by Eq. 1 [2]:

$$P_w = \frac{1}{2} C_p \rho \pi R^2 V_w^3 \tag{1}$$

Where, ρ is the air density (kg/m³), R is the turbine radius (m), V_w is the wind speed (m/s), C_p is the coefficient of the turbine.

The tip speed ratio (TSR) λ is defined as [1]:

$$\lambda = \frac{\Omega_t R}{V_w} \tag{2}$$

Where, Ω_t is the wind turbine angular shaft speed.

The torque on the wind turbine shaft can be calculated from the power expression as [1]:

$$T_w = \frac{P_w}{\Omega_t} = \frac{1}{2} C_p \rho \pi R^2 \frac{V_w^3}{\Omega_t} \tag{3}$$

C_p depends on the TSR and the pitch angle β such as [1]:

$$C_p = (0.5 - 0.00167(\beta - 2)).\sin\left[\frac{\pi(\lambda + 0.1)}{18 - 0.3(\beta - 2)}\right] - 0.00184(\lambda - 3)(\beta - 2) \tag{4}$$

2.2   Generator Model

Park representation of a PMSG model is the commonly used, which its voltage equations are expressed by [2]:

$$\begin{pmatrix} v_{sd} \\ v_{sq} \end{pmatrix} = \begin{pmatrix} -R_s - L_d s & \omega L_d \\ -\omega L_q & -R_s - L_q s \end{pmatrix} \begin{pmatrix} i_{sd} \\ i_{sq} \end{pmatrix} + \begin{pmatrix} 0 \\ \omega.\varphi_f \end{pmatrix} \tag{5}$$

Where, R_s, ω are the stator resistance and the generator electrical rotational speed respectively. v_{sdq}, i_{sdq} are the d, q stator voltages and currents. L_{dq}, φ_f are the d, q axis inductances and magnetic flux.

The relationship between the electrical speed and the mechanical speed can be expressed as:

$$\omega = \frac{p}{2} \Omega_t \tag{6}$$

p is the poles number of the PMSG.

The mechanical dynamics of the rotating parts can be given by maximum [1]:

$$J\frac{d\omega}{dt} = T_w - T_m - f_r\omega \tag{7}$$

Where, J is the inertia; T_w is the wind turbine torque and f_r is the coefficient of friction.

2.3 Wind Turbine Emulator

The wind turbine emulator is based on DC-DC converter and DC-Motor with separate excitation as shown in Fig. 2.

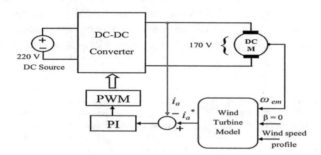

Fig. 2. Wind turbine emulator.

3 Improved Current Vector Control

Vector control technique is mainly used to maximize the extracted power from the wind [1], it has a nested-loop structure with a fast inner loop and a slow outer loop as shown in Fig. 3, where the outer loop controls the wind turbine speed (ω) and the inner loop is designed to control the currents i_d and i_q. Therefore, i_d and i_q current loop is decoupled which ensure a stable and decoupled active and reactive power.

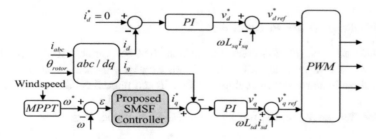

Fig. 3. Current vector control for MSC control associated with SMSF controller.

4 Design of the Proposed SMSF Controller

As many papers have highlighted [18], the most used controller in regulation loops is the conventional PI controller regarding to the simplicity of design, the good steady state and the acceptable transient state. Recently, several authors have expressed doubts about employing this kind of controllers in complicated systems that have variable parameters [17]. One of the proposed solutions is the fractional order PI controller (FO-PI). Based on fractional calculus (FC) theory and approximation methods, the FO-PI controller could be a good alternative in transients by offering fast response time and low damping but it is also limited in steady state as cited in [18]. Thus, the main idea of this paper is to propose a new controller which combines between one conventional and other one fractional then employ one of them in regulation loop by using a supervisor based on sliding mode control as shown in Fig. 4. The expected advantages of the proposed Sliding mode supervised fractional (SMSF) controller are to have a good steady state and best transient state when applying sever working conditions. The SMSF controller is employing in speed regulation loop of current vector control of Fig. 3.

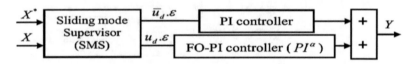

Fig. 4. Block diagram of the proposed SMSF controller.

4.1 Conventional PI Controller

In steady-state, the conventional PI controller is selected to obtain the expected control performance and its TF is given by Eq. 8 with two controlled terms (k_{pc} and k_{ic}) and an integer order of integration (s^{-1}).

$$C_c(s) = k_{pc} + \frac{k_{ic}}{s} \tag{8}$$

4.2 Fractional Order PI Controller

Fractional order PI (FO-PI) controllers are studied for variable-speed operation of WECS with a PMSG in several papers [1, 2]. However, the FO-PI controller has an additional parameter α to be developed. This will add more flexibility to the controller design [1].

In the frequency domain, the TF of the FO-PI controller is given as,

$$C_f(s) = k_{pf} + \frac{k_{if}}{s^{\alpha}} \tag{9}$$

4.3   Sliding Mode Supervisor

Sliding mode control is considered as one of the most accurate and robust control technique for several control applications. The first order sliding mode control causes the chattering problem which may damage the control performance. But, in our case, the sliding mode supervisor (SMS) of Fig. 5 employs a PI or a FO-PI controller and ignores the other one as mentioned in Eq. 10. In more details, going from up to down state and vice versa seems to be the chattering phenomena of the sliding mode control which is the main selection mechanism of the proposed SMS.

$$u_d = \begin{cases} 0 & PI \\ 1 & FO - PI \end{cases} \tag{10}$$

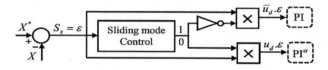

Fig. 5. Sliding mode supervisor (SMS) block diagram.

To build the SMS, firstly, the sliding surface S_x is expressed by Eq. 11.

$$S_x = |X^* - X| - m \tag{11}$$

Where, m is a real number whose values are responsible for the selection sensibility of the proper controller.

Then, the control signal can be rewritten as Eq. 12 to fulfill Eq. 10:

$$u_d = \frac{1}{2}[1 + sign(S_x)] \tag{12}$$

5   Experimental Results

In order to investigate the efficiency of the proposed SMSF controller in speed and voltage loops, a variable speed WECS is a good challenge by imposing severe wind speed profile. Figure 6 presents the experimental set-up of 6.6 kW developed in laboratory in order to examine the validity of the proposed controller integrated to the current vector control.

In order to test the efficiency of the SMSF controller integrated to current vector control, a sever wind speed profile has been applied. Actually, Figs. 7 and 8 show the response of WECS with MPPT controller based on PI and SMSF controllers respectively.

Fig. 6. Experimental setup for WECS, 1: PC with first DSPACE, 2: PC with second DSPACE, 3: DC motor, 4: DC/DC converter, 5: separate excitation, 6: PMSG, 7: machine side converter, 8: Grid side converter, 9: transformer, 10: nonlinear load, 11: numerical oscilloscope, 12: voltage sensors, 13: speed sensor, 14: currents sensors, 15: filters.

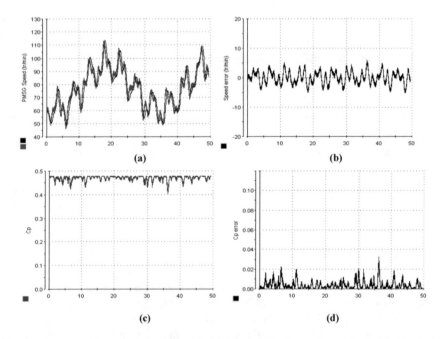

Fig. 7. Experimental results using conventional PI controller for WECS: (**a**) wind turbine speed and its reference, (**b**) error between wind speed and its reference, (**c**) power coefficient, (**d**) error in power coefficient.

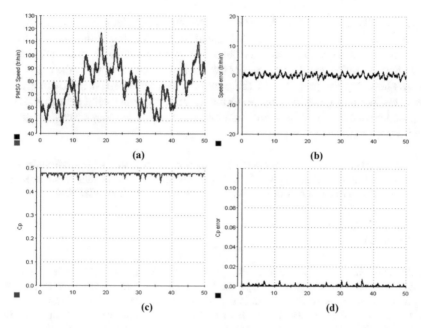

Fig. 8. Experimental results using the proposed SMSF controller for WECS: (**a**) wind turbine speed and its reference, (**b**) error between wind speed and its reference, (**c**) power coefficient, (**d**) error in power coefficient.

From Figs. 7a and 8a, it is clear that the PMSG power has the same waveform of the wind speed. But, the PMSG speed of conventional control has a considerable overshoot in tracking of its optimal value with high error between the speed and its reference (Fig. 7b), which leads to losses in the power extraction, the power coefficient C_p in this case is far from its optimum value (Fig. 7c, d). Whereas, by employing the proposed controller; it is clear that the PMSG speed follows its reference (Fig. 8a) with acceptable losses (Fig. 8b). As a result, a good C_p is recorded around the optimum values (Fig. 8c, d). Obviously, the WECS with the proposed SMSF controller presents the best tracking results in term of wind speed and power coefficient with negligible errors.

6 Conclusion

In this paper, a suitable Sliding Mode Supervised Fractional order controller (SMSF) for Wind Energy Conversion System (WECS) was developed. The potential of this approach is to behave as adaptive controller with low complexity in the design. In more details, the proposed SMSF combines between the simplicity of a conventional PI controller in steady state and the efficiency of a FO-PI controller in transient state. The SMSF controller was employed in the current vector control loop for maximum power extraction also in direct power control loop to ensure smooth injection of wind energy to the grid. The effectiveness of the proposed controller was tested experimentally in laboratory. Whereas, a wind turbine emulator based on a DC motor was realized to drive an industrial PMSG and been connected to the grid via back-to-back converters. A variety of tests have been performed to demonstrate the good tracking capabilities and the accurate power control. In transient state, the experimental results show that the proposed SMSF controller can achieve fast response time with low overshoot and good tracking performance under variable wind speed profile. The proposed controller could be a good alternative in regulation loops for variable speed wind energy conversion system.

Acknowledgements. The authors gratefully acknowledge the head of Electro-technics department and the dean of Technology faculty of University of Setif-1 for facilitates the access to the equipment. This paper is a part of the project PRFU: A10N01UN190120180002.

References

1. Afghoul, H., Krim, F., Babes, B., Beddar, A., Kihal, A.: Design and real time implementation of sliding mode supervised fractional controller for wind energy conversion system under sever working conditions. Energy Convers. Manag. **167**, 91–101 (2018)
2. Beddar, A., Bouzekri, H., Babes, B., Afghoul, A.: Experimental enhancement of fuzzy fractional order PI + I controller of grid connected variable speed wind energy conversion system. Energy Convers. Manag. **123**, 569–580 (2016)
3. Adánez, J.M., Al-Hadithi, B.M., Jiménez, A.: Wind turbine multivariable optimal control based on incremental state model. Asian J. Control **20**, 1–13 (2018)
4. Anh Tuyet, N.T., Chou, S.Y.: Maintenance strategy selection for improving cost-effectiveness of offshore wind systems. Energy Convers. Manag. **157**, 86–95 (2018)
5. Babes, B., Rahmani, L., Chaoui, A., Hamouda, N.: Design and experimental validation of a digital predictive controller for variable-speed wind turbine systems. J. Power Electron. **17**, 232–241 (2017)
6. Hoshyar, M., Mola, M.: Full adaptive integral backstepping controller for an interior permanent magnet synchronous motors. Asian J. Control **20**, 1–12 (2018)
7. Song, D., et al.: Maximum power extraction for wind turbines through a novel yaw control solution using predicted wind directions. Energy Convers. Manag. **157**, 587–599 (2018)
8. Liao, M., Dong, L., Jin, L., Wang, S.: Study on rotational speed feedback torque control for wind turbine generator system. In: International Conference on Energy and Environmental Technology (ICEET), Guilin, Guangxi, China, pp. 853–856 (2009)
9. Geng, H., Xu, D., Wu, B., Yang, G.: Comparison of oscillation damping capability in three power control strategies for PMSG-based WECS. Wind Energy **14**, 389–406 (2011)
10. Johnson, S.J., Baker, J.P., Van Dam, P., Baker, Berg: An overview of active load control techniques for wind turbines with an emphasis on microtabs. Wind Energy **13**, 239–253 (2010)
11. Mohamed, A.Z., Eskander, M.N., Ghali, F.A.: Fuzzy logic control based maximum power tracking of a wind energy system. Renew. Energy **23**, 235–245 (2001)
12. Lee S.H., Joo Y.J., Back J, Seo J.H.: Sliding mode controller for torque and pitch control of wind power system based on PMSG. In: International Conference on Control, Automation and Systems, ICCAS, pp. 1079–1084 (2010)

13. Asadollahi, M., Ghiasi, A.R., Dehghani, H.: Excitation control of a synchronous generator using a novel fractional-order controller. IET Gener. Transm. Distrib. **9**, 2255–2260 (2015)
14. Afghoul, H., Chikouche, D., Krim, F., Babes, B., Beddar, A.: Implementation of fractional-order integral-plus proportional controller to enhance the power quality of an electrical grid. Electr. Power Compon. Syst. **44**, 1018–1028 (2016)
15. Podlubny, I.: Fractional-order systems and $PI^{\lambda}D^{\mu}$. IEEE Trans. Autom. Control **44**, 208–214 (1999)
16. Afghoul, H., Krim, F., Chikouche, D., Beddar, A.: Robust switched fractional controller for performance improvement of single phase active power filter under unbalanced conditions. Front. Energy **10**, 203–212 (2016)
17. Afghoul, H., Krim, F., Chikouche, D., Beddar, A.: Design and real time implementation of fuzzy switched controller for single phase active power filter. ISA Trans. **58**, 614–621 (2015)
18. Beddar, A., Bouzekri, H., Babes, B., Afghoul, H.: Real time implementation of improved fractional order proportional-integral controller for grid connected wind energy conversion system. Rev. Roum. Sci. Tech. Ser. Électrotech. Énerg. **61**, 402–407 (2016)

Sliding Mode Control of DFIG Driven by Wind Turbine with SVM Inverter

Youcef Bekakra[✉], Djilani Ben Attous, and Hocine Bennadji

LEVRES-Research Laboratory, Department of Electrical Engineering,
University of El Oued, P.O. Box 789, El Oued, Algeria
youcef-bekakra@univ-eloued.dz

Abstract. In this paper, a direct Sliding Mode Control (SMC) of Doubly Fed Induction Generator (DFIG) driven by wind turbine with Space Vector Modulation (SVM) inverter is presented. The SMC is used to track the stator active and reactive power their references. The SVM inverter is used to improve the quality of the energy generated by DFIG which allows the minimizing of stator current harmonics and wide linear modulation range. Simulation results show that the proposed control give good performance and good quality of the energy where the THD of the stator current of the DFIG has small value.

Keywords: Doubly Fed Induction Generator · Sliding Mode Control
Wind Turbine · Space Vector Modulation · Total Harmonic Distortion

1 Introduction

One of the generation systems commercially available in the wind energy market currently is the doubly fed induction generator (DFIG) with its stator winding directly connected to the grid and with its rotor winding connected to the grid through a variable frequency converter [1, 2].

Vector control technology is used to control the generator where the rotor of DFIG is connected to an inverter of which the frequency, phase and magnitude can be adjusted. Therefore, constant operating frequency can be achieved at variable wind speeds [3].

This paper adopts the vector control method of stator oriented magnetic field to realize the decoupling control of the stator active and reactive power using sliding mode control (SMC).

The sliding mode theory, derived from the variable structure control family, was used for the induction motor drive for a long time. It performs a robust control by adding a discontinuous control signal across the sliding surface, satisfying the sliding condition [4].

This paper presents a numerical simulation study of direct sliding mode control of active and reactive power of the DFIG fed by SVM inverter to improve the quality of the energy injected into the electrical grid.

© Springer Nature Switzerland AG 2019
M. Hatti (Ed.): ICAIRES 2018, LNNS 62, pp. 192–198, 2019.
https://doi.org/10.1007/978-3-030-04789-4_21

2 DFIG Field Oriented Control

The Doubly Fed Induction Machine (DFIM) model can be described by the following state equations in the synchronous reference frame whose axis d is aligned with the stator flux vector, ($\phi_{sd} = \phi_s$ and $\phi_{sq} = 0$) [5, 6].

By neglecting resistances of the stator phases, the stator voltage will be expressed by:

$$V_{ds} = 0 \quad \text{and} \quad V_{qs} = V_s \approx \omega_s \cdot \phi_s \tag{1}$$

The reactive power is imposed by the direct component i_{rd}.

$$P_s = - V_s \frac{M}{L_s} i_{rq} \tag{2}$$

$$Q_s = \frac{V_s^2}{\omega_s L_s} - V_s \frac{M}{L_s} i_{rd} \tag{3}$$

The arrangement of the equations gives the expressions of the voltages according to the rotor currents:

$$\begin{cases} V_{rd} = R_r \cdot i_{rd} + \sigma L_r \frac{di_{rd}}{dt} - g\omega_s \sigma L_r i_{rq} \\ V_{rq} = R_r i_{rq} + \sigma L_r \frac{di_{rq}}{dt} + g \frac{M}{L_s} V_s + g\omega_s \sigma L_r i_{rd} \end{cases} \tag{4}$$

With: $T_r = \frac{L_r}{R_r}$; $T_s = \frac{L_s}{R_s}$; $\sigma = 1 - \frac{M^2}{L_s \cdot L_r}$.

The system studied in the present paper is constituted of a DFIG directly connected through the stator windings to the grid, and supplied through the rotor by a static frequency inverter as presented in Fig. 1.

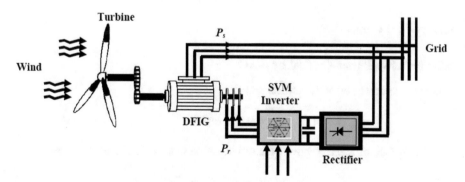

Fig. 1. Configuration of the doubly fed induction generator

3 Sliding Mode Control

3.1 Active Power Control by SMC

The active power error is defined by [4]:

$$e = P_s^* - P_s \tag{5}$$

The surface of active power control equation can be obtained as follow:

$$\sigma_s(P_s) = e = P_s^* - P_s \tag{6}$$

$$\dot{\sigma}_s(P_s) = \dot{P}_s^* - \dot{P}_s \tag{7}$$

Substituting the expression of $\dot{P}_s$ Eq. (2) in Eq. (7), we obtain:

$$\dot{\sigma}_s(P_s) = \dot{P}_s^* - \left(-V_s \frac{M}{L_s} \dot{i}_{rq}\right) \tag{8}$$

We take:

$$V_{rq} = V_{rq}^{eq} + V_{rq}^n \tag{9}$$

Where the equivalent control is:

$$V_{rq}^{eq} = R_r i_{rq} - \dot{P}_s^* \frac{L_s L_r . \sigma}{V_s M} \tag{10}$$

Therefore, the correction factor is given by:

$$V_{rq}^n = k_{Vrq} sat(\sigma_s(P_s)) \tag{11}$$

k_{Vrq}: positive constant.

3.2 Reactive Power Control by SMC

The reactive power error is defined by [4]:

$$e = Q_s^* - Q_s \tag{12}$$

The surface of reactive power control equation can be obtained as follow:

$$\sigma_s(Q_s) = e = Q_s^* - Q_s \tag{13}$$

$$\dot{\sigma}_s(Q_s) = \dot{Q}_s^* - \dot{Q}_s \tag{14}$$

Substituting the expression of $\dot{Q}_s$ Eq. (3) in Eq. (14), we obtain:

$$\dot{\sigma}_s(Q_s) = \dot{Q}_s^* - \left(\frac{V_s^2}{\omega_s L_s} - V_s \frac{M}{L_s} \dot{i}_{rd}\right) \tag{15}$$

We take:

$$V_{rd} = V_{rd}^{eq} + V_{rd}^{n} \tag{16}$$

Where the equivalent control is:

$$V_{rd}^{eq} = R_r i_{rd} - \dot{Q}_s^* \frac{L_s L_r \cdot \sigma}{V_s M} \tag{17}$$

Therefore, the correction factor is given by:

$$V_{rd}^{n} = k_{Vrd} sat(\sigma_s(Q_s)) \tag{18}$$

k_{Vrd}: positive constant.

4 Simulation Results

The DFIG used in this work is a 4 kW, whose nominal parameters are indicated in appendix.

To verify the feasibility of the proposed control scheme, computer simulations were performed using Matlab/Simulink software.

We have proposed a random variable wind speed as shown in Fig. 2 to verified the robustness of the proposed control,

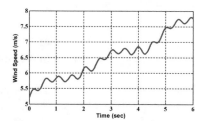
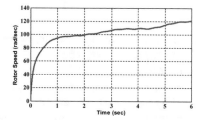

Fig. 2. Proposed wind speed   **Fig. 3.** Rotor speed

Figure 3 shows the turbine rotor speed. Figure 4 presents the power coefficient variation C_p, it is kept around its maximum value $C_p = 0.5$.

Figure 5a presents the stator active power and its reference profile injected into the grid. The stator reactive power and its reference profile are presented in Fig. 5b. After these figures a very good decoupling obtained between the stator active and reactive power. It is clear that the actual stator active power tracks its desired values using the proposed control with the presence of the oscillations produced by the chattering phenomena of the SMC, where the reactive power is maintained to zero to guarantee a unity power factor $(cos(\phi) = 1)$ at the stator side.

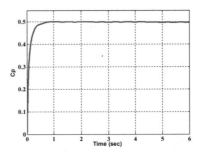

Fig. 4. Power coefficient C_p variation

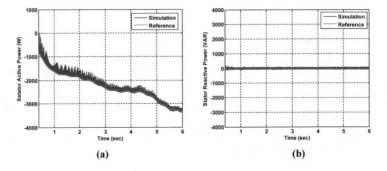

Fig. 5. (a) Stator active and (b) reactive power

Figure 6 shows the DFIG stator current changes versus time and its zoom, it shows good sinusoidal currents, where the amplitude of this current increases when the wind speed increases.

Figure 7 shows the harmonic spectrum of the output phase stator current which obtained by using Fast Fourier Transform (FFT) technique. It can be clearly observed that the stator current has a low Total Harmonic Distortion (THD) where its value is 4.38%, as indicated in the Fig. 7. Where this value is acceptable according to "IEEE Std 519-1992" which recommended by "*require AC sources that have no more than 5% Total Harmonic Distortion*" [7].

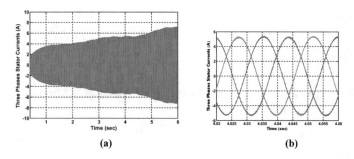

Fig. 6. (a) Stator currents with (b) a zoom

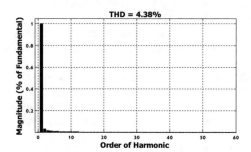

Fig. 7. Spectrum of the stator current harmonics

5 Conclusion

In this paper, a direct sliding mode control (SMC) of active and reactive power based on DFIG driven by wind turbine with space vector modulation (SVM) inverter has been studied and designed in Matlab/Simulink software.

The SMC has been used to track the stator active and reactive power their references. The stator active and reactive powers are exchanged between the stator of the DFIG and the electrical grid by the control of the rotor inverter where the simulation results show good dynamic performances and good robustness of the proposed control.

In addition, SVM technique is used for the inverter control to improve the quality of energy injected into the electrical grid, which this technique allows the minimizing of stator current harmonics and wide linear modulation range where the THD of the stator current has the value 4.38% (<5% according IEEE Std 519-1992).

Appendix

DFIG Data:

$R_s = 1.2\ \Omega, \quad R_r = 1.8\ \Omega, \quad L_s = 0.1554\ \text{H}, \quad L_r = 0.1568\ \text{H}, \quad M = 0.15\ \text{H}, \quad P = 2,$
$J = 0.2\ \text{kg.m}^2, f = 0.001\ \text{N.m.s/rad}.$

References

1. Ghedamsi, K., Aouzellag, D.: Improvement of the performances for wind energy conversions systems. Int. J. Electr. Power Energy Syst. **32**(9), 936–945 (2010)
2. Kumar, V., Pandey, A.S., Sinha, S.K.: Grid integration and power quality issues of wind and solar energy system: a review. In: International Conference on Emerging Trends in Electrical, Electronics and Sustainable Energy Systems (ICETEESES-16), IEEE, 11–12 March 2016 (2016)
3. Abniki, H., Abolhasani, M., Kargahi, M.E.: Vector control analysis of doubly-fed induction generator in wind farms. Energy Power **3**(2), 18–25 (2013)
4. Bekakra, Y., Ben Attous, D.: DFIG sliding mode control fed by back-to-back PWM converter with dc-link voltage control for variable speed wind turbine. Front. Energy **8**(3), 345–354 (2014)
5. Rahimi, M., Parniani, M.: Dynamic behavior analysis of doubly-fed induction generator wind turbines—the influence of rotor and speed controller parameters. Int. J. Electr. Power Energy Syst. **32**(5), 464–477 (2010)
6. Aydin, E., Polat, A., Ergene, L.T.: Vector control of DFIG in wind power applications and analysis for voltage drop condition. In: 2016 National Conference on Electrical and Electronics and Biomedical Engineering (ELECO) (2016)
7. IEEE Std 519-1992: IEEE recommended practices and requirements for harmonic control in electrical power systems. IEEE Industry Applications Society, New York. ISBN 1-55937-239-7, pp. 1–112 (1993)

A Hybrid of Sliding Mode Control and Fuzzy Logic Control for a Five-Phase Synchronous Motor Speed Control

Fayçal Mehedi[1,2(✉)], Lazhari Nezli[2], Mohand Oulhadj Mahmoudi[2], and Abdelkadir Belhadj Djilali[1]

[1] University of Hassiba Benbouali, 02000 Chlef, Algeria
Faycalmehedi@yahoo.fr, blhadj.prof@yahoo.fr
[2] Process Control Laboratory, National Polytechnic School, 16000 Algiers, Algeria
l_nezli@yahoo.fr, momahmoudi@yahoo.fr

Abstract. This paper introduces a new technique for controlling the speed of multi-phase permanent magnet synchronous motor (PMSM). This technique depends on two well known control methods; the first one using a nonlinear control based on sliding mode control (SMC), which has a main advantage known as a sliding mode property. In the second, the fuzzy logic control is used to overcome the occurring chattering phenomena. A combination between the two mentioned above methods is suggested in this paper. Simulations results of the proposed control theme present good dynamic and steady-state performances as compared to the classical SMC from aspects of the reduction of the torque chattering, the quickly dynamic torque response and robustness to disturbance.

Keywords: Five-phase PMSM · Sliding Mode Control (SMC)
Chattering phenomena · Fuzzy Sliding Mode Control (FSMC)
Robustness

1 Introduction

Among the AC motors, PMSM systems have been used more and more in many applications, e.g., aerospace application [1], electric-drive vehicle systems [2], and wind energy conversion systems (WECSs) [3], due to their distinctive advantages of high efficiency, high power density, and wide constant power region [4]. The multi-phase systems have more advantages compared to the three phase systems like high output power rating, low torque pulsations and stable speed response [5, 6]. Multiphase machines have gained attention in numerous fields of applications such as Aircraft, ship propulsion, petrochemical and automobiles, where high reliability is required [7, 8].

Traditional vector control structures which include proportional-integral (PI) regulator for application to an multiphase machine driven have some disadvantages such as parameter tuning complications, mediocre dynamic performances and reduced robustness.

© Springer Nature Switzerland AG 2019
M. Hatti (Ed.): ICAIRES 2018, LNNS 62, pp. 199–205, 2019.
https://doi.org/10.1007/978-3-030-04789-4_22

The SMC theory was proposed by Utkin in 1977 [9]. Thereafter, the theoretical works and its applications of the SMC were developed [10]. The SMC achieves robust control by adding a discontinuous control signal across the sliding surface, satisfying the sliding condition. Nevertheless, this type of control has an essential disadvantage, which is the chattering phenomenon caused by the discontinuous control action. A plenty of research papers focus on elimination/avoidance chattering by using different methods [11, 12].

In fuzzy logic, an exact mathematical model is not necessary because linguistic variables are used to define system behavior rapidly.

This paper presents a comprehensive study of the hybrid of SMC and fuzzy logic control for a five-phase PMSM speed control, which can assure robustness against a load torque disturbance and speed variation with chattering reduction.

The main body of this paper contains six sections. Section 2 gives the model of five-phase PMSM. Section 3 introduces the sliding control strategy. In Sect. 4, fuzzy sliding mode speed control of the five-phase PMSM is presented. Section 5 summarizes the results of this study. Finally, the main conclusions of the work are drawn.

2   Model of Five Phase PMSM

The model of the five-phase PMSM is presented in a rotating d, q frame as [14]:

$$
\begin{aligned}
v_{ds} &= R_s i_{ds} + L_d \frac{d}{dt} i_{ds} - w_r L_q i_{qs} \\
v_{qs} &= R_s i_{qs} + L_q \frac{d}{dt} i_{qs} + w_r L_d i_{ds} + \sqrt{\frac{5}{2}} w_r \phi_f
\end{aligned}
\tag{1}
$$

Where v_{ds} and v_{qs} are the stator voltages in the d-q axis, i_{ds} and i_{qs} are the stator currents in d-q, R_s is the stator resistance, L_d and L_q are inductances in the rotating frame.

The expression of electromagnetic torque is given by:

$$
T_e = p((L_d - L_q) i_{ds} i_{qs} + \sqrt{\frac{5}{2}} \phi_f i_{qs})
\tag{2}
$$

On the other hand, the mechanical equation of the machine is:

$$
J_m \frac{dw_r}{dt} = p T_e - p T_r - f_m w_r
\tag{3}
$$

With J_m is the inertia coefficient, f_m is the viscous damping, P is the number of poles pairs, and T_r is the external load torque.

3   Sliding Mode Speed Control of the Five-Phase PMSM

The design of a SMC requires mainly two stages. The first stage is choosing an appropriate sliding surface. The second stage is designing a control law.

In order to prescribe the desired dynamic characteristics of the controlled system, [15] proposes a general function of the switching surface:

$$S(x) = \left(\frac{d}{dt} + \lambda\right)^{n-1} e(x) \tag{4}$$

Here, $e(x) = x^* - x$ is the tracking error vector, λ is a positive coefficient and n is the relative degree.

The second phase consists to find the control law which meets the sufficiency conditions for the existence and reachability of a sliding mode such as [15, 16]

$$S(x)\dot{S}(x) < 0 \tag{5}$$

So that the state trajectory be attracted to the switching surface $S(x) = 0$. A commonly used from of U_n is a constant relay control.

$$U_n = K_x \text{sgn}(S(x)) \tag{6}$$

sgn(S(x)) is a sign function, which is defined as

$$\text{sgn}(S(x)) = \begin{cases} -1 & \text{if } S(x) < 0 \\ 1 & \text{if } S(x) > 0 \end{cases} \tag{7}$$

k_x is a constant.

The sliding surfaces are chosen according to (3) as follows:

$$S(w_r) = w_r^* - w_r \tag{8}$$

Based on the proposed switching surface, the speed control laws are:

$$i_q^* = \frac{J_m \dot{w}_r^* + pT_r + f_m w_r}{p\left((L_d - L_q)i_{ds} + \sqrt{\frac{5}{2}}\phi_f\right)} + K_1.sign\left(S(w_r)\right) \tag{9}$$

Where K_1 is positive constant.

Figure 1 represents the sliding mode speed control scheme of a five-phase PMSM.

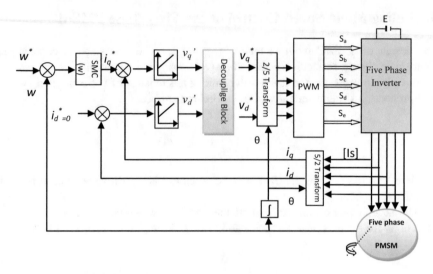

Fig. 1. Sliding mode speed control scheme of a five-phase PMSM.

4 Fuzzy Sliding Mode Speed Control of the Five-Phase PMSM

The disadvantage of SMC is the chattering effect. In this paper, the regulator of speed is substituted by a FSMC to obtain a robust performance. One part of the equivalent control (SMC) and part of fuzzy control are contained in this hybrid control (FSMC)

$$U_{FSMC} = U_{eq} + U_{fuzzzy} \tag{10}$$

The two parts are combined to provide stability and robustness of the system, the method of control by fuzzy logic approach is adopted to solve the problem of chattering [13] (Fig. 2).

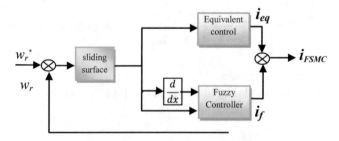

Fig. 2. Diagram of the hybrid control FSMC of a five-phase PMSM speed control.

5   Results and Discussions

To perform this tests we have introduced a load torque $T_r = 5$ Nm in the interval $t = [0.3, 0.6]$ s and a step change of the reference speed from 150 rad/s to −150 rad/s at $t = 0.9$ s. Parameters of the machine are given in Table 1. The simulation results are show in Fig. 3. These results show that the response of speed control is satisfying in all intervals of operation; the motor is running at rated speed, load impacts do not influence on its value. It can be noticed the FSMC have a nearly perfect speed disturbance rejection (less than 2%). So our control is robust to variations in load and speed, we can still note that the chattering phenomenon is reduced to the level of the electromagnetic torque. The current I_q is the image of the torque.

Table 1.   Five-phase PMSM parameters.

R_s	L_d	L_q	ϕ_f	p	J_m	f_m
3.6 Ω	0.0021 H	0.0021 H	0.12T	2	0.0011 kg/m^2	0 Nm/s

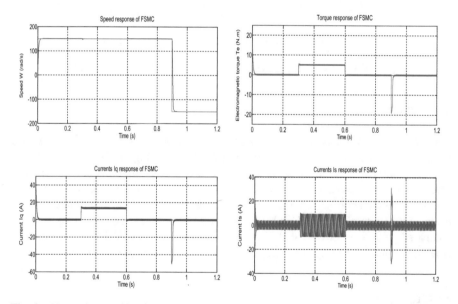

Fig. 3.   Simulation results of FSMC speed, torque, currents Iq and five-phase stator currents.

Figure 4 shows the speed response comparison and torque response comparison of the two methods. From Fig. 4a, it is obviously that the FSMC for the regulation of speed is better than the classical PI speed regulator. The start-up speed, dynamics performance and the robustness of the FSMC are all very good. The torque response comparison is shown in Fig. 4b. It can be seen that excluded the excellent steady-state and dynamics performance, the proposed drive method also restricts the torque chattering in a lower degree than the other conventional method. The torque chattering value of FSMC method and SMC method are 2 Nm and 6 Nm respectively. So the advantage of the proposed method control is very obviously.

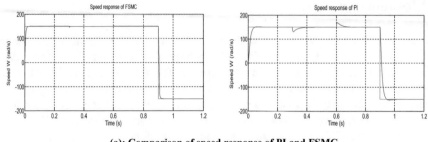

(a): Comparison of speed response of PI and FSMC

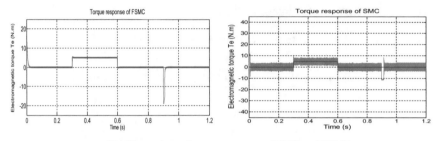

(b): Comparison of torque response of SMC and FSMC

Fig. 4. Comparison of speed and torque responses of PI, SMC and FSMC.

6   Conclusion

This paper presents the design of control system for controlling the speed of five-phase permanent magnet synchronous motor using the fuzzy sliding mode control algorithm. The important feature required in multi-phase PMSM speed control is less sensitivity to mechanical motor parameters and load variation. These advantages are provided by the developed association of the SMC controller and the fuzzy logic algorithm used to mitigate the chattering. FSMC achieves less speed oscillation and low torque ripples compared with the conventional PI controllers and the classical SMC. Simulation results are presented to demonstrate the effectiveness of the control strategy.

References

1. Guo, H., Xu, J., Chen, Y.H.: Robust control of fault tolerant permanent magnet synchronous motor for aerospace application with guaranteed fault switch process. IEEE Trans. Ind. Electron. **62**(12), 7309–7321 (2015)
2. Dai, P., Sun, W., Xu, N., Lv, Y., Zhu, X.: Research on energy management system of hybrid electric vehicle based on permanent magnet synchronous motor. In: Proceedings 11th IEEE Conference on Industrial Electronics and Applications, Hefei, China, pp. 2345–2349 (2016)
3. Chinchilla, M., Arnaltes, S., Burgos, J.C.: Control of permanent-magnet generators applied to variable-speed wind-energy systems connected to the grid. IEEE Trans. Energy Convers. **21**(1), 130–135 (2006)

4. Zhao, Y.: Position/speed sensorless control for permanent-magnet synchronous machines. Electrical Engineering Theses and Dissertations. University of Nebraska-Lincoln, Spring 4 (2014)
5. Ramana, N.V., Sastry, V.L.N.: A novel speed control strategy for five phases permanent magnet synchronous motor with linear quadratic regulator. Int. J. Comput. Electr. Eng. **7**(6), 408–416 (2015)
6. Hosseyni, A., Trabelsi, R., Iqbal, A., Mimouni, M.F.: Backstepping control for a five-phase permanent magnet synchronous motor drive. Int. J. Power Electron. Drive Syst. **6**(4), 842–852 (2015)
7. Salehifar, M., Arashloo, R.S., Eguilaz, M.M., Sala, V., Romeral, L.: Observer-based open transistor fault diagnosis and fault-tolerant control of five-phase permanent magnet motor drive for application in electric vehicles. IEEE IET Power Electron. **8**(1), 76–87 (2015)
8. Kim, H., Shin, K., Englebretson, S., Frank, N., Arshad, W.: Application areas of multiphase machines. IEEE Conference on Electric Machines & Drives, Chicago, IL, USA, pp. 172–179 (2013)
9. Utkin, V.: Variable structure systems with sliding modes. IEEE Trans. Autom. Control **22**(2), 212–222 (1977)
10. Junhui, Z., Mingyu, W., Yang, L., Yanjing, Z., Shuxi, L.: The study on the constant switching frequency direct torque controlled induction motor drive with a fuzzy sliding mode speed controller. IEEE the Natural Science Foundation of Chongqing (CSTC 2007BB3169), pp. 1543–1548 (2007)
11. Utkin, V.: Discussion aspects of high order sliding mode control. IEEE Trans. Autom. Control **61**(3), 829–833 (2016)
12. Fezzani, A., Drid, S., Makouf, A., Chrifi, L., Ouriagli, M.: Speed sensoless robust control of permanent magnet synchronous motor based on second-order sliding-mode observer. Serb. J. Electr. Eng. **11**(3), 419–433 (2014)
13. Benyattou, L., Zeghlache, S.: Adaptive fuzzy sliding mode controller using nonlinear sliding surfaces applied to the twin rotor multi-input- multi-output system. Mediterr J Meas Control **13**(1), 702–719 (2017)
14. Mekri, F., Charpentier, J.F., Benelghali, S., Kestelyn, X.,: High order sliding mode optimal current control of five phase permanent magnet motor under open circuited Phase fault conditions. In: Proceedings IEEE Conference on Vehicle Power and Propulsion, Lille, France pp. 1–6 (2010)
15. Levant, A.: Integral high-order sliding modes. IEEE Trans. Autom. Control **52**(7), 1278–1282 (2007)
16. Shah, M., Muhammad, I.: Comparative study of hierarchical sliding mode control and decoupled sliding mode control. In: 12th IEEE Conference on Industrial Electronics and Applications, Siem Reap, Cambodia, pp. 818–824 (2017)

Nonlinear Sliding Mode Control of DFIG Based on Wind Turbines

Bahia Kelkoul$^{(\boxtimes)}$ and Abdelmadjid Boumediene

LAT, Laboratoire D'Automatique de Tlemcen, Université de Tlemcen,
13000 Tlemcen, Algeria
bahiakelkoul@yahoo.com, al0boumediene@yahoo.fr

Abstract. This paper presents the modeling and control of doubly fed induction generator (DFIG) based on wind turbine systems. In order to control the stator active and reactive powers of the DFIG, a control law is synthesized using two types of controllers: a linear PI controller and nonlinear Sliding Mode Controller (SMC). Their performances are compared in terms of power reference tracking and robustness against machine parameters variations. Simulation results using Matlab/Simulink have shown good performances of the wind energy converter system operate under typical wind variations and every propose control strategies.

Keywords: Modeling · DFIG · PI controller · Sliding Mode Control (SMC)

1 Introduction

Doubly fed induction generators (DFIGs) are mostly used in high-power wind energy conversion systems because of their salient features [14, 15]. Different ways of driving the DFIG have been proposed in literature like controlling the electromagnetic torque [16], controlling the rotating speed [1, 11], and controlling the stator active power [12, 17]. The main goal of all these control methods is to track the maximum attainable power of the wind turbine as long as the wind speed is below a certain upper limit [2, 8].

This paper presents a control method for the machine inverter in order to regulate the stator active and reactive power. The active power is controlled in order to be adapted to the wind speed in a wind energy conversion system and the reactive power control allows getting a unitary power factor between the stator and the grid. Firstly, we adopt the vector transformation control method of stator oriented magnetic field to realize the decoupling control for the active power and reactive power using sliding mode control (SMC) [9].

Sliding mode theory, stemmed from the variable structure control family. it has been used for the induction motor drive for a long time. It has been known for its capabilities in accounting for modeling imprecision and bounded disturbances. It achieves robust control by adding a discontinuous control signal across the sliding surface, satisfying the sliding condition [3]. In this paper, we apply the sliding mode control (SMC) to stator active and reactive powers of a DFIG.

© Springer Nature Switzerland AG 2019
M. Hatti (Ed.): ICAIRES 2018, LNNS 62, pp. 206–215, 2019.
https://doi.org/10.1007/978-3-030-04789-4_23

2 Modeling of DFIG

DIFG electrical equations in the park frame will be written as follow [4, 5, 11]:

$$
\begin{cases}
V_{ds} = R_s I_{ds} + \dfrac{d}{dt}\phi_{ds} - \omega_s \phi_{qs} \\[2mm]
V_{qs} = R_s I_{qs} + \dfrac{d}{dt}\phi_{qs} + \omega_s \phi_{ds} \\[2mm]
V_{dr} = R_r I_{dr} + \dfrac{d}{dt}\phi_{dr} - \omega_r \phi_{qr} \\[2mm]
V_{qr} = R_r I_{qr} + \dfrac{d}{dt}\phi_{qr} + \omega_r \phi_{dr}
\end{cases}
\tag{1}
$$

Where R_s and R_r are respectively the stator and rotor phase resistances.

$$
\omega = P.\Omega_{mec}
\tag{2}
$$

Ω_{mec} is the electrical speed and P is the pair pole number.

The stator and rotor flux can be expressed as:

$$
\begin{cases}
\phi_{ds} = L_s I_{ds} + M I_{dr} \\[1mm]
\phi_{qs} = L_s I_{qs} + M I_{qr} \\[1mm]
\phi_{dr} = L_r I_{dr} + M I_{ds} \\[1mm]
\phi_{qr} = L_r I_{qr} + M I_{qs}
\end{cases}
\tag{3}
$$

Where I_{ds}, I_{qs}, I_{dr}, I_{qr} are respectively the direct and quadrature stator and rotor currents.

The active and reactive powers at the stator are defined as [11]:

$$
\begin{cases}
P_s = V_{ds} I_{ds} + V_{qs} I_{qs} \\[1mm]
Q_s = V_{qs} I_{ds} - V_{ds} I_{qs}
\end{cases}
\tag{4}
$$

DFIM torque generated based on stator currents and rotor flux is shown as follows:

$$
T_{em} = \frac{pM}{L_s}(I_{dr}\phi_{qs} - I_{qr}\phi_{ds})
\tag{5}
$$

3 Control Strategy of the DFIG

By aligning the d-axis in the direction of flux φ_s [6, 7] we have:

$$\Phi_{ds} = \Phi_s, \quad \Phi_{qs} = 0 \tag{6}$$

Therefore by using Eqs. (5) and (6) we obtain:

$$T_{em} = p\frac{M}{L_s}\left(I_{qr}\Phi_{ds}\right) \tag{7}$$

The stator flux can be expressed as:

$$\begin{cases} \phi_s = L_s I_{ds} + M I_{dr} \\ 0 = M I_{qr} + L_s I_{qs} \end{cases} \tag{8}$$

By neglecting resistances of the stator phases, with the assumption that the stator flux is also constant then the direct and quadrature stator voltage can be written as [7, 14]:

$$\begin{cases} V_{ds} = 0 \\ V_{qs} = V_s = \omega_s \phi_s \end{cases} \tag{9}$$

Stator currents are expressed in terms of rotor currents:

$$\begin{cases} I_{ds} = \frac{\phi_s}{L_s} - \frac{M}{L_s} I_{dr} \\ I_{qs} = -\frac{M}{L_s} I_{qr} \end{cases} \tag{10}$$

The orientation of the stator flux to the direct axis, and as a result, the voltage aligns with the quadrature axis. The stator active and reactive power can then be expressed only versus these rotor currents

$$\begin{cases} P_s = V_s \frac{M}{L_s} I_{qr} \\ Q_s = -V_s \frac{M}{L_s} I_{dr} + \frac{V_s \psi_s}{L_s} \end{cases} \tag{11}$$

Facts of constant stator voltage, the stator active and reactive powers are controlled by i_{qr} and i_{rd} respectively. The expressions of rotor voltages as a function of rotor currents are given as follows:

$$\begin{cases} V_{dr} = R_r I_{dr} + \sigma \cdot L_r \cdot \frac{dI_{dr}}{dt} - \sigma \cdot L_r \cdot \omega_r I_{qr} \\ V_{qr} = R_r I_{qr} + \sigma \cdot L_r \frac{dI_{qr}}{dt} + \sigma \cdot L_r \cdot \omega_r \cdot I_{dr} - g \cdot \frac{M \cdot V_s}{L_s} \end{cases} \tag{12}$$

Where: $g = \frac{\omega_s - \omega}{\omega_s} = \frac{\omega_r}{\omega_s}$ is the sliding value of the DFIG.

$\sigma = 1 - \frac{M^2}{L_s L_r}$ Dispersion coefficient.

These two controls variables V_{dr} and V_{qr} equations are coupled. The decoupling is obtained by compensation in order to ensure the control of P_s and Q_s independenly.

4   Controllers Syntheses

A. *PI regulator*

This controller is simple to elaborate. Figure 1 shows the block diagram of the system implemented with this controller. The terms K_p and K_i represent respectively the proportional and integral gains.

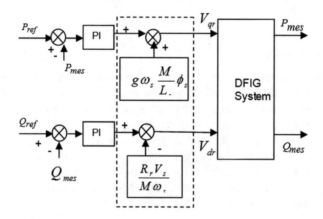

Fig. 1. PI controller of active and reactive powers of DFIG

$$C(p) = k_p + \frac{k_i}{p} \tag{13}$$

Transfer function in open loop is expressed by:

$$T(p) = \left(k_p + \frac{k_i}{p}\right)\left(\frac{MV_s}{L_s(R_r + pL_r\sigma)}\right) \tag{14}$$

$$T(p) = \frac{k_p}{p}\left(p + \frac{k_i}{k_p}\right)\frac{MV_s}{L_s L_r \sigma\left(p + \frac{R_r}{L_r\sigma}\right)} = \frac{p + \frac{k_i}{k_p}}{\frac{p}{k_p}}\frac{\frac{MV_s}{L_s L_r \sigma}}{p + \frac{R_r}{L_r\sigma}} \tag{15}$$

For eliminate zero present in transfer function, we use method of poles compensation for synthesis regulator, we put:

$$\frac{k_i}{k_p} = \frac{R_r}{L_r\sigma} \quad \text{Then:} \quad F(P) = \frac{k_p \frac{MV_s}{L_sL_r\sigma}}{p} = \frac{1}{\tau_r p} \tag{16}$$

Transfer function in closed loop:

$$F(p) = \frac{1}{1+p\tau_r} \tag{17}$$

τ_r is the time constant of the system fixed at 10 ms, corresponding at a value sufficiently rapid for application in power production of a DFIG of 7.5Kw.

Terms k_p and k_i are expressed by:

$$k_p = \frac{1}{\tau_r}\frac{L_sL_r\sigma}{MV_s}, \quad k_i = \frac{1}{\tau_r}\frac{R_rL_s}{MV_s} \tag{18}$$

B. Sliding mode control

The sliding mode control is a nonlinear control method which guarantees the control strategy despite of uncertainty. In this case the system stability is obtained by keeping the system's states on the sliding surface. A special issue about control plan is that define a sliding surface with property guaranteeing of stability and attractivity [5, 10]. This method has high flexibility, and it is very robust. The sliding mode controller doesn't need to the mathematical models accurately like classical controllers but needs to know the range of parameter changes for ensuring sustainability and condition satisfactory [5, 11]. The block diagram of the SMC in a variable speed wind turbine application in this paper is shown in Fig. 2.

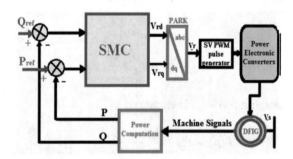

Fig. 2. Block diagram of a sliding mode control of a DFIG [5].

Active power control:
For the sliding order n = 1 the expressions of the active power control surface become [9, 13]:

$$S(P_s) = e = P_s^* - P_s \tag{19}$$

Lyapunov equation: $V(S_{Ps}) = \dfrac{1}{2} S_{Ps}^2 \tag{20}$

We have: $\dot{S}_{P_s} = \dot{P}_s^* - \dot{P}_s = \dot{P}_s^* + \dfrac{\sigma.L_s.L_r}{V_s.M_{sr}} \dfrac{d i_{rq}}{dt} \tag{21}$

$$\dot{S}_{P_s} = \dot{P}_s^* + \dfrac{V_s.M_{sr}}{\sigma.L_s.L_r} \left(V_{qr} - R_r.i_{qr} - \sigma.L_r.w_r.i_{dr} + g.\dfrac{M_{sr}.V_s}{w_s.L_s} \right) \tag{22}$$

We have: $V_{qr} = V_{qrn} + V_{qreq} \tag{23}$

$$\dot{S}_{P_s} = \dot{P}_s^* + \dfrac{V_s.M_{sr}}{\sigma.L_s.L_r} \left(V_{qrn} + V_{qreq} \right) - R_r.i_{qr} - \sigma.L_r.w_r.i_{dr} + g.\dfrac{M_{sr}.V_s}{w_s.L_s} \tag{24}$$

In sliding mode we have: $\dot{S}(P_s) = 0, \quad S(P_s) = 0, V_{qrn} = 0 \tag{25}$

$$V_{qreq} = -\dfrac{\sigma.L_s.L_r}{V_s.M_{sr}} \dot{P}_s^* + R_r.i_{qr} + \sigma.L_r.w_r.i_{dr} - g.\dfrac{M_{sr}.V_s}{w_s.L_s} \tag{26}$$

$$\dot{S}_{P_s} = \dfrac{V_s.M_{sr}}{\sigma.L_s.L_r}.V_{qrn} \tag{27}$$

On convergence mode: $S(P_s)\dot{S}(P_s) \leq 0$

$$V_{qrn} = -K_{vq}.signS(P_s) \tag{28}$$

Reactive power control:
For n = 1 surface of active stator power is defined by:

$$S(Q_s) = e = Q_s^* - Q_S \tag{29}$$

Lyapunov equation: $V(S_{Qs}) = \dfrac{1}{2} S_{Qs}^2 \tag{30}$

$$\dot{S}_{Q_s} = \dot{Q}_s^* - \dot{Q}_s = \dot{Q}_s^* + \dfrac{\sigma.L_s.L_r}{V_s.M_{sr}} \tag{31}$$

$$\dot{S}_{Q_s} = \dot{Q}_s^* + \dfrac{V_s.M_{sr}}{\sigma.L_s.L_r} \left(V_{dr} - R_r.i_{dr} + \sigma.L_r.w_r.i_{qr} \right) \tag{32}$$

$$V_{dr} = V_{drn} + V_{dreq} \tag{33}$$

$$\dot{S}_{Q_s} = \dot{Q}_s^* + \dfrac{V_s.M_{sr}}{\sigma.L_s.L_r} \left(V_{drn} + V_{dreq} \right) - R_r.i_{dr} + \sigma.L_r.w_r.i_{qr} \tag{34}$$

In sliding mode we have:

$$S(Q_s), \dot{S}_{Q_s} = 0 \text{ and } V_{drn} = 0 \tag{35}$$

$$V_{dreq} = -\frac{\sigma.L_s.L_r}{V_s.M_{sr}}\dot{Q}_s^* + R_r.i_{dr} - \sigma.L_r.w_r.i_{qr} \tag{36}$$

For assure the convergence of Lyapunov function, we put:

$$V_{drn} = -K_{vd}.sign(S_{Qs}) \tag{37}$$

$$\dot{S}(Q_s) = V_{qrn} = -K_{vq}.sign(S_{Ps}) \tag{38}$$

5   Results and Discussion

Simulation results are shown by MATLAB/Simulink software. The DFIG used in this work is a 7.5 kW, where nominal parameters are indicated in appendix.

A. *Tracking test*

We apply a stator active power reference step (−1000 W to −4000 W) at t = 1.5 s while we fixed the stator reactive power reference at zero to ensure unit power factor in the stator side. The simulation results are presented in Figs. 3, 4. The active and reactive powers track almost perfectly their references. The results show that the sliding mode control ensures a perfect decoupling between them.

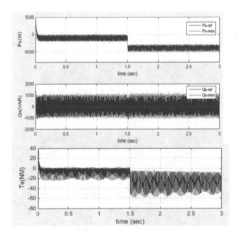

Fig. 3. Tracking test with PI controller

B. Robustness test

In this test we have changed the values the generator parameters as below:

$$R'_s = R_s * 1.4, \quad R'_r = R_r * 1.4$$

For the test of robustness, the variation of parameters (R_s, R_r) has a little influence on the response time and on the amplitude of oscillations in transient state (Figs. 5, 6).

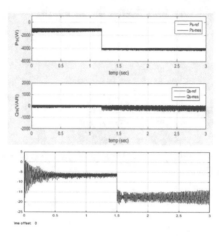

Fig. 4. tracking test Sliding mode control

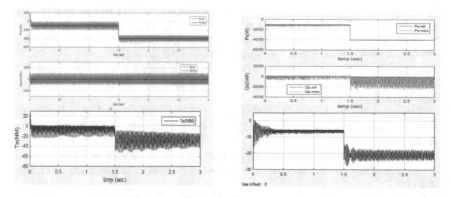

Fig. 5. Robustness test with PI controller.   **Fig. 6.** Robustness test with sliding mode control

6 Conclusion

This paper aims to control the stator active and reactive powers of a doubly fed induction generator. A modelling of this last and its linear strategy of control (vector control) are presented. Two types of controller are synthesizing and compared: proportional-integral and nonlinear sliding mode control for active and reactive power of DFIG wind turbine in order to demonstrate its performance, the controllers are tested in term of reference tracking, and robustness. The obtained results demonstrate that the sliding mode controller is very robust then the PI controller.

Appendix

Doubly fed induction generator parameters:
Rated values: 7.5 kW, 220/380 V, 15/8.6 A.
Rated parameters:
$R_s = 0.455(\Omega)$, $R_r = 0.62(\Omega)$, $L_s = 0.084(mH)$, $L_r = 0.081(mH)$, $M = 0.098(mH)$, $p = 2$, $J = 0.2(Kg.m2)$.

References

1. Zou, Y., Elbuluk, M., Sozer, Y.: Stability analysis of maximum power pointtracking (MPPT) method in wind power systems. IEEE Trans. Ind. Appl. **49**(3), 1129–1136 (2013)
2. Bo, Y., et al.: Nonlinear maximum power point tracking control and modal analysis of DFIG based wind turbine. Int. J. Electr. Power Energy Syst. **74**, 429–436 (2016)
3. Carrasco, J.M., et al.: Power-electronic systems for the grid integration of renewable energy sources: A survey. IEEE Trans. Ind. Electron. **53**(4), 1002–1016 (2006)
4. Belgacem, K., Mezouar, A., Massoum, A.: Sliding mode control of a doubly-fed induction generator for wind energy conversion. Int. J. Energy Eng. **3**(1), 30–36 (2013). https://doi.org/10.5923/j.ijee.20130301.05
5. Zadehbagheri, M., Ildarabadi, R., Nejad, M.B.: Sliding mode control of a doubly- fed induction generator (DFIG) for wind energy conversion system. Int. J. Sci. Eng. Res. **4**(11) (2013)
6. Asghar, M.: Performance comparison of wind turbine based doubly fed induction generator system using fault tolerant fractional and integer order controllers. Renew. Energy **116**, 244–264 (2018)
7. Patton, R., Frank, P., Clark, R.: Issues of Fault Diagnosis for Dynamic Systems. Springer, Berlin (2000)
8. Djoudi, A., Bachac, S., Iman-Eini, H., Rekiouae, T.: Sliding mode control of DFIG powers in the case of unknown flux and rotor currents with reduced witching frequency. Electr. Power Energy Syst. **96**, 347–356 (2018)
9. Bekakra, Y., Attous, D.B.: Active and reactive power control of a DFIG with MPPT for variable speed wind energy conversion using sliding mode control. World Acad. Sci. Eng. Technol. Int. J. Comput. Electr. Autom. Control Inf. Eng. **5**(12) (2011)

10. Munteanu, I., Bratcu, A.I., Cutululis, N.-A., Ceanga, E.: Optimal Control of Wind Energy Systems: Towards a Global Approach, 1st edn. Springer, Berlin (2008)
11. Liu, X., Kong, X.: Nonlinear model predictive control for DFIG-basedwind power generation. IEEE Trans. Autom. Sci. Eng. **11**(4), 1046–1055 (2014)
12. Moradi, H., Vossoughi, G.: Robust control of the variable speed wind turbines in the presence of uncertainties: a comparison between H∞ and PID controllers. Energy **90**, 1508–1521 (2015)
13. Kiruthiga, B.: Implementation of first order sliding mode control of active and reactive power for DFIG based wind turbine. Int. J. Inf. Futur. Res. **2**(8), 2487–2497 (2015)
14. Pinghua, X., Dan, S.: Backstepping-based DPC strategy of a wind turbine-driven DFIG under normal and harmonic grid voltage. IEEE Trans. Power Electron. **31**(6), 4216–4225 (2016)
15. Dan, S., Xiaohe, W.: Low-complexity model predictive direct power control forDFIG under both balanced and unbalanced grid conditions. IEEE Trans. Industr. Electron. **63**(8), 5186–5196 (2016)
16. Elghali, S.E.B., et al.: High-order sliding mode control of DFIG-based marine currentturbine. In: 34th Annual Conference of Industrial Electronics, IECON 2008, IEEE (2008)
17. Chen, S.Z., et al.: Integral variable structure direct torque control of doubly fed induction generator. IET Renew. Power Gener. **5**(1), 18–25 (2011)
18. MorfinOnofre, A., et al.: Torque controller of a doubly-fed induction generator im-pelled by a DC motor for wind system applications. IET Renew. Power Gener. **8**(5), 484–497 (2014)
19. Yipeng, S., Heng, N.: Modularized control strategy and performance analysis of DFIG system under unbalanced and harmonic grid voltage. IEEE Trans. Power Electron. **30**(9), 4831–4842 (2015)

New Design of Neural Direct Power Control of DSIM Fed by Indirect Matrix Converter

Y. Moati[1(✉)], K. Kouzi[1], A. Bencherif[2], and S. Bensaoucha[3]

[1] Laboratory of Semiconductors and Functional, Materials, Electrical Engineering Department, University of Laghouat, Laghouat, Algeria
y.moati@lagh-univ.dz, kouzi.univ@gmail.com
[2] Laboratory of Telecommunications, Signals and Systems, Electrical Engineering Department, University of Laghouat, Laghouat, Algeria
a.bencherif@lagh-univ.dz
[3] Laboratoire d'Etude et Dveloppement des Matriaux Semi-Conducteurs et Dilectriques (LeDMaSD), Electrical Engineering Department, University of Laghouat, Laghouat, Algeria
s.bensaoucha@lagh-univ.dz

Abstract. This paper present a new direct power control (DPC) of dual stator induction motor (DSIM) fed by indirect matrix converter (IMC) based on the neural network controller, this study aims to improve the control performance (flux, power and speed), the use of indirect matrix converter for to controlled the input power factor, reduce the power ripple and to eliminate dc-link capacitor. Simulation results obtained with matlab-simulink are presented to show the improvement in performance of the proposed method.

Keywords: Dual stator induction motor (DSIM)
Indirect matrix converter (IMC) · Space vector modulation (SVM)
Hysteresis control · Neural controller · Direct power control (DPC)

1 Introduction

In recent years, the multiphase induction machines have received a great attention in the field of power electronics because, by comparison with particularly attractive features, they offer an interesting alternative to reducing stress on switches and windings [1, 2]. DSIM is among the most used multiphase machines. Modeling, analysis and control of such machines are studied in [3, 4]. Recent research improvement in the control of DSIM have been presented in [5, 6]. Direct torque control (DTC) is the most one of these control strategies, Different types of the DTC have been introduced in the literature for DSIM [7].

The use of matrix converter in the DTC was initially presented by "Casadei" in [8], this method was extended to the indirect matrix converter (IMC) in [9]. DTC was extended to direct control of the active power in the generator [10].

© Springer Nature Switzerland AG 2019
M. Hatti (Ed.): ICAIRES 2018, LNNS 62, pp. 216–224, 2019.
https://doi.org/10.1007/978-3-030-04789-4_24

In this paper, we apply the DPC for the DSIM fed by IMC based on neural network, using this control method, it is possible to combine the advantages of IMC with that of the DPC schemes and neural controller. The simulation study clearly shows the robustness and performances of the presented DPC under unity power factor conditions and minimizes flux and power ripple while keeping constant switching frequency.

2   System Modeling

2.1   DSIM Model

The DSIM studied in this paper is a machine that includes two systems of three-phase star-coupled windings fixed to the stator and out of phase with each other at an angle γ ($\gamma = 30°$) and a mobile rotor similar to that of a conventional asynchronous machine [3]. The state-space formula is a fourth-order model [2, 3]:

$$\frac{dX}{dt} = AX + BU \tag{1}$$

With $X = \begin{bmatrix} \varphi_{ds1} & \varphi_{ds2} & \varphi_{qs1} & \varphi_{qs2} & \varphi_{dr} & \varphi_{qr} \end{bmatrix}^T$ is the case vector.
$U = \begin{bmatrix} V_{ds1} & V_{ds2} & V_{qs1} & V_{qs2} & 0 & 0 \end{bmatrix}^T$ is the control vector.

$$A = \begin{bmatrix}
-\frac{R_{s1}}{L_{s1}} + \frac{R_{s1}L_a}{L_{s1}^2} & \frac{R_{s1}L_a}{L_{s1}L_{s2}} & \omega_s & 0 & \frac{R_{s1}L_a}{L_rL_{s1}} & 0 \\
\frac{R_{s2}L_a}{L_{s1}L_{s2}} & -\frac{R_{s2}}{L_{s2}} + \frac{R_{s2}L_a}{L_{s2}^2} & 0 & \omega_s & \frac{R_{s2}L_a}{L_rL_{s2}} & 0 \\
-\omega_s & 0 & -\frac{R_{s1}}{L_{s1}} + \frac{R_{s1}L_a}{L_{s1}^2} & \frac{R_{s1}L_a}{L_{s1}L_{s2}} & 0 & \frac{R_{s1}L_a}{L_rL_{s1}} \\
0 & -\omega_s & \frac{R_{s1}L_a}{L_{s1}L_{s2}} & -\frac{R_{s2}}{L_{s2}} + \frac{R_{s2}L_a}{L_{s2}^2} & 0 & \frac{R_{s1}L_a}{L_rL_{s2}} \\
\frac{R_rL_a}{L_rL_{s1}} & \frac{R_rL_a}{L_rL_{s2}} & 0 & 0 & -\frac{R_r}{L_r} + \frac{R_rL_a}{L_r^2} & -(\omega_s - \omega_r) \\
0 & 0 & \frac{R_rL_a}{L_rL_{s1}} & \frac{R_rL_a}{L_rL_{s2}} & -(\omega_s - \omega_r) & -\frac{R_r}{L_r} + \frac{R_rL_a}{L_r^2}
\end{bmatrix}$$

$$B = \begin{bmatrix}
1 & 0 & 0 & 0 & 0 & 0 \\
0 & 1 & 0 & 0 & 0 & 0 \\
0 & 0 & 1 & 0 & 0 & 0 \\
0 & 0 & 0 & 1 & 0 & 0 \\
0 & 0 & 0 & 0 & 0 & 0 \\
0 & 0 & 0 & 0 & 0 & 0
\end{bmatrix}$$

$T_{s(1,2)} = \frac{R_{s(1,2)}}{L_{s(1,2)}}$, $T_r = \frac{R_r}{L_r}$ are the stator time constant of the two stars and rotor time constant respectively.

The mechanical equation of the machine is written:

$$\frac{d\Omega}{dt} = \frac{1}{J}(C_{em} - C_r - f_f\Omega) \tag{2}$$

Electromagnetic torque established by the DISM it given by:

$$C_{em} = P \frac{L_m}{L_m + L_r} \left(\Psi_{dr}(i_{qs1} + i_{qs2}) - (i_{ds1} + i_{ds2})\Psi_{qr} \right) \tag{3}$$

Where: P is the number of poles pairs.

2.2   Modelling of Indirect Matrix Converter (IMC)

The indirect matrix converter is an alternative converter is used for directly power conversion ac/ac, the conversion is separate by two stages: rectifier and inverter with no dc-link capacitor. The topology of this converter is illustrated in Fig. 1. Six bidirectional switches S1-S6 create rectifier stage and the inverter stage has a standard and three-phase voltages source topology based on six switches [13, 14].

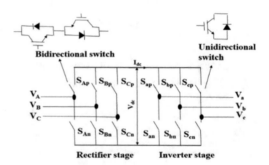

Fig. 1. Indirect matrix converter.

The transfer function of IMC is written as

$$\begin{bmatrix} V_a \\ V_b \\ V_c \end{bmatrix} = \begin{bmatrix} S_{ap} & S_{an} \\ S_{bp} & S_{bn} \\ S_{cp} & S_{cn} \end{bmatrix} \cdot \begin{bmatrix} S_{Ap} & S_{Bp} & S_{Cp} \\ S_{An} & S_{Bn} & S_{Cn} \end{bmatrix} \cdot \begin{bmatrix} V_A \\ V_B \\ V_C \end{bmatrix} == S_{inv} \cdot S_{rec} \cdot V_{in} \tag{4}$$

S_{inv}, S_{rec} are the transfer functions of two stages inversion and rectification of IMC [14] and S_{jk} defines the connection function as follow

$$S_{jk} = \begin{cases} 1 \ \text{if the switch } S_{jk} \ \text{is on} \\ 0 \ \text{if the switch } S_{jk} \ \text{is of} \end{cases} \tag{5}$$

The two space vectors of the input current I_{in} and output voltage V_o

$$I_{in} = \frac{2}{3} \left(I_A + I_B \, e^{j\frac{2\pi}{3}} + I_C \, e^{j\frac{4\pi}{3}} \right), \quad V_o = \frac{2}{3} \left(V_a + V_b \, e^{j\frac{2\pi}{3}} + V_c \, e^{j\frac{4\pi}{3}} \right) \tag{6}$$

3   Direct Power Control Strategy (DPC)

The present DPC of DSIM in this paper it is similar to DTC, and then the DTC is based on the torque and the flux control, although the DPC it based on the power and flux control [11].

$$P_{out} = T_e \, \omega_m = T_e \frac{\omega_r}{P} \tag{8}$$

The expression of the torque it given by:

$$T_e = \frac{2}{3} K_c \varphi_s \varphi_r \sin(\gamma) \tag{9}$$

Where, γ is the angle between the stator and rotor flux linkage, K_c is the constant depending on the parameter of motor. Substituting the torque in (9) with (8), the expression of the output power becomes:

$$P_{out} = \frac{2}{3} \omega_r K_c \varphi_s \varphi_r \sin(\theta_s - \theta_r) \tag{10}$$

The reference power is given by

$$P_{ref} = T_e^* \omega_{ref} \tag{11}$$

DPC is based on the orientation of the stator flux. The latter can be expressed in referential linked to the stator of the machine by the equation

$$\varphi_s = \int_0^t (V_s - R_s I_s) \, dt + \varphi_{s0} \tag{12}$$

In the case where a non-zero vector is applied during an interval $[T_0, T_e]$, we can neglect the fall of tension $(R_s \cdot I_s)$, this last equation is then written:

$$\varphi_s = V_s T_e + \varphi_{s0} \tag{13}$$

So the flux vector move with the amount $V_s T_e$

3.1   Control of Rectifier

The goal is to control the power factor of the input and to obtain the maximum value of the virtual V_{dc} bus voltage, based on SVM, the reference vector can be synthesized with these two adjacent vectors I_γ, I_δ and a null vector $I_0 \, I_0$ [12, 14].

$$I_{in} = d\gamma I\gamma + d_\delta I_\delta \qquad (14)$$

We define the cyclic ratios

$$\begin{cases} d_\gamma = m_c \sin(\theta_{in}) \\ d_\delta = m_c \sin(\frac{\pi}{3} - \theta_{in}) \\ d_0 = 1 - d_\gamma - d_\delta \end{cases} \qquad (15)$$

To maximize the voltage V_{dc}, the null vector is eliminate. The two cyclic ratios are normalized according to the following equation [12]

$$\begin{cases} d_\gamma' = \frac{d_\gamma}{d_\gamma + d_\delta} = \frac{\sin(\theta_{in})}{\cos(\frac{\pi}{6} - \theta_{in})} \\ d_\delta' = \frac{d_\delta}{d_\delta + d_\gamma} = \frac{\sin(\frac{\pi}{3} - \theta_{in})}{\cos(\frac{\pi}{6} - \theta_{in})} \\ d_0 = 0 \quad and \quad d_\gamma' + d_\delta' = 1 \end{cases} \qquad (16)$$

3.2   Control of Inverter

For the control of the inverter, we apply the table of DPC, therefore according to the error of the power and flux as well as its sector, the table of DPC chose the vector inverter, which makes it possible to ensure the respect of the bands of the errors (power and flux), is reported in switching Table 1.

Table 1. Switching table of DPC

$\Delta\varphi_s$	ΔP	S_1	S_2	S_3	S_4	S_5	S_6
1	1	V_2	V_3	V_4	V_5	V_6	V_1
	0	V_7	V_0	V_7	V_0	V_7	V_0
	− 1	V_6	V_1	V_2	V_3	V_4	V_5
0	1	V_3	V_4	V_5	V_6	V_1	V_2
	0	V_0	V_7	V_0	V_7	V_0	V_7
	− 1	V_5	V_6	V_1	V_2	V_3	V_4

4   DPC oF DSIM Obtained via the Artificial Neural Network Technique

To guarantee the speed control we used a neural controller, which is insensible to machine parametric variation and in order to reduce the response time for speed. Figure 2 shows the schematic model of neural controller based speed regulator.

The principle of neural direct power control is similar to traditional DPC. However, the classical PI controller replaced by neural controller. The general diagram Of DPC with IMC based on neural controller of DSIM is shown in Fig. 3.

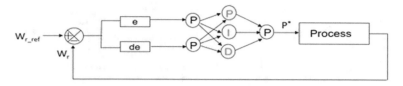

Fig. 2. Structure of neural network controller

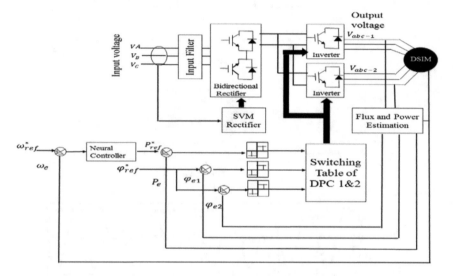

Fig. 3. Direct power control based on neural controller

5   Simulation Results

The behavior of the structure of the DPC based on neural controller applied to a 4.5 kW machine is simulated using Matlab/Simulink environment with a hysteresis band of ±0.1 W for the power comparator, ±0.01 Wb for the flux comparator.

The following figures illustrate the simulation results. The ability of the DSIM well found with the proposed DPC (Fig. 4a). Control scheme in tracking a reference power was confirmed. Control scheme in tracking a reference power was confirmed. The hysteresis band is practically respected for all the speeds considered. In steady state, the actual power is almost equal to the reference power. The speed follows its reference (Fig. 4b). The variation of the power creates a small variation in the speed then the speed follows the reference, has been shown using direct power control strategy. Good performance in terms of stator current (Fig. 4c). As it can be seen, stator current shows a sinusoidal waveform.

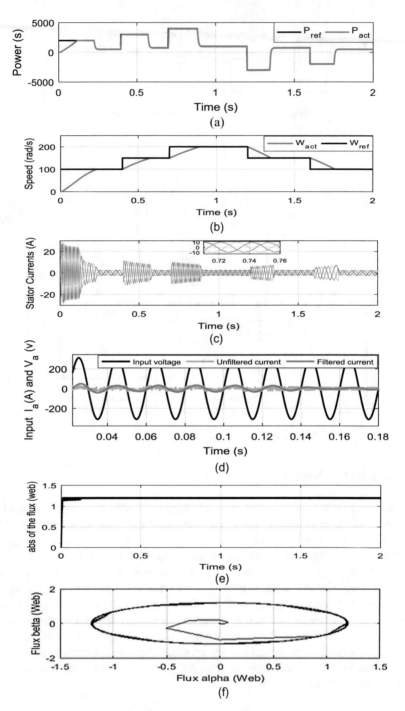

Fig. 4. (a) Power (w), (b) speed (rad/s), (c) stator currents (A), (d) input voltage and filtered current, (e) stator flux magnitude (web), (f) stator flux vector (web), (g) output phase voltage (v)

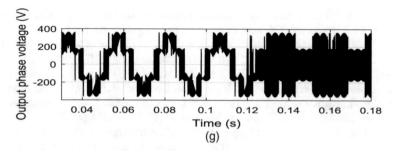

Fig. 4. (*continued*)

The input current is in phase with the input voltage (Fig. 4d), that is, the input power factor is equal to one. The flux virtually respects its hysteresis band (Fig. 4e, f) and follows its reference with some overshoot at low speeds. The output voltage and the current (Fig. 4g) vary in amplitude and frequency according to the speed set point and the power. The performance of IMC to reduced ripple in power and flux stator. The drawback of variable switching frequency in the classical inverter DPC is solved with IMC.

6 Conclusion

This paper investigated the control of IMC by using the DPC technique. This latter with its simple structure to control the DSIM presents a solution to the problems of robustness (flux, Power). Although the implementation of this method is a little more complex compared to the DPC with DMC, but it gives good performances in both transient and steady state behaviors and the best quality of the input current and the input filtered current. The control of input power factor is independent of the output current, the neural controller confirms its good performance when controlling the dual stator induction motor speed, and this is clearly shown in the obtained simulation results. The advantages provided by the proposed control scheme make this latter competitive with regard to the other proposed control techniques.

References

1. Levi, E.: Multiphase electric machines for variable-speed applications. IEEE Trans. Ind. Electron. **55**(5), 1893–1909 (2008)
2. Basak, S., Chakraborty, C.: Dual stator winding induction machine: problems, progress, and future scope. IEEE Trans. Ind. Electron. **62**(7), 4641–4652 (2015)
3. Kianinezhad, R., Nahid, B., Baghi, L., Betin, F., Capolino, G.A.: Modeling and control of six-phase symmetrical induction machine under fault condition due to open phases. IEEE Trans. Ind. Appl. **55**(5), 1966–1977 (2008)
4. Zhao, Y., Lipo, T.A.: Space vector PWM control of dual three-phase induction machine using vector space. IEEE Trans. Ind. Appl. **31**(5), 1100–1108 (1995)

5. Federico, B.: Recent advances in the design, modeling and control of multiphase machines—Part 1. IEEE Trans. Ind. Electron. (2015). https://doi.org/10.1109/TIE.2447733
6. Federico, B.: Recent advances in the design, modeling and control of multiphase machines—Part 2. IEEE Trans. Ind. Electron. (2015). https://doi.org/10.1109/TIE.2448211
7. Bojoi, R., Farina, F., Griva, G., Profumo, F., Tenconi, A.: Direct torque control for dual three-phase induction motor drives. IEEE Trans. Ind. Appl. **41**(6), 1627–1636 (2005)
8. Casadei, D., Serra, G., Tani, A.: The use of matrix converters in direct torque control of induction machines. IEEE Trans. Ind. Electron. **48**(6), 1057–1064 (2001)
9. Li, Y., Liu, W.: A novel direct torque control method for induction motor drive system fed by two-stage matrix converter with strong robustness for input voltage. In: Proceedings of the 2nd IEEE Conferences on Industrial Electronics and Applications, pp. 698–702 (2007)
10. Tremblay, E., Atayde, S., Chandra, A.: Comparative study of control strategies for the doubly fed induction generator in wind energy conversion systems: a DSP-based implementation approach. IEEE Trans. Sustain Energy **2**(3), 288–299 (2011)
11. Zolfaghari, M., Taher, S.A., Munuz, D.V.: Neural network-based sensorless direct power control of permanent magnet synchronous motor. Ain Shams Eng. J. **7**, 729–740 (2016)
12. Benachour, A.: Commande sans Capteur basée sur DTC d'une Machine Asynchrone alimentée par Convertisseur Matriciel. PhD. Thesis, Ecole Nationale Polytechnique (2017)
13. Kolar, J.W., Schafmeister, F., Round, S.D., Ertl, H.: Novel three-phase AC–AC sparse matrix converters. IEEE Trans. Power Electron. **22**(5), 1649–1661 (2007)
14. Li, Y., Liu, W.: A novel direct torque control method for induction motor drive system fed by two-stage matrix converter with strong robustness for input voltage. In: Proceedings of the 2nd IEEE Conference on Industrial Electronics and Applications, pp. 698–702 (2007)

Design of Stand-Alone PV System to Provide Electricity for a House in Adrar, Algeria

T. Touahri[(✉)], N. Aoun, R. Maouedj, S. Laribi, and T. Ghaitaoui

Unité de Recherche en Energies renouvelables en Milieu Saharien,
UERMS, Centre de Développement des Energies Renouvelables, CDER,
01000 Adrar, Algeria
Touahri01@gmail.com

Abstract. This paper presents a design of stand-alone photovoltaic (PV) system to generate electricity in a house in Adrar, 1 year recorded solar radiation is used for the design of PV solar energy system. The proposed system takes into account the meteorological data of the site of Adrar and the needs of electric charge of the house; in 2015 annual average solar energy resource available is 6.45 KWh/m^2/day and energy which are consumed by electrical house loads in a course of 1 year is 2575.07 kWh. The paper provides a general understanding of a photovoltaic system to electrify individual housing in rural areas and how can be practically applied solar energy to power the house.

Keywords: Solar energy · PV systems · Power · Stand-alone

1 Introduction

The renewable energies are developed with a strong growth these last years. In the next year, the durable energy system will be using the rational of the classic sources and on a recourse increased to renewable energies. It is interesting to exploit the electricity consumption of a location according to the needs. The provision of electricity through the use of renewable energy resource is to assure the safety of consumers and to keep the environment clean [1].

Algeria has a large stock of the sun's energy due to its geographical location. Where, the sun duration exposure exceed 2,000 h yearly across the country, in the most nation received 5 Kwh of daily energy on a horizontal surface in 1 m. That is around 1.700 kWh/m^2/year in the north and 2.263 kWh/m^2/year in the southern part of the country [2]. The high costs of investment in PV technology obstructed spread in the market, that must estimate the parameters of the sizing of PV systems is very useful to imagine PV systems optimization, economic, especially in remote areas in developing countries [3].

This paper presents a design for a stand-alone PV system to provide the required electrical energy for a single residential home in Adrar, Algeria.

© Springer Nature Switzerland AG 2019
M. Hatti (Ed.): ICAIRES 2018, LNNS 62, pp. 225–232, 2019.
https://doi.org/10.1007/978-3-030-04789-4_25

2 Composition of Stand-Alone PV System

Figure 1 show the block diagram of the study stand-alone photovoltaic system and the main components of the system are:

- The PV array: convert the sunlight directly into DC electrical power.
- The charge controller: regulates the voltage and current coming from the PV panels going to battery and prevents battery overcharging and prolongs the battery life.
- The inverter is used to convert DC currant of PV panel into AC currant for AC appliances.
- The battery: stores energy for supplying to electrical appliances when there is a demand.
- The electrical load: electrical appliances that connected to solar PV system such as lights, radio, TV, computer, refrigerator, etc.

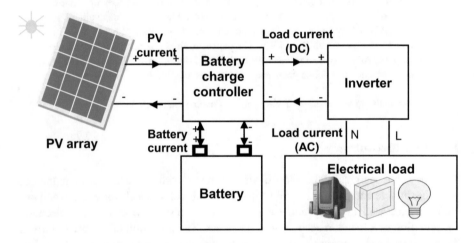

Fig. 1. Schematic diagram of a stand-alone PV system.

3 Solar Radiation Data

The monthly average solar radiation on horizontal surface and on tilted surface of the site Adrar is shown in Fig. 2. As can be seen, the solar radiation on horizontal surface is limited between 3.87–8.09 kw/m^2/day and between 5.45–7.57 kw/m^2/day for the solar radiation on tilted surface.

Figure 3 show the monthly average daylight hours, the daylight hours in Adrar reaches its minimum in December with 10.9 h/day and its maximum in June with 13.3 h/day. So, the monthly average daylight hour in the whole year is 10.9 h [5].

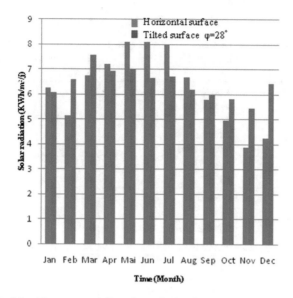

Fig. 2. Monthly average daily solar radiation for Adrar, Algeria on 2015.

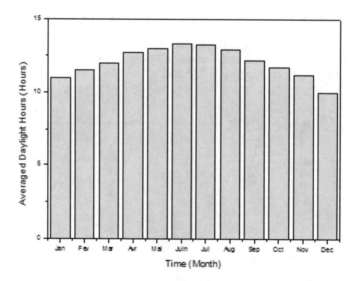

Fig. 3. The monthly average daylight hours.

4 Electrical House Loads

The house used simple and essential electrical appliances that do not consume large quantities of electrical energy, the house appliances are lamps, TV, refrigerator, computer, washing machine and water pump. The energy consumption of the electrical appliances is shown in Table 1.

Table 1. Household power consumption.

Electrical load	Number of units	Operating hours per day	Wattages (W)	Daily electricity load (KWh/day)	Annual electricity load (KWh/year)
Lamp	5	4 lamps from 18 to 24 1 lamps from 0 to 6	60	3.6	1314
TV	1	From 16 to 20	80	0.32	116.8
Refrigerator	1	From 0 to 24	100	2.4	876
Computer	1	From 20 to 23	65	0.195	71.17
Washing machine	1	From 12 to 14	150	0.3	109.5
Water pump	1	From 12 to 14	120	0.24	87.6
			Total	7.055	2575.07

The load profile for a typical day is shown in Fig. 4.The total house residential load profile statistics are:

✓ The maximum hourly power consumption = 0.37 kW,
✓ The average hourly power consumption = 0.24 kW and
✓ The total daily power consumption is 7.05 kW.

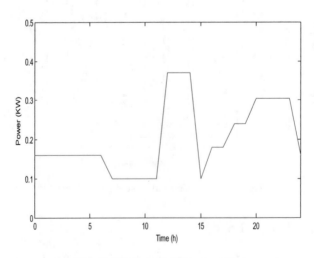

Fig. 4. The household load profile.

5 Sizing System Components

The sizing of system component required to calculate the energy consumed by each electrical appliance used in the house.

Equation (1) is used to compute the Energy consumed per day [6].

$$E_c = P * Q * h \tag{1}$$

Where

E_c: Energy consumed by the load (Wh/day)
P: Electric power (w)
Q: Number of appliance used
h: Hours Used (h)

A. *Photovoltaic Module Sizing*

Different size of PV modules will produce different amount of power. To find out the sizing of PV module, the total peak watt produced needs [7]. The peak power (Wp) of the PV generator is obtained from the Eq. (2):

$$P_p = \frac{E_c}{k.I_r} \tag{2}$$

Where the coefficient k takes into account the following factors:

- Meteorological uncertainty;
- Uncorrected tilt modules depending on the season;
- The operating point of the modules that is rarely optimal, which can be exacerbated by falling characteristics of the modules, the module efficiency loss over time (aging and dust);
- The performance of charge and discharge cycles of the battery (80%);
- Charger efficiency (90 to 95%);
- Losses in cables and connections.

P_p: The peak power (Wp)
I_r : The average daily irradiation (KWh/m^2/day)

For systems with Battery Park, the coefficient k is generally between 0.55 and 0.75. The approximate value for the systems with battery used is often 0.65 [8]. Substituting these values in the above equation, we obtain the peak power of the PV generator:

$$P_P = \frac{7.05}{0.65 * 6.4} = 1.69 \, KW_p$$

The size of PV modules (N_M) required meeting the load demand can be determined by using the following equation:

$$N_M = \frac{P_p}{P_{CM}}$$

(3)

A mono-crystalline PV module type Techno sun SYP80S-Mwith peak power = 80 Wp, Isc = 5 A, Voc = 21.6 V and area = 0.64 m^2 are used to install the power of house, the number of PV modules is obtained as:

$$N_M = \frac{1.69 * 10^3}{80} = 21.12$$

The total number of modules used to provide energy for the residential house is 21 modules, the configuration of PV array can be connected 3 modules in series and 7 ranges will be connected in parallel

B. *Battery Sizing*

The battery type recommended for using in solar PV system is deep cycle battery. The storage capacity of the battery (C_D) can be determined using the following equation [7, 8]:

$$C_D = \frac{E_C \cdot N}{D.U}$$

(4)

Where
N: Number of autonomy days.
D: Maximum permissible of discharge of the battery
U: The nominal voltage of the battery

$$C_D = \frac{7.05 * 10^3 * 3}{0.8 * 24} = 1101,56 \text{ Ah}$$

So the number of battery is 10 (2 in series and 5 in parallel) of a 12 V with capacity of 110 Ah were required. These batteries will provide adequate storage to meet the daily energy requirements.

C. *Controller Sizing*

The battery charge controller is necessary to protect the battery block against deep discharge and over charge [9]. The solar charge controller is obligatory to charge the battery with safety and preserve their longer lifetime. It has to be capable of wearing the short circuit current of the PV array [4]. The sizing of the solar charge controller is to take the Isc of the PV module and multiply it by a factor of 1.3 and the number of string, in this case, it can be chosen to handle 48 A (5 A*7*1.3) and the DC voltage about 24 V.

D. *Inverter Sizing*

An inverter is used in the system where AC power output is needed., the inverter must be sized to provide 20% higher than the rated power of the total AC loads you wish to run simultaneously at any one moment, the rated output power of the inverter becomes $(1.2 * 0.37 = 0.44$ kW). The specifications of the required inverter will be 450 W, 220 VAC, and 50 Hz.

6 Conclusion

The optimum exploitation of renewable energy sources is very important to all countries of the world, especially in the developing countries, like Algeria. Solar photovoltaic energy is one of the important renewable energy technologies, due to their high reliability and safety.

In this paper a stand-alone system is designed and analyzed for the electrification for a single residential house in a rural area located in the city of Adrar, Algeria. The proposed photovoltaic system required a minimum 21 photovoltaic modules of 80 Wp, 10 unit of battery capacity of 110 Ah and an inverter size of 450 W for the design.

Finally, we conclude that the provision of electricity to the rural site using the renewable energy sources in Adrar, Algeria creates many opportunities and new perspectives in national development through the rural Development.

Acknowledgements. The authors thank the team Potential Solar Energy and Wind, Photovoltaic Conversion Division, Research Unit in Renewable Energies in the Saharan Medium for provide the meteorological data.

References

1. Belmili, H., Ayad, M., Berkouk, E.M., Haddadi, M.: Optimisation de dimensionnement des installations photovoltaïques autonomes—Exemples d'applications, éclairage et pompage au fil du soleil. In: Revue des Energies Renouvelables CICME'08 Sousse, pp. 27–39 (2008)
2. Ministry of energy and mines.Algeria. Guidelines to renewable energies. In: *Edition 2007* (2007)
3. Abdulateef, J., Sopian, K., Kader, W., Bais, B., Sirwan, R., Bakhtyar, B., Saadatian, O.: Economic analysis of a stand-alone PV system to electrify a residential home in Malaysia. In: Heat Transfer, Thermal Engineering and Environment; Advances in fluid mechanics & Heat & Mass Transfer. by WSEAS, Greece, pp. 169–174 (2012)
4. Nafeh, E.S.A.: Design and economic analysis of a stand-alone PV system to electrify a remote area household in Egypt. Open Renew. Energy J. 33–37 (2009)
5. Bentouba, S., Hamouda, M., Slimani, A., Pere Roca, C., Bourouis, M., Coronas, A., Draoui, B., Boucherit, M.S.: Hybrid system and environmental evaluation case house in south of Algeria. Energy Procedia **36**, 1328–1338 (2013)
6. Christian, K., Emmanuel, E.: Design of a photovoltaic system as an alternative source of electrical energy for powering the lighting circuits for premises in Ghana. J. Electr. Electron. Eng. **2**(1), 9–16 (2014)

7. Leonics: How to design solar PV system, April 2016. http://www.leonics.com/support/article2_12j/articles2_12j_en.php. Accessed 3 April 2016
8. Activity Report Research: Sizing the PV system. Unit of Research in Renewable Energies in Saharan Medium Adrar (2009)
9. Marwan, M., Ibrik, I.H.: Techno-economic feasibility of energy supply of remote villages in Palestine by PV-systems, diesel generators and electric grid. Renew. Sustain. Energy Rev. **10**, 128–138 (2006)

Extremum Seeking and P&O Control Strategies for Achieving the Maximum Power for a PV Array

B. K. Oubbati[1(✉)], M. Boutoubat[2], M. Belkheiri[1], and A. Rabhi[3]

[1] LTSS Laboratory, Laghouat University, Laghouat, Algeria
i.oubbati@lagh-niv.dz, m.belkheiri@lagh-univ.dz
[2] LACoSERE Laboratory LaghouatUniversity, Laghouat, Algeria
boutoubat90@yahoo.fr
[3] MIS Laboratory, Picardie University, Amiens, France
abdelhamid.rabhi@u-picardie.fr

Abstract. The aim of this paper is to study and to compare between two control strategies for tracking the Maximum Power Point (MPP) for a photovoltaic (PV). The PV array is connected to the load through a boost converter. In fact, the Maximum Power Point Tracking (MPPT) is achieved by using two different methods for searching the Maximum Power point (MPP). These strategies methods are Perturb and observe (P&O) and Extremum Seeking Control (ESC). The aim goal of these strategies is to predict the maximum power point. Simulation results show the effectiveness of the control strategies applied to the studied system.

Keywords: Extremum Seeking Control (ESC) · Perturb and observe technique MPPT · PV array

1 Introduction

The production of energy is a challenge of great importance for the coming years. Indeed, the needed energy of industrialized societies is steadily increasing. In addition, the countries development needs more and more energy to carry out their development. Today has seen a growing interest in the research in the field of renewable energy sources, especially the solar energy which has remarkable advantages over traditional energy sources such as (oil, petrol). Integration of solar PV in the gird and in standalone generation systems has increased. Besides, the power generated by PV modules depends on solar irradiation levels and temperatures. So, a Maximum Power Point Tracking (MPPT) controller is required to ensure that the maximum power is generated to the load. In literature, There were several proposed MPPT algorithms such as: Perturbation and Observation (P&O) [1]; Incremental Conductance (IC) [2]; Fuzzy Logic controller (FLC) [3] and Fractional short-circuit voltage (FSCV) [2].

These cited algorithms of MPPT strategy have been successfully demonstrated in tracking the MPP under uniform insolation where only one maximum power point (MPP) exists in the power-against – voltage (P-V) curve [4, 5]. However, in some cases

© Springer Nature Switzerland AG 2019
M. Hatti (Ed.): ICAIRES 2018, LNNS 62, pp. 233–241, 2019.
https://doi.org/10.1007/978-3-030-04789-4_26

it can be very difficult to find an appropriate reference value witch will be used to achieve the MPPT. For illustration, the energy efficiency of photovoltaic system depends on the irradiation and the temperature. If we is desire to test the maintaining of the optimal efficiency of his system, it is necessary to change one of these perturbations. Extremum Seeking Control is a family of control design approaches whose purpose is to autonomously find optimal point of the system behavior for the closed-loop system. Besides, as it is known, the power characteristic of a photovoltaic (PV) array is nonlinear [6] ESC will be successfully used to stabilize the system at its optimum point. Therefore, the Extremum seeking control is largely used to achieve the real-time optimization for dynamic systems and tracking a varying maximum or minimum (extremum, or optimum value) of a performance fitness functions. It is a very interesting methodology in practice because it does not necessitate any knowledge of the process dynamics or model of the system [7].

The work of this paper focusses on a comparative study between two control strategies applied to PV array for achieving the MPPT. The scheme of the studied system is presented in Fig. 1.

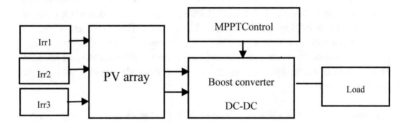

Fig. 1. The layout of the studied system.

2   Photovoltaic Array Modelling

The equivalent circuit of a classical PV cell is shown in Fig. 2.

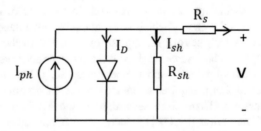

Fig. 2. PV cell equivalent circuit.

Where, I_{ph} represents the cell photo current; R_{sh} and R_s are the intrinsic shunt and series resistances of the cell, respectively.

The photo-current I_{ph} is expressed by the following equation:

$$I_{ph} = [I_{sc} + K_i(T - 298)] \times I_r/1000 \tag{1}$$

The reverse saturation current I_{rs} is given by:

$$I_{rs} = I_{sc}/\left[\exp\left(\frac{qV_{oc}}{N_s knT}\right) - 1\right] \tag{2}$$

The saturated current (I_0) that varies with the cell temperature, is expressed:

$$I_0 = I_{rs}\left[\frac{T}{T_r}\right]^3 \exp\left[\frac{qE_{g0}}{nk}\left(\frac{1}{T} - \frac{1}{T_r}\right)\right] \tag{3}$$

The output current of the PV is written as follow:

$$I = N_p \times I_{ph} - N_p \times I_0 \times \left[\exp\left(\frac{V/N_s + I \times R_s/N_p}{n \times V_t}\right) - 1\right] - I_{sh} \tag{4}$$

Where,

$$V_t = \frac{k \times T}{q} \tag{5}$$

And,

$$I_{sh} = \frac{V \times N_p/N_s + I \times R_S}{R_{sh}} \tag{6}$$

With;

Np: Number of the PV modules connected in parallel.; R_S: Serie resistance (Ω).; N_S: Number of the PV modules connected in series.; R_{Sh}: Shunt resistance (Ω).; V_t: Diode thermal voltage (V).

2.1   Characteristic of PV Array Model

The power-voltage and current-voltage characteristics for different values of temperature and irradiation are obtained from the simulation and are shown in Fig. 3.

The (I-V) and the (P-V) characteristics of the PV array for different values of temperature are presented in Fig. 4.

From Figs. 3 and 4, we can see clearly the effect of the variation of weather conditions. The PV works practically near its MPP for important value of irradiations. In addition, the PV array works also near its optimum values at low temperature values.

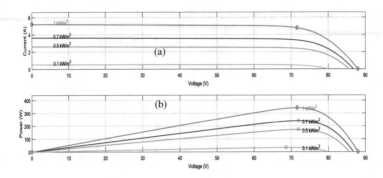

Fig. 3. PV characteristics for different values of irradiation: (a) I-V characteristics, (b) P-V characteristics

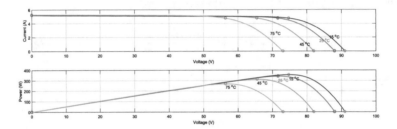

Fig. 4. PV characteristics for different values of temperature.

3   MPPT Strategies

In this study, the two MPPT strategies are fully achieved by using the Extremum seeking control and (P&O) technique. The general control scheme of the studied system is presented in (Fig. 5).

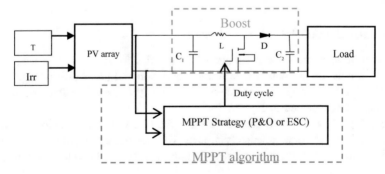

Fig. 5. Control scheme of the PV array by using the (Extremum Seeking Control) or (P&O) strategy.

The algorithms of the applied MPPT strategies (P&O) and (ESC) are developed here after:

3.1   Extremum Seeking Control

Extremum Seeking Control is a real-time optimization algorithm and an adaptive control tool [8, 9], which resolves the problem of tracking the maximum or the minimum of a given performance function. In fact, generally, it attempts to determine the extremum value of an unknown nonlinear performance function [10]. As it is known in literature, the conventional control algorithm deals with the problem of equilibrium of a system about a known reference trajectory that easily determined, while reaching certain design criteria. However, in some cases it can be very difficult to find a required reference trajectory [11, 12]. Between the important various applications of ESC we can found the Global Maximum Power Point Tracking (GMPPT) which can seek and track the MPP. The basic algorithm of the (ESC) is detailed in Fig. 6.

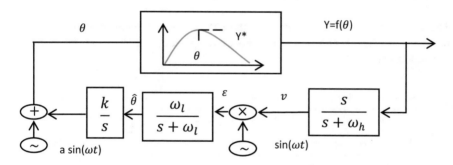

Fig. 6. Block diagram of the Extremum Seeking Control algorithm.

To obtain the optimal performances of the ESC algorithm, perturbation frequency (ω), its amplitude (a), gradient update law gain (k), and the cut-off frequencies (ωh) and (ωl) for the LPF and the HPF, must be tuned and calibrated adequately. The general model equations of the ESC are inspired from the literature and they are formulated as follow:

$$
\begin{cases}
y = f\left(\hat{\theta} + a\sin(\omega t)\right) \\
\hat{\theta} = -kv \\
\varepsilon = v * 1^{-1}\left\{\frac{\omega_l}{s + \omega_l}\right\} \\
v = \left(y * 1^{-1}\left\{\frac{s}{s + \omega_h}\right\}\right)\sin(\omega t)
\end{cases}
\tag{7}
$$

3.2   Perturb and Observe (P&O) Technique

The (P&O) algorithm is depicted in Fig. 7.

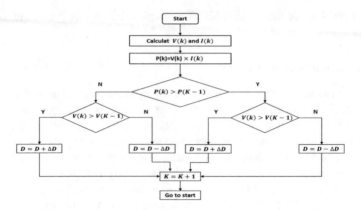

Fig. 7. Perturb and observe MPPT algorithm

4 Simulation Results and Discussions

The obtained simulation results from the use of the Extremum seeking control are presented in Figs. 8 and 9, for a fixed of temperature (25 °C) and for a different irradiation values.

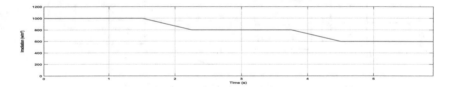

Fig. 8. The different irradiation values.

From the obtained simulation, one can that, one can remark that the maximum power is reached for different value of irradiations. In fact, the generated power to the load is equal practically to the maximum power provided by the PV array with a little difference which is due to the system losses (see Fig. 9(c)). In addition, the duty cycle value varies in such a way that the MPP is extracted from the PV (see Fig. 9(b)).

The same scenario of the irradiation variation (Fig. 8) is applied for testing the performances of the P&O technique. The obtained simulation results for fixed temperature of 25 °C and different irradiation values are presented in Fig. 10. From the simulation results that the P&O technique has permitted to the system to follow the MPP. In fact, the produced power to the load is equal to that produced by the PV array (see Fig. 10(a)) for different values of the irradiations and a fixed value of the temperature. In addition system responses track their optimum value according to the MPP technique through the use of the P&O technique. By comparing between the system responses (especially Fig. 9(c) and Fig. 1(a)) obtained by using the two studied techniques (ESC and P&O), one can remark that:

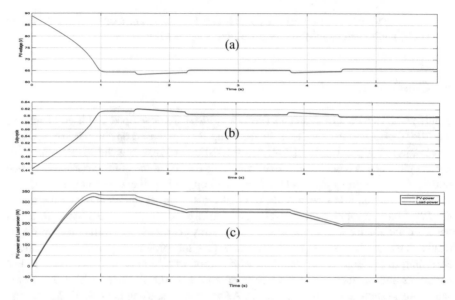

Fig. 9. Different responses of the studied system with (ESC): (a) PV-voltage (V), (b) Duty cycle, (c) PV-power (W) and Load-Power (W).

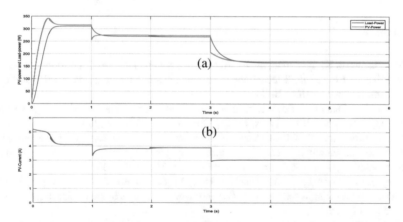

Fig. 10. Different responses of the studied system with (P&O): (a) PV-power (W) and load-power (W), (b) PV-current (A).

- From the responses profiles obtained through the ESC technique, one can remark that the produced power varies slowly according to the irradiations variation with a good transition regime (without any fast variation of the power produced by the PV array which due to the stabilizing system achieved by the ESC).
- While, the responses profiles obtained by using the P&O show that the system responses change rapidly with a good time response but with a fast variation of the produced PV array power. So and due to the fast variation of the PV array power,

the produced power at the load side cannot follow exactly its optimum reference in order to achieved exactly the best optimum value MPP.

5 Conclusion

In this work, two control strategies (P&O and ESC) are developed and used to achieve the MPPT of a PV array system. In fact, The P&O is based on the perturbation and observation of a point in the (P-V) characteristic until reaching the MPP. While, the ESC approaches is based on the stabilization of the PV at the optimum Point and then finding the optimum duty cycle value to be applied to the boost converter. From simulation results, it has been remarked that the system responses obtained the use of the P&O have a rapid dynamics change with a good time response but with a fast variation and fluctuations of the produced PV array power. But, the obtained simulation results through the use of the ESC show that the produced power varies slowly according to the irradiations variation with a good transition regime (without any fast variations and fluctuations of the power produced by the PV array which due to the stabilizing system achieved by the ESC). As perspectives, using the ESC under shading phenomena; improving the time response of the system by optimizing the filter parameters incorporated in the ESC algorithm.

References

1. Abdelsalam, A.K., Massoud, A.M., Ahmed, S., Enjeti, P.N.: High-performance adaptive perturb and observe MPPT technique for photovoltaic- based microgrids. IEEE Trans. Power Electron. **26**, 1010–1021 (2011)
2. Esram, T., Chapman, P.L.: Comparison of photovoltaic array maximum power point tracking techniques. IEEE Trans. Energy Convers. **22**, 439–449 (2007)
3. Alajmi, B.N., Ahmed, K.H., Finney, S.J., Williams, B.W.: Fuzzylogic-control approach of a modified hill-climbing method for maximum power point in microgrid standalone photovoltaic system. IEEE Trans. Power Electron. **26**, 1022–1030 (2011)
4. Jiang, L.L., Nayanasiri, D.R., Maskell, D.L., Vilathgamuwa, D.M.: A hybrid maximum power point tracking for partially shaded photovoltaic systems in the tropics. Renew. Energy **76**, 53–65 (2015)
5. Ishaque, K., Salam, Z., Lauss, G.: The performance of perturb and observe and incremental conductance maximum power point tracking method under dynamic weather conditions. Appl. Energy **119**, 119–128 (2014)
6. Tan, Y., Moase, W., Manzie, C., Nesic, D., Mareels, I.M.Y.: Extremum seeking from 1922 to 2010. In: 29th Chinese Control Conference, CCC10, pp. 14–26 (2010)
7. Ariyur, K.B., Krsti´c, M.: Real-Time Optimization by Extremum-Seeking Control. Wiley, Hoboken (2003)
8. Krstic, M., Wang, H.: Stability of extremum seeking feedback for general nonlinear dynamic systems. Automatica **36**, 595–601 (2000)

9. Dandois, J., Pamart, P.-Y.: NARX modeling and extremum-seeking control of a separation. J. Aerosp. Lab. **6**, 1–13 (2013)
10. Krstic, M.: Performance improvement and limitations in extremum seeking control. Syst. Control Lett. **39**, 313–326 (2000)
11. Yin, C., Chen, Y., MingZhong, S.: Fractional-order sliding mode based extremum seeking control of a class of nonlinear systems. Automatica **50**, 3173–3181 (2014)
12. Moballegh, S., Jiang, J.: Modeling, prediction, experimental validations of power peaks of PV arrays under partial shading conditions. IEEE Trans. Sustain. Energy **5**, 293–304 (2014)

Control of the Lateral Dynamics of Electric Vehicle Using Active Security System

K. Hartani[✉], N. Aouadj, A. Merah, M. Mankour,
and T. Mohammed Chikouche

Electrotechnical Engineering Laboratory, University of Saida, Saida, Algeria
kada_hartani@yahoo.fr, merahaek@yahoo.fr,
Nordiene.aouadj@gmail.com

Abstract. The control security of the modern vehicles is directly related to the control of the lateral dynamics of the vehicles and especially the yaw dynamics. The objective of this paper is to develop an active security system, based on the technique of sliding mode control, for the rate yaw control. The most important component of active lateral control assistance is the motor used to act on lateral dynamics. There are two principles of assistance: by differential braking of the wheels or by intervention on a steering column with a mechanical link. The approach to highlight the function to be performed to ensure the lateral stability of a road vehicle: Active Front Steering (AFS). The basic principle of this direction is to provide a steering correction with respect to the steering angle of the wheel, i.e. by adding a corrective steering angle to the driver maneuvers to facilitate the vehicle's rotational movement around its vertical axis to improve lateral vehicle stability in critical situations (skidding, turning or cornering). Several Matlab/Simulink simulation tests will be carried out to validate the effectiveness of the proposed control.

Keywords: Lateral dynamics · Lateral control · Bicycle model
Assisted steering · Active security system

1 Introduction

The motor vehicle remains one of the main causes of mortality of our modern life, and this despite the efforts made by the policies in the field of prevention, information and repression and those of the car manufacturers in terms of passive safety (air bag, …) and active safety (ABS, ESP, …) [1]. This state of affairs lies in the intervention of the driver on complex processes that govern driving (keeping the vehicle on the road, compliance with rules not always obvious, …) and that are not always adapted to its own physiological limits (sharpness visual, distance evaluation, loss of attention, nervousness, …), but that technological advances in recent years are trying to make it easier. With advances in automation, computing, telecommunications and miniaturization of instruments, researchers are now able to develop driver assistance systems that automate certain tasks by introducing new safety in order to improve safety by increasing the stability of the vehicle in cases where longitudinal or lateral accelerations occur, in which, the systems must act on the controllability of the vehicle so that the

© Springer Nature Switzerland AG 2019
M. Hatti (Ed.): ICAIRES 2018, LNNS 62, pp. 242–252, 2019.
https://doi.org/10.1007/978-3-030-04789-4_27

latter responds more quickly to the demands of the driver. According to the statistics, lane departure accidents account for 30% to 40% of claims: incorrect steering of the vehicle due to falling asleep, excessive speed, or loss of control due to inexperience, poor visibility, or reduced grip. This motivates a major research effort to help the driver and secure driving. The aids are developed on three levels: passive safety, active safety, preventive security. This work focuses on the development of one of the active safety systems for improving the stability and safety of a road vehicle [2]. In general, the control algorithm provides robust tracking of a trajectory. Simulations have been performed represent different driving situations. The various tests carried out highlight the robustness of the control law developed.

2 Modeling of the Vehicle in Its Environment

The dynamic behavior of a vehicle is the response it gives to a certain number of internal and external excitations, namely: driver maneuvers, atmospheric disturbances, the effects of the tire/road interaction and the variation of geometric characteristics of the road [3, 4]. The modeling of the vehicle is difficult task to achieve. This is due to the multiplicity of its constituents and which, by interacting, give rise to non-linearities [5]. In the next section, we present the dynamic vehicle model which is characterized by the equations taking into account the dynamic and aerodynamic aspects [6].

A. *Language Dynamic Model of the Vehicle*

Taking into account the different longitudinal and lateral forces, the overall model of the vehicle is described by the equations of motion [7]:

$$
\begin{aligned}
M_v\left(\dot{v}_x - v_y r\right) &= F_{xf} \cdot \cos\left(\delta_f\right) + F_{xr} \\
&\quad - F_{yf} \cdot \sin\left(\delta_f\right) - K_x \cdot v_x \cdot |v_x| \\
M_v\left(\dot{v}_y + v_x r\right) &= F_{xf} \cdot \sin\left(\delta_f\right) + F_{yr} \\
&\quad + F_{yf} \cdot \cos\left(\delta_f\right) - K_y \cdot v_y \cdot |v_y| + F_w
\end{aligned}
\tag{1}
$$

$$
\begin{aligned}
J_v \dot{r} &= L_f\left(F_{xf} \cdot \sin\left(\delta_f\right) + F_{yf} \cdot \cos\left(\delta_f\right)\right) \\
&\quad - L_r \cdot F_{yr} + \frac{D}{2}\left(\Delta F_x - \Delta F_y \cdot \sin\left(\delta_f\right)\right) + L_w \cdot F_w
\end{aligned}
\tag{2}
$$

With

$$
\begin{aligned}
&F_{xf} = F_{xfl} + F_{xfr} \quad F_{xr} = F_{xrl} + F_{xrr} \quad F_{yf} = F_{yfl} + F_{yfr} \quad F_{yr} = F_{yrl} + F_{yrr} \\
&\Delta F_x = \left(F_{xrr} - F_{xrl}\right) + \left(F_{xfr} - F_{xfl}\right) - \cos\left(\delta_f\right) \quad \Delta F_y = F_{yfr} - F_{yfl}
\end{aligned}
$$

The modeling of the vehicle environment is based on tire-ground contact. The slip angle α_i is expressed for a steering angle δ_i as follows:

$$\alpha_i = \beta_i - \delta_i \tag{3}$$

With β_i, steering angle of the wheel (i), and given by the following relation:

$$\beta_i = \arctan\left(\frac{V_{yi}}{V_{xi}}\right) = \arctan\left(\frac{V_y + \dot{\psi}.x_i}{V_x - \dot{\psi}.y_i}\right) \tag{4}$$

The longitudinal sliding coefficient at each wheel i $(i \in [1,2,3,4])$ is calculated as follows (Fig. 1):

$$\lambda_i = \frac{R_\omega \omega_i - u_{ti}}{\max(R_\omega \omega_i, u_{ti})} \tag{5}$$

We have,

$$\alpha = \arctan\left(\frac{V_y}{V_x}\right) \tag{6}$$

Fig. 1. Side slip of wheel

The lateral slip for the fourth wheels (front and rear) is given as follows:

$$\begin{cases} \alpha_1 = \arctan\left(\frac{1}{V_x + d}\left(V_y + rL_f\right)\right) - \delta \\ \alpha_3 = \arctan\left(\frac{1}{V_x - d}\left(V_y + rL_f\right)\right) - \delta \\ \alpha_2 = \arctan\left(\frac{1}{V_x - d}\left(V_y + rL_r\right)\right) \\ \alpha_4 = \arctan\left(\frac{1}{V_x - d}\left(V_y + rL_r\right)\right) \end{cases} \tag{7}$$

Where
L_f: Perpendicular distance between the nose gear and the center of gravity;
L_r: Perpendicular distance between the rear axle and the center of gravity;
δ: Steering angle of the front wheels.

The longitudinal adhesion factor of a wheel is defined as the ratio of the force and the vertical load F_z:

$$\mu_a = \frac{F_x}{F_z} \tag{8}$$

According to Kachroo's function, the parameter μ_a varies in a non-linear manner as a function of the slip of the wheel λ, whether it is driving or braking.

$$\mu_a = \frac{2\mu_p S_p \lambda}{S_p^2 + \lambda^2} \tag{9}$$

Taking into account the previous model and after a mathematical manipulation, one can lead to the following dynamic equations:

$$
\begin{aligned}
M_v(\dot{v}_x - v_y r) &= F_{xf} \cdot \cos(\delta_f) + F_{xr} \\
&\quad - F_{yf} \cdot \sin(\delta_f) - K_x \cdot v_x \cdot |v_x| \\
M_v(\dot{v}_y + v_x r) &= F_{xf} \cdot \sin(\delta_f) + F_{yr} \\
&\quad + F_{yf} \cdot \cos(\delta_f) - K_y \cdot v_y \cdot |v_y| + F_w \\
J_v \dot{r} &= L_f \left(F_{xf} \cdot \sin(\delta_f) + F_{yf} \cdot \cos(\delta_f) \right) \\
&\quad - L_r.F_{yr} + \frac{D}{2} \left(\Delta F_x - \Delta F_y \cdot \sin(\delta_f) \right) + L_w \cdot F_w
\end{aligned}
\tag{10}
$$

with

$$F_{xf} = F_{xfl} + F_{xfr} \quad F_{xr} = F_{rfl} + F_{xrr} \quad F_{yf} = F_{yfl} + F_{yfr} \quad F_{yr} = F_{yrl} + F_{yrr}$$

Where F_{yi} and F_{ti} are the lateral and longitudinal forces respectively and they are given by:

$$
F_x \begin{cases}
x = X + S_h \\
F_x(X) = y(x) + S_v \\
y(x) = D \sin(C \tan^{-1}(BX - E(BX - \tan^{-1}(BX))))
\end{cases}
\tag{11}
$$

$$
F_y \begin{cases}
x = X + S_h \\
F_y(X) = y(x) + S_v \\
y(x) = D \sin(C \tan^{-1}(BX - E(BX - \tan^{-1}(BX))))
\end{cases}
\tag{12}
$$

With u_{ti} the linear speeds of the vehicle brought back to the centers of the wheels and given by:

$$\begin{cases} u_{t1} = (V_x + dr)\cos(\delta) + (V_y + L_f r)\sin(\delta) \\ u_{t2} = V_x + dr \\ u_{t3} = (V_x - dr)\cos(\delta) + (V_y + L_f r)\sin(\delta) \\ u_{t2} = V_x - dr \end{cases} \tag{13}$$

The loads F_{zi} on the rear and front axles can simplify as follows (Fig. 2):

$$\begin{aligned} F_{z1} &= \frac{L_r g M_v}{2L} - \frac{h_{cg} M_v}{2L}(\dot{V}_x - rV_y) + \frac{h_{cg} M_v}{2l_w}(\dot{V}_y + rV_x) \\ F_{z3} &= \frac{L_r g M_v}{2L} - \frac{h_{cg} M_v}{2L}(\dot{V}_x - rV_y) - \frac{h_{cg} M_v}{2l_w}(\dot{V}_y + rV_x) \\ F_{z2} &= \frac{L_f g M_v}{2L} - \frac{h_{cg} M_v}{2L}(\dot{V}_x - rV_y) + \frac{h_{cg} M_v}{2l_w}(\dot{V}_y + rV_x) \\ F_{z4} &= \frac{L_f g M_v}{2L} - \frac{h_{cg} M_v}{2L}(\dot{V}_x - rV_y) - \frac{h_{cg} M_v}{2l_w}(\dot{V}_y + rV_x) \end{aligned} \tag{14}$$

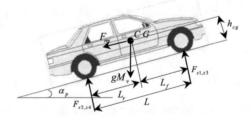

Fig. 2. Forces acting on a vehicle in a general case of movement.

B. *Electric Power Steering Column*

The studied vehicle is equipped with a conventional steering column, equipped with an electric motor, which transforms a steering angle of the steering wheel or engine (in the active assistance mode) into a steering angle of the wheels. The dynamics of the electric power steering column (Fig. 3) is described by a second order linear system [8].

The state representation of this model is as follows:

$$\begin{bmatrix} \dot{\delta}_d \\ \ddot{\delta}_d \end{bmatrix} = \begin{bmatrix} 0 & 1 \\ 0 & -\frac{B_s}{I_s} \end{bmatrix} \begin{bmatrix} \delta_d \\ \dot{\delta}_d \end{bmatrix} + \begin{bmatrix} 0 & 0 & 0 \\ \frac{1}{I_s} & \frac{1}{I_s} & -\frac{1}{I_s} \end{bmatrix} \begin{bmatrix} C_a \\ C_c \\ C_{at} \end{bmatrix}$$

$$\delta_f = \begin{bmatrix} \frac{1}{R_s} & 0 \end{bmatrix} \begin{bmatrix} \delta_d \\ \dot{\delta}_d \end{bmatrix} \tag{15}$$

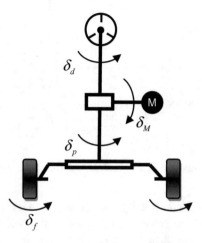

Fig. 3. Model of the steering column.

Inputs for this model are driver torque C_c, assist torque C_a, and self-aligning torque C_{at}. The latter can be modeled as the product of geometric hunting η_t and lateral force on the front wheels $F_f = -2C_y\left(\beta + \frac{L_f}{V_x}r - \delta_f\right)$. The expression obtained at the steering wheel is:

$$C_{at} = -\frac{2k_mC_y\eta_t}{R_s}\left(\beta + \frac{L_f}{V_x}r - \delta_f\right)$$

$$T_{SB} = \frac{2k_mC_y\eta_t}{R_s}, \qquad T_{Sr} = \frac{2k_mC_y\eta_t}{R_s}\frac{L_f}{V_x} \tag{16}$$

The parameter R_s is the steering reduction ratio and I_s represents the moment of inertia of the steering column and k_m is the manual steering gain.

3   Active Safety System Based on Sliding Mode Control

A. *Principle of the AFS*

Based on simplifying assumptions to pass from the nonlinear model of the vehicle to a linear model (bicycle model) exploitable in the development of the active safety system based on a robust variable structure control (sliding mode control) to generate a control signal (corrective steering angle) for proper system operation. In our study, the active front steering (AFS) uses a corrective steering angle to facilitate the vehicle's rotational movement around its vertical axis for better road holding [9].

Figure 4 shows a schematic diagram of AFS system. The control system processor (AFS Algorithm) receives signals from the various embedded sensors. Based on sensor signals and observation of status information, the processor calculates the corrective steering angle that allows for better road holding.

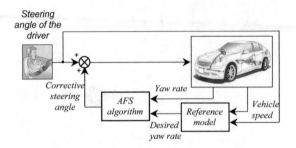

Fig. 4. Schematic of a typical AFS system for a road vehicle.

The additional simplifications will make it possible to retain the only movements of yaw and drift, and thus to arrive at the equations of the "bicycle" model.

By choosing r and v_y as state variables, the bicycle model can be written:

$$
\begin{bmatrix} \dot{V}_y \\ \dot{r} \end{bmatrix} = \begin{bmatrix} -2\frac{C_{yf}+C_{yr}}{M_v v_x} & 2\frac{-C_{yf}L_f+C_{yr}L_r}{M_v v_x} - V_x \\ 2\frac{-C_{yf}L_f+C_{yr}L_r}{J_v v_x} & -2\frac{C_{yf}L_f^2+C_{yr}L_r^2}{J_v v_x} \end{bmatrix} \begin{bmatrix} V_y \\ r \end{bmatrix}
$$
$$
+ \begin{bmatrix} \frac{2C_{yf}}{M_v} \\ \frac{2C_{yf}L_f}{J_v} \end{bmatrix} \delta_f + \begin{bmatrix} \frac{1}{M_v} \\ \frac{l_w}{J_v} \end{bmatrix} F_w
\tag{17}
$$

By defining the vehicle drift angle (β) as the angle between the vehicle heading and the velocity vector ($\beta = \arctan(V_y/V_x)$), the model can be rewritten as follows:

$$
\begin{cases}
\begin{bmatrix} \dot{\beta} \\ \dot{r} \end{bmatrix} = \begin{bmatrix} -2\frac{C_{yf}+C_{yr}}{M_v v_x} & 2\frac{-C_{yf}L_f+C_{yr}L_r}{M_v v_x^2} - 1 \\ 2\frac{-C_{yf}L_f+C_{yr}L_r}{J_v} & -2\frac{C_{yf}L_f^2+C_{yr}L_r^2}{J_v v_x} \end{bmatrix} \begin{bmatrix} \beta \\ r \end{bmatrix} \\
\qquad + \begin{bmatrix} \frac{2C_{yf}}{M_v v_x} \\ \frac{2C_{yf}L_f}{J_v} \end{bmatrix} \delta_f + \begin{bmatrix} \frac{1}{M_v v_x} \\ \frac{l_w}{J_v} \end{bmatrix} F_w \\
\begin{bmatrix} \dot{\beta} \\ \dot{r} \end{bmatrix} = \begin{bmatrix} a_{11} & a_{12} \\ a_{21} & a_{22} \end{bmatrix} \begin{bmatrix} \beta \\ r \end{bmatrix} + \begin{bmatrix} b_1 \\ b_2 \end{bmatrix} \delta_f + \begin{bmatrix} \frac{1}{M_v v_x} \\ \frac{l_w}{J_v} \end{bmatrix} F_w
\end{cases}
\tag{18}
$$

According to [10], the desired drift angle can be approached by zero and the yaw rate can be derived from the bicycle model.

$$
r_d(s) = \frac{k_r}{1+\tau s} \delta_f(s)
$$
$$
k_r = \frac{v_x}{L_f + M_v L_f L_r v_x^2 / 2L_f(L_f+L_r)C_{yr}}
\tag{19}
$$
$$
\tau = \frac{J_v v_x}{M_v L_r v_x^2 + 2C_{yf}L_f(L_f+L_r)}
$$

k_r and τ are the gain of the equilibrium state and the time constant of the yaw rate response respectively.

B. *Sliding Mode Controller Design*

Based on the mathematical relationship between yaw rate and yaw moment, an AFS method based on yaw rate is proposed. The principle of CMG is to force the trajectories of the system to reach in a finite time, and to remain there on a sliding surface.

The error between the actual vehicle yaw rate (r) and the desired yaw rate (r_d) is defined by the following equation:

$$e_r = r - r_d \tag{20}$$

We choose the sliding surface as follows:

$$S = e_r + c\dot{e}_r \tag{21}$$

The dynamics of sliding movement is governed by: $S = 0$

$$
\begin{aligned}
S = 0 &\Rightarrow e_r + c\dot{e}_r = 0 \\
&\Rightarrow (r - r_d) + c(\dot{r} - \dot{r}_d) = 0 \\
&\Rightarrow (r - r_d) + c(a_{21}v_y + a_{22}r + b_2\delta_f - \dot{r}_d) = 0
\end{aligned}
\tag{22}
$$

So, the value of the equivalent order:

$$\delta_{feq} = -\frac{1}{cb_2}\left[(r - r_d) + c(a_{21}v_y + a_{22}r - \dot{r}_d)\right] \tag{23}$$

If the states of the systems have not reached the sliding surface, the equivalent control must be reinforced by another so-called robust control, we then define the resulting steering angle by:

$$\Delta\delta_f = \delta_{feq} - \delta_{rob}.sgn(S) \tag{24}$$

To remedy the unwanted effects of "Chattering", at this level, by replacing the function Sign by the function Saturation. The steering angle becomes:

$$\Delta\delta_f = \delta_{feq} - \delta_{rob}.sat(S) \tag{25}$$

The robust control is determined using the boundary condition, so we define the steering angle such that:

$$\Delta\delta_f = -\frac{1}{cb_2}\left[(r - r_d) + c(a_{21}v_y + a_{22}r - \dot{r}_d)\right] - \frac{\eta}{b_2}.sat(S) \tag{26}$$

4 Stimulation and Discussion of Results

Series of simulations were performed under the MATLAB/Simulink environment to
verify the robustness of the proposed AFS system, Fig. 5. By comparing two control
techniques, with a conventional PI controller and with a sliding mode controller
(CMG). To clarify the effects of these two controllers, the behavior of the lateral
dynamics of the vehicle with and without controller are analyzed. The test is performed
on a test track with double lane change. In the first phase $t \in (0 \div 24)$ s, we assume
that the double lane change test of the vehicle is performed on a dry road (no slippery
$\mu = 1$) while keeping the longitudinal velocity constant $V_x = 36$ km/h (Fig. 6b). In the
second phase, $t \in (24 \div 48)$ s, we assume that the double lane change test of the
vehicle is performed on a slippery road (no slippery $\mu = 0, 6$) while increasing the
longitudinal velocity to $V_x = 54$ km/h (Fig. 6b). The controllers subtract or add a
corrective steering angle when applying a double lane change. Figure 6i shows that the
corrective control signal generated by the two controllers is smooth and less oscillations. The yaw rate velocity of vehicle follows its desired trajectory despite the disturbances and uncertainties characterizing the passage of the vehicle on a road with low
traction (the tires are found on a slippery road), Fig. 6f, j. The side angle of vehicle is
closed to zero, Fig. 6h, which increases the stability during lane change and therefore
the vehicle follows the reference path, Fig. 6j. The corrective steering angle created by
the controllers makes it possible to reduce the lateral acceleration of the vehicle to
ensure the comfort of the vehicle, Fig. 6k. According to the simulation results, we
notice that the vehicle follow exactly the driver instructions that allows the SM controller to give good dynamic performance.

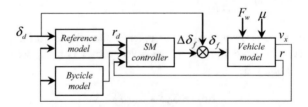

Fig. 5. Schematic diagram of the AFS with sliding mode control.

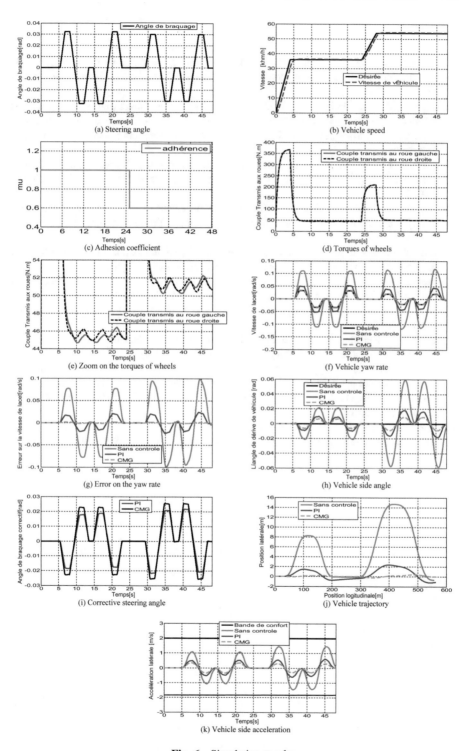

Fig. 6. Simulation results.

References

1. Gang, L., Changfu, Z., Qiang, Z.: Acceleration slip regulation control of 4WID electric vehicles based on fuzzy road identification. J. South China Univ. Technol. **40**, 99–104 (2012)
2. Ariff, M.M., Zamzuri, H., Idris, N.N., Mazlan, S., Nordin, M.: Direct yaw moment control of independent-wheel-drive electric vehicle (IWD-EV) via composite nonlinear feedback controller. In: 2014 First International Conference on Systems Informatics, Modelling and Simulation, pp. 112–117 (2014)
3. Hartani, K., Draou, A., Allali, A.: Sensorless fuzzy direct torque control for high performance electric vehicle with four in-wheel motors. J. Electr. Eng. Technol. **8**, 530–543 (2013)
4. Hartani, K., Merah, A., Draou, A.: Stability enhancement of four-in-wheel motor-driven electric vehicles using an electric differential system. J. Power Electron. **15**, 1244–1255 (2015)
5. Hartani, K., Draou, A.: A new multimachine robust based anti-skid control system for high performance electric vehicle. J. Electr. Eng. Technol. **9**, 214–230 (2014)
6. Merah, A., Hartani, K.: Shared steering control between a human and an automation designed for low curvature road. Int. J. Veh. Saf. **9**, 136–158 (2016)
7. Abe, M.: Vehicle handling dynamics: theory and application. Butterworth-Heinemann, Oxford (2015)
8. Klier, W., Reinelt, W.: Active Front Steering (Part 1): Mathematical Modeling and Parameter Estimation. SAE Technical Paper 0148-7191 (2004)
9. Jin, X., Yin, G., Zhang, N., Chen, J.: Stabilizing electric vehicle lateral motion with considerations of state delay of active front steering system through robust control. In: 2016 IEEE Conference and Expo, Transportation Electrification Asia-Pacific (ITEC Asia-Pacific), pp. 616–620 (2016)
10. Nagai, M., Yamanaka, S., Hirano, Y.: Integrated control of active rear wheel steering and yaw moment control using braking forces. JSME Int J., Ser. C **42**, 301–308 (1999)

Inverters-Converters and Power Quality

Wind Energy Conversion Systems Based on a Doubly Fed Induction Generator Using Artificial Fuzzy Logic Control

Z. Zeghdi[1]([⊠]), L. Barazane[1], A. Larabi[1], B. Benchama[1], and K. Khechiba[2]

[1] Industrial and Electrical Systems Laboratory (LSEI), Faculty of Electronics and Computer, University of Sciences and Technology Houari Boumediene, Algiers, Algeria
zoubirzeghdi@gmail.com, lbarazane@yahoo.fr,
Abdelkaderlarabi@gmail.com,
benchamma.bilel@gmail.com,
[2] Department of Electrical Engineering, Faculty of Technology, Djelfa University, Djelfa, Algeria
khechibakamel@gmail.com

Abstract. This paper presents the application of a simulation with a proposed model and control of a (DFIG) associated into a wind energy conversion system with a variable speed wind turbine using Artificial Fuzzy Logic techniques, and we are particularly interested in the application of indirect vector control by stator field orientation of DFIG. Firstly, a mathematical model of the machine written in an appropriate d–q reference frame is proposed to investigate simulations. secondly, and in order to control the power flowing between the stator of the DFIG and the power network, a control law is synthesized using two types of controllers: Proportional-Integral (PI) controller and fuzzy logic based controller. The proposed controller was tested and compared with one other technique, the PI controller. Finally, the obtained results show that the proposed controller exhibits better behaviour in terms of settling time, overshoot, robustness with respect to machine parameters variation, and good tracking references. The simulation was carried out by means of computational simulations in Matlab/Simulink Software.

Keywords: Doubly fed induction machine · Wind turbine
Proportional-integral · Fuzzy logic · Vector control

1 Introduction

Wind energy is the source of the renewable energy. It is the most promising of electrical power generation for the future. The wind power is a clean source of energy. It is non-degraded, firstly, it was exploited in mechanical applications. After that, it has been used to generate electricity [1].

Nowadays, the Doubly fed induction generator (DFIG) is one of the most popular used variable speed wind turbines, due of its advantages in better reliability, supply

© Springer Nature Switzerland AG 2019
M. Hatti (Ed.): ICAIRES 2018, LNNS 62, pp. 255–262, 2019.
https://doi.org/10.1007/978-3-030-04789-4_28

division and low maintenance cost. It is normally powered by a voltage source inverter [1, 2]. Otherwise, indirect vector control techniques proposed by Pena and al. in [2] and other researches which is based on classical PI (Proportional-Integral) controllers are traditionally used for controlling the active and reactive power of the DFIG [1–5]. The main goal of such a technique consists in dissociate the rotor current into its two corresponding (d–q) components; i.e., the active and reactive ones. Nevertheless, the computing of the gains of PI controllers are tedious and the appropriate corresponding values could be difficult to be obtained, due to the nonlinearity and the great complexity of the overall system, and also to the parameter's variations that will occur during the functioning of such a process which are considered as a major disadvantage of such a kind of regulator [6, 7]. So, and in order to handle such problems, some robust control techniques were proposed in the literature and were applied in several applications as: sliding mode control, adaptive control, predictive control and H_∞ control approach,…etc; [3–6]. On the other hand, artificial intelligence techniques in general such as: Takagi-Sugeno fuzzy method, fuzzy logic, Artificial Neural Networks, genetic algorithms and particle swarm optimization, were used and proves that they are too suitable to control systems that are characterised by variations of their parameters and ensured optimal performances [6–9]. Better solutions we can achieve by using the fuzzy logic controller since it is based on the knowledge of the expert (human knowledge) and the implicit inaccuracy [10]. In fact, in several applications, the use of an Artificial Fuzzy Logic system (controller) proved that such an approach ensures more robustness towards parameter's variations and external perturbations during the functioning of the process [9, 10]. In this paper, the authors propose to control active and reactive powers by using first in the process the Proportional -Integral (PI) controller then this latter is replaced by an AFL one respectively. After more, the two controllers are compared and results are discussed, the main objective is to show that controllers can improve performances of doubly-fed induction generators in terms of reference tracking, sensibility to perturbations and parameter's variations.

2   Modelling of System Understudy

This section briefly describes the modelling of the studied system, as depicted in Fig. 1.

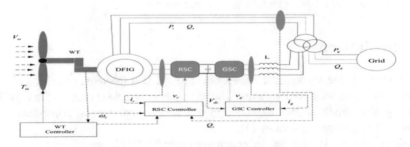

Fig. 1. A DFIG based wind energy conversion system.

2.1 Modelling of the Wind Turbine and Gearbox

The wind turbine model is given by the following system of equations [2, 4]:

$$P_{ex} = \frac{1}{2} \rho S_u C_p(\lambda, \beta) V_w^3 \tag{1}$$

$$\lambda = \frac{\Omega_t R}{V} \tag{2}$$

$$\Omega_t = \frac{\Omega_{mec}}{G} \tag{3}$$

$$C_p(\lambda, \beta) = 0.5176 \left(\frac{116}{\lambda_i} - 0.4\beta - 5 \right) \exp\left(\frac{-21}{\lambda_i} \right) + 0.0068\lambda \tag{4}$$

$$\frac{1}{\lambda_i} = \frac{1}{\lambda + 0.08\beta} - \frac{0.035}{\beta^3 + 1} \tag{5}$$

2.2 Modelling of the DFIG

The electrical equations of a DFIG are given by [2]:

$$\begin{cases} V_{ds} = R_s i_{ds} + \dfrac{d\phi_{ds}}{dt} - \omega_s \phi_{qs} \\[2mm] V_{qs} = R_s i_{qs} + \dfrac{d\phi_{qs}}{dt} + \omega_s \phi_{ds} \\[2mm] V_{dr} = R_r i_{dr} + \dfrac{d\phi_{dr}}{dt} - \omega_r \phi_{qr} \\[2mm] V_{qr} = R_r i_{qr} + \dfrac{d\phi_{qr}}{dt} + \omega_r \phi_{dr} \end{cases} \tag{6}$$

With;

$$\begin{cases} \phi_{ds} = L_s i_{ds} + L_m i_{dr} \\ \phi_{qs} = L_s i_{qs} + L_m i_{qr} \\ \phi_{dr} = L_r i_{dr} + L_m i_{ds} \\ \phi_{qr} = L_r i_{qr} + L_m i_{qs} \end{cases} \tag{7}$$

The rotor and stator angular velocities are expressed by the following relationship:

$$\omega_r = \omega_s - P\Omega_{mec} \tag{8}$$

The electromagnetic torque can be obtained using the power budget. There are several expressions of electromagnetic torque all equal:

$$C_{em} = \frac{3}{2}p\frac{L_m}{L_s}\left(\Phi_{qs}i_{dr} - \Phi_{ds}i_{qr}\right) \tag{9}$$

3 Field Oriented Control of DFIG

Among the above control techniques, the FOC is widely employed in the industrial field due to its simplicity of conception and its good performances [2, 3].

We use asynchronous reference with the q axis aligned to the computed stator flux. Assuming that the resistance of the stator windings (R_s) is neglected, the voltage and the flux equations of the stator winding can be simplified as follow:

$$V_{ds} = 0, V_{qs} = V_s = \omega_s\Phi_{ds}, \Phi_{ds} = \Phi_s, \Phi_{qs} = 0 \tag{10}$$

The active and reactive power equations of a DFIG can be expressed as

$$\begin{cases} P_s = -\frac{L_mV_s}{L_s}i_{qr} \\ Q_s = \frac{V_s^2}{\omega_sL_s} - \frac{V_sL_m}{L_s}i_{dr} \end{cases} \tag{11}$$

By exploiting Eqs. (6), (7), (10) and (11), we can get the rotor voltages as

$$\begin{cases} V_{dr} = R_rI_{dr} + \sigma L_r\frac{dI_{dr}}{dt} - g\omega_s\sigma L_rI_{qr} \\ V_{qr} = R_rI_{qr} + \sigma L_r\frac{dI_{qr}}{dt} + g\omega_s\sigma L_rI_{dr} - g\frac{L_mV_s}{L_s} \end{cases} \tag{12}$$

where: $\sigma = 1 - \frac{L_m}{L_rL_s}$ (leakage coefficient), $g = \frac{\omega_s - \omega_r}{\omega_s}$ (slip) and V_s (stator voltage).

4 Controller Synthesis

4.1 PI Controller

The PI controller is the most commonly used and the easiest to synthesise for DFIG control purposes and in many industrial control schemes. This type of regulator is a combination of both proportional and integral action. It has the effect of improving simultaneously the steady and transient states [8].

The open loop transfer function is given by

$$OLTF = \frac{p + \frac{k_i}{k_p}}{\frac{p}{k_p}} \cdot \frac{\frac{3L_mV_s}{2L_s\left(L_r - \frac{L_m^2}{L_s}\right)}}{p + \frac{L_sR_r}{L_s\left(L_r - \frac{L_m^2}{L_s}\right)}} \tag{13}$$

In order to eliminate the zero of the transfer function (the compensation method) we choose:

$$\frac{k_i}{k_p} = \frac{L_s R_r}{L_s \left(L_r - \frac{L_m^2}{L_s} \right)} \tag{14}$$

If the poles are perfectly compensated, the open loop transfer function becomes

$$OLTF = \frac{k_p \frac{3}{2} \frac{L_m V_s}{L_s \left(L_r - \frac{L_m^2}{L_s} \right)}}{p} \tag{15}$$

The closed-loop transfer function can be expressed as

$$CLTF = \frac{1}{1 + p\tau_r} \quad With: \quad \tau_r = \frac{1}{k_p} \frac{2}{3} \frac{L_s \left(L_r - \frac{L_m^2}{L_s} \right)}{L_m V_s}$$

Finally, we get:

$$k_p = \frac{1}{\tau_r} \frac{2 L_s L_r - L_m^2}{3 L_m V_s} \tag{16}$$

$$k_i = \frac{1}{\tau_r} \frac{2 L_s R_r}{3 L_m V_s} \tag{17}$$

4.2   Design of Fuzzy Feedback Controller

Fuzzy controllers have been popular especially in the industrial processes. Because of its effectiveness techniques when the mathematical model of the system is nonlinear or when there is no mathematical model [10].

The structure of the fuzzy control system consists of: Fuzzification, KnowledgeBase, Inference engine, Defuzzification

The schematic diagram of a complete fuzzy control system is given in Fig. 2. The plant control dI_{qr} is inferred from the two state variables, error E_p and change in error dE_p [9, 10].

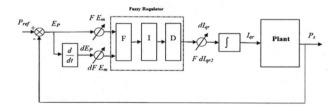

Fig. 2. Basic structure of the fuzzy control system

We can calculate the two input variables $E_P(k)$ and $dE_P(k)$ at every sampling time as:

$$E_p(k) = P_{ref}(k) - P_s(k) \tag{18}$$

$$dE_p(k) = E_p(k) - E_p(k-1) \tag{19}$$

The control signal I_{rq}^{ref} is obtained after integrating the output of the fuzzy logic regulator.

$$I_{rq}^{ref}(k) = I_{rq}^{ref}(k-1) + dI_{rq}^{ref}(k) \tag{20}$$

In this paper, the triangular membership function, the max-min reasoning technique, and the centre of gravity defuzzification method are used, as those methods are most frequently employed in many works of literature [8–10].

5 Regulators Performances

In this section, to access the effectiveness of the proposed Artificial Fuzzy Logic control strategy, simulation results are shown for different scenarios using Matlab/Simulink software.

5.1 Reference Tracking Scenario

The main goal of this scenario is to study the behavior of different control strategies while the DFIG's operates in ideal conditions, by considering different steps for active and reactive powers. Figs. 3, and 4 show the obtained simulation results. Note that as it is shown in Fig. 4, the AFL regulator offers perfectly track their references (active and reactive powers values) compared to the PI regulator.

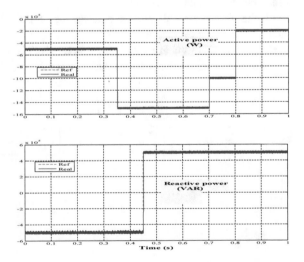

Fig. 3. Simulation results to the active and a reactive power impact (PI controller)

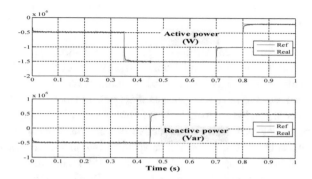

Fig. 4. Simulation results to the active and a reactive power impact (FL controller)

5.2   Robustness Scenario

In order to test the robustness of the two controllers, the value of mutual inductance L_m is decreased by 10% and 30% of its nominal value. Figures 5 and 6 show the effect of parameter's variations on the active power's responses with the two different controllers applied, respectively.

The results in Fig. 6 show a good dynamic and static performance, such as very fast response time and no overshoot and minimal static error compared to Fig. 5. From these results, we find that the Artificial Fuzzy Logic regulator (AFL) illustrates a satisfactory improvement in robustness, compared to the PI regulator.

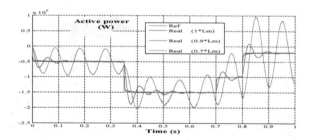

Fig. 5. Simulation results of active power to parameter's variations (PI controller)

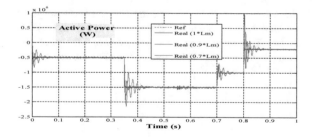

Fig. 6. Simulation results of active power to parameter's variations (AFL controller)

6 Conclusion

In this paper, we introduce an Artificial Fuzzy Logic control approach to improved power control of DFIG's used in wind turbine, after modelling the wind turbine and the DFIG in the d and q axis, we have established the indirect vector control of DFIG based on stator flux oriented. We have also presented the performance of the AFL and PI and compared between them, the robustness of the controllers is evaluated and allows us to have a decoupling between active and reactive power thus independent control. The simulation results show that the control strategy AFL is much more efficient compared to PI, it also improves the performance of the power DFIG, and ensure some important strength despite the variation of the parameters of the DFIG, fast and accurate dynamic response with an excellent steady-state performance. The use of the AFL has many advantages.

References

1. Lee, H.H., Dzung, P.Q., Phuong, L.M., Khoa, L.D., Nhan N.H.: A new fuzzy logic approach for control system of wind turbine with doubly fed induction generator. In: International Forum Strategic Technology (IFOST), pp. 135–139 (2010)
2. Pena, R., Clare, J. C., Asher, G. M.: A doubly fed induction generator using back to back converters supplying an isolated load from a variable speed wind turbine. In: IEE Proceeding on Electrical Power Applications vol. 143(September (5)) (1996)
3. Yao, X., Yi, C., Ying, D., Guo, J., Yang, L.: The grid-side PWM converter of the wind power generation system based on fuzzy sliding mode control. In: IEEE/ASME International Conference on Advanced Intelligent Mechatronics, pp. 973–978 (2008)
4. Rachidi, M., BIdrissi, B.B.: Adaptive nonlinear control of doubly-fed induction machine in wind power generation. J. Theor. Appl. Inf. Technol. **87**(1) (2016)
5. Aidoud, M., et al.: A robustification of the two degree-of-freedom controller based upon multivariable generalized predictive control law and robust H_∞ control for a doubly-fed induction generator. Trans. Inst. Meas Control **40**(3), 1005–1017 (2018)
6. Khwaldeh, A., Barazane, L., Krishan, M.M.: Robust neural network to improve hybrid control of an induction motor. Electromotion **16**(1), 28–37 (2009). ISSN 1223-057X
7. Azeem, B., et al. Robust neural network scheme for generator side converter of doubly fed induction generator. In: 2017 IEEE International Symposium on Recent Advances In Electrical Engineering (RAEE) (2017)
8. Tapia, A., Tapia, G., Ostolaza, J.X., Sáenz, J.R.: Modeling and control of a wind turbine driven doubly fed induction generator. IEEE Trans. Energy Convers. **18**(2), 194–204 (2003)
9. Abdelmalek, S., et al.: A novel scheme for current sensor faults diagnosis in the stator of a DFIG described by a TS fuzzy model. Measurement **91**, 680–691 (2016)
10. Li, X., Hui, D., Lai, X., Ouyang, M.: Control strategy of wind power output by pith angle control using fuzzy logic. In: 2010 IEEE International Symposium on Industrial Electronics (ISIE), pp. 120–124 (2010)

Control of a Dual Rotor Wind Turbine System

Housseyn Kahal[1,2(✉)], Rachid Taleb[1,2], Zinlabidine Boudjema[1,2],
Adil Yahdou[1,2], and Fayçal Chabni[1,2]

[1] Electrical Engineering Department,
Hassiba Benbouali University, Chlef, Algeria
hous.kahal@gmail.com
[2] Laboratoire Génie Electrique et Energies Renouvelables (LGEER),
Chlef, Algeria

Abstract. In this work, we try to improve the performance of the wind turbine by using a dual rotor wind system; this yield is subject to the BETZ limit of a value of 0.59 for a simple rotor wind. We use a proportional integral (PI) regulator for the generator velocity control.

Keywords: Dual rotor wind turbine · MPPT · PI regulator

1 Introduction

In recent years, wind energy has become the most widely used renewable resource and has been installed worldwide (Bu et al. 2015; Zamani et al. 2016). The reason is that the energy produced does not cause greenhouse gases or other pollutants. Modern wind turbines use advanced power electronics to ensure efficient control of the generator and ensure compatible operation with the power system (Anaya-Lara et al. 2009). Despite these advances, the wind power yield remains limited by the BETZ limit which reaches a maximum of 59% (Yahdou et al. 2013). In this paper, we are interested in improving this efficiency of the aerogenerator by using a double rotor wind turbine.

2 Dual Rotor Wind Turbine

This type of wind turbine is composed of two rotors on a horizontal axis. There are several types of dual rotor wind turbines; they are distinguished according to the direction of rotation and the size of the propellers of the two turbines. We focus in this paper on the study of contra-rotating wind turbines and rotors of different sizes. The rotor with large radius propellers is said to be main, it develops a great torque. On the other hand, the small rotor is called secondary connected to the fast shaft. He develops a couple less important than the previous one (Yahdou et al. 2016) (Fig. 1).

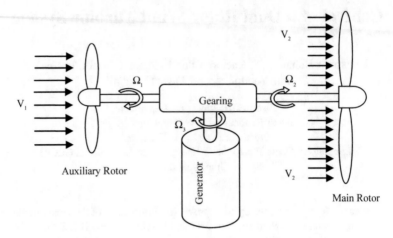

Fig. 1. The components of a dual rotor wind turbine

2.1   Modeling of the Dual Rotor Wind

The purpose of this part is to establish the dynamic model of the mechanical part of the double rotor wind turbine (Fig. 2).

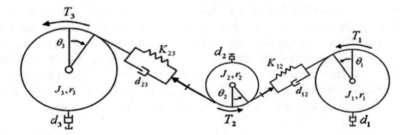

Fig. 2. Mechanical elements of a double rotor wind turbine.

Applying the superposition principle, we have the following cinematic relation:

$$\Omega_3 = \eta_1 \Omega_1 + \eta_2 \Omega_2 \qquad (01)$$

With η1 and η2 are positive.

We assume that the efficiency of the gear is 100%, we can write the conservation of energy:

$$\sum_{k=1}^{3} P_k = P_1 + P_2 + P_3 \qquad (02)$$

$$= \Omega_1 T_1 + \Omega_2 T_2 + \Gamma(\eta_1 \Omega_1 + \eta_2 \Omega_2)^2 = 0$$

The mechanical part of the counter-rotating wind turbine equations is represented by Eq. 03 (Rudaz 2009):

$$\begin{pmatrix} \dot{\Omega}_1 \\ \dot{\Omega}_2 \end{pmatrix} = \frac{1}{J_1 J_2 + n_1^2 J_2 J_3 + n_2^2 J_1 J_3} \begin{pmatrix} J_2 + n_2^2 J_3 & -n_1 n_2 J_3 & n_1 J_2 \\ -n_1 n_2 J_3 & J_1 + n_1^2 J_3 & n_2 J_1 \end{pmatrix} \begin{pmatrix} T_1 \\ T_2 \\ T_3 \end{pmatrix} \tag{03}$$

With J is the Lagrange multiplier

The kinetic power of the wind through the secondary rotor of radius R1 is given by the following relation:

$$P_1 = \frac{1}{2} \rho \pi R_1^2 V_1^3 \tag{04}$$

The aerodynamic torque developed by the secondary rotor is given by:

$$T_1 = \frac{C_p \rho \pi R_1^2 V_1^3}{2 \Omega_1} \tag{05}$$

The kinetic power of the wind through the main rotor of radius R2 is given by the following relation:

$$P_2 = \frac{1}{2} \rho \pi R_2^2 V_2^3 \tag{06}$$

The aerodynamic torque developed by the main rotor is given by:

$$T_2 = \frac{C_p \rho \pi R_2^2 V_2^3}{2 \Omega_2} \tag{07}$$

The relationship between wind speeds v1 and v2 is given by % (Yahdou et al. 2013):

$$V_2 = \frac{2}{3} V_1 \tag{08}$$

For our study, we assume that the generator can be modeled by a pure dissipative load:

$$T_3 = \Gamma \Omega_3 = \Gamma(\eta_1 \Omega_1 + \eta_2 \Omega_2) \tag{09}$$

Where Γ is negative.

2.2   Control MPPT of the Dual Rotor Wind

The most classic approach used in our work is to control the speed of rotation by a PI regulator. The Fig. 3 shows a part of our system looped and corrected by a PI regulator whose transfer function is of the form Kp + Ki/p (Fig. 4).

$$\begin{cases} K_{i\Omega} = \omega_n^2 J_d \\ K_{p\Omega} = 2\zeta\omega_n J_d - f_v \end{cases} \quad \text{Whose} \quad J_d = \left(\frac{J_1}{G_1^2} + \frac{J_2}{G_2^2} + J_g \right) \tag{10}$$

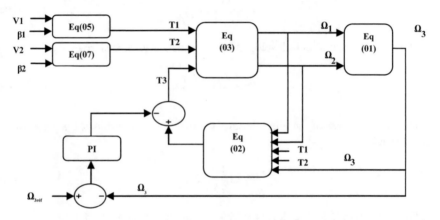

Fig. 3. Control by a PI regulator of the dual rotor wind turbine

3   Simulation Result and Discussions

The simulation is done by MATLAB/Simulink software. The simulation of the dual rotor wind turbine is based on the diagram presented in Fig. 3, the characteristics of which are given in Table 2. It can be seen in Fig. 5 that the speed of the generator shaft follows perfectly reference. We also see in Fig. 6 that the power on the generator shaft equal to the sum of the powers captured by the two main and auxiliary turbines. We see that the power will increase by 20% using a double rotor wind turbine compared to a single rotor wind turbine.

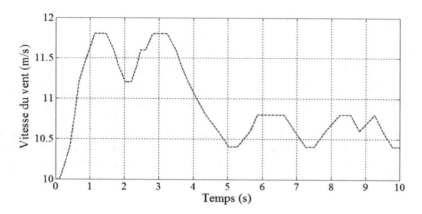

Fig. 4. Profile of the wind speed.

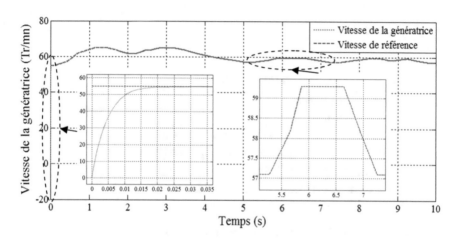

Fig. 5. Speed Ω3 of the generator

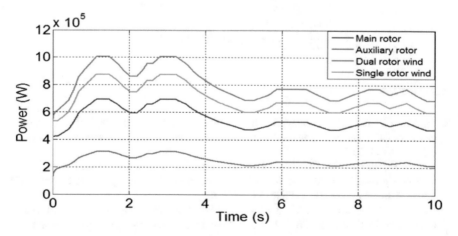

Fig. 6. Powers of both turbines and power on the generator shaft

4 Conclusion

In this article, we present a simple technique for controlling a twin rotor wind turbine. We used a MPPT to control the rotational speed and the power of the generator. The results obtained after simulation show the efficiency of the PI regulator. It can be deduced that dual rotor wind turbines are more efficient than single rotor wind turbines.

Acknowledgements. This work was financially supported by the Electrical Engineering Department. University Hassiba Benbouali, Chlef, Algeria. The data collection and theory in this paper are completed by the help of Mr. R Taleb, sam Department and university, to whom we express our heart feelings.

Annex

See Tables 1 and 2.

Table 1. Nomenclature

Symbol	Signification
J	Inertia
V1, V2	Wind captured by auxiliary and main turbines respectively
$\Omega 1, \Omega 2$	Auxiliary and main turbine rotation speed respectively
Cp	Power coefficients
ρ	Air density

Table 2. The parameters of the dual rotor wind turbine

Parameter	Value	Units
Rated power	1.5	MW
Main rotor diameter R1	51	m
Auxiliary rotor diameter R2	26.4	m

References

Bu, F., Hu, Y., Huang, W., Zhuang, S., Shi, K.: Wide speed rang operation dual stator winding generator DC generating system for wind power applications. IEEE Trans. Power Electron. **30** (2), 561–573 (2015)

Zamani, M.H., Riahy, G.H., Abedi, M.: Rotor speed stability improvement of dual stator winding induction generator based wind farms by control windings voltage oriented control. IEEE Trans. Power Electron. **31**(8), 5538–5546 (2016)

Anaya-Lara, O., Jenkins, N., Ekanayake, J., Cartwright, P., Hughes, M.: Wind Energy Generation: Modeling and Control. Wiley, New York (2009). ISBN 978-0-470-71433-1

Yahdou, A., Hemici, B., Boudjema, Z.: Commande Robuste d'un Système Eolien double Axe à base d'une Génératrice asynchrone à double alimentation. In: 2nd International Conference on Electronics, Electrical and Automatic, Oran, Algeria (2013)

Yahdou, A., Hemici, B., Boudjema, Z.: Second order sliding mode control of a dual-rotor wind turbine system by employing a matrix converter. J. Electr. Eng. **16**(3), 89–100 (2016)

Rudaz, S.: Double Rotor Wind Turbine Control and Optimization, Master Project Section of Mechanical Engineering. Ecolepolytechnique fédérale de Lausanne, Lausanne (2009)

Fractional Order PI Controller Design for Control of Wind Energy Conversion System Using Bat Algorithm

Maroufi Oussama$^{(\boxtimes)}$, Abdelghani Choucha, and Lakhdar Chaib

Laboratoire d'Analyse, de Commande des Systèmes d'Energie et Réseaux électriques (LACoSERE), Université Amar Telidji de Laghouat, Laghouat, Algeria
ou.maroufi@lagh-univ.dz

Abstract. This study presents intelligent control of the global wind energy conversion system (WECS), through a Permanent Magnet Synchronous Generator based variable speed Wind Turbine (PMSG-WT). The proposed control design is touched many parts in PMSG-WT, which are control of Maximum Power Point Tracking, Pitch Angle controller, and control of PMSG. The proposed controller is designed using Fractional PI controller. Where the controller parameters are successfully tuned by a new metaheuristic optimization Bat algorithm (BA). To highlight and compare the performances of this controller, it is employed under changes wind speed condition and compared with the conventional PI controller. Simulations results show clearly effectiveness of the proposed controller. Moreover, the PMSG-WT plant is effectively controlled at different operating conditions by the proposed scheme.

Keywords: Wind energy conversion system · PMSG
Fractional order PI controller · Bat algorithm

1 Introduction

Today renewable energies becoming more and more popular, with the increase of energy demand and the green house effect also pollution are caused by conventional energies (petrol-gas-carbon), moreover a green energy presents a solution for isolated sites [1]. The wind energy conversion system (WECS) using variable speed wind turbine (WT) with a permanent magnet synchronous generator (PMSG) is one of the most promising ones due to its advantages of high power, no gearbox, high precision, except initial installation costs. However, its control is a difficult task, due to the inherent nonlinearities [2].

The WT characterized by three different regions. The first is the low-speed region, where the turbine should be stopped and disconnected from the grid to prevent it from being driven by the generator. The second is the moderate-speed region that is bounded by the cut-in speed at which the turbine starts working, and it wants to be controlled with Maximum power point tracking mode. In the high-speed region, the wind turbine must be protected with pitch angle control.

Thus, many control strategies have been proposed to overcome the various difficulties in the control and design of PMSG-WT, e.g. conventional PI control [3], fuzzy

© Springer Nature Switzerland AG 2019
M. Hatti (Ed.): ICAIRES 2018, LNNS 62, pp. 270–278, 2019.
https://doi.org/10.1007/978-3-030-04789-4_30

logic [4], and Fuzzy-Sliding mode [5]. There is a certain difficulty to find the controller parameters, for this cause the control objective can be formulated as an optimization problem. The last one can be solved using metaheuristic optimization methods. Some of these approaches include Genetic Algorithm (GA) [6], gray Wolf technique [5] and In our study we using Bat Algorithm optimization technique.

In this contribution, a fractional PI controller tuning with Bat Algorithm optimization technique has applied for WT-PMSG, where the parts touched are MPPT control, Pitch angle control and control of the PMSG. The proposed controller is able to drive the state of the PMSG-WT system to track its optimal trajectory. The present paper is structured as follows In Sect. 2, the proposed WECS is presented. Then in Sect. 3 the proposed controller and optimization technique are detailed. Finally, a comparative study between the optimized controllers FOPI and the conventional is illustrated in Sect. 4.

2 Wind Energy Conversion Model

The conversion chain consists of a variable-pitch turbine coupled with a PMSG in order to capture the maximal wind energy, it is necessary to install the power electronic devices between the wind turbine generator (WTG) and the grid where the frequency is constant. The input of a wind turbine is the wind and the output is the mechanical power turning the generator rotor. In this work, we study the generator side as mentioned in the overall WECS layout shown in Fig. 1.

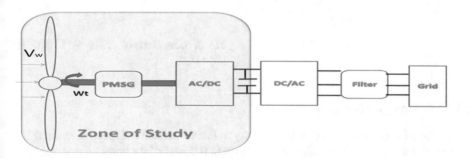

Fig. 1. Structural diagram of WECS.

The power captured from a wind turbine is given by [2]:

$$Pt = \frac{1}{2}\rho\pi R^2 V_w^3 C_P(\lambda, \beta) \tag{1}$$

$$\lambda = \frac{\omega_t R}{V_w} \tag{2}$$

Where C_p is the power coefficient, ω_t is the mechanical rotation speed, λ is the tip speed ratio, V_w is the wind speed, ρ is the air density and R is the blade length.

2.1 MPPT Control

In the low-speed case, it must capture a maximum power from wind, the power coefficient should be maintained at its maximum ($C_{P\max}$) which is achieved by making the tip speed ratio λ equal to the optimal value λ_{opt} at a fixed value. The mechanical rotation speed requires to follow its optimal reference ω_{opt} as:

$$\omega_{opt} = \frac{\lambda V_w}{R} \tag{3}$$

From the Fig. 2 The optimal tip speed ratio is $\lambda_{opt} = 8.1$ and $C_{P\max} = 0.48$.

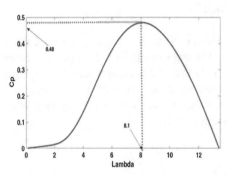

Fig. 2. Power coefficient.

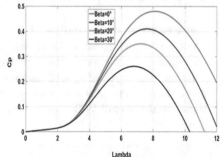

Fig. 3. Characteristic of the WEC different pitch angle.

2.2 Pitch Angle Control

In the case of high wind region, it is necessary to limit the rotational speed to avoid the damage of the turbine and in other words, the output power of the wind turbine can be regulated by pitch angle control Fig. 3.

The control strategy implemented is as follows [7]:

$$\begin{cases} \beta_{ref} = \beta_0 = 0 & \text{for} \quad 0 < P < Pn \\ \beta_{ref} = \frac{\Delta\beta}{\Delta P}(P_t - P_n) & \text{for} \quad P > Pn \end{cases} \tag{4}$$

2.3   Modeling of the PMSG

The voltage dynamic equations tutoring reference frame are expressed as [2]

$$\begin{cases} V_d = R_s i_d + L_d \frac{di_d}{dt} - L_q \omega i_q \\ V_q = R_s i_q + L_q \frac{di_q}{dt} + \omega (L_d i_d + \psi) \end{cases} \tag{5}$$

Then the mechanical system is represented by the following equation:

$$j \frac{d\omega_t}{dt} = \Gamma_m - \Gamma_e - f\omega_t \tag{6}$$

Where (V_d, V_q) are the stator voltages in the d-q axis, (i_d, i_q) are the currents in the d-q axis, (L_d, L_q) are the d-q axis inductances in the d-q axis, $\omega = p\omega_t$ is the electrical rotation speed, R_s is the stator resistance, ψ is the flux linkage of permanent magnets and is p the number of pole, The smooth-air-gap of the synchronous machine are considered, $L_d = L_q = L$. The electromagnetic torque in the d-q reference frame is given by:

$$\Gamma_e = \frac{3}{2}p((L_d - L_q)i_d i_q + \psi i_q) = \frac{3}{2}p\psi i_q \tag{7}$$

Finally Fig. 4 presents the global control system.

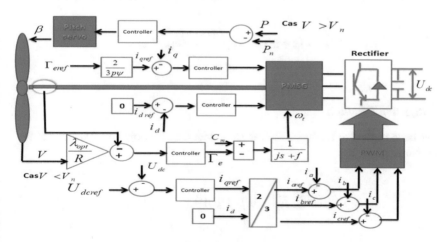

Fig. 4. Strategie scheme control.

3 Control Strategy

3.1 Fractional-Order Controllers

In the last decades, the applications of Fractional Calculus (FC) attract more attention of scientists in several domains. First, FC is an area of mathematics which generalizes the order of integration and derivation from integer to real value. The operator of fractional order differential-integral is given [8, 9].

$$
\begin{cases}
\frac{d^{\alpha}}{dt^{\alpha}} & R(\alpha) > 0 \\
1 & R(\alpha) = 0 \\
\int_0^t d(\tau)^{-\alpha} & R(\alpha) < 0
\end{cases}
\tag{8}
$$

One of the most commonly used definitions is the Caputo definition [13]

$$
{}_0D_{\alpha}^t f(t) = \frac{1}{\Gamma(n-\alpha)} \int_{\alpha}^t \frac{f^{(n)}(\tau)}{(t-\tau)}
\tag{9}
$$

Where $\Gamma_n(n) = \int_0^{+\alpha} x^{n-1} e^{-x} dx$ is the gammas function based on the Riemann–Liouville, the fractional order.

PI controller transfer function can be rewritten as follows in the time domain:

$$
U(t) = k_p + e(t) + K_i \int_0^t \frac{(t-\tau)}{\Gamma(\alpha)} e(\tau) d(\tau)
\tag{10}
$$

3.2 Objective Function

BA have been applied to minimize the values provided by the objective functions of the system that is given by:

$$
IAE = \int_0^{tsim} |e(t)| dt
\tag{11}
$$

3.3 Bat Algorithm

The bat-inspired metaheuristic algorithm, called the Bat algorithm (BA), was newly implemented by Yang [10], inspired by the echolocation of microbat. In nature, echolocation can have just a few thousandths of a second (up to about 8–10 ms) with a changing frequency in the area of 25–150 kHz, matching to the wavelengths of

2–14 mm in the air. Microbats usually utilize echolocation for searching forprey. During roaming, microbats produce short pulses, but, their emitted pulse rates augment and the frequency is tuned, when a potential prey is nearby. The augment of the frequency, called frequency-tuning, together with the acceleration of pulse emission will shorten the wavelength of echolocations and therefore augment precision of the detection. The echolocation characteristics of microbats can be idealized as the following rules:

(a) All bats utilize echolocation to sense distance, as well as they also recognize the difference between prey/food and background barriers in a few magical manners;
(b) Bats fly randomly with velocity v_i at position x_j with an unchanging frequency f_{min}, varying loudness A_o and wave length k to look for prey. They can routinely tune the rate of pulse emission $r \in [0, 1]$ and adjust the wavelength (or frequency) of their emitted pulses, depending on the proximity of their aim, and
(c) While the loudness can change in many manners, we suppose that the loudness varies from a great (positive) A_0 to a least constant value A_{min}. One must identify for every bat (j), its velocity vj and position x_j in a d-dimensional search space, the novel solutions velocities v_j^t and x_j^t at time step t can be written as follows:

$$f_j = f_{min} + (f_{max} - f_{min})\alpha \tag{12}$$

$$v_j^t = v_j^t + (x_j^{t-1} - x^*)f_j \tag{13}$$

$$x_j = x_j^{t-1} + v_j^t \tag{14}$$

where α in the range of [0,1] is a random vector drawn from a uniform distribution and is the current global best location, after comparing all the solutions among all the n bats at the current iteration x^* is located. As the product $k_j f_j$ is the velocity increment, one can utilize either f_j (or k_j) while fixing the other factor, to tune the velocity change. For implementation, every bat is randomly assigned a frequency which is drawn uniformly from (f_{min}, f_{max}). The local search is principally a random walk around the current best solutions.

4 Results and Discussions

In order to analyze the performance of the proposed Fractional Order PI Controller. Simulation tests have been carried out on PMSG-WT with the parameters being given in Table 1. Numerical values of FOPI and PI controllers tuning with Bat Algorithm are given Table 2. The proposed control strategy is employed under changes wind speed condition Fig. 5.

Figures 6 and 7 shows the variation of the rotor speed and mechanical power controlled by PI and proposed FOPI controllers compared with a reference generator speed. Note that the proposed controller has the lower time response of the tracking Furthermore, the less speed tracking errors and it highlights the good performances than the baseline PI regulator.

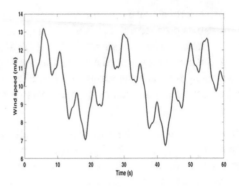

Fig. 5. Wind speed variation.

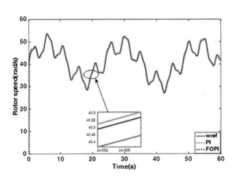

Fig. 6. Rotor speed.

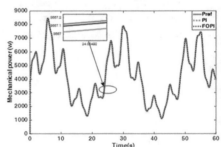

Fig. 7. Mechanical power.

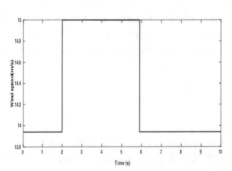

Fig. 8. Wind speed profile.

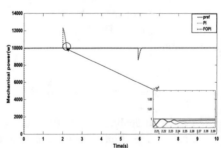

Fig. 9. Mechanical power.

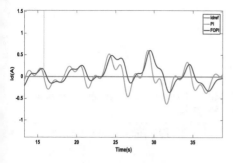
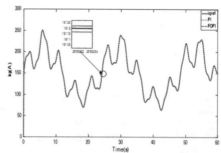

Fig. 10. Current d-axis.   **Fig. 11.** Current q-axis.

In the case of pitch angle control. The high wind speed profile has been chosen as illustrated in Fig. 7 to show how the controllers do, when the wind speed atteint the rated value (13.94 m/s). We note that the results were positive for tow controllers in Fig. 8, with superiority of proposed controller, This shows clearly in sudden change of wind speed at t = 2 s the FOPI regulator has a good and robustness response than PI controller. Where maintains the rated power of the WT Pn = 10000w.

In Figs. 9 and 10 we apply optimized FOPI to control the i_d and i_q currents of PMSG, it is clear that the currents well their optimal references for the two controllers. But the response time for the proposed controller is less than the conventional controller (Fig. 11).

5 Conclusion

This paper has presented a new meta-heuristic called Bat Algorithm to find the optimal parameters of Fractional order PI controller applied to a Permanent Magnet Synchronous Generator used in a wind energy conversion system. The different parts of the proposed WECS have been modeled separately. The objectives of this study are: ensuring maximum power generation in the low-speed region, protect the wind turbine in case of the high wind speed region using Pitch angle control, control of the generator currents. All of this is tested in the variable wind speed conditions. Simulations results and comparison with baseline PI controller show that the PMSG-WT plant can be controlled with more efficiency and more stability using proposed optimized FOPI controller.

Appendix

See Tables 1 and 2.

Table 1. Wind conversion system parameters.

Parameter	Value
Blad length (R)	2 m
Air density (R)	1.225 Kg/m^3
Stator inductance (L)	0.174 mH
Friction coefficient (f)	0.005 N · m
Moment of inertia (J)	0.089 Kg m^2
Stator resistance (R_s)	0.00829 Ω
Magnetic flux (ψ)	0.071 wb
Number of poles (p)	6

Table 2. Optimal controllers parameters given by the (BA).

		MPPT	Pitche angle	Vector control i_d	i_q
PI	P	76.2	91.67	0.4	0.39
	I	87.8	14.33	59.64	98.47
FOPI	P	98.7	69.93	0.94	0.89
	I	97.2	3.24	93.88	44.72
	α	0.86	0.5	0.89	0.5

References

1. Hong, C.M., Chen, C.H.: Intelligent control of a grid-connected wind-photovoltaic hybrid power systems. Int. J. Electr. Power Energy Syst. **55**, 554–561 (2014)
2. Ounnas, D., Ramdani, M., Chenikher, S., Bouktir, T.: Optimal reference model based fuzzy tracking control for wind energy conversion system. Int. J. Renew. Energy Res. (IJRER) **6** (3), 1129–1136 (2016)
3. Mansour, M., Mansouri, M.N., Mmimouni, M.F.: Study and control of a variable-speed wind-energy system connected to the grid. Int. J. Renew. Energy Res. **1**(2), 96–104 (2011)
4. Meghni, B., Dib, D., Azar, A.T.: A second-order sliding mode and fuzzy logic control to optimal energy management in wind turbine with battery storage. Neural Comput. Appl. **28** (6), 1417–1434 (2017)
5. Kahla, S., Soufi, Y., Sedraoui, M., Bechouat, M.: Maximum power point tracking of wind energy conversion system using multi-objective grey wolf optimization of fuzzy-sliding mode controller. Int. J. Renew. Energy Res. (IJRER) **7**(2), 926–936 (2017)
6. Bechouat, M., Soufi, Y., Sedraoui, M., Kahla, S.: Energy storage based on maximum power point tracking in photovoltaic systems: a comparison between GAs and PSO approaches. Int. J. Hydrog. Energy **40**(39), 13737–13748 (2015)
7. Smida, M.B., Sakly, A.: Pitch angle control for variable speed wind turbines. Renew. Energy Sustain. Dev. **1**(1), 81–88 (2015)
8. Chaib, L., Choucha, A., Arif, S.: Optimal design and tuning of novel fractional order PID power system stabilizer using a new metaheuristic Bat algorithm. Ain Shams Eng. J. **8**(2), 113–125 (2017)
9. Delassi, A., Arif, S., Mokrani, L.: Load frequency control problem in interconnected power systems using robust fractional PIλD controller. Ain Shams Eng. J. **9**, 77–78 (2015)
10. Yang, X.S.: A new metaheuristic bat-inspired algorithm. In: Nature Inspired Cooperative Strategies for Optimization (NICSO 2010) (pp. 65–74). Springer, Berlin (2010)

Breaking Rotor Bars and Ring Segments Faults on Squirrel Induction Motor: Analytical Study and Comparison of Its Spectral Signatures on Stator Currents

Thamir Deghboudj[1(✉)], Abdellah Derghal[2],
and Abdelhamid Tlemçani[1]

[1] Laboratory of Electrical Engineering and Automatic,
Yahia Fares University, Médéa, Algeria
de.thamir@yahoo.fr
[2] LGEA Laboratory, Oum el Bouaghi University, Oum el Bouaghi, Algeria
derghal01@yahoo.fr

Abstract. In this paper, we attempt to apply breaking bars and broken ring segments faults to a squirrel cage induction motor, which involve the same conditioning. The principle objective in this manuscript is how to clarify the difference in terms of amplitude and frequency, the outcomes of this test system examined by comparing with that of a healthy induction motor, these comparisons reveal the efficient of the PSD analysis.

Keywords: Induction motors · Squirrel cage · Ring segment
Broken bars · PSD

1 Introduction

The Squirrel-cage induction motors are widely used in many industrial applications because they are cost effective and mechanically robust. However, production will stop if these motors fail. Therefore, early detection of motor faults is highly desirable. Induction motor faults summarized in (Thomson and Fenger 2001), and rotor failures account for approximately 10% of the total induction motor failures.

Numerous fault detection methods have been proposed to identify the above faults. The fault detection methods involved several different types of fields of science and technology. They performed by mechanical and/or electrical monitoring.

The most frequent used detection methods are:

- Motor current signature analysis (MCSA),
- Acoustic noise measurements,
- Model, artificial intelligence and neural network based techniques, …, etc.

For the detection of the induction motor's rotor faults here, the motor current signature analysis method applied.

M. Hatti (Ed.): ICAIRES 2018, LNNS 62, pp. 279–289, 2019.
https://doi.org/10.1007/978-3-030-04789-4_31

2 Multiple-Coupled Circuit Model

We use the multiple-coupled circuit model described in (Filipetti et al. 1995) to analyze the harmonic content of the stator currents when a motor has broken bars. The rotor cage portion of the model for a healthy motor consists of *n* identical and equally spaced current loops formed by two rotor bars and two end ring segments (see Fig. 1a). Rb and Lb represent the resistance and leakage inductance of each bar. Re and Le represent the resistance and leakage inductance of each end ring segment between adjacent bars. This model does not consider saturation, parasitic currents, and inter bar currents but still shows the effect of the broken bars on the stator currents. Figure 1b shows the Squirrel Cage equivalent circuit for a motor with one broken bar, as described in (Noureddine and Touhami 2015). This circuit eliminates the loop n and modifies the self-inductance and mutual inductance associated with the loop $n - 1$. Figure 1c shows the Squirrel Cage equivalent circuit for a motor with one broken ring segments.

2.1 Stator Voltage Equations

An induction machine stator comprises of 3 phase concentric windings. Each of these windings is treated as a separate coil (Noureddine and Touhami 2015). The stator equations for induction machine can be written in vector matrix form as (Toliyat and Lipo 1995):

$$[V_s] = [R_s][I_s] + \frac{d[\psi_s]}{dt} \tag{1}$$

where:

$$[\psi_s] = [\mathcal{L}_s][I_s] + [M_{sr}][J_r] \tag{2}$$

The matrix [Rs] is a 3×3 diagonal matrix, which consists of resistances of each coil

$$[R_s] = \begin{bmatrix} R_s & 0 & 0 \\ 0 & R_s & 0 \\ 0 & 0 & R_s \end{bmatrix} \tag{3}$$

The matrix $[\mathcal{L}_s]$ is a symmetric 3×3 matrix.

$$[\mathcal{L}_s] = \begin{bmatrix} L_s & M_{ss} & M_{ss} \\ M_{ss} & L_s & M_{ss} \\ M_{ss} & M_{ss} & L_s \end{bmatrix} \tag{4}$$

The mutual inductance matrix [Msr] is a $3 \times (n + 1)$ matrix comprised of the mutual inductances between the stator coils and the rotor loops.

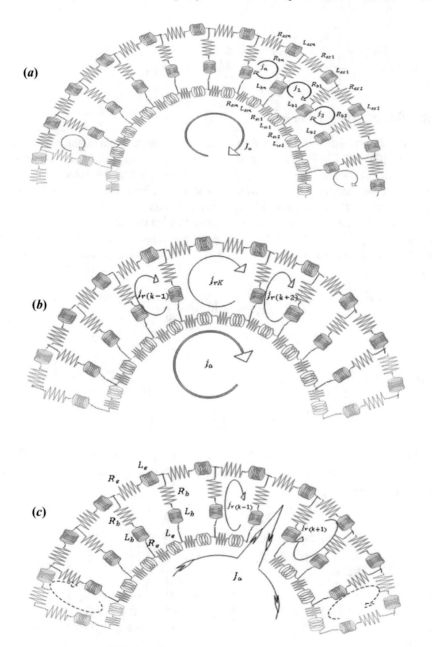

Fig. 1. Squirrel Cage IMs Equivalent for: (**a**) healthy, (**b**) one broken bar, (**c**) one broke ring segments.

$$[M_{sr}] = \begin{bmatrix} M_{ar1} & M_{ar2} & \cdots & M_{ark} & \cdots & M_{ar(n-1)} & M_{arn} & M_{are} \\ M_{br2} & M_{br2} & \cdots & M_{brk} & \cdots & M_{br(n-1)} & M_{brn} & M_{bre} \\ M_{cr1} & M_{cr2} & \cdots & M_{crk} & \cdots & M_{cr(n-1)} & M_{crn} & M_{cre} \end{bmatrix} \tag{5}$$

2.2 Rotor Voltage Equations

A cage rotor consists of n bars can be treated as n identical and equally spaced rotor loop. As illustrated in Fig. 1, each loop formed by two adjacent rotor bars and the connecting portions of the end-ring between them. Hence, the rotor circuits have (n + 1) independent currents as variables. The n rotor loop currents coupled to each other and to the stator windings through mutual inductances. The end-ring loop does not couple with the stator windings; it however couples the rotor currents only through the end leakage inductance and the end-ring resistance. From Fig. 1a, the voltage equations for the rotor loops can written in vector matrix form (Munoz and Lipo 1999) as:

$$[V_r] = [R_r][I_r] + \frac{d[\psi_r]}{dt} \tag{6}$$

where:

$$\left. \begin{array}{l} [V_r] = \begin{bmatrix} V_{r1} & V_{r2} & V_{r3} & \cdots\cdots\cdots & V_{rk} & \cdots\cdots\cdots\cdots & V_{r(n-1)} & V_m & V_{ra} \end{bmatrix}^t \\ = \begin{bmatrix} 0 & 0 & 0 & \cdots\cdots & 0 & \cdots\cdots\cdots\cdots & 0 & 0 & 0 \end{bmatrix}^t \end{array} \right\} \tag{7}$$

$$[\psi_r] = [\mathcal{L}_r][J_r] + [M_{rs}][I_s] \tag{8}$$

Since each loop is assumed to be identical, the equation (Meshgin et al. 2004) is valid for every loop. Therefore, the resistance matrix [Rr] is a symmetric $(n+1)^2$ matrix given by:

$$[R_r] = \begin{bmatrix} R_r & -R_b & 0 & \cdots & 0 & -R_b & -R_e \\ -R_b & R_r & -R_b & \cdots & 0 & 0 & -R_e \\ 0 & \vdots & \vdots & \vdots & \vdots & \vdots & \vdots \\ \vdots & \vdots & \vdots & \vdots & \vdots & \vdots & \vdots \\ 0 & 0 & 0 & \cdots & R_r & -R_b & -R_e \\ -R_b & 0 & 0 & \cdots & -R_b & R_r & -R_e \\ -R_e & -R_e & -R_e & \cdots & \cdots & -R_e & n*R_e \end{bmatrix} \tag{9}$$

where: $R_r = 2(R_b - R_e)$.

The matrix $[M_{rs}]$ is the transpose of the matrix $[M_{sr}]$. Due to the structural symmetry of the rotor, $[\mathcal{L}_r]$ can be written in matrix form (Toliyat and Lipo 1995), where L_r is the self-inductance of the kth rotor loop, L_b is the rotor bar leakage inductance, L_e is the rotor end-ring leakage inductance and M_{rr} is the mutual inductance between two rotor loop.

$$[\mathcal{L}_r] = \begin{bmatrix} L_r & M_r & M_{rr} & \cdots & M_{rr} & M_r & -L_e \\ M_r & L_r & M_r & M_{rr} & \cdots & M_{rr} & -L_e \\ 0 & \vdots & \vdots & \vdots & \vdots & \vdots & \vdots \\ \vdots & \vdots & \vdots & \vdots & \vdots & \vdots & \vdots \\ M_{rr} & \cdots & \cdots & \cdots & L_r & M_r & -L_e \\ M_r & M_{rr} & \cdots & M_{rr} & M_r & L_r & -L_e \\ -L_e & -L_e & -L_e & \cdots & \cdots & -L_e & n*L_e \end{bmatrix} \tag{10}$$

where: $M_r = M_{rr} - L_b$, $L_r = M_{rr} - L_b$.

The global equation of voltage is given by

$$[V_G] = [R_G][I_G] + [L_G]\frac{d[I_G]}{dt} + \frac{d[L_G]}{dt}[I_G] \tag{11}$$

The modified Park transformation is applied on the system (11) we find

$$[V_t] = [R_t][I_t] + [\mathcal{L}_t]\frac{d[I_t]}{dt} \tag{12}$$

What gives

$$\frac{d[I_t]}{dt} = [\mathcal{L}_t]^{-1}\{[V_t] - [R_t][I_t]\} \tag{13}$$

Where

$$[V_t] = \begin{bmatrix} V_{os} & V_{ds} & V_{qs} & \vdots & 0 & 0 & \cdots & \cdots & 0 & \cdots & \cdots & \cdots & \cdots & 0 & 0 & \vdots & 0 \end{bmatrix}^t$$

$$[I_t] = \begin{bmatrix} I_{os} & I_{ds} & I_{qs} & \vdots & j_{r_1} & j_{r_2} & j_{r_3} & j_{r_4} & \cdots & \cdots & \cdots & \cdots & \cdots & j_{r_{(n-1)}} & j_{r_n} & \vdots & j_e \end{bmatrix}^t$$

$$[R_t] = \begin{bmatrix} [R_s] + \frac{d\theta}{dt}[F][\mathcal{L}_{sp}] & [P(\theta_s)]^{-1}\frac{d[M_{sr}]}{dt} \\ \underbrace{[M_{rs}]\frac{d[P(\theta_s)]}{dt} + \frac{d[M_{rs}]}{dt}[P(\theta_s)]}_{[0]_{(n+1)*3}} & [R_r] \end{bmatrix}_{(n+4)^2} \tag{14}$$

$$[\mathcal{L}_t] = \begin{bmatrix} [\mathcal{L}_{sp}] & \vdots & [M_{srp}] \\ \cdots & \vdots & \cdots \\ [M_{srp}]^t & \cdots & [\mathcal{L}_r] \end{bmatrix}_{(n+4)^2} \tag{15}$$

$$[M_{srp}] = [P(\theta_s)]^{-1}[M_{sr}] \tag{16}$$

where: $[P(\theta_s)][\mathcal{L}_s][P(\theta_s)]^{-1} = [\mathcal{L}_{sp}]$.

The modified Park transformation is $[P(\theta_s)]$.

The matrix of the currents of the bars and the rotor segments given respectively by the following equations

$$
[I_b] = \begin{bmatrix} I_{b1} \\ I_{b2} \\ \vdots \\ I_{bk} \\ \vdots \\ I_{b(n-1)} \\ I_{bn} \end{bmatrix} = \begin{bmatrix} j_{r1} - j_{r2} \\ j_{r2} - j_{r3} \\ \vdots \\ j_{rk} - j_{r(k+1)} \\ \vdots \\ j_{r(n-1)} - j_{rn} \\ j_{rn} - j_{r1} \end{bmatrix} ; \quad [I_{ri}] = \begin{bmatrix} I_{ri1} \\ I_{ri2} \\ \vdots \\ I_{rik} \\ \vdots \\ I_{ri(n-1)} \\ I_{rin} \end{bmatrix} = \begin{bmatrix} j_{r1} - j_a \\ j_{r2} - j_a \\ \vdots \\ j_{rk} - j_a \\ \vdots \\ j_{r(n-1)} - j_a \\ j_{rn} - j_a \end{bmatrix} \quad (17)
$$

j_{r_k} The current flowing in the kth maille.
j_{r_a} The current flowing in the short-circuit ring.
I_{bk} The current flowing in the kth bar.
I_{re} The current flowing in the kth ring segment.

The electromagnetic torque equation

$$
C_e = \sqrt{\frac{3}{2}} P M \left[I_{qs} \sum_{k=1}^{N} j_{rk} \cos(k'\alpha') - I_{ds} \sum_{k=1}^{N} j_{rk} \sin(k'\alpha') \right] \quad (18)
$$

where: $k' = ((2k-1)/2); \, k = 1 \ldots\ldots\ldots\ldots n$.

2.3 System Resolution

We can re-write the obtained $(n + 4)$ equations system's in a condensed matrix form as fellow

$$
\frac{d[I_t]}{dt} = [\mathcal{L}_t]^{-1} \{ [V_t] - [R_t][I_t] \} \quad (19)
$$

Coupled with the following two mechanical equations

$$
\frac{d\omega}{dt} = \left(\frac{P}{J}\right)(C_e - C_r - f\omega); \quad \frac{d\theta}{dt} = \omega \quad (20)
$$

The total is $(n + 6)$ differential equations of the first order solved by the implicit method of Rung–Kutta.

3   Broken Rotor Bars and Broken Ring Segments

An induction motor has two parts – stator and rotor. The rotor has bars with slots for the rotor windings and end rings to short the ends of the windings. The rotor bars may crack or break due to many reasons, which gives rise to broken rotor bars. A broken bar causes several effects in induction motors. A well-known effect of a broken bar is the appearance of the socalled side-band components in the frequency spectrum of the stator current. These founded on the left and right sides of the fundamental frequency component. The frequencies of these sidebands are given by: $(1 \mp 2.k.g) * fs$, and $k = 1,2,3\ldots$ (Sahar et al. 2018; Roman 2012).

Where g the slip in per unit and fs is is the fundamental frequency of the stator current other electric effects of broken bars used for motor fault classification purposes including speed oscillations, torque ripples (Filipetti et al. 1996).

4   Simulation Broken Bars and Broken Ring Segments

We modeled a faulty motor with one, two, and three adjacent broken bars or adjacent broken ring segments to analyze the harmonic content of the stator currents for these operating conditions. The simulation method of the broken bar: is to replace the value of the resistance of the broken bar by a sufficiently large value; the aim is to make the current passing through this bar is equivalent to zero. As a remark for broken ring segments, it used the same method of simulation.

4.1   Broken Bar

The figures represented below identify the analysis of stator current via PSD method in two cases: healthy induction motor and broken induction motors (Fig. 2).

The magnitude of the harmonics of the absorbed current increases each time there is an increase in the number of broken bars. The nature of the broken influences the magnitude of the harmonics components $(1 \mp 2.k.g) * fs$ there is a greater value in the spectrum of the absorbed current when the number of bars broken. The slip is larger than those of the broken bars are when the number of faults is one, two or three.

4.2   Broken Ring Segments

When a break occurs in two adjacent rings segments, the current passing through the rod that falls between them is zero, i.e., a break in two adjacent rings segments causes a break in bar between them (Fig. 3).

For a break in three, adjacent rings segments causing a fracture between the bars. This explains the difference between components of $(1 \mp 2kg) * fs$ in both figures.

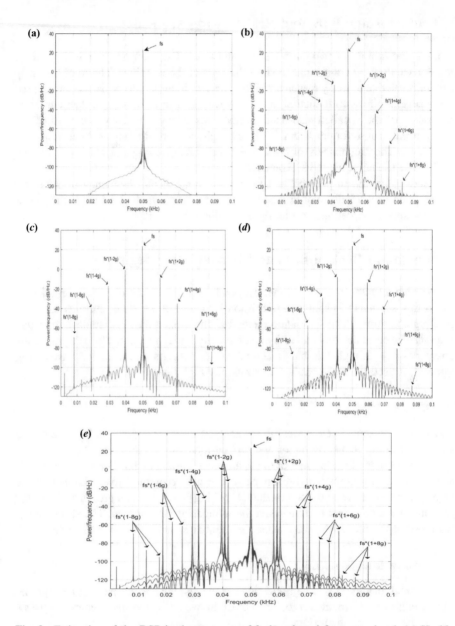

Fig. 2. Estimation of the DSP in the presence of fault, selected frequency band. (**a**) Healthy, (**b**) 1 broken bar, (**c**) 2 broken bars, (**d**) 3 broken bars and (**e**) broken bars comparison.

4.3   Comparison of Broken Rotor Bars and the Broken Ring Segments

In Fig. 4a, the frequency $(1 \mp 2.k.g) * fs$ is identical in frequency that is to say that they have the same slip. Due to the magnitude of the harmonics $(1 \mp 2.k.g) * fs$, broken in the ring segments has a greater effect than the broken bars.

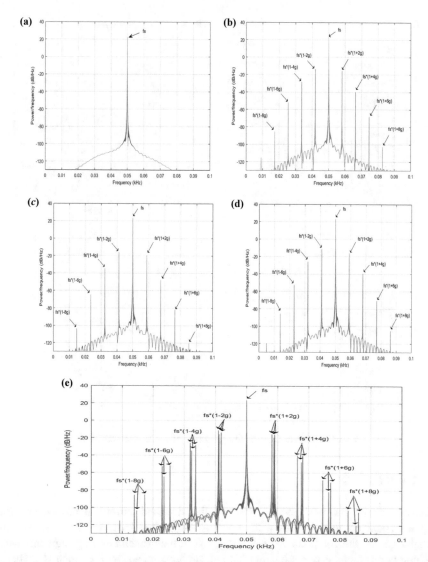

Fig. 3. Estimation of the DSP in the presence of fault, selected frequency band. (**a**) Healthy, (**b**) 1 broken ring segment, (**c**) 2 broken rings segments, (**d**) 3 broken rings segments, (**e**) broken rings segments comparison.

In both figure (Fig. 4b and c), the frequency is not the same and this means that they do not have the same slip, the magnitude of the components in the break $(1 \mp 2.k.g) * fs$ in two bars or three adjacent larger than in two or three adjacent ring segments.

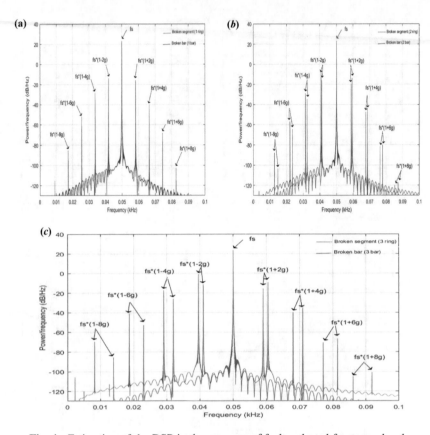

Fig. 4. Estimation of the DSP in the presence of fault, selected frequency band.

5 Conclusions

This manuscript has analyzed the PSD of the absorbed current in broken end ring segments and compared them with those of the broken bars. We have verified by simulation that the components $(1 \mp 2.k.g) * fs$ of the broken end ring segments when the number of faults is one are larger than of the broken bars, and the opposite when it is the number of faults is two or three. In addition, when the number of faults is two or three the slip became larger than of these broken end ring segments did, it remains constant when the number of faults is one. Moreover, the effect of the number of broken bars and broken end-ring segments on the motor performance has clarified with these outcomes.

Acknowledgements. The authors gratefully acknowledge the helpful suggestions of the reviewers, which have improved the manuscript.

References

Bonnett, A.H., Soukup, G.C.: Rotor failures in squirrel cage induction motors. IEEE Trans. Ind. Appl. **22**(6), 1165–1173 (1986)

Thomson, W.T., Fenger, M.: Current signature analysis to detect induction motor faults. IEEE Ind. Appl. Magazine **7**, 26–34 (2001)

Noureddine, L., Touhami, O.: Diagnosis of wind energy system faults Part I: modeling of the squirrel cage induction generator. Int. J. Adv. Comput. Sci. Appl. **6**(8), 46–53 (2015)

Kelk, M.H., Milimonfared, J., Toliyat, H.A.: Interbar Currents and Axial Fluxes in Healthy and Faulty Induction Motors. IEEE Trans. Ind. Appl. **40**(1), 128–134 (2004)

Roman M (2012) Analysis of squirrel cage motor with broken bars and rings. Trans. Electr. Eng. **1**(2)

Filipetti, F., Franceschinzi, G., et al.: Impact of speed ripple on rotor fault diagnosis of induction machine. In: International Conference on Electrical Machines, vol. 2. Vigo (1996)

Filipetti, F., Franceschini, G., Vas, P.: Transient analysis of cage induction machines under stator, rotor bar and end ring faults. IEEE Trans Energy Conv **10**(2), 241–247 (1995)

Munoz, A.R., Lipo, T.A.: Complex vector model of the squirrel-cage induction machine including instantaneous rotor bar currents. IEEE-IAP **35**(6), 1332–1340 (1999)

Toliyat, H.A., Lipo, T.A.: Transsient analysis of cage induction machines under stator rotor bar and endring faults. IEEE Trans. Energy Conv. **10**(2), 241–247 (1995)

Sahar, Z., Samsul, B.N., et al.: Broken rotor bar fault detection and classification using wavelet packet signature analysis based on fourier transform and multi-layer perceptron neural network. Appl. Sci. **8**, 25 (2018)

Field Analysis of Dual Stator Winding Induction Machine

H. Hammache[✉] and D. Aouzellag

Electrical Engineering Department, Bejaia University, Béjaïa, Algeria
hammache.hakim@gmail.com, aouzellag@hotmail.com

Abstract. Basic concepts of six-phase ac motor drives have existed for a number of years ago and were considered extensively in the eighties for safety-critical and/or high power applications. There has been an upsurge in the interest in these drives in recent times, initiated by various application areas, such as 'more-electric' aircraft, electric ship propulsion and EV/HVs. In our study, an ac drive with dual stator windings fed by dual six step converters is proposed. Two sets of electrically isolated windings are placed in the stator slots, and one set is shifted from the other by 30° in space. Two converters are used to fed the dual windings, and the phase voltages from the two converters are also shifted 30° from each other in time. The 5th and 7th harmonics of the air-gap flux and rotor currents are reduced dramatically.

However, for each individual converter, the 5th and 7th harmonic currents are found to be substantial and the functioning of the power converter evidently is deteriorated. In this paper, the computer-simulation of the dual-stator-winding induction machine using FLUX2D software in which the space harmonics of the stator windings and those of the rotor circuits are accounted for has been presented.

Keywords: Dual-stator induction machine · Finite elements method Harmonics analysis

1 Introduction

Three-phase induction machines are today a standard for industrial electrical drives. Cost, reliability, robustness, and maintenance-free operation are among the reasons these machines are replacing dc drive systems. The development of power electronics and signal processing systems has eliminated one of the greatest disadvantages of such ac systems, that is, the issue of control. With modern techniques of field-oriented vector control, the task of variable-speed control of induction machines is no longer a disadvantage.

In a multiphase system, here assumed to be a system that comprises more than the conventional three phases, the machine output power can be divided into two or more solid-state inverters that could each be kept within prescribed power limits. In addition, having additional phases to control means additional degrees of freedom available for further improvements in the drive system. With split-phase induction machines, and appropriate drive system, the sixth harmonic torque pulsation, typical in a six-step

© Springer Nature Switzerland AG 2019
M. Hatti (Ed.): ICAIRES 2018, LNNS 62, pp. 290–296, 2019.
https://doi.org/10.1007/978-3-030-04789-4_32

three-phase drive, can be eliminated (Bakhshai et al. 1998), (Nelson and Krause 1974). Also, air-gap flux created by fifth and seventh harmonic currents in a high-power six-step converter-fed system is dramatically reduced with the penalty of increased converter harmonic currents (Xu and Ye 1995).

Dual-stator machines are similar to split-phase machines with the difference that the stator groups are not necessarily equal. A dual-stator machine with different numbers of poles in each three-phase group has been proposed in (Lipo 1980) to obtain controllability at low speeds. Two independent stator windings are used in (Ojo and Davidson 2000) for an induction generator system. One set of windings is responsible for the electromechanical power conversion while the second one is used for excitation purposes. A PWM converter is connected to the excitation windings and the load is connected directly to the power windings (Fig. 1).

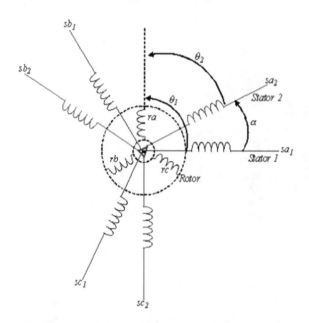

Fig. 1. Windings of the Double-star induction machine (DSIM).

2 Motor Design

Standard squirrel cage induction motors, rated at 5.5 kW, was used to investigate the new drive configurations. the motors was wound in a 6-pole concentric arrangement with six slots per-phase per-pole in a thirty six slot stator. At least two differently wound motors can be used.

(1) 2-Layer Wound Motor: This motor had two windings per-phase arranged in the slots in two layers. The separation of the windings in the slot allows slot leakage flux paths to increase the inter-winding leakage inductance. The separation also minimizes inter-winding capacitive effects.

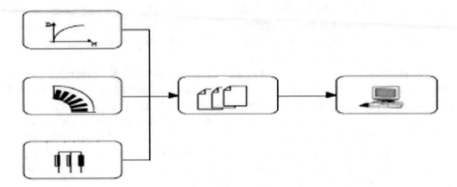

Fig. 2. Pre-processing steps.

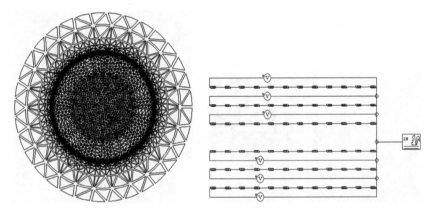

Fig. 3. Finite-element mesh.   **Fig. 4.** Coupled circuits

(2) 3-Layer Wound Motor: This motor was wound with three windings per phase
    placed in the stator slots in three layers. The winding separation increases the inter-
    winding leakage inductance between the unipolar windings by allowing the slot
    leakage flux paths to contribute. The effective stator resistance is low in this
    winding. The use of only two unipolar winding sections means that this motor can
    only be used in a star connection. For our study, we choose the first case.

3 Field Calculation

In the following, the results concerning two kinds of six-phase asynchronous machines are presented, it depend of the angle between winding. The rated values of these machines are summarized in Table 1. According to the previous section, the three stages of the whole procedure have to be performed. Figures 2 and 3 illustrate respectively the structure of FLUX2D and mesh generation in the case of the DSIM; Fig. 4 represents the external coupled circuits.

Table 1. Machine characteristics

Rated power Sn (kVA)	7.346
Rated voltage Un (V)	380
Power factor cos φ	0.75
Pole number	6
Rated frequency fn (Hz)	50
Phase number	6
Stator notches number	36
Rotor notches number	28

Where D is the stator inner diameter, ls the stator length, P the number of poles, and Bg1 the peak fundamental air-gap flux. Figures 5 and 6 give, for null shift between the two winding or $\alpha = 0°$, the field and current distribution, it's the same obtained for conventional three phases induction machine. Figure 7 give the harmonics analysis of magnetic field in air-gap. We can observe the presence of space harmonic 5 and 7.

3.1 Result for $\alpha = 0°$

Finite-element analysis is conducted in the six-phase machine to calculate the air gap flux distribution and to demonstrate the influence of different harmonics like harmonic 5, 7, 11 and 13; other harmonics are not studied because, for symmetrical structure and balanced feeding, they don't exist. An asymmetric six-phase induction motor was designed using a conventional three-phase motor as baseline. For the baseline machine, from the nameplate and geometrical data, the air-gap flux is calculated. The stator phase voltage of the machine is calculated as

$$Vsf = \omega e \ k1Ns \ \Phi_p \tag{1}$$

Where

$$\Phi_p = \frac{2Dis \ ls}{p} Bgl \tag{2}$$

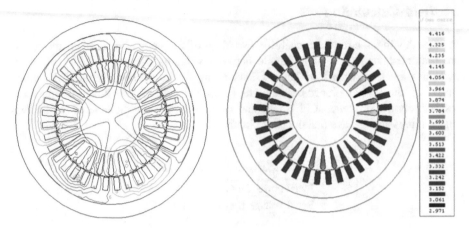

Fig. 5. Field distribution   **Fig. 6.** Current density distribution

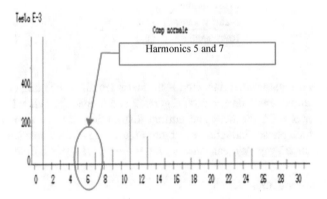

Fig. 7. Spectral analysis of normal induction ($\alpha = 0°$)

3.2   Result for $\alpha = 30°$

Figures 8 and 9 shows that the space harmonic 5 and 7 does not exist in this case. This means that these harmonics does not contribute to the electromechanical conversion of energy.

The power supply of the machine by the harmonics of higher row revealed us the influence of this one on the behavior of the machine (Fig. 10):

We can see on the Figs. 11 and 12 that when the machine is supplied by harmonic 5 and 7 (same results obtained for harmonics 11 and 13), the lines of field do not cross the air-gap. As result they do not contribute to the electromechanical energy conversion. Whereas the density of the currents which circulates in the windings stator is very significant, it is almost zero in the rotor bars.

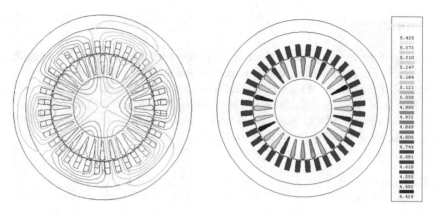

Fig. 8. Field distribution

Fig. 9. Current density distribution

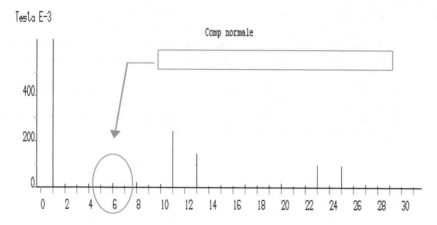

Fig. 10. Spectral analysis of normal induction ($\alpha = 30°$)

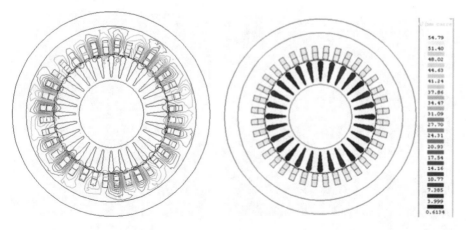

Fig. 11. Field distribution (fh = 5 fs)

Fig. 12. Current density distribution

4 Conclusion

In this paper, the results of simulations obtained by Flux2D made it possible to analyze and quantify the performances of our prototype and to realize of specificity functional of the DSIM. The study of the machine response for a supply with a source of tension containing harmonics 1, 5, 7, 11 and 13 approached, we showed the effect of each one of these harmonics on the magnetic behavior of the machine for the two principal configurations of windings $\alpha = 0°$ and $\alpha = 30°$.

References

Bakhshai, A.R., Joos, G., Jin, H.: Space vector PWM control of split-phase induction machine using the vector classification technique. In: Proceedings of IEEE APEC 1998, vol. 2, pp. 802–808, February 1998

Nelson, R.H., Krause, P.C.: Induction machine analysis for arbitrary displacement between multiple windings. IEEE Trans. Power App. Syst. **PAS-93**, 841–848 (1974)

Xu, L., Ye, L.: Analysis of a novel stator winding structure minimizing harmonic current and torque ripple for dual six-step converter-fed high power ac machines. IEEE Trans. Ind. Appl. **31**, 84–90 (1995)

Lipo, T.A.: A d–q model for six phase induction machines. In: Proceedings of ICEM 1980, Athens, Greece, pp. 860–867, September 1980

Ojo, O., Davidson, I.E.: PWM-VSI inverter-assisted stand-alone dual stator winding induction generator. IEEE Trans. Ind. Appl. **36**, 1604–1611 (2000)

Investigate the Performance of an Optimized Synergetic Control Approach of Dual Star Induction Motor Fed by Photovoltaic Generator with Fuzzy MPPT

Hossine Guermit$^{(\boxtimes)}$ and Katia Kouzi

Laboratory of Semi-conductors and Functional Materials, University Amar
Telidji Laghouat, Laghouat, Algeria
{ho.guermit,k.kouzi}@lagh-univ.dz

Abstract. The aim of this work is to investigate the performance of robust vector control based on synergetic theory of dual star induction motor (DSIM) fed by two inverters coupled in Photovoltaic Generator (PVG). To do this, in first stage in order to improve the performance of vector control of DSIM, we propose a novel control scheme based on synergetic control theory newly integrated in the control of DSIM. The main advantage of synergetic control is fast response, asymptotic stability of the closed-loop system in the all range of admissible operating condition, and robustness of the system to the variation parameter. On the other one, to overcome the problem of synergetic controller parameter tuning, it is proposed in this study, the Particle Swarm Optimization (PSO) algorithm which allow obtaining the optimal parameter of suggested controller and consequently improving the performance of control system. In the second stage, to couple the DSIM to a photovoltaic generator, we have use the model with two exponential models. Then we have presented the control of this latter by an algorithm called MPPT, based on fuzzy logic theory. The obtained simulation results illustrate clearly that the suggested scheme control provides high performance in all range of operating conditions.

Keywords: Dual Star Indication Motor (DSIM) · PV generator
Fuzzy MPPT · Synergistic control · PSO · Optimization

1 Introduction

The demand on the electrical energy is increase with very fast way these last years as well as the constraints related to its production, such that the effect of pollution and global warming, lead the research toward the development of renewable sources of energy. The advent of new so-called renewable sources has facilitated the introduction of DSIM in energy production and especially for electric traction. Among these energies, photovoltaic energy which gives an efficient system by associating it with DSIM. This combination gives rise to various new, powerful, reliable and robust applications. In this context, the photovoltaic (PV) systems offer a solution very competitive. To overcome the problem of performance of solar panels and obtain a

© Springer Nature Switzerland AG 2019
M. Hatti (Ed.): ICAIRES 2018, LNNS 62, pp. 297–310, 2019.
https://doi.org/10.1007/978-3-030-04789-4_33

maximum performance, it is necessary to optimize the design of all parties of the PV system [1].

In this context, the photovoltaic (PV) systems offer a solution very competitive. To overcome the problem of performance of solar panels and obtain a maximum performance, it is necessary to optimize the design of all parties of the PV system. In addition, it is necessary to optimize the converters DC/DC employees as interface between the PV generator and the load in order to extract the maximum of power, and thus operate the generator PV generator to its maximum power point (MPP) using a MPPT controller (maximum power point tracking), consequently, get a maximum electric current that flows under the variation of the load and the atmospheric conditions (brightness and temperature. The growth of the consumption of electric energy and electrical applications of high power, have led to use Dual Star Indication Machine (DSIM) for segment the power. In addition to this advantage, the DSIM present several other benefits such as the segmentation of power without increasing the currents by phase and the minimization of losses iron, robustness, the simplicity of its structure, its low cost and the no need a regular maintenance [1–3]. Through these benefits, the DSIM is used in several applications, especially in the area of high powers such as naval, railway propulsion systems and renewable energy [4]. However, its control remains complex and difficult to implement. This is due in particular to the problem of close coupling between the various variables and the strong non-linearity is present in the machine model, which may lead to poor dynamic performances compared to those obtained with the DC motor. One of the first commands to solve this problem is the vector control, also called field-oriented control (FOC). The principle of this control method consists in making the performances of the DSIM similar to those of the DC machine, by orienting the flux vector on an axis of the reference frame linked to the rotating field. This made it possible to eliminate the coupling problem between the two variables "torque and flux" [4]. However, this type of control has a major disadvantage, which lies in its sensitivity to the parameters of the machine, in particular the rotor time constant. Moreover, the classical regulators of the PID family rely on a state model with constant coefficients where the parameters of the system are supposed to be known with precision [2, 4, 5]. What is no longer the case of DSIM where some of its state variables are inaccessible to direct measurements (rotor flux) and its parameters (in particular rotor resistance) are affected by the thermal effect, leading to limited performances [6]. Actually, because of their success, intelligent controllers such as Artificial Neural Network intelligent (ANN) and Fuzzy Logic Controller (FLC) have become one of the most favorable areas of research for controlling nonlinear systems [1]. However, membership function type, number of rules and correct selection of parameters of FLC are very important to obtain desired performance in the system. The selection of suitable fuzzy rules, membership functions, and their definitions in the universe of discourse invariable involves painstaking trial-error. To solve these problems, the use of non-linear ordering techniques was timely and justified. Known for its simplicity, speed and robustness has been widely adopted and has proven to be effective in many applications. It consists in changing the control structure as a function of the state of the system, ensuring good system performance and robustness to external disturbances and parametric variations [2, 4, 5]. The steady state of the system in this case is called the sliding mode. That is, the state trajectory of the system is brought to a hyper surface called the sliding surface and switches around this surface to the point of equilibrium [4–8]. Despite these numerous advantages, this type of controller with variable structures suffers from a major disadvantage which is the phenomenon of chattering. Indeed, in the theory of systems with variable structures, it is assumed that the switching between the control structures will take place instantaneously [4–6]. However, the physical limitations of the semiconductor components of the inverter, which cannot follow the switching imposed by the control, causes the system to oscillate around the surface instead of sliding over that surface, which may excite non-modeled dynamics of the system, this is called chattering.

In order to reduce this phenomenon, a technique consisting in making continuous an approximation of the discontinuous function was used. A new approach for controlling non-linear systems is the synergistic approach, presented in [9, 10]. The work on the application of the synergistic controller, showed that it offers a better robustness with respect to possible parametric variations as well as level of yield High design simplicity and flexibility of synergistic controllers. The theory of the synergistic controller has several advantages and is widely used in the case of numerical controls. It therefore ensures a reduction of chattering with in addition of robustness [9–16]. The synergistic approach can help not only reduce the size of a modeled system, but also ensure the stability of the power system in general. Theoretically, the synergetic regulator shows a great capacity to ensure the robustness of control system presence of various disturbances. However, this technique has been used only for nonlinear systems whose dynamic model is perfectly known, which is rarely the case. Moreover, the parameters of this regulator are not optimal, which has been proposed to be remedied. In the literature, different approaches using Particle Swarm Optimization (PSO) algorithm have been proposed for the optimal adjustment of regulators in control systems [17]. The main advantage of this optimization approach compared to traditional techniques is the robustness, and flexibility.

The paper is organized as follows, in Sect. 2, the study of photovoltaic generator system PVG is presented. The fuzzy MPPT strategy is explained in Sect. 3. In Sect. 4 the modeling of DSIM is described. Section 5 deals with indirect vector control of DSIM. The design of proposed synergetic controller of DSIM is developed in Sect. 6. The tuning the parameters of the suggested controller by PSO is proposed in Sect. 7. The performance and robustness of the suggested controller is illustrated by simulation results in Sect. 8. Finally, conclusion is presented in Sect. 9.

2 Study of Photovoltaic Generator System "PVG"

Connecting the PV generator to a load requires adapting the generator to operate at maximum power. And so that the power supplied by the generator to the load is maximum, it is necessary permanently to follow the point called Maximum Power Point (MPP). Our objective is to adapt the photovoltaic generator (GPV) to the DSIM. The adaptation between source and load is achieved by inserting a DC/DC converter (chopper) controlled by a MPPT mechanism. The two diode model or still known as the exponential two diode model is widely used in literature [1]. It is gaining importance because it gives results closest to those obtained by a real GPV. The model with two diodes is shown by this figure (Fig. 1).

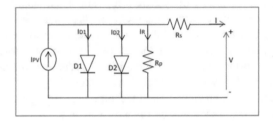

Fig. 1. Equivalent circuit of a photovoltaic cell.

The following equation describes the output current of the photovoltaic cell for the two diode model.

$$I = I_{PV} - I_{01}\left[\exp\left(\frac{V + IR_s}{a_1 V_{T1}}\right) - 1\right] - I_{02}\left[\exp\left(\frac{V + IR_S}{a_2 V_{T2}}\right) - 1\right] - \left(\frac{V + IR_S}{R_P}\right) \quad (1)$$

3   Fuzzy MPPT Mechanism

As conventional methods of MPPT shown their limitations in the face of sudden changes in climatic conditions and the load connected to the generator, several methods have emerged to mitigate these failures and improve the operation of these generators. The fuzzy logic approach is used to improve MPPT control performance. Fuzzy logic algorithm has proposed to obtain faster MPP control and more stable output power under transient and steady state conditions. Two input variables, which are error and change in error, and the duty cycle is considered as the output linguistic variable. The inputs fuzzy E and CE variables are given by:

$$E = \frac{P(K) - P(K - 1)}{V(K) - V(K - 1)} \quad (2)$$

$$CE = E(K) - E(K - 1) \quad (3)$$

The five fuzzy sets are namely: NB: Negative Big; NS: Negative Small; EZ: Zero; PS: Positive Small; PB: Positive Big. Hence, 25 fuzzy rules were created. We have chosen the MAX–MIN inference algorithm to complete the fuzzy procedure. The defuzzification process employs the gravity center method (Table 1).

Figures 2 and 3 show the Rules View and Surface View of fuzzy MPPT.

The Fig. 4 represent the output power of the PV generator with and without fuzzy MPPT and with two different irradiation.

Table 1. Inference rules

CE	E				
	NB	NS	ZE	PS	PB
NB	NB	NB	NB	NS	ZE
NS	NB	NS	NS	ZE	PS
ZE	NB	NS	ZE	PS	PB
PS	NS	ZE	PS	PS	PB
PB	ZE	PS	PB	PB	PB

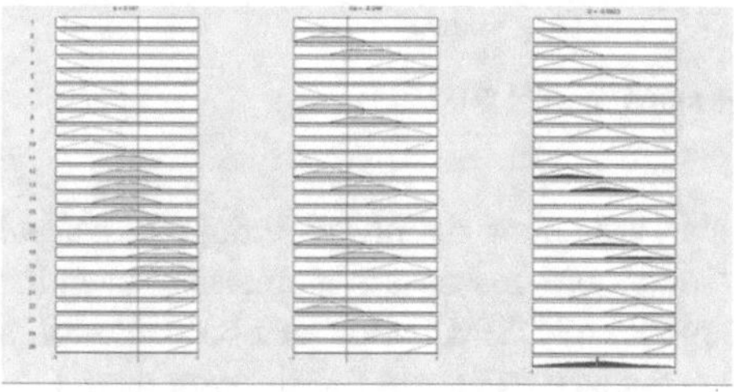

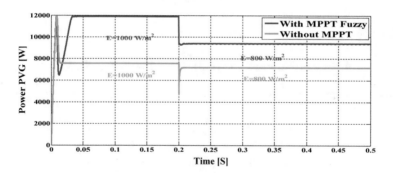

Fig. 2. Rules view of fuzzy MPTT. 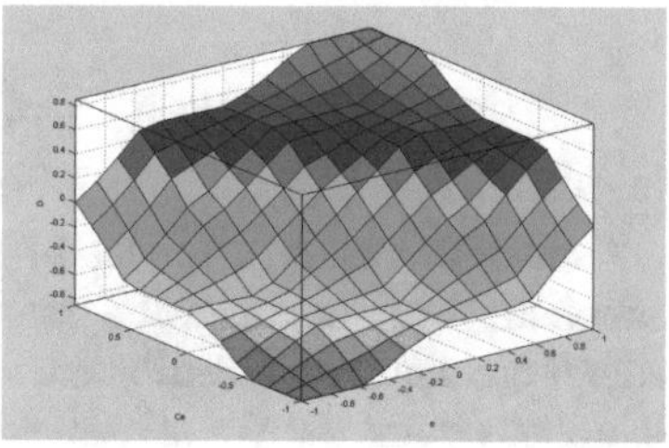   **Fig. 3.** Surface view of fuzzy MPTT.

Fig. 4. Output power of PVG with and without fuzzy MPTT.

4 Dual Star Induction Motor Model

The DSIM includes in the stator two systems of 3 phase windings shifted between them with α electrical angle (in this modeling we take $\alpha = 30°$) and a rotor is coil or squirrel cage. To simplify the study, we consider the electrical circuits of the rotor as equivalent to a three-phase winding short-circuit. The Fig. 5 shown up the position of the axes of coiling of the nine phases constituting the machine.

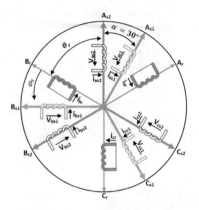

Fig. 5. Vector representation of DSIM.

In order to obtain a model of dual star induction motor, we adopt the usual assumptions i.e.: the magnetic saturation and core losses are neglected, machine windings are sinusoidal distributed, so, the model of DSIM motor, can be written in a synchronous frame (d, q) and expressed in state-space form, is a fourth-order model [2, 3] as:

$$\dot{x} = Ax + BU \tag{4}$$

Where:

$$X = \left[\phi_{ds1}\,\phi_{qs1}\,\phi_{ds2}\,\phi_{qs2}\,\phi_{dr}\,\phi_{qr}\right]^{T}; U = \left[v_{ds1}\,v_{qs1}\,v_{ds2}\,v_{qs2}\,0\,0\right]^{T};$$

The system matrices given by:

$$A = \begin{bmatrix}
\frac{L_a-L_{s1}}{T_{s1}L_{s1}} & \omega_s & \frac{L_a}{T_{s1}L_{s2}} & 0 & \frac{L_a}{T_{s1}L_r} & 0 \\
-\omega_s & \frac{L_a-L_{s1}}{T_{s1}L_{s1}} & 0 & \frac{L_a}{T_{s1}L_{s2}} & 0 & \frac{L_a}{T_{s1}L_r} \\
\frac{L_a}{T_{s2}L_{s1}} & 0 & \frac{L_a-L_{s2}}{T_{s2}L_{s2}} & \omega_s & \frac{L_a}{T_{s2}L_r} & 0 \\
0 & \frac{L_a}{T_{s2}L_{s1}} & -\omega_s & \frac{L_a-L_{s2}}{T_{s2}L_{s2}} & 0 & \frac{L_a}{T_{s2}L_r} \\
\frac{L_a}{T_rL_{s1}} & 0 & \frac{L_a}{T_rL_{s2}} & 0 & \frac{L_a-L_r}{T_rL_r} & \omega_{gl} \\
0 & \frac{L_a}{T_rL_{s1}} & 0 & \frac{L_a}{T_rL_{s2}} & -\omega_{gl} & \frac{L_a-L_r}{T_rL_r}
\end{bmatrix}$$

And

$$B = \begin{bmatrix} 1 & 0 & 0 & 0 & 0 & 0 \\ 0 & 1 & 0 & 0 & 0 & 0 \\ 0 & 0 & 1 & 0 & 0 & 0 \\ 0 & 0 & 0 & 1 & 0 & 0 \\ 0 & 0 & 0 & 0 & 0 & 0 \\ 0 & 0 & 0 & 0 & 0 & 0 \end{bmatrix}$$

The mechanical modeling part of the system given by:

$$J \frac{d\Omega_r}{dt} = T_{em} - T_l - k_f \Omega_r \tag{5}$$

Moreover, the electromagnetic torque given by:

$$T_{em} = P \frac{L_m}{L_m + L_r} \left[(i_{qs1} + i_{qs2}) \phi_{dr} - (i_{ds1} + i_{ds2}) \phi_{qr} \right] \tag{6}$$

5 Rotor Field Oriented Control

The basic idea of vector control is to assimilate the behavior of the DSIM to that of separately excited DC machine, hence to create a linear and decoupled model between electromagnetic torque and flux, which allows improving its dynamic behavior. By considering the following conditions $\phi\_qr = 0$ and $\phi\_dr = \psi\_r$, the drive behavior can be expressed by simplified model as follow [3]:

$$T_{em} = P \frac{L_m}{L_m + L_r} \left[(i_{qs1} + i_{qs2}) \cdot \phi_{dr} \right] \tag{7}$$

The rotor currents given as:

$$i_{dr} = \frac{1}{L_r + L_m} \left[\psi_r - L_m \cdot (i_{ds1} + i_{ds2}) \right] \tag{8}$$

$$i_{qr} = \frac{L_m}{L_r + L_m} (i_{qs1} + i_{qs2}) \tag{9}$$

Moreover, the sliding pulsation given by:

$$w_{sl} = \frac{R_r \cdot L_m}{(L_m + L_r) \cdot \phi_r} (i_{qs1} + i_{qs2}) \tag{10}$$

6 Synthesis of Synergetic Control

Generally, nonlinear dynamic system represented by the following equation [10, 12, 15]:

$$\dot{x} = f(x, u, t) \tag{11}$$

Where x represents the system state vector, and u represents the control vector.

The first step in the design of a synergetic control resides in the training of macro-variables defined as a function of the state variables of the system in the form of algebraic relationship between these variables that reflect the characteristics of the requirements of the design. In the simple case, these macro-variables can be defined in the form of linear combinations of these variables of state and determine the properties of the movement of the system [13, 14] from an initial state of any kind to a state of desired balance said: the manifold.

$$\Psi = \psi(x, t) \tag{12}$$

Where Ψ is the macro-variable and $\psi(x, t)$ a function defined by the user. Each macro-variable Ψ, presents a new constraint on the system in the state space, as well its order reduced to a unit, by forcing it to assess toward a global stability in the Desired State $\Psi = 0$.

The designer can choose the characteristics of these macro-variables according to the requirements or limitations on some of the variables of state.

In the second step, the fixing of the dynamic evolution of macro-variables to the manifolds ($\psi = 0$) by an equation, the functional equation, defined by the following general form [13, 16]:

$$T\dot{\psi} + \psi = 0 \tag{13}$$

With $T > 0$.

T is the control parameter, which indicates the speed of convergence of the system in a closed loop to the area indicated.

The solution of the Eq. (13) gives the following function:

$$\psi(t) = \psi_0 e^{-\frac{t}{\tau}} \tag{14}$$

Taking into account the chain of the differentiation, which is given by [13, 14, 16]:

$$\frac{d\psi(x, t)}{dt} = \frac{d\psi(x, t)}{dx} \frac{dx}{dt} \tag{15}$$

Substitution of the (12) and (13) in (15) allows us to write [10, 14]:

$$T\frac{d\psi(x, t)}{dx} f(x, u, t) + \psi(x, t) = 0 \tag{16}$$

Solving the Eq. (16) for "u", the control law is then expressed as follows [13, 14, 16]:

$$u = g(x, \psi(x,t), T, t) \tag{17}$$

From the Eq. (17), it is apparent that the control depends not only of the state variables of the system, but also of the macro-variable and the control parameter t. In other words, the designer can choose the characteristics of the controller by choosing macro-variable appropriate and specific control parameters T. In the synthesis of the synergistic controller shown above, it is clear that the latter acts on the non-linear system and a linearization or a simplification of the model is not necessary, as is often the case for the approaches traditional control.

6.1   Synergetic Controller Design for DSIM

Generally, the control laws are a function of the parameters, state variables and the convergence time of the system. If we limit our search to a macro-variable that is a linear function of the mechanical state variables, in general it has the following form:

$$\psi_1 = \alpha x_1 + \beta x_2 \tag{18}$$

Where:

$$x_1 = \omega_{r-ref} - \omega_r$$
$$x_2 = \phi_{r-ref} - \phi_r$$

ψ_1 Must satisfy the following equation:

$$T\dot{\psi}_1 + \psi_1 = 0 \quad T > 0 \tag{19}$$

By introducing the Eq. (19) in the first functional Eq. (20), we get:

$$T\left(\alpha \dot{x}_1 + \beta \dot{x}_2\right) + \alpha x_1 + \beta x_2 = 0$$
$$-T\left(\alpha \frac{d\omega_r}{dt} + \beta \frac{d\phi_r}{dt}\right) + \alpha(\omega_{r-ref} - \omega_r) + \beta(\phi_{r-ref} - \phi_r) = 0$$

Where:

$$\begin{cases} \frac{d\omega_r}{dt} = \frac{1}{J}(T_{em} - T_L - k_f\omega_r) \\ \frac{d\phi_r}{dt} = -\frac{R_r}{L_r + L_m}\phi_r + \frac{R_r L_m}{L_r + L_m}(i_{ds1} + i_{ds2}) \end{cases} \tag{20}$$

Replacing $\frac{d\omega_r}{dt}$ *and* $\frac{d\phi_r}{dt}$ by their expressions in the system (20) we get:

$$-T\left(\alpha\frac{1}{J}(T_{em} - T_L - k_f\omega_r) + \beta\frac{d\phi_r}{dt}\right) + \alpha\omega_r + \beta\phi_r = 0 \tag{21}$$

By Solving (21) for T_{em}, we obtained the following control law:

$$T_{em} = \frac{J}{T\alpha}\left[\alpha x_1 + \beta x_2 - T\beta\frac{d\phi_r}{dt}\right] + T_L + k_f\omega_r = 0$$

$$T_{em} = \frac{J}{T\alpha}\left[\alpha(\omega_{r-ref} - \omega_r) + \beta(\phi_{r-ref} - \phi_r) - T\beta\frac{d\phi_r}{dt}\right] + T_L + k_f\omega_r = 0 \tag{22}$$

Where α, β and T are the controller parameters.

7 Tuning of Synergetic Controller Parameters by PSO Technique

Particle Swarm Optimization (PSO) is a robust stochastic optimization method introduced by Kennedy and Eberhart the early 1995 [17]. It is a meta-heuristic inspired by social behavior of bird flocking and fish schooling when searching for food. A brief description of PSO is presented in this section while a detailed explanation can be found in the references [18]. In PSO, through the search process in the problem space, each individual (particle) will adjust its flying velocity and position according to its self-flying experience as well as the experiences of the other companion particles of the swarm. The velocity and position of each particle can be determined by the following equations [17]:

$$v_i(t+1) = w \cdot v_i(t) + \varphi_1 \cdot r_1(t) \cdot (p_{bi}(t) - x_i(t)) + \varphi_2 \cdot r_2(t) \cdot \left(p_{gi}(t) - x_i(t)\right) \tag{23}$$

$$x_i(t+1) = x_i(t) + v_i(t) \tag{24}$$

Where *vi* and *xi* are the current velocity and position of particle at the kth iteration of each particle i; p_{bi} is the best position; g_{bi} is the best swarm position; r1 and r2 are random variables; $\varphi1$ and $\varphi2$ are two accelerations constants; and w is the inertial weight used as a compromise between the local and global exploration capabilities of the swarm. In optimizing the parameters of synergetic controller using PSO algorithm, each particle is requested to represents a potential solution, comprised of a vector of combination (K1, K2, K3). In control system of DSIM, the objective is to improve its performance, specifically, reduce the errors between the reference speed and the real speed of DSIM. Consequently, the fitness function of each possible solution was evaluated using Integral Time Absolute of Error (ITAE) (see Fig. 6) expressed by:

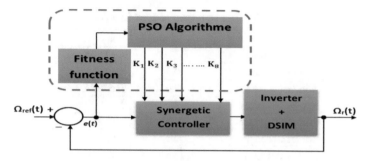

Fig. 6. An optimized synergetic controller by PSO algorithm.

$$J = \text{ITAE} = \int_0^t |e(t)| dt \tag{25}$$

The error e of Eq. (22) represents the function to be minimized is given by:

$$e(t) = \Omega_{\text{ref}} - \Omega_r \tag{26}$$

The good adjustment of scaling factor K1, K2, K3 of synergetic controller can improve greatly the performance of control system. The main steps of the proposed PSO-Synergetic Controller are the following:

Step 1. Initialize randomly the swarm such particle population size, dimension of the search space and specify the search boundaries of controller parameters.

Step 2. Evaluate the fitness of each search particle and determine the best position from the particle with the minimal fitness in the swarm.

Step 3. Update the velocity and position of each particle.

Step 4. Check stopping criteria, if the maximal number of iterations is not yet reached, return to step 2. Otherwise go to following sub step.

Step 5. Terminate the algorithm and give the best controller-parameters.

8 Simulation Results and Discussion

In order to show the performance of the optimized synergetic control of DSIM fed by two inverters coupled in Photovoltaic Generator (PVG) with fuzzy MPPT, a numerical simulation was carried out under different conditions. The parameters of the test DSIM used in the simulation are given in Table 2. From Fig. 7, one can notice that the application of the synergetic control to the DSIM has made it possible to demonstrate its simplicity of design and the superiority of the performances obtained, relative to those obtained with the conventional regulation, or a variable-structure regulator. In fact, it is observed that the orientation condition of the flux of the proposed synergetic

Table 2. Parameters of DSIM model

Parameters	Symbol	Value
Nominal power	Pn	1.5 MW
Nominal voltage	Vn	220/380 V
Nominal rotor speed	ωr	315 rad/s
Number of poles	P	2
Stators resistances	Rs1, Rs2	3.72 Ω
Rotor resistance	Rr	2.12 Ω
Stators leakages inductances	Ls1, Ls2	0.022 H
Rotor leakages inductances	Lr	0.006 H
Mutual leakages inductances	Lm	0.3672 H
Inertial moment	J	0.0662 kg . m^2
Viscous coefficient	Kf	0.001 N.m.s/rad

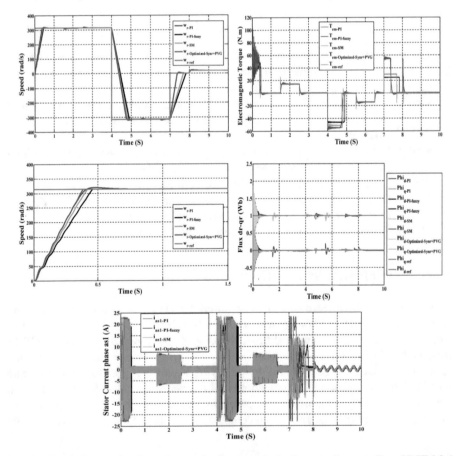

Fig. 7. Simulation results of speed control using an optimized synergetic controller of DSIM fed by PVG with fuzzy MPTT.

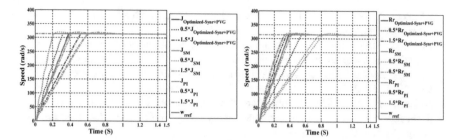

Fig. 8. Simulation results of speed control using an optimized synergetic controller of DSIM fed by PVG with fuzzy MPTT under some key parameters variation: J and Rr.

control is demonstrated by the cancellation of the quadrature flux ϕ_{qr} and by similar performances to those of the separately excited DC machine.

Also the synergetic control gives satisfactory results even with sudden reversal of the speed reference. Besides, the synergetic controller shows a good performance in the rejection of the torque of the load, and the reduction of the vibrations of the torque introduced by the chattering. Moreover, in order to highlight the robustness of suggested controller against key parameter variations of DSIM (J), and Rr, simulation works have been carried out considering increase of this parameter. The obtained results are shown in Fig. 8.

It can be seen that the speed dynamic is unaffected by the key parameter variations. In fact, one can notice that a rapid and no overshoot speed responses are achieved against large variation of inertia and rotor resistance.

9   Conclusion

In this research work, an optimized synergetic control of dual star induction motor (DSIM) fed by two inverters coupled in Photovoltaic Generator (PVG) is presented. By using fuzzy MPPT, solar power is efficiently extracted and fed DSIM controlled by synergetic vector control. To solve the problem of scaling factor determination of synergetic controller, it was proposed in this study, the (PSO) algorithm which can provide the optimal parameter of suggested controller and consequently improve the performance of control system.

From the obtained results, it can be say: that an optimized synergetic control can increase the robustness of the drive speed control. In fact, the s optimized synergistic controller demonstrates best performance in the load torque rejection, and reduction of the vibration torque introduced by the chattering phenomenon.

References

1. Pakkiraiah, B.: Investigations on performance improvement of solar photovoltaic system fed asynchronous motor drive. Thesis doctor of University, Andhra Pradesh, India (2017)

2. Khedher, A., Mimouni, M.F.: Sensorless-adaptive DTC of double star induction motor. Energy Convers. Manag. **51**(12), 2878–2892 (2010)
3. Kouki, H., Ben Fredj, M., Rehaoulia, H.: Modeling of double star induction machine including magnetic saturation and skin effect. In: 10th International Multi-Conferences on Systems, Signals & Devices 2013 (SSD13), pp. 1–5. IEEE (2013). https://doi.org/10.1109/SSD.2013.6564045
4. Amimeur, H., Aouzellag, D., Abdessemed, R., Ghedamsi, K.: Sliding mode control of a dual-stator induction generator for wind energy conversion systems. Int. J. Electr. Power Energy Syst. **42**(1), 60–70 (2012)
5. Yan, Z., Jin, C., Utkin, V.I.: Sensorless sliding-mode control of induction motors. IEEE Trans. Ind. Electron. **47**(6), 1286–1297 (2000)
6. Fnaiech, M.A., Betin, F., Capolino, G.-A., Fnaiech, F.: Fuzzy logic and sliding-mode controls applied to six-phase induction machine with open phases. IEEE Trans. Ind. Electron. **57**, 354–364 (2010)
7. Sabanovic, A., Bilalovic, F.: Sliding mode control of AC drives. IEEE Trans. Ind. Appl. **25** (1), 70–75 (1989)
8. Chang, T.Y., Hong, C.M., Pan, C.T.: No chattering discrete-time sliding mode controller for field acceleration method induction motor drives. In: PESC Record—IEEE Annual Power Electronics Specialists Conference, vol. 00, pp. 1158–1164 (1993)
9. Zhang, Y., Jiang, Z., Yu, X.: Indirect field-oriented control of induction machines based on synergetic control theory. In: 2008 IEEE Power and Energy Society General Meeting - Conversion and Delivery of Electrical Energy in the 21st Century, pp. 1–7. IEEE (2008). https://doi.org/10.1109/PES.2008.4596187
10. Yu, X., Jiang, Z., Zhang, Y.: A synergetic control approach to grid-connected, wind-turbine doubly-fed induction generators. In: 2008 IEEE Power Electronics Specialists Conference, pp. 2070–2076, IEEE (2008). https://doi.org/10.1109/PESC.2008.4592248
11. Kolesnikov, A., Veselov, G., Kolesnikov, A., et al.: Modern applied control theory: synergetic approach in control theory. TRTU, Moscow, Taganrog (2000)
12. Son, Y.-D., Hen, T.-W., Santi, E., Monti, A.: Synergetic control approach for induction motor speed control. In: 30th Annual Conference of IEEE Industrial Electronics Society, 2004. IECON, vol. 1, pp. 883–887. IEEE (2004)
13. Medjbeur, L., Harmas, M.N., Benaggoune, S.: Robust induction motor control using adaptive fuzzy synergetic control. JEE., 1–6 (2012)
14. Davoudi, A., Bazzi, A.M., Chapman, P.L.: Application of synergetic control theory to non-sinusoidal PMSMs via multiple reference frame theory. In: 2008 34th Annual Conference of IEEE Industrial Electronics, pp. 2794–2799. IEEE (2008). https://doi.org/10.1109/IECON.2008.4758401
15. Kolesnikov, A. et al.: A synergetic approach to the modeling of power electronic systems. In: 7th Workshop on Computers in Power Electronics COMPEL 2000, Proceedings (Cat. No.00TH8535), pp. 259–262. IEEE (2000). https://doi.org/10.1109/CIPE.2000.904726
16. Nusawardhana, A., Zak, S.H., Crossley, W.A.: Nonlinear synergetic optimal controllers. J. Guid. Control. Dyn. **30**, 1134–1147 (2007)
17. Eissa, M.M., Virk, G.S., AbelGhany, A.M., et al.: Optimum induction motor speed control technique using particle swarm optimization. Int. J. Energy Eng. **3**, 65–73 (2013)
18. Kennedy, J., Eberhart, R.C.: Swarm Intelligence. Morgan Kaufman Publishers, Burlington (2000)

MPPT and Optimization

Optimization of the Preventive Maintenance for a Multi-component System Using Genetic Algorithm

Zakaria Dahia$^{(\boxtimes)}$, Ahmed Bellaouar, and Soulmana Billel

LITE Laboratory, University of Constantine 1, Constantine, Algeria
zaki.dahia@umc.edu.dz, bellaouar_ahmed@yahoo.fr

Abstract. The main goal of the manufacturers is to efficiently exploit their technological systems to improve their agility and productivity by using all available resources optimally in minimum time. For this, it is necessary to develop effective maintenance strategies to ensure the continuity of production and machine availability while minimizing the overall cost.

Our work describes a policy adopted for the determination of the minimum cost of preventive maintenance of a multi-component system with respect to an availability constraint. For this, we have proposed a model that examines the situation of maintenance decision in which three actions, a minimal repair, a periodic overhaul and a complete renewal, the improvement of the system due to the maintenance action of a revision differs of the virtual age approach by considering a direct reduction of the failure rate.

The genetic algorithm (GA) was used as a technique to optimize the cost function with respect to an availability constraint, i.e. to determine the optimum couples of periodicity of the preventive maintenance and the revision number.

The results obtained considerably improved the preventive maintenance plan.

Keywords: Preventive maintenance · Cost · Multi-components system
Optimization · Genetic algorithm · Preventive maintenance plan

1 Introduction

Optimizing Preventive Maintenance is a process of improving their efficiency and effectiveness. This process tries to balance the requirements of the preventive maintenance (economic, technical, etc.) and the resources used for the realization of their program (manpower, spare parts, articles of consumption, equipment, etc.).

The goal of optimizing preventive maintenance is to choose the appropriate policy for each system and the identification of the periodicity of this policy should be carried out to achieve the objectives regarding safety, reliability of equipment and system availability. The effective management of maintenance requires the use of multi-criteria optimization procedures, based on this procedure; it is up to the decision-maker to choose the optimal preventive maintenance [2].

The criteria for efficiency measures for optimization policies are based solely on maintenance cost measures, such as the expected cost per unit of time and total

© Springer Nature Switzerland AG 2019
M. Hatti (Ed.): ICAIRES 2018, LNNS 62, pp. 313–320, 2019.
https://doi.org/10.1007/978-3-030-04789-4_34

discounted costs [3]. A small part of the maintenance models is based solely on reliability measures and availability [4, 8].

In the literature, the basic assumptions about maintenance efficiency are known as perfect maintenance, which restores the system to its new state. Minimal repair does not change the failure rate. The most realistic hypothesis is the imperfect maintenance that places the system between the two extreme situations [4–7].

In this paper, we used the quasi-renewal process to model maintenance efficiency by considering a direct reduction in the system failure rate due to the maintenance action of a revision [1]. The improvement model assumes that each revision makes the system failure rate "bad as old" and "satisfactory as the previous revision period" with a fixed degree, the measure of effectiveness that will be considered in this work are

- Maintenance costs.
- The availability of system.

Both criteria are used simultaneously; in other words, the model is used to determine the optimal couples (x*, n*) that minimizes the maintenance cost function with respect to an availability constraint.

2   Preventive Maintenance Policy

This strategy assumes that the system undergoes preventative maintenance, which is considered a revision at the end of each time interval x, and whenever a failure occurs, it is repaired and restored to an operational state without changing the failure rate of the system. After a number of overhauls, the system receives a complete renewal that brings the system to a new state; the cycle is repeated over an infinite time horizon [1, 7].

A. System Improvement Model

A system is improved if its failure rate is reduced. Here, the improvement due to an overhaul is defined as follows:

$$\lambda_k(t) = p\lambda_{k-1}(t - X) + (1 - p)\lambda_{k-1}(t) \tag{1}$$

$\lambda_{k-1}(t)$ is the system's failure rate function just before the overhaul, $\lambda_k(t)$ is the failure rate function right after the overhaul, x is the overhaul interval and p improvement degree $p \in [0, 1]$.

Notation
$\lambda(t)$: Original failure rate of the system (without PM);
$\hat{\lambda}(t)$: Actual failure rate of the system (with periodic overhauls);
$H(t)$: Originally expected number of failures in the interval [0, t);

$$H(t) = \int_0^t \lambda(x)dx$$

$\hat{H}(t)$: Actual expected number of failures in the interval [0, t);

$$\hat{H}(t) = \int_0^t \hat{\lambda}(x)dx$$

p, q: p is the improvement degree and $q = 1 - p$;
X: PM (overhaul) interval;
n: Number of overhauls in a renewal cycle;
C_m, C_P, C_r: Costs of minimal repair, PM and renewal respectively;
T_m, T_P, T_r: Downtime of minimal repair, PM and renewal respectively;
$C(n, X)$: The expected unit-time cost when the system is receives $(n - 1)$ overhauls with interval X in a renewal cycle.
A: Threshold value of Availability of the system (Availability constraint).

Assumptions
(1) A minimal repair does not change the failure rate.
(2) In this paper, each component is assumed to follow a Weibull distribution with known beta (β) and Etta (η) values.
(3) An overhaul (PM) improves the system with a fixed degree p.
(4) All the renewal cycles have the same length ns.
(5) $C_m, C_P, C_r, T_m, T_P, T_r\, \lambda(t)\, and\, H(t)$ are known.

3 Optimization Model

A. Cost Model

The total expected cost in a renewal cycle $C(n, s)$ is given by:

$$C(n, X) = \frac{Expected\ cost\ incurred\ during\ a\ cycle\ E(c)}{Expected\ length\ of\ cycle\ E(L)} \tag{2}$$

$$C(n, X) = \frac{C_r + C_P(n - 1) + C_m\hat{H}(ns)}{ns} \tag{3}$$

Where

$$\hat{H}(nX) = \sum_{i=0}^n \binom{n}{i} p^{n-1}q^{i-1}H(iX) \tag{4}$$

B. Availability Model

The other measure of effectiveness is the availability function $A(n, X)$ is given by:

$$A(n, X) = 1 - \frac{downtime}{downtime + uptime} \tag{5}$$

The downtime of the system is given by: $T_r + T_P(n - 1) + T_m \hat{H}(nX)$
The uptime of the system is given by: nx

$$A(n, X) = \frac{nX}{nX + T_r + T_P(n - 1) + T_m \hat{H}(nX)} \tag{6}$$

C. Problem Formulation

We used the cost model as an objective function to be minimized and the availability model as a constraint, so the problem can be formulated as:

$$\text{Min } C(n, X)$$

Subject to

$$A(n, X) \leq A$$

- For Weibull failure rate:

$$\lambda(t) = \frac{\beta}{\eta} \left(\frac{t}{\eta}\right)^{\beta - 1} \quad \text{With } \beta > 0$$

The originally expected number of failures in a renewal cycle

$$H(iX) = \int_0^{iX} \lambda(x) dx$$

By introducing in Eq. (4) the actual expected number of failures becomes

$$\hat{H}(nX) = \left(\frac{X}{\eta}\right)^{\beta} \sum_{i=0}^{n} \binom{n}{i} p^{n-1} q^{i-1} i^{\beta}$$

D. Solution Procedure

The above model is used to determine the optimal number and intervals overhauls for each subsystem according to the following procedure: first, estimate the interval in which the optimal number of revisions is found, then find x, n which minimizes the objective function (cost) for each fixed n in this range, and finally to select X, n such that the objective function is minimized. The optimal couple is the one that gives a minimum of maintenance costs.

In the multi-components of the system, the solutions provided by this model remain unrealistic, because we cannot perform maintenance at any time (resource constraint), for an effective maintenance strategy and to avoid frequent maintenance shutdowns on different components. In this case, the constraint where the maintenance intervals must be a positive integer multiple is implemented to round the intervals.

- For overhaul interval

$$X* = a * T \text{ a is an positive integer}$$

T: is constant for realistic maintenance time step determined on the basis of the availability of technicians and resources.

- For the number of overhauls

To create the opportunity to renew components simultaneously:

1. We determine the component that seems critical
2. Determine the optimal number for this component (n1)
3. For the number of revisions for the other components n must be chosen so that all the instants of renewal of the components become divisible by an integer (ni = b n1 *)

In our case we assume that the first component is critical.
Finally, we distinguish two cases:

1. Optimization without implantation of rounding constraint of the intervals (type X = aT)
2. Optimization with implantation of rounding constraint.

Optimization by genetic algorithms is appropriate to be used in the search for optimal couples.

Following this technique, we realized a program in Matlab in order to minimize the cost compared to a availability constraint.

4   Numerical Example

As numerical case, our system essentially composed of three subsystems in series, in this paper, each component is follow a Weibull distribution with known (β) and (η) values; Tables 1 and 2 gives the Weibull parameters, data costs and constraints (Availability).

Table 1. Input parameters of the subsystems

	β	η	Minimal repair cost $C_m(\$)$	Preventive cost (PM) $C_p(\$)$	Renewal cost $C_r(\$)$
Component 1	2	2300	190	7000	280000
Component 2	2.5	6500	300	5000	200000
Component 3	1.6	1500	200	9500	300000

Table 2. Availability constraint

	Availability constraint (%)
Component 1	80
Component 2	90
Component 3	90

Table 3. Optimization results for first component

n	A (%)	C ($/hour)	PM interval (hour)
1	83.23	6.34	88293.77
2	84.96	5.56	51609.70
5	87.62	4.41	27920.94
6	87.78	4.34	24180.24
7	87.75	4.34	21172.10
8	87.86	4.29	19141.63
9	87.84	4.29	17371.60
10	**87.89**	**4.26**	**16065.61**
11	87.86	4.27	14881.91
12	87.78	4.30	13825.58
15	87.65	4.33	11615.81

5   Optimization Results

A. Without Constraint Of Rounding

At first, each individual component is taken into account and the optimal numbers and time intervals for each component are determined. The expected cost is calculated for each component, independently, the following Table 3 shows the results for the first component.

According to the results, the optimal couples (n*, x*) obtained give the minimum cost per unit of time is 4.26, so in this case for the first component, the objective value is

C (n*, X*) = C (10, 16065.61) = 4.26 ($/hour) and the availability is 87.89%.

Table 4 summarizes the optimal results for the other components.

Table 4. Optimization results for system

Components	n*	A (%)	C* ($/hour)	PM interval (hour) X*	Renewal cycle (year) (n*x)
C1	10	87.89	4.26	16065.61	18.33
C2	11	94.37	2.83	13344.10	16.75
C3	8	91.46	2.93	41599.8	37.99

This model leads to system availability 75.90% and each component has a renewal time, we can see that there is no opportunity to renew the subsystems simultaneously.

B. With Constraint of Rounding

This constraint makes it possible to coordinate the maintenance operations of the overall system in order to better manage the times between the maintenance operations of the subsystems. The Table 5 gives the results after the integration of this constraint.

Table 5 gives the new optimum couples (n*, X*) after the application of this constraint to permit coordination of subsystem overhauls and their renewal. We can see that this constraint leads to small additional maintenance cost, but at the same time, it leads to an increase in the availability (A before − A after = 78-75.90). So the availability improvement is about 2.1% and less downtime due to non-joint maintenance operations.

For the renewal cycle the first and second subsystems undergo a renewal of 17.7 years, so we can see that there is a simultaneous renewal, the third subsystem undergoes a renewal every 35.47 year at the second renewal of the two subsystems.

Table 5. Optimization results for system

Components	n*	A (%)	C* ($/hour)	PM interval (hour) X*	Renewal cycle (year) (n*x)
C1	9	90.66	4.28	17260.00	17.70
C2	12	94.06	2.85	12900.00	17.67
C3	6	91.38	2.97	51800.00	35.47

6 Conclusion

The purpose of maintenance policies is to improve system availability and reduce the frequency of failures to minimize overall costs.

The approach proposed by Zhang is used to determine the periodicity and number of maintenance reviews of individual components in serial multi-component systems, based on periodic reviews with a minimum repair policy in case of failure.

This article is considered to be an aid tool for engineers and maintenance managers to manage maintenance operations for all components with different management requirements, namely the stress for availability.

As in multi-component systems in series, the shutdown of any subsystem causes the shutdown of the whole system and the production losses are very high, a constraint being exerted on the rounding of the intervals between the components to achieve a more realistic solution allowed us to adjust all maintenance intervals of the components for the asset they manage, according to the needs of the company.

The results come up:

1. The applicability and high efficiency of the genetic algorithm optimization technique in the field of optimizing preventive maintenance
2. That this model in multi-component systems can be considered as a decision support tool for the management and organization of maintenance operations.

References

1. Zhang, F., Jardine, A.K.S.: Optimal maintenance models with minimal repair, periodic overhaul and complete renewal. IIE Trans. **30**, 1109–1119 (1998)
2. Chareonsuk, C., Nagarur, N., Tabucanon, M.T.: A multicriteria approach to the selection of preventive maintenance intervals. Int. J. Prod. Econ. **49**(1), 55–64 (1997)
3. Pham, H., Wang, H.: Imperfect maintenance. Eur. J. Oper. Res. **94**, 425–438 (1996)
4. Zhao, M.: Availability for repairable components and series system. IEEE Trans. Reliab. **43**, 329–334 (1994)
5. Laggoune, R., Chateauneuf, A., Aissani, D.: Preventive maintenance scheduling for a multi-component system with non negligible replacement time. Int. J. Syst. Sci. **41**, 747–761 (2010)
6. Phelps, R.I.: Optimal policy for minimal repair. J. Oper. Res. Soc. **34**, 425–427 (1983)
7. Pongpech, J., Murthy, D.N.P.: Optimal Periodic Preventive Maintenance Policy for Leased Equipment, pp. 1–6. Elsevier, Amsterdam (2005)
8. Wang, H., Pham, H.: Some maintenance models and availability with imperfect maintenance in production systems. Ann. Oper. Res. **91**, 305–318 (1999)

Optimization of a Photovoltaic Pumping System by Applying Fuzzy Control Type-1 with Adaptive Gain

Samia Bensmail[1(✉)], Djamila Rekioua[2], and Chafiaa Serir[2]

[1] Electrical Engineering Department, University of Bouira, Bouria, Algeria
ben_sam68@yahoo.fr
[2] Laboratory LTII, Department of Electrical Engineering,
University of Bejaia, Béjaïa, Algeria
dja_rekioua@yahoo.fr, Serir.chafia@gmail.com

Abstract. In this paper a robust intelligent controller based on the theory of fuzzy logic is adopted in order to fulfil maximum power point tracking (MPPT) algorithm for a photovoltaic generator, on the one hand and to control the asynchronous motor used in the photovoltaic pumping system on the other hand. In this context, the modelization and simulation of the various constituents of the installation (photovoltaic cell, converters, asynchronous motor and pump) is done. To follow the point of maximum power, an adaptation stage equipped with an MPPT algorithm is inserted by introducing two fuzzy algorithms (classical and adaptive). In the second part we will be interested in the control of the asynchronous motor by afield oriented control (FOC, using three controllers, the simulation results obtained on Matlab/Simulink, shows the importance of the techniques implemented in particular on the efficiency, robustness and response time.

Keywords: Photovoltaic · PV pumping · MPPT · Fuzzy logic
Adaptive fuzzy logic

1 Introduction

Solar photovoltaic electricity is the most elegant way to produce electricity without moving parts, gas and noise, while converting abundant, non-exhaustible sunlight into useful electrical energy. Driven by this context the use of energy spreads in several areas (lighting, pumping …) especially in remote areas where the power network is absent [1, 2]. Unfortunately, the nonlinear characteristic of the photovoltaic generator only makes it possible to obtain the maximum power in a single point (MPP), but this point varies according to several parameters such as solar irradiation, temperature and the nature of the charge [3, 4]. These types of variations are random and very difficult to control effectively to remedy this problem several algorithms (MPPT) have been developed and widely adapted to determine the maximum power point. In this paper, we present two MPPT methods that use fuzzy logic theory, which aim to optimize the amplitude of the perturbation in order to minimize the oscillations and increase the speed of convergence then we used these two techniques to control the speed of the asynchronous machine, doing a comparative study with a classical PI regulator, it has

© Springer Nature Switzerland AG 2019
M. Hatti (Ed.): ICAIRES 2018, LNNS 62, pp. 321–328, 2019.
https://doi.org/10.1007/978-3-030-04789-4_35

been possible to show the efficiency and the robustness of the fuzzy controllers, in particular the AFLC algorithm.

2 System Description

Figure 1 describes the photovoltaic pumping system used in this article. It consists of a photovoltaic generator (GPV) that provides power to the installation, DC/DC converter to adapt the output of the GPV to the load by ensuring maximum power operation, induction motor feed via inverter and a centrifugal pump. The system objectives are to ensure maximum photovoltaic generator operation and to improve the dynamic performance of the photovoltaic pumping system by using speed controllers.

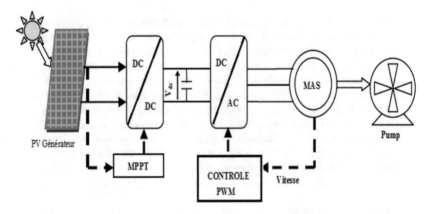

Fig. 1. Overall structure of photovoltaic pumping.

3 MPPT Controllers

3.1 Fuzzy Control [5]

Fuzzy logic is a mathematical description of a process, based on the theory of fuzzy sets, its mechanism is articulated around three main stages namely: fuzzification, fuzzy inference and deffuzification (Table 1).

- *Fuzzification* [6, 7]

Fuzzification or the interface with the fuzzy, establishes an adequate representation of the knowledge. During this step the input numeric variables are converted to linguistic variables based on the membership functions (Fig. 2).

Table 1. Fuzzy controller rules

Error(e)	Change of error(Ce)						
	N_B	N_M	N_S	Z_E	P_S	P_M	P_M
N_B	N_B	N_B	N_B	N_B	N_M	N_S	Z_E
N_M	N_B	N_B	N_B	N_M	N_S	Z_E	P_S
N_S	N_B	N_B	N_M	N_S	Z_E	P_S	P_M
Z_E	N_B	N_M	N_S	Z_E	P_S	P_M	P_B
P_S	N_M	N_S	Z_E	P_S	P_M	P_B	P_B
P_M	N_S	Z_E	P_S	P_M	P_B	P_B	P_B
P_B	Z_E	P_S	P_M	P_B	P_B	P_B	P_B

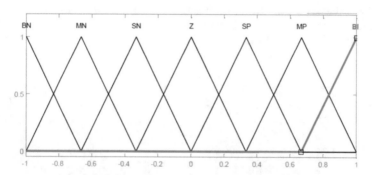

Fig. 2. Fuzzy controller membership function (inputs, output).

- *Inférence*

The objective of fuzzy inference is to construct rules of decision and to find for each of them the function of belonging to the conclusion. In this work the resulting membership functions of the linguistic variable dαi to each of its classes (Table 2).

Table 2. Rule base for computation of β.

Erreur(e)	La variation d'erreur(Ce)						
	N_B	N_M	N_S	Z_E	P_S	P_M	P_M
N_B	N_B	N_B	N_M	Z_E	Z_E	Z_E	Z_E
N_M	N_B	N_M	N_M	Z_E	N_M	P_S	P_S
N_S	N_B	N_B	N_B	N_B	P_M	P_S	P_M
Z_E	N_B	N_B	N_S	Z_E	P_S	P_M	P_B
P_S	N_M	N_S	Z_E	P_S	P_M	P_B	P_B
P_M	N_S	P_B	P_B	P_B	P_B	P_B	P_B
P_B	Z_E	P_B	P_B	P_B	P_B	P_B	P_B

- *Défuzzification*

Défuzzification uses the center of gravity method to calculate the FLC algorithm output which is the change in the duty cycle.

$$d\alpha = \frac{\sum_{j=1}^{n} \Gamma(\alpha_j) * \alpha_j}{\sum_{j=1}^{n} \Gamma(\alpha_j)} \tag{1}$$

3.2   Adaptive Fuzzy Control [8]

We call an adaptive fuzzy controller if one of its adjustable parameters (membership functions, fuzzy rules…) changes while the controller is in use. Figure 3 shows the principle diagram of the AFLC.

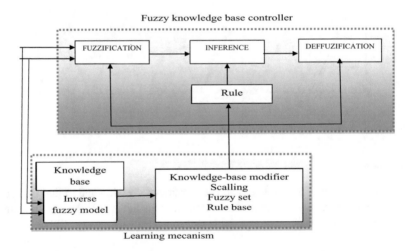

Fig. 3. Structure of an adaptive fuzzy controller

As shown in Fig. 3, the AFLC method is composed of two parts:

a. **Fuzzy basic learning controller**: represents the classic fuzzy controller
b. **Learning Mechanism:** The purpose of this part is to study the environmental parameters and to modify the FLC accordingly, so that the response of the global system is near the optimum point.

In our study we propose to modify the out scaling factor of a conventional fuzzy controller according to Eq. (2).

$$\Delta\alpha_{FA} = \beta \cdot \Delta\alpha_{FC} \tag{2}$$

With:

β: the gain adapting factor, $\Delta\alpha_{FA}$: the output of the adaptive fuzzy controller, $\Delta\alpha_{FC}$: the output of the classic fuzzy controller

The rule base for the calculation of β is given in Table 2 and the membership functions of the error (e), for change of the error (Ce) we will keep the same ones presented on (Fig. 2).

4   Numerical Simulation

To verify the execution of the methods implemented, we will perform two tests, we take the reference values of the flux and the DC bus voltage are given respectively: $\Phi d_{-ref} = 0.7 Wb$ and $V_{dcref} = 460$ V.

In a first place we simulate the global system under the STC conditions using the three controllers; the results obtained are given by (Figs. 4 and 5).

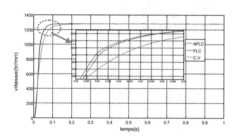

Fig. 4. Rotation speed   **Fig. 5.** Pump flow for standard conditions

The simulation results show that the response of the fuzzy controllers is faster, especially that of the AFL controller, which increases the pumping time and the flow of water. Table 3 gives the response time (τ) for each controller.

Table 3. Response time for each regulator

Controler	$\tau(s)$
PI-classic	0.0705
PI-fuzzy	0.0612
PI-adaptive fuzzy	0.056

After having tested the system under stable conditions, we will execute it under variable climatic conditions (Figs. 6 and 7). The simulation results are given in Figs. 8, 9, 10, 11 and 12.

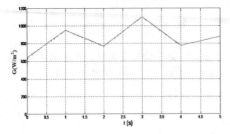

Fig. 6. Solar radiance

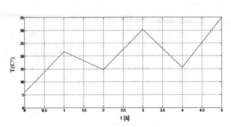

Fig. 7. Temperature

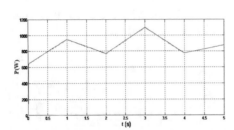

Fig. 8. Maximal photovoltaic power

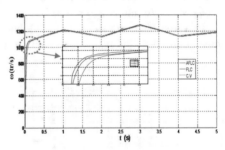

Fig. 9. Rotation speed

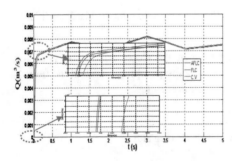

Fig. 10. Flow rate of the pump.

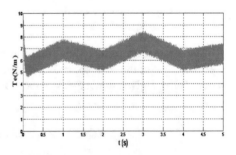

Fig. 11. Electromagnetic couple

It is clear that fuzzy controllers have a faster response, particularly the adaptive fuzzy controller; this allows pumping to begin and therefore the pumping time will be improved.

In Fig. 8 the AFLC algorithm follows the MPP and gives the maximum power at every moment. It can be clearly seen in Fig. 9 the variation of the speed depends on the variation of the photovoltaic power.

The flow pumped by the pump is given in Fig. 10, it is the image of the speed of rotation of the motor and this is explained by the principle of converting mechanical

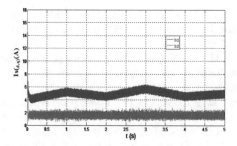

Fig. 12. stator currents (I_{sd}, I_q).

energy into hydraulic energy. It can be noted that the pumping time is improved in the case where the adaptive fuzzy regulator is used.

The form of I_{sd} remains constant independently of I_{sq} which has the same shape as the electromagnetic couple which responds well to the principle of vector control.

5 Conclusion

This work is dedicated to the optimization of a photovoltaic pumping system, to be done we divide the work into two parts.

The first part consists in maximizing the power produced by the photovoltaic generator using two controllers, the first based on the theory of classical fuzzy logic and the other adaptive. The comparison between the two methods shows a slight difference in response time.

In the second part we applied a vector control with rotor flow orientation to the asynchronous machine using three regulators namely PI-classic, PI-fuzzy and adaptive PI-fuzzy. The results obtained in the three cases are satisfactory, but comparing the dynamic performance of the system with different controllers, we find that the controller based on the adaptive fuzzy logic gives better results in terms of speed and efficiency which allowed increasing the pumping time.

References

1. Bentaillah, A.: étude expérimental et de simulation des performances d'une installation PV de faible puissance. Memory of Magister in energy physics Tlemcen (1994)
2. Rekioua, D., Bensmail, S., Bettar, N.: Development of hybrid photovoltaic-fuel cellsystem for stand-alone application. Int. J. Hydrogen Energy **39**, 1604–1611 (2014)
3. Karbakhsh, M., Aburotabi, H., Khazaee, A.: An enhanced MPPT fuzzy control of a wind turbine equipped with permanent magnet synchronous generator. In: 2nd International Conference on Computer and Knowledge Engineering (ICCKE), October 18–19, IRAN (2012)
4. Gules, R., Pellegrin, J.D.P., Hey, H.L., Imhoff, J.: A maximum power point tracking system with parallel connection for PV stand-alone applications. IEEE Trans. Ind. Electron. **55**, 2674–2683 (2008)

5. Patcharaprakiti, N., Premrudeepre-echacharn, S., Sriuthaisiriwong, Y.: Maximum power point tracking using adaptive fuzzy logic control for grid-connected photovoltaic system. Renew. Energy **30**, 1771–1788 (2005)
6. Bensmail, S., Rekioua, D.: Optimisation d'un système de pompage photovoltaïque en utilisant la commande vectorielle basée sur la logique flou adaptative. In: Conférence ICEE 2014, October 21–23, Annaba, algeria (2014)
7. Bensmail, S., Rekioua, D., Azzi, H.: Study of hybrid photovoltaic/fuel cell system for stand-alone applications. Int. J. Hydrogen Energy **40**, 13820–13826 (2015)
8. Bensmail, S., Rekioua, D., Serir, C.: Energy management and optimization of a hybrid system (wind/photovoltaic) with battery storage for water pumping. In: 6th European Conference on Renewable Energy Systems, 25–27 June, Istanbul, Turkey (2018)

Comparative Study of Maximum Power Point Tracking Algorithms for Thermoelectric Generator

Abdelkader Belboula[1(✉)], Rachid Taleb[2], Ghalem Bachir[1],
and Fayçal Chabni[2]

[1] LDEE Laboratory, Electrical Engineering Department,
USTO-MB University, Oran, Algeria
aek.belboula@gmail.com
[2] LGEER Laboratory, Electrical Engineering Department,
Hassiba Benbouali University, Chlef, Algeria

Abstract. Variations in load and temperature can cause a thermoelectric generator (TEG) to operate at a voltage that does not produce the maximum possible power for a given temperature difference. Therefore a maximum power point tracker (MPPT) is used to force the generator to a voltage that produces maximum power. This paper presents a comparative simulation study of two important MPPT algorithms specifically perturb and observe and incremental conductance. The Matlab Simulink environment is used to analyze and interpret the simulation results of these algorithms, and therefore show the performance and limitations of each algorithm. As a result, the Incremental conductance method has shown promise as a suitable MPPT algorithm for a TEG subjected to steady state conditions.

Keywords: Maximum power point tracking · Thermoelectric generator TEG
INC MPPT · P&O MPPT

1 Introduction

In the last decade, problems related to energy factors (oil crisis), ecological aspects (climatic change), electric demand (significant growth) and financial/regulatory restrictions of wholesale markets have arisen worldwide. These difficulties, far from finding effective solutions, are continuously increasing, which suggests the need of technological alternatives to ensure their solution. One of these technological alternatives is known as distributed generation (DG), and consists of generating electricity as near as possible of the consumption site, in fact like it was made in the beginnings of the electric industry, but now incorporating the advantages of the modern technology [1]. Here it is consolidated the idea of using clean non-conventional technologies of generation that use renewable energy sources (RESs) that do not cause environmental pollution [2].The thermoelectric generators (TEG) perfectly fit into this category [3, 4].

The TEGs are solid-state devices engineered to generate electricity directly from heat, what is known as Seebeck effect [5]. TEGs, which use the thermoelectric or

© Springer Nature Switzerland AG 2019
M. Hatti (Ed.): ICAIRES 2018, LNNS 62, pp. 329–338, 2019.
https://doi.org/10.1007/978-3-030-04789-4_36

Seebeck effect of semiconductors to convert heat energy to electrical energy, have existed for many years with the initial discovery of the thermoelectric effect being made in 1821 by Thomas J. Seebeck. Due to the relatively low efficiency and high costs associated with the technology it has been limited to specialized military, medical, space and remote applications [6, 7]. However, the recent need to new energy sources at all scales, and the technological development of new generation of power processing devices and circuits, has put the TEG again on the list as viable energy source to be exploited and improved for commercial use, and to diversify its applications [3].

The efficiency changes according to the TE material used in the manufacture of TEGs. In 1995, only materials with maximum efficiency of 5% were available, and after that, new TE materials that provide efficiency greater than 15% were discovered. Scientists believe that in a near future it is going to be possible to have strongly doped semiconductor materials that have efficiency greater than 25% [8, 9]. The output power of the TEG module strongly depends on the temperature gradient applied to the TEG module to maximize the power output of the TEG module, an MPPT algorithm should be used to keep the operating point of the TEG to in the optimum location. Several MPPT algorithms have been applied for TEG systems, most of these algorithms have been originally developed for photovoltaic (PV) systems [10, 11]. Different methods are used for maximum power point tracking. Currently, Incremental Conduction, Perturb and Observe and Ripple Correlation Control are the most frequently discussed and analysed MPPT algorithms in literature [12, 13]. All these maximum power point tracking algorithms are rather slow to respond to the fast-changing weather conditions. Furthermore, most of them can not accurately detect the maximum power point [14].

2   Model of Thermoelect Generator

The Thermoelectric module (TEM) is based on the Seeback effect, which states that an electromotive force is introduced between two semiconductors when a temperature difference exists [15].

Several models have been proposed to describe the operation of the thermoelectric module and its behavior under different conditions (temperature gradient and load). The model chosen is the Thevenin Equivalent Circuit of Thermoelectric module (TEM).

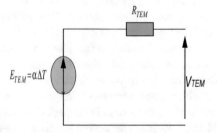

Fig. 1. Diagram electrical equivalent of a Thermoelectric module (TEM) (Thevenin Equivalent Circuit).

This model is known as a Thevenin Equivalent Circuit with voltage source as a function of the temperature gradient and a serial resistance equivalent to the internal resistance. The equivalent electrical diagram of the thermoelectric module as shown in the figure below (Fig. 1).

where, the TEM electromotive force characteristic of is described by the following expression:

$$E_{TEM} = \alpha_{np}\Delta T \tag{1}$$

E_{TEM}: electromotive force produced by the TEM;
α_{np}: the Seeback coefficients;
ΔT: Temperature gradient;

where, the TEM voltage-current characteristic is described by the following expression:

$$V_{TEM} = \alpha\Delta T - R_s I_{TEM} \tag{2}$$

V_{TEM}: the Output voltage delivered by the TEM;
α_{np}: the Seeback coefficient V/K;
ΔT: the Temperature gradient K;
I_{TEM}: Current produced by the TEM;
R_s: the equivalent serial Resistance of TEM;

A TEG consists of several TEMs, which are electrically connected in a series–parallel arrangement.

The Output voltage V_{TEG} and generated output power are expressed as:

$$V_{TEG} = N_S\alpha\Delta T - R_{TEG} I_{TEG} \tag{3}$$

$$P_{TEG} = N_S N_P V_{TEM} I_{TEM} \tag{4}$$

N_S and N_P: the number of TEMs in series and in parallels.
R_{TEG}: the equivalent serial Resistance of TEG;
V_{TEM}: the Output voltage delivered by the TEG;
I_{TEG}: Current produced by the TEG;
where:

$$R_{TEG} = \frac{N_S}{N_P} R_s \tag{5}$$

$$I_{TEG} = N_P I_{TEM} \tag{6}$$

3 Maximum Power Point Tracking Principle (MPPT)

A. *Principle*

The analytical definition of the optimum of a function is the point through which its derivative with respect to a given variable is zero. All algorithms for calculating the maximum power point consulted are based on this principle.

There is an operating point where the power delivered is at a maximum (Fig. 2). The optimization consists in performing this permanently acting automatically on the load seen by the thermoelectric generator, for this adaptation the principle is carried out in general by means of a static converter where the losses should be as low as possible and which can also ensure a shaping according to an outcomes, different attitudes may be considered as to control the adapter.

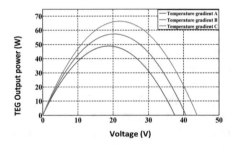

Fig. 2. Characteristic of power delivered by TEG.

This kind of control is often called "Maximum Power Point Search" or "Maximum Power Point Tracking" (MPPT). Figure 3 shows a basic chain of elementary Power conversion associated with an MPPT control. To simplify the operating conditions of this command, a DC load is chosen friendly. As we can see in this chain, in the case of TEG conversion.

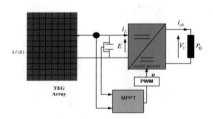

Fig. 3. Schematic diagram of the MPPT converter.

The adapter can be achieved using a DC-DC converter so that the power supplied by the thermoelectric generator corresponds to the maximum power (P max) that generates and it can then be transferred directly to load.

B. *DC-DC Boost Converter*

The DC-DC converter circuit used in this paper is a boost chopper circuit with the advantage that the boost circuit is more efficient than the buck circuit and its energy efficiency can vary with the duty cycle. When the converter runs stably, mean of the induced voltage in one switching cycle is zero.

Consider the dynamic model of a thrust power converter that has been depicted in Fig. 4. The converter status equations are:

$$\begin{cases} L\frac{di_L}{dt} = E - (1-u)V_C \\ C\frac{dV_C}{dt} = (1-u)i_L - \frac{V_C}{r_0} \end{cases} \tag{7}$$

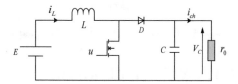

Fig. 4. Schematic boost converter.

4   Incremental Conductance (INC) Algorithm

This algorithm is selected because of its simplicity and the ability to detect and track maximum power point keeping the operation point of solar power plants at it [14].

In this algorithm, calculating the derivative of the panel output power; this derivative is zero at maximum power point, positive and negative to the left to right point MPP [16]. The panel output power P given by:$P = VI$

$$\begin{cases} \frac{dP}{dV} = 0 \Rightarrow P = P_{\text{max}} \\ \frac{dP}{dV} \langle 0 \Rightarrow P \langle P_{\text{max}} \\ \frac{dP}{dV} \rangle 0 \Rightarrow P \rangle P_{\text{max}} \end{cases} \tag{8}$$

The partial derivative is $\frac{dP}{dV}$ given by:

$$\frac{dP}{dV} = I + V\frac{dI}{dV} \cong I + V\frac{\Delta I}{\Delta V} \tag{9}$$

Then, (9) can be described as follows:

$$\begin{cases} \frac{\Delta I}{\Delta V} = -\frac{I}{V} \Rightarrow P = P_{\text{max}} \\ \frac{\Delta I}{\Delta V} \langle -\frac{I}{V} 0 \Rightarrow P \langle P_{\text{max}} \\ \frac{\Delta I}{\Delta V} \rangle - \frac{I}{V} \Leftarrow P \rangle P_{\text{max}} \end{cases} \tag{10}$$

A flowchart of the INC algorithm is shown in Fig. 5.

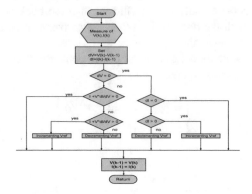

Fig. 5. The flow chart of the INC method.

5   Perturb and Observe (P&O) Algorithm

The P&O method is widely used in commercial products and is the basis of the largest part of the most sophisticated algorithms presented in the literature [17]. It is widely employed in practice, due to its low-cost, simplicity and ease of implementation [18, 19].

The principal of this algorithm is to compare the actual power with the previous one, if the difference is positive the voltage is increased by given perturbation step size, else the voltage is decreased by the same given perturbation step size, the MPP is reached when this difference is zero

A flowchart of the P&O algorithm is shown in Fig. 6.

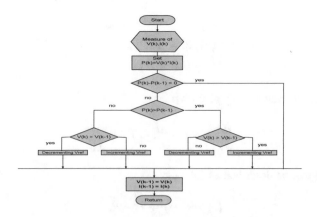

Fig. 6. The flow chart of the P&O method.

6 Simulation Results and Discussion

A. *Robustness Study of INC and P&O Applied for TEG System*

To validate the algorithm operation INC MPP and P&O MPP the TEG, is performed by introducing variations on the different intervening variables on the operational MPPT. Furthermore, ranks for some variables in t = 2.5 s, is introduced.

– *Temperature gradient variation*

Assuming a rise in temperature gradient from 60 °C to 65 °C at time t = 2.5 s, the simulation results are shown in Fig. 7.

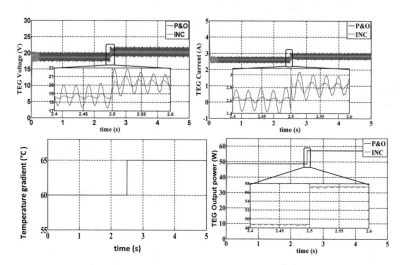

Fig. 7. Outcomes INC MPPT algorithms and P&O MPPT for a temperature gradient increase of 60 °C to 65 °C whit load 10 Ohm

– *Load variation*

Assuming a load increase from 10 to 20 Ohm at time t = 2.5 simulation results are shown in Fig. 8.

– *Temperature gradient and load changes*

Here, it's exposed both MPPT algorithms to a change in various parameters temperature gradient and the load at the same time, simulation results are shown in Fig. 9.

B. *Discussion of the Results*

In Fig. 7 shows the effect of increasing in power, caused by an increase of the temperature gradient, which causes a deviation of the maximum power point MPP for both algorithms with increased current and voltage. Once the temperature gradient stabilizes, the power returns to its steady state with less disruption to INC MPPT. This

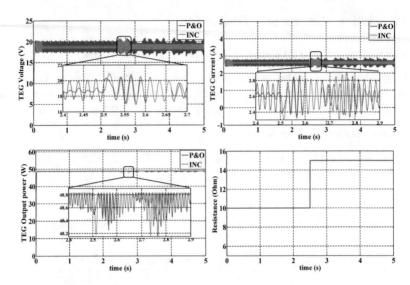

Fig. 8. Responses INC and IDA-PBC algorithms for load increase 10 to 20 Ohm with temperature gradient 60 °C

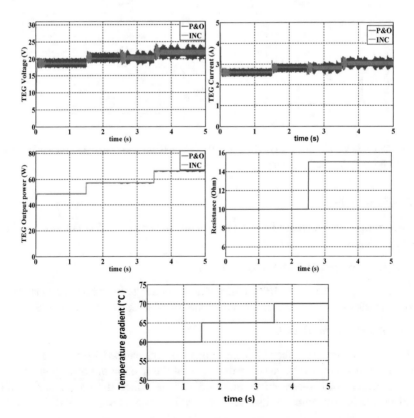

Fig. 9. Outcomes of MPPT algorithms INC and P&O achievement for a parameters variations.

has resulted a very short response time and better dynamic performance with negligible disruption over INC MPPT.

It's also noted that in Fig. 8, despite the change in the load, both MPPT algorithms have retained the optimal values of the power of TEG generator, and negligible disruption to INC MPPT, consequently a good income.

In Fig. 9 shows the performance of two algorithms in the case of variation in temperature gradient and load the results of this test show the good pursuit of both algorithms but with speed and higher steadiness and less disruption as of INC MPPT controller responses compared to P&O MPPT controller.

7 Conclusion

In this paper a mathematical model of a thermoelectric generator (TEG) has been developed using MATLAB Simulink. This model is used for the maximum power point tracking algorithms. The P&O and Incremental conductance MPPT algorithms are discussed and their simulation results are presented. It is proved that Incremental conductance method has better performance than P&O algorithm. These algorithms improve the dynamics and steady state performance of thermoelectric system as well as it improves the efficiency of the dc-dc converter system.

References

1. Willis, H.L., Scott, W.G.: Distributed Power Generation—Planning and Evaluation, 1st edn. Marcel Dekker, New York (2000). ISBN 0-8247-0336-7
2. Rahman, S.: Going green: the growth of renewable energy. IEEE Power Energy Mag. 1(6), 16–18 (2003)
3. Rowe, D.: Thermoelectric waste heat recovery as a renewable energy source. Int. J. Innov. Energy Syst. Power 1, 13–23 (2006)
4. Dalala, Z.M., Zahid, Z.U.: New MPPT algorithm based on indirect open circuit voltage and short circuit current detection for thermoelectric generators. In: Energy Conversion Congress and Exposition (ECCE), 2015 IEEE, pp. 1062–1067 (2015)
5. Riffat, S.B., Ma, X.: Thermoelectrics: a review of present and potential applications. Appl. Therm. Eng. 23, 913–935 (2003)
6. Rowe, D.: Thermoelectrics, an environmentally-friendly source of electrical power. Renew. Energy 16, 1251–1256 (1999)
7. Phillip, N., Maganga, O., Burnham, K.J., Dunn, J., Rouaud, C., Ellis, M.A., Robinson, S.: Modelling and simulation of a thermoelectric generator for waste heat energy recovery in Low Carbon Vehicles. In: 2012 2nd International Symposium on Environment. Friendly Energies and Applications (EFEA), pp. 94–99 (2012)
8. Hendricks, T., Choate, W.T.: Engineering Scoping Study of Thermoelectric Generator Systems for Industrial Waste Heat Recovery. Pacifc Northwest National Laboratory, Richland (2006)
9. Fernandes, A.E.S.S.: Conversão de energia com células de Peltier. Dissertação (Mestrado). Universidade Nova de Lisboa, Lisboa (2012)
10. Kasa, N., Iida, T., Liang, C.: Flyback inverter controlled by sensorless current MPPT for photovoltaic power system. IEEE Trans. Ind. Electron. 52, 1145–1152 (2005)

11. Rae-young, K., Jih-Sheng, L.: A seamless mode transfer maximum power point tracking controller for thermoelectric generator applications. IEEE Trans. Power Electron. **23**, 2310–2318 (2008)
12. Esram, T., Chapman, P.L.: Comparison of photovoltaic array maximum power point tracking methods. IEEE Trans. Energy Convers. **22**(2), 439–449 (2007)
13. Dolara, A., Faranda, R., Leva, S.: Energy comparison of seven MPPT techniques for PV systems. J. Electromagn. Anal. Appl. **3**, 152–162 (2009)
14. Pikutis, M., Vasarevicius, D., Martavicius, R.: Maximum power point tracking in solar power plants under partially shaded condition. Elektronika ir Elektrotechnika **20**(4), 49–52 (2014)
15. Lineykin, S., Ben-yaakov, S.: Modeling and analysis of thermoelectric modules. IEEE Trans. Ind. Appl. **43**(2), 505–512 (2007)
16. Josephine, R.L., Padmabeaula, A., Raj, A.D.: Simulation of incremental conductance mppt with direct control and fuzzy logic methods using SEPIC converter. J. Electr. Eng. **13**(3), 91–99 (2013)
17. Femia, F., Petrone, G., Spagnuolo, G., Vitelli, M.: Power Electronics and Control Techniques for Maximum Energy Harvesting in Photovoltaic Systems, pp. 35–84. CRC Press, Boca Raton (2013)
18. Piegari, L., Rizzo, R.: Adaptive perturb and observe algorithm for photovoltaic maximum power point tracking. IET Renew. Power Gener. **4**, 317–328 (2010)
19. Liu, F., Kang, Y., Zhang, Y., Duan, S.: Comparison of P&O and hill climbing MPPT methods for grid-connected PV generator. In: 3rd IEEE Conference on Industrial Electronics and Applications" 3–5 June 2008; Singapore. IEEE, New York, pp 804–807

A New MPPT Technique Sinusoidal Extremum-Seeking Control

M'hamed Sekour[(✉)] and Mohamed Mankour

Electrotechnical Engineering Laboratory,
University Tahar Moulay of Saida, Saida, Algeria
sekourmohamed@yahoo.fr

Abstract. A photovoltaic generator can operate over a wide range of voltage and output current, but it can deliver a maximum power only for particular values of the current and tension. Indeed characteristic I (V) of the generator depend on solar illumination and the temperature. These climatic variations involve the fluctuation of the point of maximum power. Because of this fluctuation, one often intercalates between the generator and the receiver one or more static inverters ordered allowing continuing the point of maximum power. In this paper, a maximum power tracking controller Sinusoidal Extremum-Seeking Control (ESC) is designed in an attempt to improve the system stability and robustness. Relative to the P&Q method, it demonstrates a superior overall efficiency [8] and well maintained robustness in the rapidly varying atmospheric.

Keywords: Photovoltaic (PV) · Maximum power point tracking (MPPT)
Extremum-seeking control · Perturb and observe (P&O)
Maximum power point (MPP)

1 Introduction

In the literature, several algorithms have been developed and introduced to achieve this objective. Among all these algorithms, the Perturb and Observe (P&O) and incremental conductance (INC) are widely used and are different from each other by their cost, complexity, benefits and drawbacks [1–5]. Since, the perturbation and observation method (P&O), is a method perturbing the cell load so as to vary the operating point toward the MPP. A resulting disadvantage is the oscillation in the vicinity of MPP, resulting in a power loss and degraded solar energy conversion efficiency. The incremental conductance method, is a method that intends to locate the maximum power operating point such that the condition dP/dV l = 0 is satisfied. A major drawback in practical applications is that an error while locating the MPP is inevitably encountered as a consequence of the low precision sensors used. In this paper, we propose a Sinusoidal Extremum-Seeking Control (ESC) algorithm for optimum maximum power point tracking. For achieving this purpose, the theoretical analysis and the design principle of the proposed method are described in detail and implemented on this PV system under gradually and rapidly changing insulation levels, (between 250 and 1000 W/m^2) to study their dynamic response for tracking the maximum power point and obtain high levels of efficiency, reliability and flexibility [10, 11].

M. Hatti (Ed.): ICAIRES 2018, LNNS 62, pp. 339–345, 2019.
https://doi.org/10.1007/978-3-030-04789-4_37

2 Description of the Photovoltaic Generator

$$I = I_{ph} - I_0 \left[\exp\left(\frac{q(V_{pv} - IR_S)}{nkT} \right) - 1 \right] - \left(\frac{V_{pv} - IR_S}{R_P} \right) \tag{1}$$

where Vpv represents the solar cell output voltage, Iph the output current, Id the diode current, I sat the photo current at specified irradiance and temperature, n the idea parameter, k the Boltzmann constant, T the reference temperature of the solar cell, q an electron charge, Rs the equivalent series resistor and Rsh the shunt resistor caused due to a flawed PN junction. In this study, the SUN POWER SPR-305 WHT PV module 1 in [9] is used. In detail, Table 1 shows its characteristics at the standard test condition (Fig. 1).

Table 1. Electrical characteristics of the solar module

Parameter	Value
Number of cells in series	$nCells = 96$
Maximum power (W)	$Pmp = 305.2$
Maximum power voltage (V)	$Vmp = 54.70$
Maximum power current (A)	$Imp = 5.58$
Open circuit voltage (V)	$Voc = 64.20$
Short circuit current (A)	$Isc = 5.96$

The simulation results of i-v curve and p-v curve of PVmodel for different solar irradiation and constant temperature are shown in Fig. 2a, b. These curves are non linear and are crucially influenced by solar radiation, it is very clear that current generated increases with increasing solar irradiance and maximum output power (Pm) also increases. In Fig. 2, the Fig. 2c, d show The three-dimensional V-I and P-I curves is changed according to solar radiation. It has two inputs the irradiance and the voltage, one output which represent the current and one output which represent the power.

3 MPPT Techniques

MPPT is an electronic system that operates PV modules in a way, so that modules are capable of producing maximum power for which they are designed for [11, 13]. Maximum Power Point Tracking is an adaptive structure that is used to control the converter between load and the solar panel. Over the years, most of the MPPT techniques are being studied, analyzed and are further defined to bring quality output. Different techniques have different algorithm and convergence speed which is further discussed below.

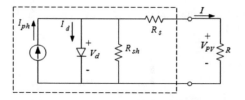

Fig. 1. The equivalent circuit of a solar cell.

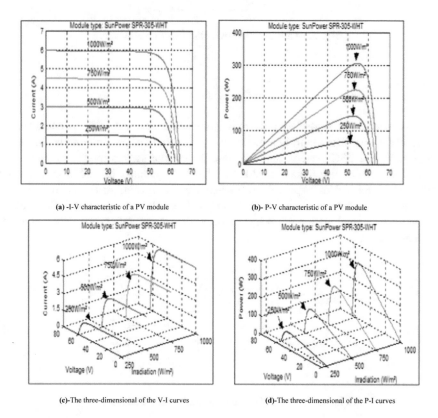

(a) -I-V characteristic of a PV module

(b)- P-V characteristic of a PV module

(c)-The three-dimensional of the V-I curves

(d)-The three-dimensional of the P-I curves

Fig. 2. I-V and P-V characteristics of PV module at various sun radiations

3.1 Sinusoidal Extremum-Seeking Control

As illustrated in Fig. 3, a small amount of sinusoidal perturbation is introduced into the system in the vicinity of MPPT. The perturbation frequency, identical to the central frequency of the filter, must be made lower than that of the triangular wave in the control system. It is noted from Fig. 3 that in the case of a negative (positive) epsilon;, the operation point lies on the right (left) of MPPT, respectively. When a small perturbation signal a: sin(w0t), where a denotes the perturbation amplitude and w0 the frequency w0, is introduced into the objective function y = f(x), the resultant output is represented as [11, 12]:

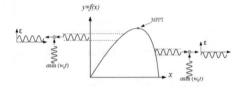

Fig. 3. An illustration of a sinusoidal perturbation.

$$y = f(x + a\sin(w_0 t)) \tag{2}$$

Illustrated in Fig. 4 [4–8] is a block diagram of a sinusoidal ESC system, a system applicable to nonlinear control problems. A small amount of perturbation is introduced into a stabilized ESC system, thus affecting the overall system dynamics. Hence, an external sinusoidal disturbance is added to the P-V dynamics in such a way that the MPP can be located as expected via filtered signals. Assuming that there exists an extremum on a concave objective function y = f (x), then the goal of MPPT can be realized with a combination of an integrator, a filter, a multiplier, an adder and a sine wave generator.

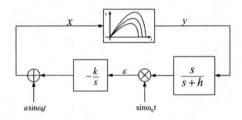

Fig. 4. A block diagram of a sinusoidal SESC system.

3.2   Perturb and Observe (P&O) Algorithm

This technique is the most commonly used out of all MPPT methods due to its simplicity of execution and implementation. (P&O) technique can be easily programmed and it provides satisfactory results within the convergence. (P&O) algorithms work by periodically perturbing (i.e. increasing or decreasing) the array terminal voltage and compares the PV output power with that of the previous perturbation cycle. The algorithm keeps incrementing or decrementing the reference voltage based on the previous value of power until it reaches the maximum power point (MPP) [5–8]. The perturbation moves the operating point towards the MP P if dP/dV is greater than zero as illustrated in Fig. 5. It continues to perturb the PV voltage in the same direction until it reaches required MPP. If dP/dV is less than zero, it is assumed that the perturbation moves the operating point away from MPP; hence, (P&O) technique reverses the direction of perturbation. In steady state operation the output power oscillates around the MPP and results in voltage oscillations. Hence, when multiplied to the current, these voltage oscillations cause power fluctuations which results in power losses. (P&O) technique can also fail under rapidly changing atmospheric conditions [11].

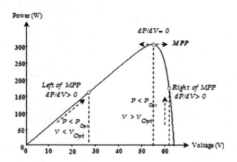

Fig. 5. Perturb and observe (P&O)

4 Comparison Study and Simulation Results

MPPT algorithm was design and simulated in MATLAB/SIMULINK environment. The proposed SESC technique was then compared with the techniques P&O MPPT. Simulation was carried out by giving a constant temperature of 25 °C under rapidly changing insolation levels, (between 200, 600, and 1000 W/m^2). Simulation results for output voltage, current and power are shown on Figs. 6, 7. A polycrystalline silicon panel of 305.2 W was used for experimental purpose. Table 1 shows the solar panel specifications [8].

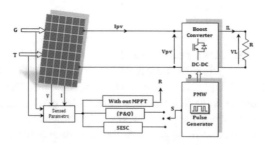

Fig. 6. Schematic diagram of various MPPT methods assisted PV system

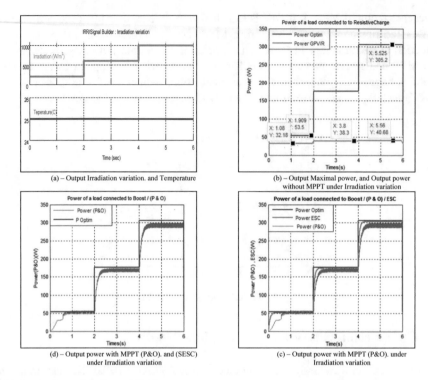

(a) – Output Irradiation variation. and Temperature

(b) – Output Maximal power, and Output power without MPPT under Irradiation variation

(d) – Output power with MPPT (P&O). and (SESC) under Irradiation variation

(c) – Output power with MPPT (P&O). under Irradiation variation

Fig. 7. Comparison of sinusoidal extremum-seeking control (SESC) to Perturb & Observe (P&O). under Irradiation variation

It is observed that simulated results of ESCS control algorithm show the improved convergence speed of output power and also reduces oscillation around the maximum power point thus avoiding power loss and hence better tracking efficiency.

5   Conclusion

The research was targeted to analyse the MPPT implementation using the ESEC technique. The ESEC technique was compared with the P&O. Simulink software was used to simulate the existing and proposed MPPT techniques. The simulation results illustrated that the proposed MPPT technique has better performance than P&O as it yield maximum power from the PV systems. The proposed method also reduces oscillation around the maximum power point thus avoiding power loss. The drawback of proposed technique is that it has higher settling time when compared to the P&O method. However, this can be reduced by having further research on filters or tuners to increment or decrement the duty cycle to achieve the optimum output.

References

1. for general nonlinear dynamic systems. Automatica **36**(4), 595–601 (2000)
2. DeHaan, D., Guay, M.: Extremum-seeking control of state constrained nonlinear systems. Automatica **41**(9), 1567–1574 (2005)
3. Chang, Y.A., Moura, S.J.: Air flow control in fuel cell systems: an extremum seeking approach. In: 2009 American Control Conference, 2009, St. Louis, MO, USA
4. Creaby, J., Li, Y., Seem, S.: Maximizing wind energy capture via extremum seeking control. In: 2008 ASME Dynamic Systems and Control Conference, 2008, Ann Arbor, MI, USA
5. Ariyur, K., Krstić, M.: Real-Time Optimization by Extremum-Seeking Control. Wiley-Inter science, Hoboken (2003)
6. Krstić, M.: Performance improvement and limitations in extremum seeking control. Syst. Control Lett. **39**, 313–326 (2000)
7. Laboratory, Technical Report 2005 [Online]
8. http://www.nrel.gov/analysis/docs/cost_curves_2005.ppt
9. Dave, F.: Introduction to Photovoltaic Systems Maximum Power Point Tracking. Texas Instruments. Report number: 2010: 446
10. Hansen, A., Sorensen, P., Hansen, L., Bindner, H.: Models for Stand-Alone PV System. Technical University of Denmark. Report number: 2001: 1219
11. Mohammed, Y., Mohammed, F., Abdelkrim, M.: A neural network based MPPT technique controller for photovoltaic pumping system. Int. J. Power Electron. Drive Syst. **4**(2), 241–255 (2014)
12. Mamatha, G.: Perturb and observe MPPT alogrithm implementation for PV applications. Int. J. Comput. Sci. Inf. Technol. **6**(2), 1884–1887 (2015)
13. Islam, F.R., Pota, H.: Impact of dynamic PHEV loads on photovoltaic system. Int. J. Electr. Comput. Eng. **2**(5), 644–654 (2012)

The MPPT Command for a PV System Comparative Study: Fuzzy Control Based on Logic with the Command "P&O"

Aicha Djalab[✉], Mohamed Mounir Rezaoui, Ali Teta,
and Mohamed Boudiaf

Applied Automation and Industrial Diagnostics Laboratory,
Djelfa University, Djelfa, Algeria
{A.djalab,mm.rezaoui,a.teta,m.boudiaf}@univ-djelfa.dz

Abstract. The photovoltaic system is able to provide a maximum power to the load to a point of particular operation which is usually called as the maximum point of power (MPP); thus, it is important to use a maximum power point trackers (MPPT) to achieve the photovoltaic maximum power nevertheless the unsteady environmental conditions. The (P&O) algorithm has used to be a classical solution for the purpose of tracking the maximum power, however, there are several drawbacks in this technique such as it causes an oscillation nearby the maximum power point. To overcome the existing problems in the classical method a lot of researches have been done.

This article provides an intelligent method to improve and optimize the performance of the maximum power point tracker associated with the PV array with the help of fuzzy logic technology, as well as compare its behavior by report has other techniques (P&O) used in the photovoltaic systems controls. The simulation results are developed under MATLAB/Simulink software. An extensive simulation have been done and compared with the traditional (P&O) technique under various climatic conditions to provide the effectiveness of the proposed controller.

Keywords: The PV system · Maximum power point tracking MPPT
The perturbation and observation (P&O) · Fuzzy logic controller (FLC)
Matlab/Simulink

1 Introduction

Currently, energy consumption is increased more and more as the development of societies industrialized. The issues caused by the warming world and the effect of pollution become very important problems to be investigated. Renewable energy sources are considered to be a perfect option for generating sustainable, clean and inexhaustible energy. There are several renewable energy sources such as solar energy, wind energy, etc. Due to the cleanliness, the inexhaustibility and the advanced technology of the photovoltaic energy, it is classified as the most ecological type of usable energy, which provides a huge opportunity for a future based nonpolluting technology. The PV array transforms solar energy directly into electric energy. The PV module

© Springer Nature Switzerland AG 2019
M. Hatti (Ed.): ICAIRES 2018, LNNS 62, pp. 346–354, 2019.
https://doi.org/10.1007/978-3-030-04789-4_38

energy conversion efficiency is rather low. To solve this issue and reach the maximum efficiency, all the elements of the PV system design should be improved. The utilization of the MPPT controller is very important due to their direct impact on the PV system efficiency. There are many algorithms have been developed for the maximum power point (MPPT) control. One of the most well-known MPPT techniques is P&O (Perturb and Observe) technique; however, this technique suffers from several drawbacks as the convergence issue and oscillation problems during the tracking. To improve the performance of the P&O algorithm, this paper presents an MPPT controller that benefits from Fuzzy Logic Control (FLC). The simulation study in this paper is done in MATLAB/Simulink software. The fuzzy logic controller based results are compared with the conventional technique such as P&O method.

2 The Photovoltaic System

A PV system is composed of four blocks as illustrated in (Fig. 1). Block 1 is the source of energy (solar panel), block 2 is a boost converter, the load is represented in the third block while the fourth block shows the control system. The objective of the boost converter is to achieve the MPP voltage defined by the MPPT controller [1].

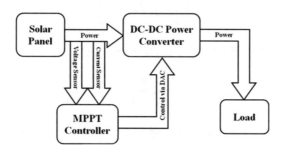

Fig. 1. Photovoltaic system.

3 Modeling of the Photovoltaic System

To obtain a desired output such as the power, the output current and output voltage [3, 4], the photovoltaic system (SPV) which consist of a set of basic photovoltaic cells can be connected in series and/or parallel, where the Photovoltaic cells are the main components of the module.

A. Model of a photovoltaic cell

The simplest circuit of the photovoltaic cell model is represented by a current source in parallel with an ideal diode, which will be adopted in this study [5]. The equivalent circuit of a solar cell is given in (Fig. 2). This equivalent circuit is composed of a current source controlled models which the photovoltaic effect (the generated current is controlled by the Sun's rays). The led represents the effect of the junction semiconductor of the cell (model use two diodes in parallel for more precision). Two resistors (Rs, Rp) represent the mass and the effect of resistivity in the cell respectively [2].

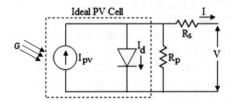

Fig. 2. Equivalent circuit of a solar cell

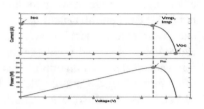

Fig. 3. I-V and P-V characteristics of a photovoltaic cell

The equivalent circuit gives the following relationship:

$$I_{ph} = I_d + I_p + I \tag{1}$$

The current in the diode Id is given by [6]:

$$I_d = I_0 \left[\exp\left(\frac{V + R_s I}{V_t a}\right) - 1 \right] \tag{2}$$

The current in the RP resistance is given by:

$$I_p = \left(\frac{V + R_s I}{R_p}\right) \tag{3}$$

From Eq. (1), we obtain the expression of current I:

$$I = I_{ph} - I_d - I_p \tag{4}$$

Replacing (4) in the Eqs. (2) and (3), the characteristic equation becomes:

$$I = I_{ph} - I_0 \left[\exp\left(\frac{V + R_s I}{V_t a}\right) - 1 \right] - \left(\frac{V + R_s I}{R_p}\right) \tag{5}$$

Where:
V: The cell voltage
R_s: The resistance series cell [Ω]
R_p: is the parallel resistance
I_0: Saturation current (A)
Vt: the thermal voltage of the module: Vt = Ns kT/q, with
Ns the number of cells connected in series
T: The temperature of the cell [°K]
q: electron's charge e = 1.6 * 10^{-19} C

K: The Boltzmann constant $(1.3854 * 10^{-23}$ J $°K^{-1)}$

a: a constant of the diode,

Figure 3 shows a nonlinear characteristic of the photovoltaic cell. This characteristic varies with the change in metrological terms. As the optimal power point varies broadly according to weather conditions, a power converter switch should be controlled by a specific algorithm to track the maximum power point [7].

B. Pv array characteristics

The characteristics of the PV array used in this paper are presented in Table 1:

Table 1. Electrical characteristics of the SPR-305-WHT PV module

PV Characteristics	
Voltage in open circuit (Vco)	64.2 V
Optimal operation voltage (Vmp)	54.7 V
Short circuit current (Isc)	5.96 A
Optimal operating current (Imp)	5.58 A
Maximum power on STC (Pmax)	305 Wp
Operating temperature	De - 40 C° à +85 C°
Power tolerance	± 5%

4 Maximum Power Point Tracking Controllers

The photovoltaic system to work at maximum power points of their characteristics, there are specific laws that meet this need. This command is named in the "Maximum Power Point Tracking" (MPPT) literature. The principle of these commands is to seek the maximum power point (MPP) by keeping a good adaptation between the generator and the load to ensure the transfer of maximum power. The technique of control so to act on the duty cycle in an automatic way to bring the point of operation of the generator at its optimum value whatever weather instabilities or brutal changes in load [8]. For such reason, three MPPT control techniques will be discussed.

A. P&O Controller

Among the MPPT technique the perturbation and observation (P&O) algorithm is considered as one of the most used techniques, owing to its easy implementation [9]. As it is suggested by its name, the PO technique is based on the increasing and decreasing in Vref by the continuous adjustment of the DC-DC converter duty cycle, then observes of the variation of the PV array output power. If the current value of the power P(k) array is bigger than the previous value P(k-1) is then keep the same direction of previous perturbation or reverse the disruption of the previous cycle [2]. The PO technique flowchart is illustrated in Fig. 4:

B. Fuzzy logic controller

Fuzzy logic allows to convert the linguistic description any process to a control laws of the adopted control strategy. Fuzzy logic based controller is a rule-based controller; it is composed of an input, processing, and output stages [10]. Figure 5 shows the fuzzy logic process structure, which confirms the basic components of a fuzzy controller: a fuzzification interface, a knowledge base, a data base, inference procedure, and a defuzzification interface.

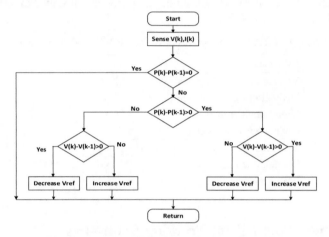

Fig. 4. Flowchart Perturb and Observe Algorithm

The fuzzification allows conversion of physical variables of entry into fuzzy sets. In the studied case, there are two inputs the error E and the variation in the error ΔE defined as follows [11]:

$$E(n) = \frac{P(n) - p(n-1)}{V(n) - V(n-1)} \tag{6}$$

$$\Delta E(n) = E(n) - E(n-1) \tag{7}$$

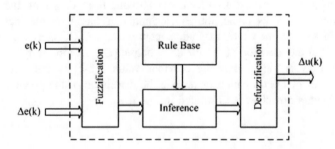

Fig. 5. Basic structure of fuzzy logic control

We assign to these variables linguistic variables: NB (Negative Big), NS (Negative Small), Z (Zero), PS (Positive Small), PB (Positive Big): These two input variables and the output duty cycle D used in this controller are illustrated in (Fig. 6).

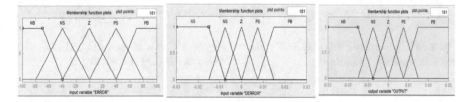

Fig. 6. Membership function of FLC

In the step of inference, we make decisions. Indeed, it establishes logical relationships between the inputs and the output setting membership rules. Subsequently, it paints the picture of inference rules (Table 2). Finally, in defuzzification, converting the fuzzy subassemblies of output to a numeric value.

Table 2. Fuzzy table rules

E	CE				
	NB	NS	ZE	PS	PB
NB	ZE	ZE	PB	PB	PB
NS	ZE	ZE	PS	PS	PS
ZE	PS	ZE	ZE	ZE	NS
PS	NS	NS	NS	ZE	ZE
PB	NB	NB	NB	ZE	ZE

5 Results and Simulation

In this section, we begin by assessing the photovoltaic system by simulation with MATLAB/Simulink simulation tool. Then, two further MPPT methods are studied: the method (P&O) and the fuzzy controller. The two systems are simulated under standard environmental conditions and many changes of weather conditions.

A. Operation under standard test conditions

In this test the irradiance and temperature are held constant. It takes the values of the standard conditions: the temperature T = 25 °C and irradiance = 1000 W/m². The purpose of these simulations is to view the offset of the point of operation compared to the MPP point. Figures 7 and 8 shows the results of the output power and voltage under standard conditions:

It has seen clearly how the FLC is faster than the P&O tracker of photovoltaic system. It is further observed that the P&O MPPT technique is further express more

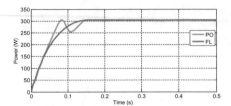

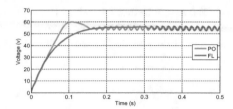

Fig. 7. PV power by P&O and FLC Controller in standard conditions

Fig. 8. PV voltage by P&O and FLC Controller in standard conditions

power loss in the energy. It has been noticed the continues oscillation of operation point for the P&O conventional technique. It is a result of the continuous increasing and decreasing of the operating voltage in order to achieve the MPP. Whereas this phenomenon of oscillation it doesn't observed in FL based MPPT technique, where signals of power, voltage, current, and duty ratio remain almost constant. This has as result a power losses reduction.

B. The system behavior under unsteady irradiance and temperature

In order to evaluate the response time of the three MPP trackers, they will be tested under the following tests: first a constant temperature and unsteady irradiance, then the fixed value (1000 W/m²) of the irradiance and variable temperature. A rapid increase in irradiance from 300 W/m² to 1000 W/m² within a time period of 1 s was simulated. The cell temperature was kept at a constant value of 25 °C, while Figs. 9 and 10 illustrate the evolution of the power and the voltage at the terminals of the PV Panel using the two controllers. Under these operating conditions the FL based MPPT method is more effective. Figures 9 and 10 show how the power output and voltage of the FL based MPPT increases linearly, whereas the conventional P&O MPPT technique experiences a vast deviation from the MPP.

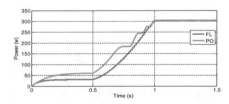

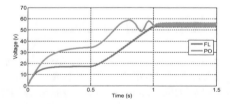

Fig. 9. PV power by P&O and FLC Controller in unsteady conditions

Fig. 10. PV voltage by P&O and FLC Controller in unsteady conditions

In the next part of simulation, the PV array is under fast temperature variation (increasing the temperature of 25 °C to 455°C in 0.5 s).

Figures 11 and 12 show that the output pv power and voltage are decreased with the augmentation of the temperature from 0.5 S to 1 S, however, all the two trackers keep tracking the MPP. The fuzzy logic based tracker gives better results compared to the P&O tracker under changing irradiance and temperature.

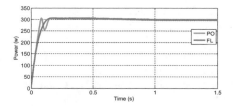

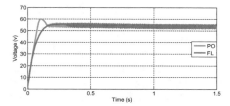

Fig. 11. PV power by P&O and FLC Controller in unsteady temperature

Fig. 12. PV voltage by P&O and FLC Controller in unsteady temperature

6 Conclusion

MPPT or maximum power point tracking is algorithm that covered in charge controllers used for extracting maximum available energy from PV module under certain conditions. In this paper, a new method for MPPT based on fuzzy logic controller was presented and compared with the conventional P&O MPPT method. The models were tested under changing irradiances and temperature degrees. Simulation results showed that the proposed method effectively tracks the maximum power point under different ambient conditions. In addition to comparing the tracking efficiency of both methods indicates that the proposed method had higher efficiency than the conventional P&O MPPT method. The fuzzy logic controller had proved that it had better performance, fast time response and the error was very low in the permanent state, and it was robust in different variations of atmospheric conditions.

References

1. Pradeep Kumar Yadav, A., Thirumaliah, S., Haritha, G.: Comparison of MPPT algorithms for DC–DC converters based PV systems. **1**(1) (2012). ISSN: 2278–8875
2. Drir, N., Barazane, L., Loudini, M.: Fuzzy logic for tracking maximum power point of photovoltaic generator. Revue des energies Renouvelables **16**(1), 1–9 (2013)
3. Abbes, H., Abid, H., Loukil, K., Toumi, A., Abid, M.: Comparative study of five algorithms of MPPT control for a photovoltaic system, (CIER'13) Sousse, Tunisie (2013). ISSN 2356-5608
4. Takun, P., Kaitwanidvilai, S., Jettanasen, C.: Maximum Power Point Tracking using Fuzzy Logic Control for Photovoltaic Systems. IMECS, Hong Kong (2011)
5. Balasubramanian, G., Singaravelu, S.: Fuzzy logic controller for the maximum power point tracking in photovoltaic system. Int. J. Comput. Appl. **41**(12), 22–28 (2012)
6. Karthika, S., Velayutham, K., Rathika, P., Devarai, D.: Fuzzy logic based maximum power point tracking designed for 10 kW solar photovoltaic system with different membership functions. World Acad. Sci. Eng. Technol. Int. J. Electric. Comput. Eng. **8**(6), 1013–1018 (2014)
7. Robles Algarín, C., Taborda Giraldo, J., Rodríguez Álvarez, O.: Fuzzy logic based MPPT controller for a PV system. Energies **10**, 2036 (2017). https://doi.org/10.3390/en10122036
8. Balamurugan, T., Manoharan, S., Sheeba, P., Savithri, M.: Design photovoltaic array with boost converter using fuzzy logic controller **3**(2), 444–456 (2012)

9. Harsha, P.P., Dhanya, P.M., Karthika, K. (2013) Simulation & proposed hardware implementation of MP controller for a solar PV system. Int. J. Adv. Electric. Electron. Eng. (IJAEEE) **2**
10. Abbes, H., Abid, H., Loukil, K.: An improved MPPT incremental conductance algorithm using T-S fuzzy system for photovoltaic panel. IJRER **5**(1), 160–167 (2015)
11. Terki, A., Moussi, A., Betka, A., Terki, N.: An improved efficiency of fuzzy logic control of PMBLDC for PV pumping system. Appl. Math. Model. **36**(3), 934–944 (2012)

Impact of Load Characteristics on PV Solar System with MPPT Control

Saidi Ahmed$^{(\boxtimes)}$, Benoudjafer Cherif, and Chellali Benachaiba

Electrical Engineering Department,
Tahri Mohamed University, 08000 Béchar, Algeria
ahmedsaidi@outlook.com, benoudjafer@gmail.com,
chellali99@yahoo.fr

Abstract. In the last decade, the use of renewable energy resources instead of fossil fuels pollutants has increased exponentially. Photovoltaic energy generation is ever more important as a renewable resource since it does not cause in fuel costs, pollution, maintenance, and emitting noise compared to other renewable resources as more accessibility of solar irradiation.

This paper presents impact load characteristics on MPPT controller of R, RC, RL and RLC circuit load with Perturb and Observe Maximum Power Point Tracking (MPPT) algorithm for a stand-alone photovoltaic system and sees the comparison of each circuit load on the system and the algorithm of control. The different results of power, voltage and current are discussed and shown that the inductor it has a capital effect on Maximum power point (MPP) and system in general. The simulation results and a comparative analysis are discussed in this paper.

Keywords: tPV array · MPPT · P&O algorithm · Load characteristics
DC-DC converter

1 Introduction

Photovoltaic energy generation is ever more important as a renewable resource since it does not cause in fuel costs, pollution, maintenance, and emitting noise compared to other renewable resource as more accessibility of solar irradiation.

Recently, solar energy or photovoltaic energy applications are getting increased especially in a stand-alone configuration. It is one of the most promising sources of renewable energy [1]. The limitations of PV energy system such as the low efficiency and the non-linearity of the output characteristics, make it necessary to obtain an MPP operation. Variations on solar irradiance levels, ambient temperatures and dust accumulation on the surface of the PV panel affect the output of the PV system [2].

The MPPT working principle is based on the maximum power transfer theory. The power delivered from the source to the load is maximized when the input resistance seen by the source matches the source resistance. Therefore, in order to transfer maximum power from the panel to the load, the internal resistance of the panel has to equal the resistance seen by the PV panel. For a fixed load, the resistance seen by the

© Springer Nature Switzerland AG 2019
M. Hatti (Ed.): ICAIRES 2018, LNNS 62, pp. 355–368, 2019.
https://doi.org/10.1007/978-3-030-04789-4_39

panel can be adjusted by changing the charger (DC/DC converter) duty cycle [3]. Further illustrates the PV system input resistance concept [4].

In order to improve the performance of the PV system and to extract the maximum power point under any environmental condition is necessary to track the maximum power point using control methods. The MPPT algorithm calculates the MPP in each instant of time for any irradiance and temperature [5].

MPPT are used for operating PV array at the point of maximum power irrespective of irradiance, temperature and load current variation. In literature, different MPPT techniques have been proposed but their suitability largely depends on factors like the end application, dynamic of irradiance, design simplicity, convergence speed, hardware implementation and the cost [6]. The available MPPT methods range from simple voltage relationships to complex multiple sample-based analysis which includes but not limited to constant voltage method, short current pulse method, open voltage method, perturb and observe method, incremental conductance method, and temperature method [7]. The design and implementation simplicity coupled with good performance have make perturb and observe (P&O) MPPT to be one of the most widely used MPPT techniques for solar PV applications [6, 8].

This paper presents studies on solar PV model module, the energy pattern was investigated under different load and irradiance conditions and MPPT was incorporated to maximize energy harvesting. A model solar PV panel intended to modelled in Matlab-Simulink environment. The characteristic behaviour of the PV panel is analysed considering different operating conditions of irradiance level and different circuit load with a temperature constant and P&O MPPT algorithm was used for maximum power tracking. The simulation results show that system performed satisfactorily as the maximum output power from solar PV panel with MPPT show a close relationship with the maximum power available from the PV under test at different irradiance levels and characteristic load.

2   PV System Modelling

Figure 1 shows a simplified scheme of a standalone PV system with DC–DC buck converter.

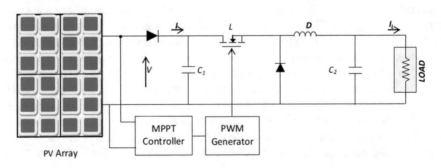

Fig. 1. A PV system with a DC–DC buck converter

This section is devoted to PV module modelling which is a matrix of elementary cells that are the heart of PV systems. The modelling of PV systems starts from the model of the elementary PV cell that is derived from that of the P–N junction [9].

A. Ideal photovoltaic cell

The PV cell combines the behavior of either voltage or current sources according to the operating point. This behavior can be obtained by connecting a sunlight-sensitive current source with a P–N junction of a semiconductor material being sensitive to sunlight and temperature. The dot-line square in Fig. 2 shows the model of the ideal PV cell. The DC current generated by the PV cell is expressed as follows

$$I = I_{PV,Cell} - I_{s,Cell}\left(e^{\frac{V}{aVt}} - 1\right)$$ (1)

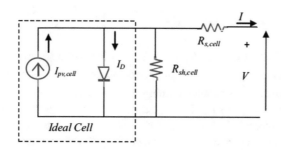

Fig. 2. The Equivalent circuit of an ideal and practical PV cell.

The first term in Eq. (1), that is $I_{PV,Cell}$, is proportional to the irradiance intensity whereas the second term, the diode current, expresses the non-linear relationship between the PV cell current and voltage. A practical PV cell, shown in Fig. 1, includes series and parallel resistances [10]. The series resistance represents the contact resistance of the elements constituting the PV cell while the parallel resistance models the leakage current of the P–N junction.

This model is known as the single diode equivalent circuit of the PV cell. The larger number of diodes the equivalent circuit contains, the more accurate is the modelling of the PV cell behavior, however, at the expense of more computation complexity. The single diode model is shown in Fig. 2 is adopted for this study, due to its simplicity.

B. PV module modelling

Commercially photovoltaic devices are available as sets of series and/or parallel-connected PV cells combined into one item, the PV module, to produce a higher voltage, current and power, as shown in Fig. 3.

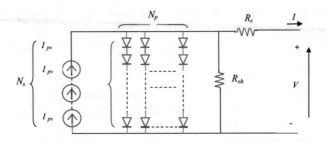

Fig. 3. Equivalent circuit of PV module.

The equation of the I–V characteristic of the PV module is obtained from Eq. (1) by including the equivalent module series resistance, shunt resistance and the number of cells connected in series and in parallel.

$$I = N_p \left(I_{PV} - I_s \left(e^{\frac{q(V + I.R_s)}{aN_sKT}} - 1 \right) \right) - \frac{(V + I.R_s)}{R_{sh}} \qquad (2)$$

Where V_t the PV cell thermal voltage in Eq. (1) is substituted by that of the module thermal voltage given by $V_t = \frac{N_sKT}{q}$ and N_s and N_p are respectively the number of cells connected in series and in parallel forming the PV module.

The constant an expressing the degree of ideality of the diode may be arbitrary chosen from the interval (1, 1.5) [11]. The light generated current of PV cell depends linearly on the irradiance and is also influenced by the temperature:

$$I_{PV} = \left(\frac{G}{G_{STC}} \right) (I_{PVn} + K_i(T - T_{STC})) \qquad (3)$$

I_{PVn} is the nominal light-generated current provided at GSTC, TSTC which refers to the values at nominal or Standard Test Conditions (1 kW/m^2, 25 °C). The nominal light-generated current is not available in the datasheet of the PV panel but estimated as [11]:

$$I_{PVn} = \left(\frac{R_s + R_{sh}}{R_s} \right) I_{scn} \qquad (4)$$

-The second term in Eq. (2) is the diode current that is function of the voltage and current coefficients given by the equation below:

$$I_s = \frac{I_{scn} + K_I \Delta T}{e^{\frac{V_{ocn} + K_V \Delta T}{aV_t}} - 1} \qquad (5)$$

Where I_{scn} is the nominal short-circuit current or the maximum current available at the terminals of the practical device at nominal conditions.

C. I–V and P–V characteristics

A PV module can be modeled as a current source that is dependent on the solar irradiance and temperature. The complex relationship between the temperature and irradiation results in a non-linear current–voltage characteristics. A typical I–V and P–V curve for the variations of irradiance and temperature is shown in Fig. 4 (a) and (b), respectively. As can be observed, the MPP is not a fixed point; it fluctuates continuously as the temperature or the irradiance does. Due to this dynamics, the controller needs to track the MPP by updating the duty cycle of the converter at every control sample. A quicker response from the controller (to match the MPP) will result in better extraction of the PV energy and vice versa [12].

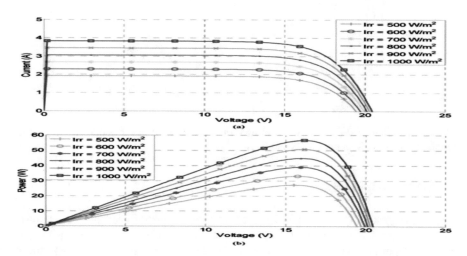

Fig. 4. Solar cell characteristics (a) voltage-current characteristics and (b) voltage-power characteristics

D. Partial shading

The MPP tracking becomes more complicated when the entire PV array does not receive uniform irradiance. This condition is known as partial shading. Typically, it is caused by the clouds that strike on certain spots of the solar array, while other parts are left uniformly irradiated [13]. Another source of partial shading-like characteristics is exhibited by module irregularities; a common example would be the presence of cracks on one or more modules of the PV array. Figure 5 (a) shows a PV array in a typical series–parallel configuration. Commonly, a bypass diode is fitted across the module to ensure that hot spot will not occur if that module is shaded. In this example, three modules are connected in a single string. In a normal condition, i.e. when the solar irradiance on the entire PV array is uniform, the P–V curve exhibits a unique maximum power point (curve 1 of Fig. 5 (c)). However, during partial shading in Fig. 5 (b), the difference in irradiance between two modules activates the bypass diode. As a result, two stairs current waveform is created on the I–V curve, while the P–V curve is characterized by multiple maxima points, as depicted by curve 2 of Fig. 5 (c). The MPPT needs to ensure that the tracked maximum point is the true global peak, not one of the local maxima. If the algorithm is trapped at the local peak, significant loss in power incurs.

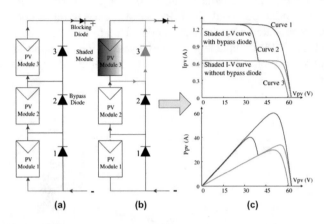

Fig. 5. Operation of PV array (a) under uniform irradiance (b) under partial shading (c) the resulting I-V and P-V curve for (a) and (b).

3 MPPT Algorithm

By adjusting the duty cycle of DC-DC converter the source impedance can be made equal to output impedance and the peak power is reached. So, the question is reduced only to variation of duty cycle in a particular direction to reach to the peak power. There are various algorithms to perform this duty and these are:

- Perturb and Observe (P & O)
- Incrémentale Conductance (IC)
- Fractional Open Circuit Voltage
- Fractional Short Circuit Current
- Fuzzy Logic Control

Each of the algorithms has its own merits and demerits. Out of these algorithms one is most popular, i.e., Perturb and Observe (P&O) is discussed here in detail.

E. Perturbation and Observation (P & O) Method

The problem considered by MPPT methods is to automatically find the voltage VMPP or current IMPP at which a PV array delivers maximum power under a given temperature and irradiance. In P&O method, the MPPT algorithm is based on the calculation of the PV output power and the power change by sampling both the PV Array current and voltage. The tracker operates by periodically incrementing or decrementing the solar array voltage [14, 15]. If a given perturbation leads to an increase (decrease) in the output power of the PV, then the subsequent perturbation is generated in the same (opposite) direction. The duty cycle of the dc chopper is varied and the process is repeated until the maximum power point has been reached. Actually, the system oscillates about the MPP. Reducing the perturbation step size can minimize the oscillation. However, small step size slows down the MPPT. For different values of irradiance and cell temperatures [16], the PV array would exhibit different characteristic curves. Each curve has its maximum power point [17]. It is at this point, where the corresponding maximum voltage is supplied to the converter [18].

A DC-DC boost converter is used to achieve the MPPT power stage owing to the advantage of high reliability [20], less component parts to reduce implementation cost. The boost converter configuration comprises of power MOSFET as the switching transistor with input inductor (L) placed in series with the PV voltage (Vin) as shown in Fig. 6. The boost converter parameters used in this paper is shown in Table 1.

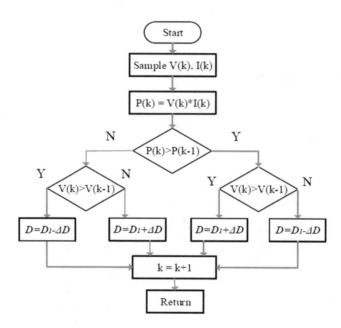

Fig. 6. Flow chart of P&O MPPT method

Table 1. Boost converter parameters

Parameters	Symbol	Value
Input voltage	Vin	20.97 [V]
Output voltage	Vout	51.76 [V]
Load resistance	RL	30 [Ω]
Inductor	L	1.5 [mH]
Output capacitor	C	100 [μF]
Switching frequency	fs	5 [KHz]

The steady state conversion ratio (input-output voltage) of the converter is given by [21–23]:

$$V_{out} = \frac{V_{in}}{1 - D}$$ (6)

The magnitude of peak-to-peak inductor current ripple is given by [24–26]:

$$\Delta I_L = \frac{V_{in}D}{f_sL}$$ (7)

And, also the output capacitor voltage ripple is [27]:

$$\Delta V_c = \Delta V_{out} = \frac{I_oD}{f_sC}$$ (8)

4 Simulation Results

In this part, the PV system simulation is realized to research the effects of the sliding mode MPPT controller parameters on the system performance. The PV system presented by Fig. 7

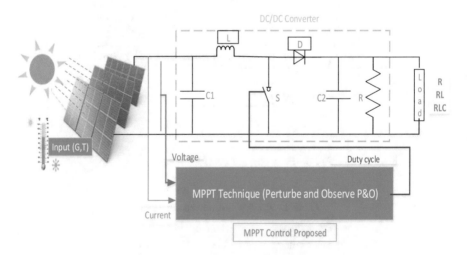

Fig. 7. Photovoltaic system Proposed

In the following, we will discuss reliability of the MPPT controller based on Perturb and Observe. So for the evaluation of the proposed MPPT controller performance, simulation results are presented in four subsections [28–30]:

1. The reliability of the MPPT controller with changing irradiations,
2. The robustness of the MPPT controller against load variations,
3. The reliability of the MPPT controller with simultaneous change of irradiation and load.

F. R, RL, RC loads

Figures 8, 9, 10, illustrates the irradiation and load variations used in this simulation work. Indeed, we suppose that the irradiations trajectory mentioned takes this form: it decreases irradiation from 1 000 to 800 to 600 W/m^2 in steps of 200 units and we consider that the temperature is a constant (T = 25 °C).

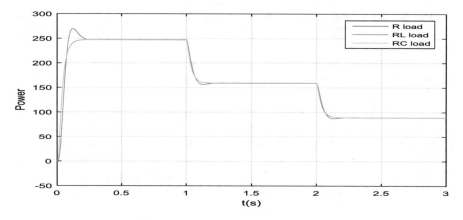

Fig. 8. Output Power of the solar array

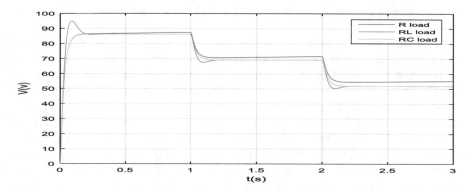

Fig. 9. Output Voltage of the solar array

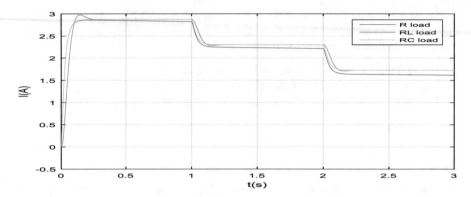

Fig. 10. Output current of the solar array

The simulation results show that the characteristic load it has a major effect on MPPT, moreover on Perturb and Observe controller.

The following results are shown, with a variation of irradiance levels. The irradiance levels change at fixed intervals of time while maintaining the ambient temperature constant:

Using purely resistive load circuit to the PV system keep the system very stable and efficient, Moreover, using the proposed MPPT controller (P&O) more performants.

Using RC load circuit to the PV systems keep the power stable compared the resistive load but in voltage and current curve is up and down respectively than resistive load curve this is represents that the capacitor circuit provoke the system to have a higher voltage supply compared to the resistive load therefore the decrease current called.

Using RL load circuit to the PV systems disrupting the system and the MPPT controller creates disruption, oscillation and overtaking time to change irradiation caused by inductive load.

G. RL, RC, RLC loads:

Figures 11, 12, 13, illustrates the irradiation and load variations used in this simulation work. Indeed, we suppose that the irradiations trajectory mentioned take same of first simulation.

This simulation is presented to mention the principal factor in the exceeding and the effect of each element of the charge and its effect on MPPT controller.

The results of RL, RC, RLC load with changes irradiation are indicated in Figs. 11, 12, 13. If we examine Fig. 8, 9, 10 that presents the power generated by the PV panel under different types of abrupt changes, we could remark that the power in Fig. 8 is increasing with the high value of irradiations and decreasing if the irradiation is diminishing. We note that the MPPT controller offers a rapid response against the irradiations change in the RC and RLC load. So changing the light intensity incident on a solar cell changes all solar cell parameters, including the short-circuit current, the open-circuit voltage, and the efficiency. This is clear if we examine the power response.

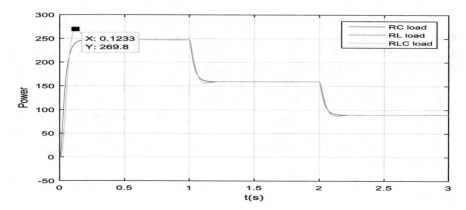

Fig. 11. Output Power of the solar array

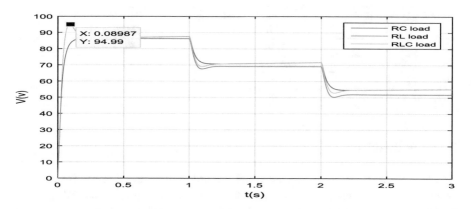

Fig. 12. Output Voltage of the solar array

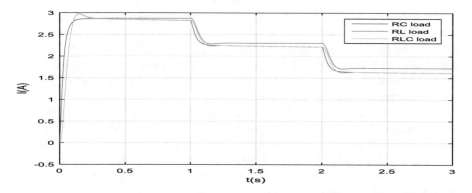

Fig. 13. Output current of the solar array

The simulation of the PV array with a deference load R, RC, RL, and RLC shows the negative effect of the inductor load on maximum power point of the system and the response of MPPT controller, and we can see that the load has a large importance to determine the maximum power point and a direct influence on the MPPT controller.

So the proposed siding mode controller under variable irradiations and characteristic load conditions are exhibiting an accurate control signal. This is clear if we examine Figs. 11 and 12, the sliding controller is adjusting in every case of the corresponding value of the control signal that extracts the maximum power from the PV panel. In Fig. 13, the proposed Perturb and Observe controller is proving its robustness against load change.

Table 2 presents typical peak of Power, voltage and current for each characteristic circuit load.

Table 2. Simulation results for different load

Load	Typical peak power (P) [W]	Voltage at peak power (VMPP) [V]	Current at peak power (IMPP) [A]
R	249	87.74	2.81
RC	249.12	86.37	2.83
RL	269.8	94.99	2.91
RLC	269.8	94.99	2.91

5 Conclusions

A solar PV panel has been modelled and perturb and observe MPPT technique is developed for maximising its operating efficiency. Studies on the solar PV Simulink model shows the nonlinear I-V and P-V characteristics of the PV panel to different solar irradiance of 1000 W/m^2, 800 W/m^2, 600 W/m^2 and different characteristics circuit loads R, RC, RL, RLC with fixed temperature of 25 °C.

The programmes implemented in the MPPT technique achieve the maximum power point. It has been shown that for the particular irradiance levels the maximum power delivered by the PV is delivered to the load. It is a simple MPPT setup resulting in a highly efficient system. In conclusion, non-conventional energy sources will dominate the conventional sources of energy in the near future and here one uses the greatest renewable energy of all, the sun's energy.

The techniques have been implemented in standalone PV system as shown in the simulation model to study their individual efficiency and response towards the variation in solar irradiance and load. The comparison proves that the proposed MPPT technique has better tracking ability and speed of response towards the PV system irrespective of irradiance and load variation. Moreover, the optimal operating point does not oscillate around the MPP in case of R and RC Load. But in the case of RL and RLC load overtaking time to change irradiation caused by inductive.

The capacitive component it has a very important role in the compensation of voltage and stability of current generated.

Acknowledgment. The authors thank ENERGARID, SimulIA,SGRE Laboratorys and Tahri Mohammed University for all helps and supports.

References

1. Alsumiri, M.A., Jiang, L., Tang, W.H.: Maximum power point tracking controller for photovoltaic system using sliding mode control. In: Renewable Power Generation Conference (RPG 2014), 3rd, 24–25 Sept 2014
2. Mahdi, A., Tang, W., Wu, Q.: Improvement of a MPPT algorithm for pv systems and its experimental validation. In: International Conference on Renewable Energies and Power Quality, Granada, Spain (2010)
3. Aranda, E., Galán, J.A., Cardona, M., Rquez, J.: Measuring the I-V curve of PV generators. IEEE Ind. Electron. Mag. **3**(3), 4–14 (2009)
4. Scarpa, V.V.R., Buso, S., Spiazzi, G.: Low-complexity MPPT technique exploiting the PV module MPP locus characterization. IEEE Trans. Ind. Electron. **56**(5), 1531–1538 (2009)
5. Martin, A.D., IEEE Student Member, Jesus R. Vazquez, IEEE Member: MPPT Algorithms Comparison in PV Systems P&O, PI, Neuro-fuzzy and Back Stepping Controls. IEEE. ISBN: 978-1-4799-7800-7/15/$31.00 ©2015 (2015)
6. Esram, T., Chapman, P.L.: Comparison of photovoltaic array maximum power point tracking technique. IEEE Trans. Energy Convers. **22**, 439 (2007)
7. Freeman, D.: Introduction to photovoltaic systems maximum power point tracking. Texas instruments Rapport D'application SLVA446-November (2010)
8. Jain, S., Agarwal, V.: Comparison of the performance of maximum power point tracking schemes applied to single-stage grid-connected photovoltaic systems. IET Electric Power Appl. **1**, 753–762 (2007)
9. Chao, P.C.-P., Chen, W.-D., Chang, C.-K.: Maximum power tracking of a generic photovoltaic system via a fuzzy controller and a two-stage DC–DC converter. Microsyst. Technol. **18**, 1267 (2012)
10. Villalva, M.G., Gazoli, J.R.: Comprehensive approach to modeling and simulation of photovoltaic arrays. IEEE Trans. Power Electron. **24**, 1198–1208 (2009)
11. Bennett, T., Zilouchian, A., Messenger, R.: Photovoltaic model and converter topology considerations for MPPT purposes. Sol. Energy **86**, 2029–2040 (2012)
12. Salam, Z., Ahmed, J., Merugu, B.S.: The application of soft computing methods for MPPT of PV system: a technological and status review. Appl. Energy **107**, 135–148 (2013)
13. Di Piazza, M.C., Vitale, G.: Photovoltaic field emulation including dynamic and partial shadow conditions. Appl. Energy **87**, 814–823 (2012)
14. Oi, A.: Design and simulation of photovoltaic water pumping system. Thesis (2005)
15. Elgendy, M.A., Zahawi, B., Atkinson, D.J.: Assessment of perturb and observe MPPT algorithm implementation techniques for PV pumping applications. IEEE Trans. Sustain. Energy **3**, 21–33 (2012)
16. Femia, N., Petrone, G., Spagnuolo, G., Vitelli, M.: Optimization of perturb and observe maximum power point tracking method. IEEE Trans. Power Electron. **20**, 963–973 (2005)
17. Dorofte, C., Borup, U. and Blaabjerg, F. (2005) A Combined two-method MPPT control scheme for grid-connected photovoltaic systems. In: 2005 European Conference on Power Electronics and Applications, Dresden, 11–14 Sept 2005, p. 10
18. Liu, C., Wu, B., Cheung, R.: Advanced algorithm for MPPT control of photovoltaic systems. In: Canadian Solar Buildings Conference, Montreal, 20–24 Aug 2004 (2004)

19. Rosu-Hamzescu, M., Oprea, S.: Practical guide to implementing solar panel MPPT algorithms. Microchip Technology Inc. AN1521 (2013)
20. Ibrahim, O., Yahaya, N.Z., Saad, N., Umar, M.W.: Matlab/Simulink model of solar PV array with perturb and observe MPPT for maximising PV array efficiency. In: 2015 IEEE Conference on Energy Conversion (CENCON) 19–20 Oct 2015
21. Saidi, A.: Comparative study of different load under P&O MPPT algorithm for PV systems. EEA Electroteh. Electron. Autom. **64**(4), 24–27 (2016)
22. Saidi, A., Benachaiba, C.: Comparison of IC and P&O algorithms in MPPT for grid connected PV module. In: Proceedings of 2016 8th International Conference on Modelling, Identification and Control, ICMIC 2016 (2017)
23. Benoudjafer, C., Benachaiba, C., Saidi, A.: Hybrid shunt active filter: impact of the network's impedance on the filtering characteristic. WSEAS Trans. Power Syst. **9**, 171–177 (2014)
24. Saidi, A., Cherif, B., Chellali, B.: Fuzzy intelligent control for solar/wind hybrid renewable power system. EEA Electroteh. Electron. Autom. **65**(4), 128–136 (2017)
25. Ahmed, S., Benoudjafer, C., Benachaiba, C.: MPPT technique for standalone hybrid PV-wind using fuzzy controller. In: Hatti, M. (ed.) ICAIRES 2017. LNNS, vol. 35, pp. 185–196. Springer, Cham (2018). https://doi.org/10.1007/978-3-319-73192-6_19
26. Saidi, A., Chellali, B.: Simulation and control of Solar Wind hybrid renewable power system. In: 2017 6th International Conference on Systems and Control, ICSC 2017 (2017)
27. Harrou, F., Sun, Y., Saidi, A.: Online model-based fault detection for grid connected PV systems monitoring. In: 2017 5th International Conference on Electrical Engineering—Boumerdes, ICEE-B 2017 (2017)
28. Ahmed, Saidi, Benoudjafer, Cherif, Benachaiba, Chellali: Modeling and operation of PV/fuel cell standalone hybrid system with battery resource. In: Hatti, Mustapha (ed.) ICAIRES 2017. LNNS, vol. 35, pp. 299–307. Springer, Cham (2018). https://doi.org/10.1007/978-3-319-73192-6_31
29. Harrou, F., Sun, Y., Taghezouit, B., Saidi, A., Hamlati, M.-E.: Reliable fault detection and diagnosis of photovoltaic systems based on statistical monitoring approaches. Renew. Energy **116**, 22–37 (2018)
30. Saidi, A.: Solar-wind hybrid renewable energy systems: evolutionary technique. Electrotehn. Electronica Autom. **64**(4), 24–27 (2016)

Innovative Stateflow Models Assessment of P&O and IC PV MPPTs

Abdelghani Harrag[1,2(✉)] and Sabir Messalti[3]

[1] Optics and Precision Mechanics Institute, Ferhat Abbas University - Setif 1,
Cite Maabouda (ex. Travaux), 19000 Setif, Algeria
a.harrag@univ-setif.z
[2] CCNS Laboratory, Department of Electronics, Faculty of Technology,
Ferhat Abbas University - Setif 1, Campus Maabouda, 19000 Setif, Algeria
[3] Electrical Engineering Department, Faculty of Technology,
Mohamed Boudiaf University, Route BBA, 28000 M'Sila, Algeria
messalti.sabir@yahoo.fr

Abstract. In this paper, P&O and IC MPPT controllers have been proposed and investigated using the Matlab Stateflow models. The proposed MPPT controllers have been tested and validated using Matlab/Simulink models under different scenario tests including irradiation change, temperature change and step size change. Results and analysis show that the proposed stateflow models produce the correct output power regarding all considered scenarios tests.

Keywords: MPPT · Perturb and observe · Incremental conductance
Stateflow

1 Introduction

Renewable energies will inevitably dominate the world's energy supply system in the long run. The reason is both very simple and imperative: there is no alternative. Mankind cannot base its life on the consumption of finite energy resources indefinitely. Today, the world's energy supply is largely based on fossil fuels. These sources of energy will not last forever and have proven to be one of the main causes of our environmental problems. Environmental impacts of energy use are not new but they are increasingly well known. As links between energy use and global environmental problems such as climate change are widely acknowledged, reliance on renewable energy is not only possible, desirable and necessary, it is an imperative (Renewable Energy in Europe Markets 2010).

Among renewable energy, solar energy is growing rapidly, PV is the second-most deployed renewable technology in terms of global installed capacity, after wind. Solar PV represented about 44% of the new installed renewable power capacity, followed by wind and hydropower contributing about 33% and 18%, respectively. The increase in the global energy demand and decreasing system costs are the primary growth drivers for the PV industry. Additionally, climate and environmental concerns

© Springer Nature Switzerland AG 2019
M. Hatti (Ed.): ICAIRES 2018, LNNS 62, pp. 369–375, 2019.
https://doi.org/10.1007/978-3-030-04789-4_40

and government support are likely to further propel growth in the market. During the 2017–2025 forecast period, the global PV capacity is expected to increase from 387.3 FW to 969 GW (Rekioua and Matagne 2012; IRENA 2016).

PV power systems offer many unique benefits above and beyond simple energy delivery. PV also brings important social benefits in terms of:

- PV fuel is free;
- PV produces no noise, harmful emissions or polluting gases;
- PV systems are extremely safe and highly reliable. The estimated lifetime of a PV module is 30 years;
- PV modules can be recycled;
- PV requires virtually no maintenance.
- PV brings electricity to remote rural areas.

However, the power generated by PV systems is an irregular energy source presents two major problems: low efficiency and the output characteristic is not linear and varies with the conditions atmospheric such as temperature and irradiance (Kok Soon and Mekhilef 2014; Isaldo and Amiri 2016). Therefore, in order to improve the output power of PV system, maximum power point tracking controller is applying to adapt constantly the generator through the converter used acting as an adaptive impedance to maximize the power transferred to the load (Babaa et al. 2014).

In the last two decades, several MPPT techniques have been proposed, namely: perturb and observe (Femia et al. 2005), hill climbing method (Xiao and Dunford 2004), incremental conductance method (Kok Soon and Mekhilef 2014), fractional short circuit current (Sher et al. 2015), fractional open circuit voltage (Noguchi et al. 2000, fuzzy logic (Harrag and Messalti 2018), neural networks (Messalti et al. 2017), genetic algorithm (Harrag and Messalti 2015), particle swarm optimization (Mirbagheri et al. 2015), ant colony optimization (Oshaba et al. 2015), ...etc. Among several techniques mentioned, the perturb and observe (P&O) method and the incremental conductance (IC) algorithms are the most commonly applied algorithms.

This paper proposes an innovative implementation of P&O and IC MPPTs using Matlab/Simulink stateflow models to track the maximum power of PV system composed of Solarex MSX-60W PV panel operating at variable atmospheric conditions and DC-DC boost converter powering resistive load. The performance of the proposed stateflow MPPT models has been evaluated under different scenarios test considering irradiation, temperature and step size variation, respectively. Simulation results show that the proposed stateflow MPPT models can effectively track the maximum power transfer. The remainder of the paper is organized as follows. In Sect. 2, the photovoltaic cell modelling as well as the P&O and IC MPPTs are presented. Section 3 presents the simulations results and discussions. Section 4 stated the main conclusions of this study.

2   Materials and Methods

2.1   PV Cell Modeling

The well-known and widely used model based on the well-known Shockley diode equation is presented below Fig. 1 (Shockley 1949).

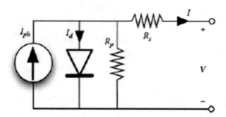

Fig. 1. Solar cell single-diode model.

The output current i can be expressed by:

$$I = N_p I_{ph} - N_p I_{rs}\left[e^{\left(\frac{(q(V+R_sI))}{A.k.T.N_s}\right)} - 1\right] - N_p\left(\frac{(q(V+R_sI))}{N_s.R_p}\right) \tag{1}$$

where V is the cell output voltage; q is the electron charge ($1.60217646 \times 10^{-19}$ C); k is the Boltzmann's constant ($1.3806503 \times 10^{-23}$ J/k); t is the temperature in Kelvin; I_{rs} is the cell reverse saturation current; a is the diode ideality constant and N_p and Ns are the number of PV cells connected parallel and in series, respectively.

The generated photocurrent I_{ph} is related to the solar irradiation by the following equation:

$$I_{ph} = [I_{sc} + k_i(T - T_r)]\frac{s}{1000} \tag{2}$$

where k_i is the short-circuit current temperature coefficient; s is the solar irradiation in w/m^2; T_r is the cell reference temperature and I_{sc} is the cell short-circuit current at reference temperature.

The cell's saturation current varies with temperature according to the following equation:

$$I_{rs} = I_{rr}\left[\frac{T}{T_r}\right]^3 exp\left(\frac{q.E_G}{k.A}\left[\frac{1}{T_r} - \frac{1}{T}\right]\right) \tag{3}$$

where E_g is the band-gap energy of the semiconductor and I_{rr} is the reverse saturation at T_r.

2.2 PV MPPT

The Perturb and Observe (P&O) and the Incremental Conductance (IC) MPPTs are the most used algorithms for their simplicity and easy implementation (Seyedmahmoudian et al. 2015).

2.2.1 P&O MPPT

Perturb and Observe MPPT involves perturbation of the operating voltage or the duty cycle based on a comparison of the generated power to ensures maximum power point. The principle is very simple: in the ascending phase of the P-V curve and considering a positive change of the voltage, the tracker generates a positive voltage increasing as a consequence the delivered PV power and change of the operating point. In this case, the PV voltage and power increase up to a new point. Similar steps with opposite direction can be done in the case of a decrease in the PV power; the instantaneous PV voltage follows the maximum power point according to a predetermined PV voltage and power values. Under these conditions, the tracker seeks the MPP permanently. At specified PV voltage, the desired power is the solution of the nonlinear equation given by (dP/dV = 0) (Ishaque et al. 2015).

2.2.2 IC MPPT

The Incremental Conductance focuses directly on power variations. The output current and voltage of the photovoltaic panel are used to calculate the conductance and the incremental conductance. The basic equations of this method are as follows (Putria et al. 2015):

$$\frac{dP}{dV} = 0 \tag{4}$$

Equation (4) can be rewritten as:

$$\frac{dP}{dV} = \frac{d(I.V)}{dV} = I + V\frac{dI}{dV} = 0 \tag{5}$$

$$\frac{dI}{dV} = -\frac{I}{V} \quad atMPP \tag{6}$$

$$\frac{dI}{dV} > -\frac{I}{V} \quad \text{at the left of MPP} \tag{7}$$

$$\frac{dI}{dV} < -\frac{I}{V} \quad \text{at the right of MPP} \tag{8}$$

3   Results and Discussion

The proposed stateflow MPPT controllers have been investigated by implementing under Matlab/Simulink software the MPPTs stateflow models driving the boost DC-DC converter to transfer the maximum power from the PV module MSX-60 W PV panel operating at variable atmospheric conditions and the resistive load.

Figure 2 shows the output power using the stateflow P&O MPPT.

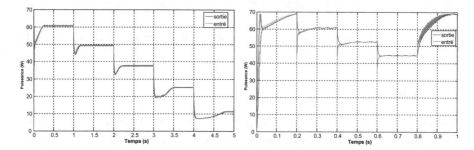

Fig. 2. Output power variation using stateflow P&O MPPT under varying: (a) irradiation [1000 800 600 400 200]W/m², (b) temperature [25 50 75]°C.

Figure 3 shows the output power using the stateflow IC MPPT.

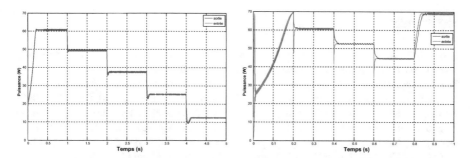

Fig. 3. Output power variation using stateflow IC MPPT under varying: (a) irradiation [1000 800 600 400 200]W/m², (b) temperature [25 50 75].

Figure 4 shows the output power using the stateflow MPPTs in case of step size variation.

From Figs. 2, 3 and 4 we see clearly that the proposed stateflow MPPT models reacts correctly and provides PV characteristics accordingly to irradiation, temperature or step size varying corresponding to the theoretical values.

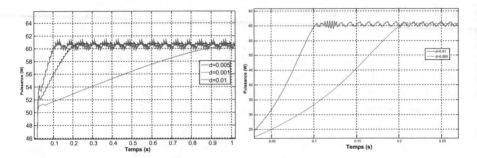

Fig. 4. Output power variation using stateflow MPPTs under varying: P&O step size [0.001 0.005 0.01], (b) IC step size[0.005 0.01].

4   Conclusion

This paper addresses the assessment of the two well known MPPTs controllers P&O and IC using the Matlab/Simulink stateflow models used to track the maximum power of PV generator system composed of Solarex msx-60 W PV panel operating at variable atmospheric conditions and DC-DC boost converter controlled using the proposed MPPTs. The simulation results obtained using the proposed stateflow MPPT models prove that the proposed models can effectively track the maximum power in case of irradiation, temperature and step size varying, respectively.

References

Babaa, S.E., Armstrong, M., Pickert, V.: Overview of maximum power point tracking control methods for PV systems. J. Power Energy Eng. **2**, 59–72 (2014)

Femia, G.N., Petrone, G., Spagnuolo, G., Vitelli, M.: Optimization of perturb and observe maximum power point tracking method. IEEE Trans. Power Electron. **20**, 963–973 (2005)

Harrag, A., Messalti, S.: Variable step size modified P&O MPPT algorithm using GA-based hybrid offline/online PID controller. Renew. Sustain. Energy Rev. **49**, 1247–1260 (2015)

Harrag, A., Messalti, S.: Improving PV performances using fuzzy-based MPPT. In: Hatti, M. (ed.) ICAIRES 2017. LNNS, vol. 35, pp. 236–244. Springer, Cham (2018). https://doi.org/10.1007/978-3-319-73192-6_24

IRENA_Roadmap: http://www.irena.org/DocumentDownloads/Publications/IRENA_REmap_2016_edition_report.pdf (2016)

Isaldo, B.A., Amiri, P.: Improved variable step size incremental conductance MPPT method with high convergence speed for PV systems. J. Eng. Sci. Technol. **11**, 516–528 (2016)

Ishaque, K., Salam, Z., Lauss, G.: The performance of perturb and observe and incremental conductance maximum power point tracking method under dynamic weather conditions. Appl. Energy **119**, 228–236 (2015)

Kok Soon, T., Mekhilef, S.: A fast-converging MPPT technique for photovoltaic system under fast varying solar irradiation and load resistance. IEEE Trans. Ind. Inform. **11**(1), 176–186 (2015)

Messalti, S., Harrag, A., Loukriz, A.: A New variable step size neural networks MPPT controller: review, simulation and hardware implementation. Renew. Sustain. Energy Rev. **68**, 221–233 (2017)

Mirbagheri, S.Z., Aldeen, M., Saha, S.: A PSO-based MPPT re-initialised by incremental conductance method for a standalone PV system. In: 2015 23rd Mediterranean Conference on Control and Automation (MED), Torremolinos, Spain, 16–19 June, pp. 293–303 (2015)

Noguchi, T., Togashi, S., Nakamoto, R.: Short-current pulse based adaptive maximum power point tracking for photovoltaic power generation system. In: IEEE International Symposium on Industrial Electronics, pp. 157–162 (2000)

Oshaba, A.S., Ali, E.S., Abd Elazim, S.M.: PI controller design using ABC algorithm for MPPT of PV system supplying DC motor pump load. Neural Comput. Appl. **28**, 353–364 (2015). https://doi.org/10.1007/s00521-015-2067-9

Putria, R.I., Wibowo, S., Rifai, M.: Maximum power point tracking for photovoltaic using incremental conductance method. Energy Procedia **68**, 22–30 (2015)

Rekioua, D., Matagne, E.: Optimization of Photovoltaic Power Systems: Modelization, Simulation and Control. Springer, London (2012)

Renewable Energy in Europe Markets: Trends and Technologies - European Renewable Energy Council (EREC). Earthscan, London (2010)

Seyedmahmoudian, M., Rahmani, R., Mekhilef, S., Maung Than Oo, A., Stojcevski, A., Kok Soon, T., Safdari Ghandhari, A.: Simulation and hardware implementation of new maximum power point tracking technique for partially shaded PV system using hybrid DEPSO method,". IEEE Trans. Sustain. Energy **6**, 850–862 (2015)

Sher, H.A., Murtaza, A.G., Noman, A., Addoweesh, K.E.: A new sensorless hybrid MPPT algorithm based on fractional short-circuit current measurement and P&O MPPT. IEEE Trans. Sustain. Energy **6**(4), 1426–1434 (2015)

Shockley, W.: The theory of p-n junctions in semiconductors and p-n junction transistors. Bell Syst. Tech. J. **28**, 435–489 (1949)

Tey, K.S.S., Mekhilef, S.: Modified incremental conductance MPPT algorithm to mitigate inaccurate responses under fast-changing solar irradiance level. Sol. Energy **101**, 333–342 (2014)

Xiao, W., Dunford, W.G.: A modified adaptive hill climbing MPPT method for photovoltaic power systems. In: Proceedings on 35th Annual IEEE Power Electronic Specialists Conference 1957–1963 (2004)

New Combined Fuzzy-IC Variable Step Size MPPT Reducing Steady State Oscillations

Abdelghani Harrag[1,2(✉)] and Sabir Messalti[3]

[1] Optics and Precision Mechanics Institute,
Ferhat Abbas University - Setif 1,
Cite Maabouda (ex. Travaux), 19000 Setif, Algeria
a.harrag@univ-setif.z
[2] CCNS Laboratory, Department of Electronics, Faculty of Technology,
Ferhat Abbas University - Setif 1, Campus Maabouda, 19000 Setif, Algeria
[3] Electrical Engineering Department, Faculty of Technology,
Mohamed Boudiaf University, Route BBA, 28000 Msila, Algeria
messalti.sabir@yahoo.fr

Abstract. In this paper, a combined fuzzy-IC variable step size MPPT controller has been proposed. The proposed MPPT uses the classical IC algorithm where the step size is drived using a fuzzy logic controller. The MPPT controller has been tested and validated under Matlab/Simulink environment where the model of the system including a Solarex MSX-60W module fed by DC-DC boost converter drived using the proposed combined fuzzy-IC controller has been implemented. The simulation results prove the performance of the proposed algorithm to reduce the steady state oscillation leading to the reduction of energy losses between 23.08% and 93.15%. In addition, the proposed MPPT controller improves the dynamic performances by reduction the response time between 54.39% and 82.64% as well as a reducing the overshoot between 10.47% and 33.53% which will reduce effectively the energy losses.

Keywords: IC · MPPT · Fuzzy logic · PV cell modeling · Variable step size

1 Introduction

In past years it has become increasingly clear that the present method of generating energy has no future due to finiteness of resources reflected in the rising prices of oil and gas as well as the first effects of burning fossil fuels. The melting of the glaciers, the rise of the ocean, levels and the increase of the weather extremes, the Fukushima nuclear catastrophe, all show that nuclear energy is not the path to follow in the future. Fortunately, there is a solution with which a sustainable energy supply can be assured: renewable energy sources. These use infinite sources as a basis for energy supplies and can ensure a full supply with a suitable combination of different technologies such as biomass, photovoltaic, wind power, etc. a particular role in the number of renewable energies is played by PVs. They permit an emission-free conversion of sunlight into electrical energy and will be an important pillar in future energy systems (Nam and Pardo 2011).

© Springer Nature Switzerland AG 2019
M. Hatti (Ed.): ICAIRES 2018, LNNS 62, pp. 376–383, 2019.
https://doi.org/10.1007/978-3-030-04789-4_41

Despite the large number of photovoltaic advantages, it still presents some drawbacks comparing to conventional energy resources especially its low conversion efficiency which is only in the range of 9–17%, high fabrication cost, and nonlinear characteristics. As consequence, the energy harvesting at maximum efficiency is not enough (Martens 2014; IEA 2014; IRENA 2016). To overcome these problems, a maximum power point tracking controller is required to adjust continuously the duty cycle of the power interfaces to assure the transfer of the maximum power available from a photovoltaic array to the load at any given time and under variable conditions. The nonlinear P-V characteristic presents a unique operating point called the maximum power point (MPP). The role of the maximum power point tracking controllers is to operate by sensing at periodical time the current and voltage of the PV array and compute the power in order to adjust the duty cycle of the power converter to track the maximum power point (Rekioua and Matagne 2012; Piegari and Rizzo 2010; Eltawil and Zhao 2013; Bhatnagar and Nema 2013; Abu Eldahab et al. 2007; Esram and Chapman 2007; Liu et al. 2015; El-Khozondar et al. 2016; Pakkiraiah and Durga Sukumar 2016).

In the last two decades, vast number of maximum power point tracking controllers have been proposed: perturbation and observation (P&O) (Farahat et al. 2015; Femia et al. 2005), incremental conductance (IC) (Harrag et al. 2017), fractional open-circuit voltage (FOCV) (Safari and Mekhilef 2011), fractional short-circuit current (FSCC) (Masoum et al. 2002), hill climbing (HC) (Noguchi et al. 2000), neural network (Messalti et al. 2017) (Pakkiraiah and Durga Sukumar 2016), fuzzy logic (FL) (Harrag and Messalti 2018), genetic algorithms (GA) (Harrag and Messalti 2015), particle swarm optimization (PSO) (Letting et al. 2012), teaching-learning-based optimization (TLBO) (Chao and Wu 2016), etc. Among all the previous maximum power point tracking strategies, the perturb and observe (P&O), incremental conductance (IC) and hill climbing algorithms are widely employed due to easy implementation and simplicity. Conversely the performance depends essentially on the fixed step size, a faster dynamics with large oscillations around the maximum power point is obtained using a large step size, a slow tracking speed and less oscillations around the maximum power point is obtained using a small step size. Hence, the tradeoff between the dynamics and steady state accuracy must be established by the corresponding design. To overcome this problem, variable step-size maximum power point tracking is required.

In this paper, a combined fuzzy-IC variable step size maximum power point tracking controller has been proposed to provide the duty cycle under different atmospheric conditions. The proposed maximum power point tracking uses the classical IC maximum power point tracking algorithm where the step size is controlled using a fuzzy logic controller. The proposed fuzzy-IC variable step size maximum power point tracking controller has been tested and validated using Matlab/Simulink model for different atmospheric conditions. The remainder of the paper is organized as follows. In Sect. 2, the photovoltaic cell modeling is presented. Section 3 describes the proposed fuzzy-IC variable step size maximum power point tracking. Section 4 presents the simulations results and discussions. In Sect. 5, the conclusions are stated.

2 PV Cell Modeling

The well-known and widely used model based on the well-known one diode has been used in this study. The output current i can be expressed by:

$$I = N_p I_{ph} - N_p I_{rs} \left[e^{\left(\frac{q(V + R_S I)}{A.k.T.N_S} \right)} - 1 \right] - N_p \left(\frac{q(V + R_S I)}{N_S.R_p} \right) \tag{1}$$

where v is the cell output voltage; q is the electron charge ($1.60217646 \times 10^{-19}$ c); k is the Boltzmann's constant ($1.3806503 \times 10^{-23}$ j/k); t is the temperature in Kelvin; Irs is the cell reverse saturation current; a is the diode ideality constant and Np and Ns are the number of PV cells connected parallel and in series, respectively.

The generated photocurrent I_{ph} is related to the solar irradiation by the following equation:

$$I_{ph} = [I_{sc} + k_i(T - T_r)] \frac{s}{1000} \tag{2}$$

where k_i is the short-circuit current temperature coefficient; s is the solar irradiation in W/m^2; T_r is the cell reference temperature and I_{sc} is the cell short-circuit current at reference temperature.

The cell's saturation current varies with temperature according to the following equation:

$$I_{rs} = I_{rr} \left[\frac{T}{T_r} \right]^3 \exp \left(\frac{q.E_G}{k.A} \left[\frac{1}{T_r} - \frac{1}{T} \right] \right) \tag{3}$$

where E_g is the band-gap energy of the semiconductor and I_{rr} is the reverse saturation at Tr.

3 Proposed Variable Step Size MPPT

3.1 Classical Fixed Step Size IC MPPT

The incremental conductance is widely used maximum power point tracking methods for its simplicity and ease of implementation, high tracking speed and better efficiency (Kok Soon and Mekhilef 2014). This method focuses directly on power variations. The output current and voltage of the photovoltaic panel are used to calculate the conductance and the incremental conductance. The basic equations of this method are as follows:

$$\frac{dP}{dV} = 0 \tag{4}$$

$$\frac{dP}{dV} = \frac{d(I.V)}{dV} = I + V\frac{dI}{dV} = 0 \tag{5}$$

$$\frac{dI}{dV} = -\frac{I}{V} \ at \ \text{maximum power point} \tag{6}$$

$$\frac{dI}{dV} > -\frac{I}{V} \ \text{at the left of maximum power point} \tag{7}$$

$$\frac{dI}{dV} < -\frac{I}{V} \ \text{at the right of maximum power point} \tag{8}$$

3.2 Fuzzy Logic Controller

Fuzzy logic have the benefits of operating with general inputs instead of a correct mathematical model and handling nonlinearities mathematical logic management typically consists of four stages: fuzzification, rule base, inference method, and defuzzification (Pedrycz 1993). In this study, the proposed fuzzy maximum power point tracking requires as inputs the error E and the change in error ΔE defined by:

$$E_k = \frac{P(k) - P(k-1)}{V(k) - V(k-1)} \tag{9}$$

$$\Delta E_k = E_k - E_{k-1} \tag{10}$$

where E_k and E_{k-1} are the error at instant k and k − 1 and $\Delta E(k)$ is the change of error at instant k. The output is the fuzzy scaling (FS) of the pulse width modulation duty cycle step variation Δd. Table 1 gives the rule base.

Table 1. Rules base.

$\Delta E_k/E_k$	NL	NS	ZE	PS	PL
NB	ZE	ZE	PS	NS	NB
NS	ZE	ZE	ZE	NS	NB
ZE	PB	PS	ZE	NS	NB
PS	PB	PS	ZE	ZE	ZE
PB	PB	PS	NS	ZE	ZE

3.3 Proposed Combined Fuzzy-IC Variable Step Size MPPT

As mentioned previously, the performances of pv systems depends mainly on the step size. Therefore, a good calculation of step size provides a high performance of pv systems. The proposed variable step size is given as follows:

$$D(k) = D(k-1) + FS * \Delta D \qquad (11)$$

where D(k) and D(k−1) are the duty cycle at instants k and k − 1; Fs is the scaling factor adjusted at the sampling period and Δd is the fixed step size.

4   Results and Discussion

The proposed combined fuzzy-IC variable step size maximum power point tracking controller has been investigated by implementing the model of the whole system using Matlab/Simulink software, composed of Solarex MSX-60W PV panel (Table 2) operating at variable atmospheric conditions supplying resistive load via a DC-DC boost converter driven using the propose fuzzy-IC maximum power point tracking controller.

Table 2. PV Module parameters.

Description	MSX-60
Maximum Power (P_{MPP})	60 W
Voltage P_{max} (V_{MPP})	17.1 V
Current at P_{max} (I_{MPP})	3.5 A
Short Circuit current (I_{SC})	3.8 A
Open Circuit voltage (Voc)	21.1 V
Temperature coeff of Voc	− (80 ± 10) mV/°C
Temperature coeff of Isc	(0.065 ± 0.01)%°C
Temperature coeff of power	(− 0.5 0.05)%°C
NOCT	47 ± 2 °C

Figure 1.left shows the corresponding output power using both fixed step size IC and fuzzy-IC variable step size maximum power point tracking controllers. From Fig. 1.left, we can see that both IC algorithm track properly the maximum available power with less steady state oscillations for the proposed fuzzy-IC maximum power point tracking.

4.1   Static Performances; Steady State Oscillation

Table 3 give the output power ripple and steady state oscillations.

From Table 3, it's clear that the proposed fuzzy-IC maximum power point tracking controller outperforms the classical fixed step size IC maximum power point tracking controller showing a reduction ratio of steady state oscillation between 23.08% and 93.15% which will reduce effectively the energy losses.

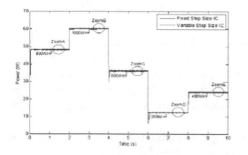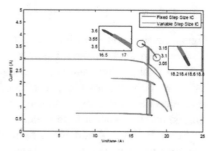

Fig. 1. left) Output Power tracking, right) I-V characteristic.

Table 3. Output power ripple and steady state oscillations.

Point	Fixed Step	fuzzy-IC	Red. Ratio (%)
Zoom A (800 W/m^2)	0.73 W	0.006 W	91.78
Zoom B (1000 W/m^2)	0.44 W	0.10 W	72.27
Zoom C (600 W/m^2)	0.73 W	0.05 W	93.15
Zoom D (200 W/m^2)	0.56 W	0.20 W	64.29
Zoom E (400 W/m^2)	0.65 W	0.15 W	23.08

4.2 Dynamic Performances - Response Time and Overshoot

Table 4 summarizes the dynamic performances.

From Table 4, the proposed fuzzy-IC (FZ-IC) maximum power point tracking controller performs better than the classical fixed step size IC maximum power point tracking controller (FS IC) showing a reduction ratio of response time between 54.39% and 82.64% as well as a reduction ratio of overshoot between 10.47% and 33.53%.

Table 4. response time and overshoot.

Point	Resp. time (ms)		Red. Ratio (%)	Overshoot (W)		Red. Ratio (%)
	FSIC	FZIC		FSIC	FZIC	
Point A	392.50	83.32	78.77	–	–	–
Point B	61	15	75.41	–	–	–
Point C	104	30	71.15	10.05	6.68	33.53
Point D	265	121	54.39	6.84	6.124	10.47
Point E	311	54	82.64	–	–	–

4.3 Efficiency - Tracking Accuracy

Figure 1.right shows the I-V characteristic. We can see that the proposed fuzzy-IC variable step size maximum power point tracking controller presents the best and minimal tracking time and course compared to conventional fixed step size IC maximum power point tracking. This property reduces energy losses and ensures the maximum impedance adaptation by transferring the maximum available power from the photovoltaic system to the load.

From Tables 3, 4 and Fig. 1.right, the proposed combined fuzzy-IC variable step size maximum power point tracking outperforms the classical fixed step size IC maximum power point tracking considering the irradiation test pattern used regarding all performance measures in static, dynamic and efficiency performances.

5 Conclusion

In this work, a combined fuzzy-IC variable step size has been proposed for reducing the steady state oscillation in photovoltaic systems power point tracking. The fuzzy logic controller has been used to scale the variable step size of the classical IC maximum power point tracking controller needed for generating the pulse width modulation duty cycle to drive DC-DC boost converter. The proposed maximum power point tracking controller addresses the challenges associated with rapidly changing insolation levels and oscillation around the maximum power point. The simulation results done using Matlab/Simulink environment prove the performance and functionality of the proposed algorithm to reduce the steady state oscillation between 23.08% and 93.15%. In addition, the proposed maximum power point tracking controller improves the dynamic performances by reduction the response time between 54.39% and 82.64% as well as a reducing the overshoot between 10.47% and 33.53% which will reduce effectively the energy losses. As future works, we plan to validate the developed fuzzy-IC variable step size maximum power point tracking controller on an experimental platform in the hardware-in-the-loop mode.

References

Abu Eldahab, Y.E., Saad, N.H., Zekry, A.: Enhancing the maximum power point tracking techniques for photovoltaic systems. Renew. Sustain. Energy Rev. **40**, 505–514 (2014)

Bhatnagar, P., Nema, R.K.: Maximum power point tracking control techniques: state-of-the-art in photovoltaic applications. Renew. Sustain. Energy Rev. **23**, 224–241 (2013)

Chao, K.H., Wu, M.C.: Global maximum power point tracking (MPPT) of a photovoltaic module array constructed through improved teaching-learning-based optimization. Energies **9**(12), 986 (2016)

El-Khozondar, R.J., Matter, K., Suntio, T.: A review study of photovoltaic array maximum power tracking algorithms. Renew. Wind Water Solar **3**, 3 (2016)

Eltawil, M.A., Zhao, Z.: MPPT techniques for photovoltaic applications. Renew. Sustain. Energy Rev. **25**, 793–813 (2013)

Esram, R., Chapman, P.L.: Comparison of photovoltaic array maximum power point tracking techniques. IEEE Trans. Energy Convers. **22**, 439–449 (2007)

Farahat, M.A., Enany, M.A., Nasr, A.: Assessment of maximum power point tracking techniques for photovoltaic system applications. J. Renew. Sustain. Energy **7**, 042702 (2015)

Femia, G.N., Petrone, G., Spagnuolo, G., Vitelli, M.: Optimization of perturb and observe maximum power point tracking method. IEEE Trans. Power Electron. **20**, 963–973 (2005)

Harrag, A., Messalti, S.: Variable step size modified P&O MPPT algorithm using GA-based hybrid offline/online PID controller. Renew. Sustain. Energy Rev. **49**, 1247–1260 (2015)

Harrag, A., Messalti, S.: Improving PV performances using fuzzy-based MPPT. In: Artificial Intelligence in Renewable Energetic Systems, Lecture Notes in Networks and Systems, vol. 35. https://doi.org/10.1007/978-3-319-73192-6_24 (2018)

Harrag, A., Titraoui, A., Bahri, H., Messalti, S.: Photovoltaic pumping system - comparative study analysis between direct and indirect coupling mode. AIP Conf. Proc. **1814**, 020002 (2017)

IEA: Technology Roadmaps Bioenergy for Heat and Power. http://www.iea.org/publication/ (2014)

IRENA_Roadmap. http://www.irena.org/DocumentDownloads/Publications/IRENA_REmap_2016_edition_report.pdf (2016)

Kok Soon, T., Mekhilef, S.: A fast-converging MPPT technique for photovoltaic system under fast varying solar irradiation and load resistance. IEEE Trans. Ind. Inf. **11**(1), 176–186 (2015)

Letting, L.K., Munda, J.L., Hamam, Y.: Optimization of a fuzzy logic controller for PV grid inverter control using S-function based PSO. Sol. Energy **86**, 1689–1700 (2012)

Liu, Y.H., Chen, J.H., Huang, J.W.: A review of maximum power point tracking techniques for use in partially shaded conditions. Renew. Sustain. Energy Rev. **41**, 436–453 (2015)

Martens, K.: Photovoltaics - fundamentals, technology an practice. Wiley, Chichester (2014)

Masoum, M.A.S., Dehbonei, H., Fuchs, E.F.: Theoretical and nexperimental analyses of photovoltaic systems with voltage and current based maximum power-point tracking. IEEE Trans. Energy Convers. **17**(4), 514–522 (2002)

Messalti, S., Harrag, A., Loukriz, A.: A new variable step size neural networks MPPT controller: review, simulation and hardware implementation. Renew. Sustain. Energy Rev. **68**, 221–233 (2017)

Nam, T., Pardo, T.A. Conceptualizing smart city with dimensions of technology, people, and institutions. In: 12th Annual International Digital Government Research Conference: Digital Government Innovation in Challenging Times, pp. 282–291 (2011)

Noguchi, T., Togashi, S., Nakamoto, R.: Short-current pulse based adaptive maximum power point tracking for photovoltaic power generation system. In: IEEE International Symposium on Industrial Electronics, pp. 157–162 (2000)

Pakkiraiah, B., Durga Sukumar, G.: Research survey on various maximum power point tracking performance issues to improve the solar PV system efficiency. J. Sol. Energy **2016**, 1–20 (2016)

Pedrycz, W.: Fuzzy Control and Fuzzy Systems. Research Studies Press Ltd (1993)

Piegari, L., Rizzo, R.: Adaptive perturb and observe algorithm for photovoltaic maximum power point tracking. IET Renew. Power Gener. **4**, 317–328 (2010)

Rekioua, D., Matagne, E.: Optimization of Photovoltaic Power Systems: Modelization, Simulation and Control. Springer, London (2012)

Safari, A., Mekhilef, S.: Simulation and hardware implementation of incremental conductance maximum power point tracking with direct control method using cuk converter. IEEE Trans. Ind. Electron. **58**, 1154–1161 (2011)

A Set of Smart Swarm-Based Optimization Algorithms Applied for Determining Solar Photovoltaic Cell's Parameters

Selma Tchoketch Kebir[✉], Mohamed Salah Ait Cheikh,
and Mourad Haddadi

Electronic Department, Ecole Nationale Polytechnique, ENP, Algiers, Algeria
selma.tchoketch_kebir@g.enp.edu.dz

Abstract. This paper presents the study and the use of two smart swarm-based optimization methods for determining the electrical unknown parameters of solar photovoltaic cells. These two methods are the well-known Particle Swarm Optimization (PSO) and a recent smart swarm-based method named, Whale Optimization Algorithm (WOA). This last one is inspired by the hunting behaviour of humpback whales in nature. The best parameters determination values are essential for the accuracy of the solar photovoltaic characteristics. The non-linear parameters determination problem is formulated mathematically as a multi-parameters or as a multi-objective optimization problem. The two swarm-based optimization methods are first described, explaining every step, and then validated using solar photovoltaic manufacturers' data sheets information. The performance of each approach is evaluated in terms of chosen criteria. The results show that the WOA method outperforms the PSO.

Keywords: Optimization · Swarm-based intelligence · Nature-inspired
PSO algorithm · WOA algorithm · Photovoltaic cells

1 Introduction

Optimization techniques are widely used in different engineering topics, such as modelling, identification, optimization, prediction, and control of complex systems. Nowadays, there is a trend, in the scientific community, to model and solve difficult optimization problems by the use of creatures inspired behaviours in nature. This is mainly due to the poor results of classical optimization algorithms in solving combinatorial and nonlinear functions. The process of optimization techniques usually leads to the best possible solution for a particular problem. As the complexity of systems has increased, over the last few decades, the need for new optimization methods has become evident. Since classical methods [1, 2], lead to poor results, a solution to these complex systems may be found through the application of meta-heuristic optimization algorithms. These methods are known for their simplicity, flexibility, derivation free process and the ability to find the global optimal solution. They are also applicable to diverse problems without changing their basic structure. Meta-heuristics are classified

© Springer Nature Switzerland AG 2019
M. Hatti (Ed.): ICAIRES 2018, LNNS 62, pp. 384–399, 2019.
https://doi.org/10.1007/978-3-030-04789-4_42

as in Fig. 1 bellow, into four main classes such as swarm-based, physics-based, evolution-based, and human-based approaches [2].

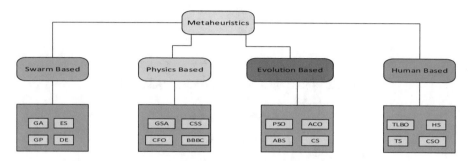

Fig. 1. Classification of metaheuristic's optimization methods.

Swarm intelligence (SI) meta-heuristic's optimization methods are based on the natural collective behaviour of decentralized and self-organized systems. Among these SI meta-heuristic methods, the Particle Swarm Optimization (PSO), which is known as a swarm-based algorithm that mimics the natural flocking birds' behaviour [3] and the Whale Optimization Algorithm (WOA) which mimics the hunting whales' humpback behaviour [4, 5]. These two optimization algorithms are applied to control Photovoltaic (PV) generation systems in order to determine the unknown electrical parameters of a solar photovoltaic cell. Also, to draw the most precise PV model characteristics [6]. To do so, different approaches exist, which are categorized as analytic [7], numeric [8] and optimization based methods [9–16]. This paper is structured as follow, the next section gives a talk about a modelling section in which the parameters determination process is described. It is followed by a description of smart swarm-based optimization methods, with details about the two developed methods. Then, the two swarm-based methods are applied and tested. In the next section, the two methods are compared and discussed, after having some simulation results. Finally, this paper ends with a conclusion.

2   Photovoltaic Parameters Determination Process

In this section, a modelling step is given, then, the problem's formulation is discussed.

2.1   Modelling

There are several electrical models to describe the physical behaviours of solar PV cells. In this work, we have chosen the researchers most used model, known as the single diode model, with series and shunt resistances. This model contains five unknown parameters, cited below, represented in Fig. 2 [17, 18].

The mathematical expression which governs the Current-Voltage (I-V) relationship of this elementary electrical PV cell's model is given below.

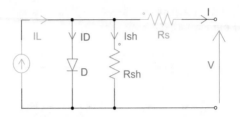

Fig. 2. Solar PV cell's electrical equivalent circuit.

$$I = I_L - I_D - I_{sh} \tag{1}$$

$$I = I_L - I_{ds} \cdot \left(exp\left(\frac{V + R_s \cdot I}{n \cdot V_t}\right) - 1 \right) - \frac{V + R_s \cdot I}{R_{sh}} \tag{2}$$

Where the PV unknown parameters are

- I_L: Light current,
- I_{ds}: Diode saturation current,
- n: Diode ideality factor,
- R_s: Series resistance and
- R_{sh}: Shunt resistance.

With, $V_t = K_B \cdot T_c/q$: Thermal voltage, K_B: Boltzmann's constant (1.380650 * 10^{-23} J/K), q: Electronic charge (1.6021764 * 10^{-19} C) and T_c: Cell's temperature.

The above mathematical non-linear equation, with its five unknown parameters, which cannot be directly measurable and are also, not given in the PV manufacturers' data-sheet. Hence, it is necessary to accurately determine these unknown parameters by an appropriate method before designing the PV system.

2.2   Problem's Formulation

Many methods allow the formulation of the optimal non-linear PV parameters determination problem. These methods can be under the mono-objective (convex or non-convex) case [19] or under multi-objectives optimization case [20], as detailed below. To do so, we first formulate the optimization problem.

(a) *Mono-objective case:*

For the description of the proposed mono-objective problem, the objective (fitness) function is obtained by the derivative of the power-voltage characteristic at the maximum power point, through the following expressions, as presented in Fig. 3.

$$Pmax \rightarrow \left.\frac{dP}{dV}\right|_{V=Vm} = 0 \tag{3}$$

$$Objective = \frac{dP}{dV} = I + V * \frac{dI}{dV} = 0 \tag{4}$$

In the example of choosing a single diode model representing the PV cell's mathematical behavior (Fig. 2), the objective function will be expressed as follow [11]:

- Convex (without shading)

$$Objective = \frac{dP}{dV} = I - V \cdot \left[\frac{\frac{1}{R_{sh}} + \frac{1}{n*Vt} * \left(I_L + I_{ds} - I - \left(\frac{V + R_s \cdot I}{R_{sh}} \right) \right)}{1 + \frac{R_s}{R_{sh}} + \frac{R_s}{n*Vt} \left(I_L + I_{ds} - I - \left(\frac{V + R_s \cdot I}{R_{sh}} \right) \right)} \right] = 0 \quad (5)$$

This objective function is subject to a set of constraints, for the five parameters range as mentioned in what follow.

$$\begin{cases} 0.9 * I_{sc\_min} < I_L < 1.1 * I_{sc\_max} \\ I_{ds\_min} \leq I_{ds} \leq I_{ds\_max} \\ n_{min} \leq n \leq n_{max} \\ 0 \leq R_s \leq R_{s\_max} \\ R_{sh\_min} \leq R_{sh} \leq R_{sh\_max} \end{cases} \quad (6)$$

The boundaries conditions for evaluating the constraints are obtained as in [19] and are presented in the Table 1.

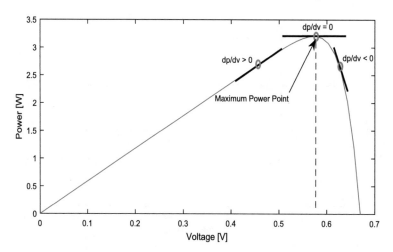

Fig. 3. Power-voltage curve characteristic and its derivative at the maximum point.

Table 1. Maximum and minimum boundaries of PV cell's parameters determination values.

	Parameters				
	I_{ds}	n	I_L	R_s	R_{sh}
Search range	$[1e^{-10}, 1e^{-7}]$	$[1, 1.5]$	$[0.9*I_{sc}, 1.1*I_{sc}]$	$[0, ((V_{oc} - V_m)/ I_m)]$	$[(V_m/(I_{sc} - I_m)) - ((V_{oc} - V_m)/ I_m), 3* (V_m/ (I_{sc} - I_m)) - ((V_{oc} - V_m)/I_m)]$

As shown in Fig. 3 there exist only one optimum (maximum) point, therefore the problem is classified under the convex optimization case, where the use of Newton's approach [20], gives also good results.

- Non-convex (with shading)

The above curve of Fig. 3 contains an optimum (maximum) point. The problem can be classified under the convex optimization case [19]. In our proposed work if we take into consideration the shading phenomenon, which can occur on an array of cells [19]. Since many optimum points can be observed on the curve, the problem is, then, classified under a non-convex optimization case which allows the use of the smart swarm-based approaches in order to achieve the Global Maximum Power Point (GMPP), Fig. 4. So, when one peak appears is considered as a particular case of the non-convex system.

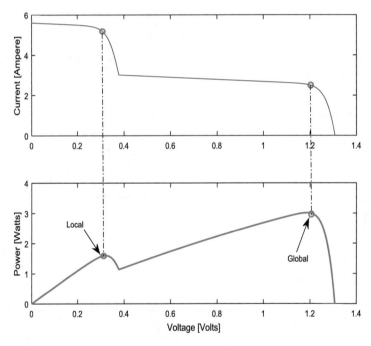

Fig. 4. Current-voltage and power-voltage curve characteristics of two PV cells with shading phenomenon (many local maximum points).

The PV cell's electrical model was also extended to the shading phenomenon, so it includes the diode characteristic's negative section. The electrical model of PV cells with consideration of shading phenomenon is presented as follow (Fig. 5) [22]:

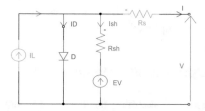

Fig. 5. A solar cell's electrical equivalent circuit considering the avalanche effect.

When shading occurs, since many optimum points can be observed on the curve, the objective function becomes.

$$Obj = I - V \cdot \left[\frac{\frac{1}{N_s \cdot R_{sh}} + b + \frac{1}{n \cdot N_s \cdot V_t} \cdot \left(I_L + I_{ds} - I - \left(\frac{V + R_s \cdot N_s \cdot I}{N_s \cdot R_{sh}}\right)\right) - b \cdot \left(\frac{V + R_s \cdot N_s \cdot I}{V_b}\right)^{-m}}{1 + \frac{R_s}{R_{sh}} + b \cdot R_s N_s + \frac{R_s}{n \cdot N_s V_t} \left(I_L + I_{ds} - I - \left(\frac{V + R_s \cdot N_s \cdot I}{N_s \cdot R_{sh}}\right)\right) + b \cdot R_s \cdot N_s \cdot \left(\frac{V + R_s \cdot N_s \cdot I}{V_b}\right)^{-m}} \right] = 0 \quad (7)$$

Where V_b is break through voltage $(-10, -50)$, m is an exponent $(1, 10)$ and α is a correction factor in $(0, 1)$.

(b) Multi-objectives case

The multi-functions problem has been formulated as in [20], so the multi-objectives optimization problem is given by the following expressions.

$$f1 : I_L + \left(1 - e^{\frac{I_{sc} \cdot R_s}{n \cdot V_T}}\right) \cdot I_{sd} - \frac{I_{sc} \cdot R_s}{R_{sh}} - I_{sc} = 0 \quad (8)$$

$$f2 : I_L + \left(1 - e^{\frac{V_{oc}}{n \cdot V_T}}\right) \cdot I_{sd} - \frac{V_{oc}}{R_{sh}} = 0 \quad (9)$$

$$f3 : I_L + \left(1 - e^{\frac{I_{mpp} \cdot R_s + V_{mpp}}{n \cdot V_T}}\right) \cdot I_{sd} - \frac{I_{mpp} \cdot R_s + V_{mpp}}{R_{sh}} - I_{mpp} = 0 \quad (10)$$

$$f4 : \frac{-n \cdot V_T - I_{sd} \cdot R_{sh} \cdot e^{\frac{I_{mpp} \cdot R_s + V_{mpp}}{n \cdot V_T}}}{n \cdot V_T \cdot (R_s + R_{sh}) + I_{sd} \cdot R_s \cdot R_{sh} \cdot e^{\frac{I_{mpp} \cdot R_s + V_{mpp}}{n \cdot V_T}}} + \frac{I_{mpp}}{V_{mpp}} = 0 \quad (11)$$

$$f5 : \frac{V_{mpp} \cdot n \cdot V_T \left(I_{sd} \cdot R_{sh} + I_L \cdot R_{sh} - 2V_{mpp}\right) + I_{sd} \cdot R_{sh} \cdot e^{\frac{I_{mpp} \cdot V_{mpp} \cdot R_s + V_{mpp}^2}{V_{mpp} \cdot n \cdot V_T}} \left[I_{mpp} \cdot V_{mpp} \cdot R_s - V_{mpp} \left(n \cdot V_T + V_{mpp}\right)\right]}{V_{mpp} \cdot \left[n \cdot V_T \cdot (R_s + R_{sh}) + I_{sd} \cdot R_s \cdot R_{sh} \cdot e^{\frac{I_{mpp} \cdot V_{mpp} \cdot R_s + V_{mpp}^2}{V_{mpp} \cdot n \cdot V_T}}\right]} = 0$$

$$\quad (12)$$

The final objective function to the problem is formulated by the expression bellow.

$$Obj(I_L, I_{sd}, n, R_s, R_{sh}) = \sum_1^5 f_i^2(I_L, I_{sd}, n, R_s, R_{sh}) \quad (13)$$

To optimize the above-mentioned objective functions (in mono or multi) many meta-heuristic algorithms can be employed.

3 Swarm-Based Optimization Methods

Optimization is the process of obtaining the best result under given situations. This process is based on a fitness function, which describes the problem under a set of constraints, representing a set of solutions for the problem [1]. Novel optimization methods, which have drawn researchers' attention recently, are the swarm-based meta-heuristics. The basic concepts of the swarm-based methods were first proposed in 1993 [23]. Swarm-based intelligence has been inspired by the collective behaviour in self-organizing systems. Swarm intelligence, which is based on the nature-inspired behaviour, is successfully applied into optimization problems in a variety of fields. These algorithms almost mimic the social natural swarm behaviour. The two algorithms, chosen in this work are the Particle Swarm Optimization and the recent Whale Optimization Algorithm. These methods are presented with more details in the following subsections.

3.1   Particle Swarm Optimization "PSO"

Proposed by *Eberhart* in 1995 [9, 10], this algorithm is based on the social behaviour of animals such as the flock of birds while they are communicating and sharing individual knowledge information during their migration way search. The equations bellow describe the PSO's mathematical model where the population is chosen as the swarm and the individuals as particles.

The position equation is,

$$X(i+1) = X(i) + V(i+1) \tag{14}$$

The velocity equation is,

$$V(i+1) = w * V(i) + c_1 * (XL_{best} - X(i)) + c_2 * (XG_{best} - X(i)) \tag{15}$$

The best local position equation is,

$$XL_{best}(i) = \begin{cases} XL_{best}(i-1) & \text{if } F(X(i)) \geq F(XL_{best}(i-1)) \\ X(i) & \text{if } F(X(i)) < F(XL_{best}(i-1)) \end{cases} \tag{16}$$

The best global position equation is,

$$XG_{best} = min\{F(XG_{best}1), F(XG_{best}2), \ldots, F(XG_{best}N)\} \tag{17}$$

The major steps of the PSO's procedures are presented as in the next pseudo code of the following Fig. 6:

Step 1: Initialization
Generate randomly a swarm of N particles
Initial set of position X and velocity V
Initial set of personnel XL_{best} and global positions XG_{best}
Set the control parameters of the PSO
Step 2: Evaluation
For each particle:
Calculate and evaluate fitness value
Step 3: Update
If the fitness value is better than the best fitness value using Eq. (16)
        Set current value as the new XL_{Best}
End
For each particle:
Find in the particle neighborhood, the particle with the best fitness using Eq. (17)
Calculate particle velocity according to the velocity Eq. (15)
Apply the velocity constriction
Update particle position according to the position Eq. (14)
Apply the position constriction
End
Step 4: Check for stopping criterion
While maximum iterations or minimum error criteria is not attained
Iter=Iter+1
Return XG_{best}

Fig. 6. Flowchart of the general steps of the PSO algorithm.

3.2   Whale Optimization Algorithm "WOA"

Developed by Mirjalili in 2016 [4], this algorithm is inspired by the bubble-net hunting strategy, observed from the natural behavior of humpback whales. Whales are fancy creatures and are considered the biggest mammals in the world. There are seven different main species of this giant mammal, among them the humpback whales. They are mostly observed in groups. They prefer to hunt, small fishes, close to the surface by creating distinctive bubbles along a circle or a '9'- shaped path as shown in the Fig. 7 below.

The Whale Optimization Algorithm (WOA) is a new swarm-based optimization method. It employs a unique and complex foraging behavior bubble-netting that involves expelling air underwater to form a vertical cylinder-ring of bubbles around the prey [24]. In the former maneuver, humpback whales dive around 12 m down and then start creating a bubble in a spiral shape around the prey and swim up toward the surface. The later maneuver includes three different stages: coral loop, lobtail, and capture loop [5]. In this work, the whale's feeding process is mathematically modeled,

Fig. 7. Bubble-net foraging process of humpback whales.

which contains the encircling prey, the spiral bubble-net attacking process and the search for prey. These models are presented, with details, in the following subsections.

3.2.1 Encircling Prey

The equations below describe the encircling humpback whales and prey (small fish) position search.

$$\vec{D} = \left| C \cdot \overrightarrow{X_P}(i) - \vec{X}(i) \right| \tag{18}$$

$$\vec{X}(i+1) = \overrightarrow{X_P}(i) - A \cdot \vec{D} \tag{19}$$

Where: i represents the current iteration. D represents distance between the whale and the prey. $\vec{X}$ and $\vec{XP}$ represent the position vectors of the whales and the prey respectively. A and C are coefficients which are calculated as follow.

$$A = 2 \cdot a \cdot r_1 - a \tag{20}$$

$$C = 2 \cdot r_2 \tag{21}$$

Where A is linearly decreasing from 2 to 0 over the course of iterations and r_1, r_2 are random in an interval from 0 to 1.

3.2.2 Bubble-Net Attacking Process

The bubble-net strategy attack method is mathematically formulated as follows:

(a) *Shrinking encircling process*

By decreasing the values of A from 2 to 0 during the optimization process, which simulates the prey encircling approach, it provides the exploration ability of the algorithm. As for the exploitation ability of the WOA, it comes from the random variable, r_2, of C.

(b) *Spiral updating position*

This approach first calculates the distance between the whale, located at $\vec{X}$ and the prey located at $X\vec{P}$. A spiral equation is then established, see below, to mimic the position between the humpback whale and the prey as a helix-shaped movement.

$$\vec{D} = \left| \vec{X_P}(i) - \vec{X}(i) \right| \tag{22}$$

$$\vec{X}(i+1) = \vec{D} \exp(b \cdot l) \cdot \cos(2 \cdot \Delta \cdot l) + \vec{X_p}(i) \tag{23}$$

Where: b is a constant, defining the shape of the logarithmic spiral, l is a random number in $(-1, 1)$, p is a random number in $(0, 1)$.

Note: It is supposed that there is 50-50% probability that whale either follow the shrinking encircling or logarithmic path during optimization process. It is mathematically, modelled as follows:

$$\vec{X}(i+1) = \begin{cases} \vec{X_P}(i) - A \cdot \vec{D} & \text{if } p \leq 0.5 \\ \vec{D} \exp(b \cdot l) \cdot \cos(2 \cdot \Delta \cdot l) + \vec{X_p}(i) & \text{if } p > 0.5 \end{cases} \tag{24}$$

3.2.3 The Prey's Attack

Finally, the humpback whales attack the prey in a random way, which is mathematically modelled as follow.

$$\vec{D} = \left| C \cdot \vec{X_{rand}}(i) - \vec{X}(i) \right| \tag{25}$$

$$\vec{X}(i+1) = \vec{X_{rand}}(i) - A \tag{26}$$

According to these conditions:

When $|A| > 1$ it enforces exploration to WOA algorithm for finding global optimum and avoiding local optima.

When $|A| < 1$ it updates the position of the current search agent when a best solution is selected.

The major steps of the WOA's procedures is shown in the pseudo code of the Fig. 8 below.

Step 1: Initialization

Initialize a swarm of N whales,

Calculate fitness of each search agent

X_p the best search agent

Step 2: Update

While (Iter < Iter_Max)

for each search agent

update a, A, c, l and p

if (p<0.5)

  if ($|A| < 1$)

Update the position of the current search agent by the Eq. (5) & (6)

  elseif ($|A| \geq 1$)

  Select a random search agent X_{rand}

Update the position of the current search agent by the Eq. (12) & (13)

  end if

 elseif (p $\geq$0.5)

Update the position of the current search by the Eq. (9) & (10)

 end if

 end for

Check if any search agent goes beyond the search space and amend it

Step 3: Evaluation

Calculate the fitness of each search agent

Update X_p if there is a better solution

Iter=Iter+1

end while

Step 4: Check for stopping criterion

return X_p

Fig. 8. Pseudo-code of the WOA algorithm.

3.3 Problem's Resolution

In this subsection, the two above presented swarm-based methods are applied. Figure 9 below shows the organigram process of Mono-crystalline silicon-based solar PV cell used for the test and the validation [25]. Table 2 gives information about the PV manufacturer's datasheet of the mono-crystalline silicon-based cell employed in this work [26].

The flowchart of the PV parameters determination procedures is represented as follows in Fig. 9.

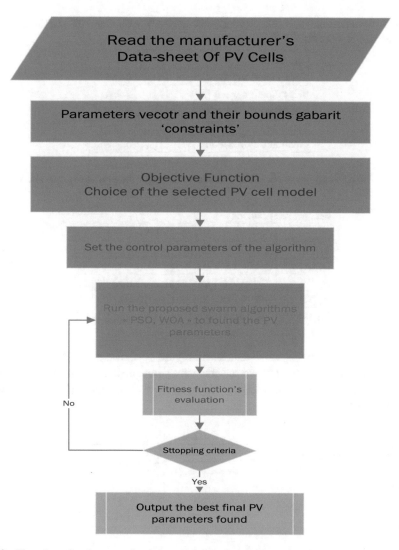

Fig. 9. Flowchart for the use of smart swarm-based algorithms for the PV cell's parameters determination values.

4   Results and Discussion

Table 3 shows the results obtained, using the two previous parameters determination methods, namely the PSO and WOA, for the unknown PV cell electrical parameters. A mono-crystalline silicon based solar PV cell is used for the test and validation of the approaches [25].

The PV parameters values obtained have been compared to those obtained as in the reference [25] ($I_L = 5.807$ A, $I_{sd} = 2.946 \times 10^{-10}$ A, $n = 1.102$, $R_s = 0.00625$ Ω, $R_{sh} = 4.957$ Ω). It can be seen that our proposed methods presents better values based

Table 2. Specification of A-300 SUNPOWER solar PV cell, from PV datasheets at STC (T = 25°C & W = 1 000 W/m²).

Parameters		Cell
Open circuit voltage [V]	V_{oc}	0.665 V
Short circuit current [A]	I_{sc}	5.9 A
Maximum power voltage [V]	V_{mpp}	0.560 V
Maximum power current [A]	I_{mpp}	5.54 A
Maximum Power Point [W]	P_{mpp}	3.1 W

Table 3. PV cell parameters simulation results values with the proposed methods: WOA and PSO.

Parameters	Methods	
	PSO	WOA
I_L [A]	5.350047694294241	5.692500000000000
I_{ds} [A]	0.000000000100000	0.000000000100000
n	1.219348238858283	1.200000000000000
R_s [Ω]	0.001000000000000	0.001000000000000
R_{sh} [Ω]	1.452541918855772	2.908079402754119

on the closest curve to that of the data sheet. Figure 10 shows the cell's current-voltage (I-V) characteristics with the parameters values obtained using the two mentioned methods. These characteristics are compared to the real measured (I-V) characteristic reference [26].

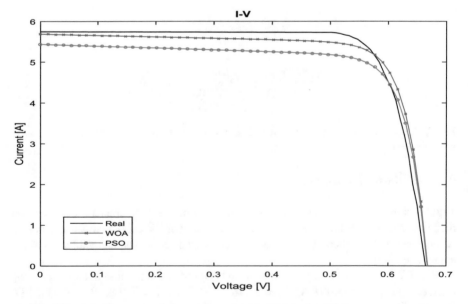

Fig. 10. PV cell I-V characteristics compared with the obtained parameters using the two proposed methods.

Figure 11 shows the absolute errors between the real and the experimental I-V characteristics obtained using results from the above obtained parameters values. The computed error is done using the following expression.

$$Error = abs(I_{Real} - I_{Model}) \tag{27}$$

These errors allow the analysis of these methods and provide information on their accuracy.

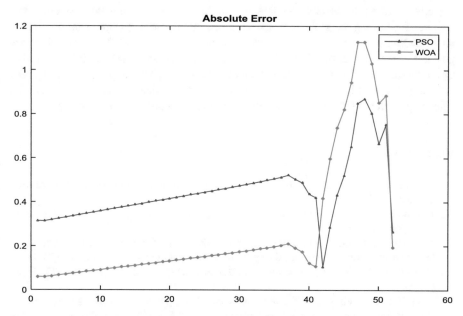

Fig. 11. Current absolute errors between I-V characteristics of the two methods and the I-V real characteristic.

Figure 11 shows that the WOA algorithm gives a smaller error then the PSO. The two algorithms are affected by a set of control parameters defined below. They have to be well adjusted in order to get to the best optimal solutions.

The control parameters values of the algorithm, used in Eq. (2) for the PSO algorithm, are taken as follow:

- Cognitive component, C_1: 2 & Social component, C_2: 2.
- Inertia constant, W: 2.
- Number of population or swarm: 100.

As for, the WOA its control parameters value are taken as:

- Number of search agents: 100.
- Maximum iteration number: 500.
- Variable a, decreases linearly from 2 to 0.

We notice that the PSO is related to more control parameters algorithm. It can be deduced that the PSO algorithm need much more operation time than the WOA.

5 Conclusion

In the present study, the two smart swarm-based WOA and PSO algorithms have been used for the determination of the electrical unknown parameters of solar PV cells. The five-parameters non-linear determination problem has been modelled as a mono-objective or in multi-objective optimization problem. The PSO and WOA are effective methods for finding the global optimal solutions in a non-convex optimization method for a shading problem. The different steps in the development of these algorithms are given. The application of WOA has shown a better precision than the PSO algorithm and outperforms it. It is simple, accurate and converges rapidly to the optimum in every test. In addition, it has fewer parameters to set. The obtained results demonstrate the efficiency of the WOA approach compared to the PSO under the case study.

References

1. Tu-Ilmenau, F., Geletu, A.: Solving Optimization Problems Using the Matlab Optimization Toolbox—A Tutorial (2018)
2. Brownlee, J.: Clever Algorithms—Nature-Inspired Programming Recipes. Lulu.com, Morrisville (2011)
3. Talukder, S.: Mathematical modelling and applications of particle swarm optimization. Independent thesis advanced level (degree of master (two years)) student thesis (2011)
4. Mirjalili, S., Lewis, A.: The whale optimization algorithm. Adv. Eng. Softw. **95**, 51–67 (2016)
5. Trivedi, I.N., Pradeep, J., Narottam, J., Arvind, K., Dilip, L.: Novel adaptive whale optimization algorithm for global optimization. Indian J. Sci. Technol. (2016). https://doi.org/10.17485/ijst%2F2016%2Fv9i38%2F101939
6. El-Fergany, A.: Efficient tool to characterize photovoltaic generating systems using mine blast algorithm. Electr. Power Compon. Syst. **43**, 890–901 (2015)
7. Carrero, C., Ramirez, D., Rodriguez, J., Platero, C.: Accurate and fast convergence method for parameter estimation of PV generators based on three main points of the I/V curve. Renew. Energy **36**, 2972–2977 (2011)
8. Ghani, F., Rosengarten, G., Duke, M., Carson, J.K.: The numerical calculation of single-diode solar-cell modelling parameters. Renew. Energy **72**, 105–112 (2014)
9. Wang, X., Xu, Y., Ye, M.: Parameter extraction of solar cells using particle swarm optimization. J. Appl. Phys. **105**, 094502 (2009)
10. Soon, J.J., Low, K.S.: Photovoltaic model identification using particle swarm optimization with inverse barrier constraint. IEEE Trans. Power Electron. **27**, 3975–3983 (2012)
11. Askarzadeh, A., Rezazadeh, A.: Parameter identification for solar cell models using harmony search-based algorithms. Sol. Energy **86**, 3241–3249 (2012)
12. Zagrouba, M., Sellami, A., Bouaïcha, M., Ksouri, M.: Identification of PV solar cells and modules parameters using the genetic algorithms: application to maximum power extraction. Sol. Energy **84**, 860–866 (2010)

13. Ishaque, K., Salam, Z.: An improved modeling method to determine the model parameters of photovoltaic (PV) modules using differential evolution (DE). Sol. Energy **85**, 2349–2359 (2011)
14. Kebir, S.T., Cheikh, M.S.A., Haddadi, M.: A detailed step-by-step electrical parameters identification method for photovoltaic generators using a combination of two approaches. Adv. Sci. Technol. Eng. Syst. J. **3**(4), 44–52 (2018)
15. Hamid, N.F.A., Rahim, N.A., Selvaraj, J.: Solar cell parameters identification using hybrid Nelder–Mead and modified particle swarm optimization. J. Renew. Sustain. Energy **8**, 015502 (2016). https://doi.org/10.1063/1.4941791
16. Mughal, M.A., Ma, Q., Xiao, C.: Photovoltaic cell parameter estimation using hybrid particle swarm optimization and simulated annealing. Energies **10**, 1213 (2017). https://doi.org/10.3390/en10081213
17. Izadian, A., Pourtaherian, A., Motahari, S.: Basic model and governing equation of solar cells used in power and control applications. IEEE Energy Convers. Congr. Expos. (ECCE) **2012**, 1483–1488 (2012)
18. Ishaque, K., Salam, Z., Taheri, H.: Simple, fast and accurate two-diode model for photovoltaic modules. Sol. Energy Mater. Sol. Cells **95**, 586–594 (2011)
19. El-Fergany, A.: Efficient tool to characterize photovoltaic generating systems using mine blast algorithm. Electr. Power Compon. Syst. **43**, 890–901 (2015)
20. Lo Brano, V., Ciulla, G.: An efficient analytical approach for obtaining a five parameters model of photovoltaic modules using only reference data. Appl. Energy **111**, 894–903 (2013)
21. Seyedmahmoudian, M., Mekhilef, S., Rahmani, R., Yusof, R., Renani, E.: Analytical modeling of partially shaded photovoltaic systems. Energies **6**, 128 (2013)
22. Bidram, A., Davoudi, A., Balog, R.S.: Control and circuit techniques to mitigate partial shading effects in photovoltaic arrays. IEEE J. Photovolt. **2**, 532–546 (2012)
23. Beni, G., Wang, J.: Swarm intelligence in cellular robotic systems. In: Dario, P., Sandini, G., Aebischer, P. (eds.) Robots and biological systems: towards a new bionics, pp. 703–712. Springer, Berlin (1993)
24. Wiley, D., Ware, C., Bocconcelli, A., Cholewiak, D., Friedlaender, A., Thompson, M., et al.: Underwater components of humpback whale bubble-net feeding. Behaviour **148**, 575–602 (2011)
25. Van Overstraeten, R.: Crystalline silicon solar cells. Renew. Energy **5**, 103–106 (1994)
26. (2016). http://us.sunpowercorp.com
27. Tian, H., Mancilla-David, F., Ellis, K., Muljadi, E., Jenkins, P.: Detailed performance model for photovoltaic systems. Preprint (2012)

Amelioration of MPPT P&O Using Fuzzy-Logic Technique for PV Pumping

K. Nebti[(✉)] and F. Debbabi

Laboratory Electrotechnique of Constantine LEC, Constantine, Algeria
idor2003@yahoo.fr, debbabi.fares@gmail.com

Abstract. This paper presents an amelioration of P&O maximum power point tracking (MPPT) technique by using fuzzy logic method, when the main goal is to extract the maximum of power to supply a pumping system in an isolated area. The role of the MPPT is to force the system for working at the maximum point for each change of the illumination or the temperature. We present in first the classical technique, by explaining how we can obtain the maximum power under a variable meteorological condition. In P&O strategy, for big value of disturbance step we can get quickly the desired point but with a large oscillation. Small value of disturbance step makes very slow system and affects the responding time. By using fuzzy logic technique the appropriate disturbance step is produced in order to obtain a fast system with an acceptable precession. The simulation of the photovoltaic pumping chain is constructed under Matlab/Simulink, when the effectiveness of the fuzzy MPPT strategy is shown by the obtained results, which makes its application for controlling solar panels very interesting.

Keywords: Solar panel · MPPT · P&O · Fuzzy logic · PV pumping

1 Introduction

Renewable energies represent an attractive solution as replacement or complement of the conventional sources. Among renewable energies, are those resulting from the sun, wind, heat of the ground, water or of the biomass. With the difference in fossil energies, renewable energies are unlimited resource. Renewable energies are divided in a certain number of technological fields according to the developed energy source and useful energy obtained. The field studied in this paper is photovoltaic solar. Photovoltaic pumping is one of the promising applications of the photovoltaic energy source. The system of PV pumping as shown in Fig. 1 is composed of a PV generator, an electrostatic converter, a control strategy which creates the commutation states of the converter switches, and a motor-pump system.

The system of pumping with direct coupling is a simple, reliable. Then in the direct couplings of the loads, the photovoltaic panels are often oversized to ensure a sufficient power to provide the load, this led to an excessively expensive system. The operation point system is obtained by the intersection of characteristics I(V) of the generator and that of the motor-pump group. So that, the motor-pump system is always optimized and works at maximum power point, it is necessary to integrate a tracking MPPT system,

© Springer Nature Switzerland AG 2019
M. Hatti (Ed.): ICAIRES 2018, LNNS 62, pp. 400–410, 2019.
https://doi.org/10.1007/978-3-030-04789-4_43

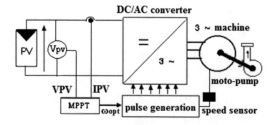

Fig. 1. Scheme of global PV pumping system

which has a role of detecting this point, and forces the system to works precisely on this point, whose operation is optimal [3–6]. The proposed solution is to carry out a vector control of such kind to force induction motor to run at an optimal speed, that last is according to the maximum values of the current and the voltage (power) which vary mainly with illumination and temperature, so we must find these maximum values through the MPPT technique, which then allow us to find the optimum value of speed reference. Several methods for seeking the maximum power point exist in the literature, we will use in our work: in first time we treat P&O method. This type of control is a largely widespread approach in the research of the MPPT because it is simple and requires only measurements of voltage and current of the photovoltaic panel V_{PV} and I_{PV} respectively (two sensors necessary), It should be known that this type of control imposes a permanent oscillation around the MPP [7]. In second time we treat the fuzzy MPPT, in order to make the disturbance adapted by the fuzzy system according to controller inputs. The theory of fuzzy logic was developed in the middle of the Sixties by Professor LOTFI A. ZADEH. The text "Fuzzy Sets" appeared in 1965 in the review "Information and Control" [8].

2   Modeling of Chaine Elements

2.1   Electric Model of a PV Cell

The simplified equivalent circuit of the photovoltaic cell with junction PN (Fig. 2) includes a current source I_{PV}, who gives the photoelectric current model, associated with a diode in parallel which gives the junction PN model, whose polarization determines voltage. And also a series resistance which is the internal resistance of the cell, it depends mainly on the resistance of the semiconductor used, it is also affected by temperature influence [1].

$$I = I_{ph} - I_D \tag{1}$$

$$I_{ph} = I_{ph}(T_1) \times [1 + K_0 \times (T - T_1)] \tag{2}$$

$$I_{ph}(T_1) = I_{cc}(T_1) \times \left(\frac{G}{G_0}\right) \tag{3}$$

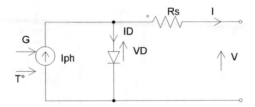

Fig. 2. Equivalent model of the photovoltaic cell

And the relationship between current and voltage of a solar cell is then becomes:

$$I = I_{ph} - I_s \left(e^{\left(\frac{q(V + R_s I)}{nKT} \right)} - 1 \right) \tag{4}$$

2.2   DC-AC Converter Modeling

The voltage inverter is composed of three arms, each one has two switching cells in series and that do not work simultaneously. To simplify the modeling of the inverter, it is assumed that the switches are ideal [2, 10]. The objective of the control is to find the right combination of control switches. C_K, control signals of the arms K of the inverter; with $K \in \{1, 2, 3\}$.

$C_K = 1$ if the switch in top of an arm is closed and that in bottom is open;

$C_K = 0$ if the switch in top is open and that in bottom is closed. The output voltages of the inverter are given by:

$$[V_{sabc}] = \frac{1}{3} U_{pv} \begin{bmatrix} 2 & -1 & -1 \\ -1 & 2 & -1 \\ -1 & -1 & 2 \end{bmatrix} \begin{bmatrix} C_1 \\ C_2 \\ C_3 \end{bmatrix} = U_c[T_s][C_k] \tag{5}$$

2.3   Induction Machines Model

The induction machine can be modeled in a two-phase reference (d, q) in rotating field $((\theta_s = \theta + \theta_r) \Rightarrow (\omega_s = \omega + \omega_r))$ frame by the following equations:

$$\begin{cases} \frac{d}{dt} I_{sd} = \frac{1}{\sigma L_s} \left[\begin{aligned} -\left(R_s + \frac{M_{sr}^2 R_r}{L_r^2} \right) I_{sd} + (\omega_s \sigma L_s) I_{sq} + \left(\frac{M_{sr} R_r}{L_r^2} \right) \Phi_{rd} \\ + \left(\frac{M_{sr}}{L_r} \omega \right) \Phi_{rq} + V_{sd} \end{aligned} \right] \\[3ex] \frac{d}{dt} I_{sq} = \frac{1}{\sigma L_s} \left[\begin{aligned} -(\omega_s \sigma L_s) I_{sd} - \left(R_s + \frac{M_{sr}^2 R_r}{L_r^2} \right) I_{sq} - \left(\frac{M_{sr}}{L_r} \omega \right) \Phi_{rd} \\ + \left(\frac{M_{sr} R_r}{L_r^2} \right) \Phi_{rq} + V_{sq} \end{aligned} \right] \\[3ex] \frac{d}{dt} \Phi_{rd} = \left(\frac{M_{sr} R_r}{L_r} \right) I_{sd} - \left(\frac{R_r}{L_r} \right) \Phi_{rd} + (\omega_s - \omega) \Phi_{rq} \\[2ex] \frac{d}{dt} \Phi_{rq} = \left(\frac{M_{sr} R_r}{L_r} \right) I_{sq} - \left(\frac{R_r}{L_r} \right) \Phi_{rq} - (\omega_s - \omega) \Phi_{rd} \end{cases} \tag{6}$$

Such as:

$$\sigma = 1 - \frac{M_{sr}^2}{L_s L_r}, \quad \omega = p\Omega, \quad \omega_r = \omega_s - \omega$$

Moreover, the expression of the electromagnetic torque can be expressed according to the stator currents and rotor fluxes as follows:

$$C_{em} = p\frac{M_{sr}}{L_r}\left(\Phi_{rd}I_{sq} - \Phi_{rq}I_{sd}\right) \tag{7}$$

And the mechanical equation becomes:

$$\frac{d\omega}{dt} = \frac{1}{J}\left[p\frac{M_{sr}}{L_r}\left(\Phi_{rd}I_{sq} - \Phi_{rq}I_{sd}\right) - C_r - f_V\Omega\right] \tag{8}$$

2.4   Model of the Centrifugal Pump

A pump is a turbine that provides a hydraulic energy to a fluid by transforming the mechanical power applied in its axe by electrical machine. The centrifugal pump applies a load torque proportional to the square of the rotational speed:

$$C_r = k_r\omega^2 + C_s \tag{9}$$

The mechanical power equation of the pump is given according to speed by the following relation:

$$P_{méc} = k_p\omega^3 \tag{10}$$

3   Rotor Flux Oriented Control (FOC)

The induction machine control with flux orientation consists in placing the reference (d, q) such as the axis d coincides with the oriented flux; which imply to impose that the components $\Phi_{rq} = 0$ and $\Phi_{rd} = \Phi_r$ (Fig. 3) in order to have a similar behavior as the DC current machine.

The expression of torque becomes:

$$C_{em} = p\frac{M_{sr}}{L_r}\left(\Phi_{rd}I_{sq}\right) \tag{11}$$

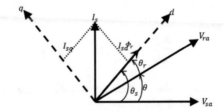

Fig. 3. Principle of the rotor flux orientation

3.1   Control Parameters Estimation

Since the flux Φ_r is oriented on the axis d, we can express I_{sd} and ω_s, with:
$\Phi_{rq} = 0 \Rightarrow \frac{d\Phi_{rq}}{dt} = 0$

$$\begin{cases} I_{sd} = \left(\frac{T_r S + 1}{M_{sr}}\right)\Phi_r \\ \omega_s = \left(\frac{M_{sr}}{T_r}\right)\frac{I_{sq}}{\Phi_r} + \omega \end{cases} \tag{12}$$

$$T_r = \frac{L_r}{R_r}, \ \omega_r = \left(\frac{M_{sr}}{T_r}\right)\frac{I_{sq}}{\Phi_r} \text{ and } \omega_s = \frac{d}{dt}\theta_s \tag{13}$$

4   P&O Algorithm

The principle of P&O MPPT control is to disturb the voltage V_{PV} with a low amplitude around its initial value and analyze the behavior of the power variation P_{PV} resulting. So, as shown in Fig. 4, we can deduce if a positive increment of the voltage V_{PV} generates increased power PPV that means that the operating point is left of MPP. If not the system has exceeded the MPP. Similar reasoning can be made when the voltage decreases [5, 6, 11].

The disadvantage of this type of control is that if quick change of the illumination such as a mobile cloud, this command has more losses, Generated by the long response time of the control to reach the new MPP.

4.1   Optimal Reference Estimation

The research algorithm of the maximum operation point will be done as follow:

- Measuring the current and the voltage of the PV generator, then the determination of the maximum power point Vpm, Ipm and Ppm using MPPT algorithm.
- Determination of the optimal reference speed of the machine according to Vpm, Ipm through following procedure:

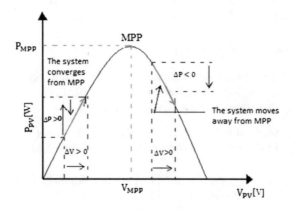

Fig. 4. P&O principle

The mechanical equation of the pump power according to speed is given by the Eq. (10). And it is known that:

$$P_{méc} = \eta_p.P_m \tag{14}$$

The power of the motor can be given according to his efficiency as follows:

$$P_m = \eta_m.P_c \tag{15}$$

In the same way

$$P_c = \eta_c.P_{pm} = \eta_c.V_{pm}.I_{pm} \tag{16}$$

Thus

$$P_{mec} = k_p\omega^3 = \eta_p.\eta_m.\eta_c.V_{pm}.I_{pm} \tag{17}$$

Finally optimal speed according to the maximum values of the current and voltage of the photovoltaic generator is:

$$\omega_{opt} = \sqrt[3]{\frac{\eta_p.\eta_m.\eta_c.V_{pm}.I_{pm}}{k_p}} \tag{18}$$

This speed will be the reference of the speed regulation loop as shown in Fig. 1.

5 Fuzzy Logic MPPT

5.1 Presentation of Fuzzy System

The majority of the developed controllers use the simple diagram suggested by MAMDANI for the single-input/single-output system [9, 12].

According to this diagram, fuzzy system includes:

- A calculation block of the error and its variation over time (CE(k), ΔCE(k)).
- Scaling factors associated with the error, in its variation and the output variation (dD).
- The fuzzification corresponds to the process of determining the degree of member ship to each fuzzy partition.
- Fuzzy rules (Inference) indicate the use of the rules started by the various fuzzified input.
- Defuzzification block corresponds to the passage of fuzzy values of outputs (of linguistic form) to a net final value (ΔV).

S_E, S_{CE}: Inputs Gains, S_{dD}: Output Gain they are a scale factors (Fig. 5).

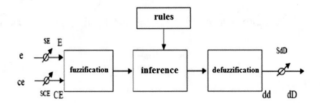

Fig. 5. General structure of a fuzzy controller

By combining these rules, we can trace decision tables to give the values of controller output corresponding to situations of interest.

The scale factors should be selected based on the study of the system such that, when the small transitory phenomenon, the permissible range for the error and its variation are not exceeded.

5.2 Description of the Used System

- **Error E:**

The error E is defined as the error between $\frac{dP}{dV}$ and the seeking values $\frac{dP}{dV} = 0$. The latter value corresponds to the unique extreme value of the curve P(V). This extreme point is a maximum. More E is positive; more the value of P increase. Conversely, more E is negative, more the value of P decreases. Finally when E tends to 0, the value of P tends towards its maximum, the MPP.

- **Variation of the error ΔE:**

The change in the error ΔE indicates in which direction and in what proportion the error changes in proportion as the algorithm is running. So when ΔE tends to 0, the system stabilizes (but not necessarily MPP).

- **Definition of output criteria (disturbance):**

The disturbance or increment corresponds to the adjustment value added to the voltage for each iteration of the algorithm.

5.3 Fuzzy System Operating

The fuzzy rules allow determining and connecting the output of the controller to input signals by linguistic terms taking into account the experience acquired by a human operator.

- **Rules Table:**

After having done some tests by varying the number of output classes and rules allocation, we get the following rules (Table 1).

Table 1 Fuzzy rules

ΔE	E						
	NG	NM	NP	Z	PP	PM	PG
NG	NG	NG	NG	NG	NM	NP	Z
NM	NG	NG	NG	NM	NP	Z	PP
NP	NG	NG	NM	NP	Z	PP	PM
Z	NG	NM	NP	Z	PP	PM	PG
PP	NM	NP	Z	PP	PM	PG	PG
PM	NP	Z	PP	PM	PG	PG	PG
PG	Z	PP	PM	PG	PG	PG	PG

6 Simulation Results of P&O MPPT

We apply two fast change of illumination, the first at time t = 0.5 s, from 1000 W/m^2 to 400 W/m^2 and the second at time 1 s, from 400 W/m^2 to 800 W/m^2.

In Fig. 6 represents the evolution of the optimal power, this curve contains fluctuations due to the oscillation of P&O around MPP with fixed step. This fluctuation, influence also on speed performance, as shown in Fig. 7. The electromagnetic torque follows the load torque with a small overshooting for each change of condition

The optimum speed (reference speed of the control loop for vector control (FOC)) is proportional to the optimum power delivered by the panel PV; we notice that whenever the sun changes the rotation speed of the machine perfectly follows this reference. P&O technique is based on applying a disturbance to the voltage (positive or negative) to reach the maximum power point. The problem with this technique is that

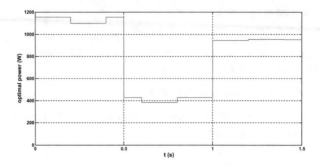

Fig. 6. Optimum power (after MPPT)

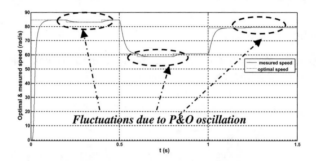

Fig. 7. Optimal (ref) and mechanical speed

the disturbance is fixed. In order to reach the point max quickly, we must apply a large disturbance, but this causes ripples, unlike that, if we apply a small perturbation, we will have a good precision but a very long response time.

7 Simulation Results of Fuzzy Logic MPPT

It is noticed the amelioration in the evolutions of the various curves comparing to the classic P&O MPPT. No fluctuation in voltage, in power or in speed evolution, this confirms the effectiveness and the superiority of the fuzzy system applied (Figs. 8, 9).

The advantage of this technique compared to the P&O MPPT is that the step of disturbance is adapted by the fuzzy system according to the variation of inputs, thus a transitory mode is shorter than in the traditional MPPT.

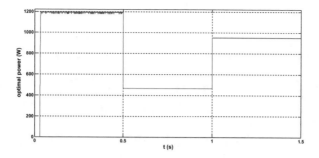

Fig. 8. Optimum power (fuzzy MPPT)

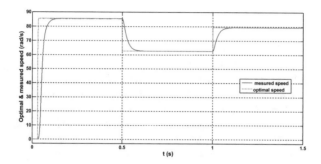

Fig. 9. Optimal (ref) and mechanical speed (fuzzy MPPT)

8 Conclusion

In this paper, in order to optimize performance of photovoltaic system of power production, we have applied flux oriented control structure, combined with two MPPT techniques: P&O and Fuzzy P&O. A comparative study was conducted. The first part concerns the classic technique P&O which give acceptable performances but it present disturbance in the optimal power and speed. The second part concerns the fuzzy MPPT strategy. Simulation results show a superiority of fuzzy MPPT compared to the P&O MPPT. With the fuzzy version, we achieve an adaptation of disturbance step in order to optimize response time and precision. We obtain also a less ripples in speed, torque and voltage responses.

Appendix

Photovoltaic panel TE600 composed of 36 series cells (ns = 36), 60 W

P (kW)	Rs (Ω)	Rr (Ω)	Ls (H)	Lr (H)	Msr (H)	η_m	J (kg/m^2)	F (Nm/m^2)	Wn (rad/s)	P
Induction machine parameters										
1.5	5.72	4.2	0.462	0.462	0.44	0.92	0.0049	0.0098	150	2
Centrifugal pump parameters										
$Kp = 3.3e{-}4$[W/(rad s^{-1})^3], $kr = 3.3e{-}4$[Nm/(rad s^{-1})^2], $\eta_p = 0.74$										

References

1. Meekhun, D., Boitier, V., Dilhac, J.M., Blin, G.: An automated and economic system for measuring of the current-voltage characteristics of photovoltaic cells and modules. In: IEEE International Conference on Sustainable Energy Technologies, Singapore (2008)
2. Bouzelata, Y., Kurt, E., Chenni, R., Altın, N.: Design and simulation of a unified power quality conditioner fed by solar energy. Int. J. Hydrog. Energy **41**(29), 12485–12496 (2016)
3. Madaci, B., Chenni, R., Kurt, E., Hemsas, K.E.: Design and control of a stand-alone hybrid power system. Int. J. Hydrog. Energy **41**(29), 12485–12496 (2016)
4. Qi, J., Zhang, Y., Chen, Y.: Modeling and maximum power point tracking (MPPT) method for PV array under partial shade conditions. Renew. Energy **66**, 337–345 (2014)
5. Kollimalla, S.K., Mishra, M.K.: Variable perturbation size adaptive P&O MPPT algorithm for sudden changes in irradiance. IEEE Trans. Sustain. Energy **5**(3), 718–728 (2014)
6. Altin, N., Yildirimoglu, T.: Labview/matlab based maximum power point substitutable photovoltaic system simulator. J. Polytech. **14**(4), 271–280 (2011)
7. El Basri, Y., Petibon, S., Estibals, B.: New P&O MPPT algorithm for FPGA implementation. In: IECON 2010-36th Annual Conference on IEEE Industrial Electronics Society, AZ, USA (2010)
8. Wu, X., Shen, J., Li, Y., Lee, K.Y.: Fuzzy modeling and stable model predictive tracking control of large-scale power plants. J. Process Control **24**(10), 1609–1626 (2014)
9. Lacrose, V., Titli, A.: Fusion and hierarchy can help fuzzy logic controller designers, fuzzy systems. In: Proceedings of the Sixth IEEE International Conference on, Barcelona, Spain (1997)
10. Labrique, F., Seguier, G., Bausier, R.: The electronic power converters. Vol 4: DC-AC Conversion, Lavoisier (1995)
11. Eskander, M.N., Zaki, A.M.: A maximum efficiency-photovoltaic-induction motor pump system. Renew. Energy Journal **10**(1), 53–60 (1997)
12. Chouder, A., Guijoan, F., Silvestre, S.: Simulation of fuzzy-based MPP tracker and performance comparison with perturb and observe method. Rev. Renew. Energy **11**(4), 577–580 (2008)

Renewable Materials and Devices

Experimental Validation of a Seven Level Inverter with Reduced Number of Switches Using Harmony Search Technique

Fayçal Chabni[✉], Rachid Taleb, Abderrahmen Benbouali,
Ismail Bouyakoub, and Ahmed Derrouazin

Electrical Engineering Department, Laboratoire Génie Electrique et Energies
Renouvelables (LGEER), Hassiba Benbouali University, Chlef, Algeria
chabni.fay@gmail.com

Abstract. Selective harmonic elimination pulse width modulation (SHEPWM) is modulation strategy particularly used to control high-power converters. The SHEPWM strategy allows the elimination of a specific order of harmonics and also controls the amplitude of the fundamental component of the output voltage. In this paper harmony search algorithm is used to determine optimum switching angles for a seven level inverter with reduced number of switches (asymmetrical inverter). Harmony search is an optimization algorithm inspired from the music improvisation process. The algorithm and the mathematical model of the power converter are developed in MATLAB. Simulation results are validated by an experimental setup using STM32F407 high-performance microcontroller.

Keywords: Multilevel inverter · Harmonic elimination · Harmony search
Optimization

1 Introduction

Inverters play a critical role in Machine drives, power transmission systems, induction heating, electric vehicles, and other technologies. The recent advancement in semiconductor technology and the increasing demand on electrical energy, have led to the development of multiple configurations of DC to AC converters. The multilevel DC to AC cascade inverter is one of the most used topologies in High and medium power applications, the advantages provided by the cascade inverter such as higher output waveform quality than most inverter topologies and low voltage stress on the power switches made it a very attractive topology. Cascade DC to AC multilevel inverters are suitable for high power applications they can withstand a huge amount of voltage stress; they are also very easy to make and to maintain due to their modular structure. The conventional multilevel cascade configuration can be achieved by connecting multiple H-bridge modules in series; this configuration will be briefly covered in this work. The harmonic content in an AC voltage waveform generated by an inverter can affect significantly the performance of AC machines. Several modulation strategies have been proposed and studied for the control of multilevel inverters such as Sinusoidal Pulse width modulation (SPWM) and space vector pulse width modulation

© Springer Nature Switzerland AG 2019
M. Hatti (Ed.): ICAIRES 2018, LNNS 62, pp. 413–420, 2019.
https://doi.org/10.1007/978-3-030-04789-4_44

(SVPWM) [1]. A more efficient method called selective harmonic elimination pulse width modulation (SHE-PWM) is also used; the method offers a lot of advantages such as operating the inverters switching devices at a low frequency which extends the lifetime of the switching devices. The main disadvantage of selective harmonic elimination method is that a set of non-linear equations extracted from the targeted system model must be solved to obtain the optimal switching angles to apply this strategy. Multiple computational methods have been used to calculate the optimal switching angles such as Newton-Raphson (N-R) [2], this method dependents on initial guess of the angle values in such a way that they are sufficiently close to the global minimum (desired solution). And if the chosen initial values are far from the global minimum, non-convergence can occur. Selecting a good initial angle, especially for a large number of switching angles can be very difficult. Another approach is to use optimization algorithms such as genetic algorithm (GA) [3, 4], particle swarm optimization (PSO) [5] and differential evolution (DE) [6]. The main advantage of these methods is that they are free from the requirement of good initial guess. This work discusses the possibility of using the harmony search algorithm to solve the selective harmonic elimination problem. The Harmony search (HS) is an optimization algorithm inspired form the music improvisation process, it was first introduced in 2001 by Geem [7]. In this work the HS algorithm is used to compute the optimal switching angles necessary for the SHEPWM method, in the case of a uniform step seven level waveform, only two harmonics are eliminated and the fundamental component is controlled. This work is organized as follows the next section will present briefly the SHEPWM for multi-level inverters and the harmony search optimization method. The obtained simulation and experimental results are presented in the last section.

2 Shepwm for Multilevel Inverts and Optimization Method

2.1 Proposed Converter and the SHEPWM Strategy

The left side of Fig. 1 presents the topology of an asymmetrical three phase cascaded seven level inverter. Each phase consists of H-bridge cells connected in series, each bridge is powered by its own isolated direct current power source u_{dc1} and u_{dc2} with $u_{dc2} = 2 \times u_{dc1}$, in this particular configuration can generate seven voltage levels per phase the right side of Fig. 1 illustrates a generalized form of a uniform stepped voltage waveform with θ_1, θ_2 and θ_3 are the optimal angles to be computed in order to eliminate the undesired harmonics and control the fundamental component simultaneously.

The number of voltage levels that can be generated by Cascade multilevel inverters is generally presented by $2P + 1$ where P represents the number of voltage levels or switching angles in a quarter waveform of the signal, and $P - 1$ is the number of undesired harmonics that can be eliminated from the generated waveform. In a seven level inverter with uniform step voltage waveform, the number of voltage levels generated in quarter waveform is three plus the zero level which means only two harmonics can be eliminated. To control the peak value of the output voltage and

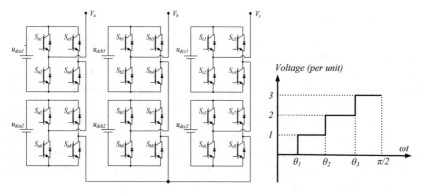

Fig. 1. Schematic of the proposed multilevel converter (left), quarter waveform of a seven-level inverter (right).

eliminate any harmonic, with quarter and half wave symmetry characteristics of the voltage waveform are taken in consideration, the Fourier series expansion is given as:

$$V(\omega t) = \sum_{n=1,5,7,\dots}^{\infty} \left[\frac{4V_{dc}}{n\pi} \sum_{i=1}^{p} \cos(n\theta_i) \right] \sin(n\omega t) \tag{1}$$

where n is rank of harmonics, $n = 1, 5, 7, \dots$, and $p = (N - 1)/2$ is the number of switching angles per quarter waveform, and θ_i is the ith switching angle, and N is the number of voltage levels per half waveform. The optimal switching angles θ_1 and θ_2 can be determined by solving the following system of non-linear equations:

$$\begin{cases} H_1 = \cos(\theta_1) + \cos(\theta_2) + \cos(\theta_3) = M \\ H_5 = \cos(5\theta_1) + \cos(5\theta_2) + \cos(5\theta_3) = 0 \\ H_7 = \cos(7\theta_1) + \cos(7\theta_2) + \cos(7\theta_3) = 0 \end{cases} \tag{2}$$

where $M = (((N - 1)/2)r/4)$, r is the modulation index and. The obtained solutions must satisfy the following constraint:

$$0 < \theta_1 < \theta_2 < \theta_3 < \pi/2 \tag{3}$$

An objective function is necessary to perform the optimization operation, the function must be chosen in such way that allows the elimination of low order harmonics while maintaining the amplitude of the fundamental component at a desired value Therefore the objective function is defined as:

$$f(\theta_1, \theta_2, \theta_3) = \left(\sum_{n1}^{2} (\cos(\theta_n) - M) \right)^2 + \left(\sum_{n1}^{p} (\cos(n\theta_n)) \right)^2 \tag{4}$$

The optimal switching angles are obtained by minimizing Eq. (4) subject to the constraint Eq. (3). The main problem is the non-linearity of the transcendental set of Eq. (2), the harmony search is used to overcome this problem.

2.2   Harmony Search

As mentioned before the harmony search (HS) is an optimization algorithm, it draws the inspiration from harmony improvisation, when an artist tries to create a harmony; he usually tries different combinations of music pitches on his instrument to obtain better harmony. Harmony search depends on two important parameters, harmony memory considering rate (HMCR) and pitch adjusting rate (PAR). The algorithm is composed of three main steps: harmony memory generation, harmony improvisation and harmony memory update. The left side of Fig. 2 shows Harmony search flowchart for solving SHEPWM for different values of r (modulation index), the harmony search algorithm and its use for solving the optimal switching problem for multilevel inverters are well explained in [8–10].

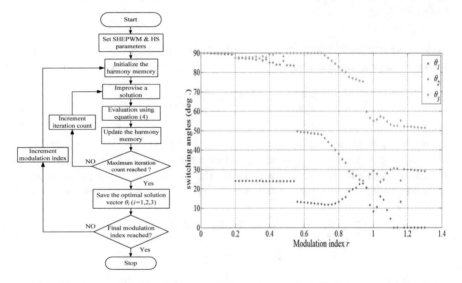

Fig. 2. Harmony search flowchart for solving SHEPWM problem (left), switching angles versus modulation index generated by HS algorithm (right)

In order to prove the theoretical predictions and to test the effectiveness of the proposed algorithm, the control method and the mathematical model of the proposed inverter were developed and simulated using MATLAB/SIMULINK software. The right side of Fig. 2 shows the computed optimal switching angles versus modulation index generated by HS algorithm.

3   Simulation and Experimental Results

The proposed method was validated by building a small scale laboratory prototype, Fig. 3 shows the experimental setup used in this work and its block diagram. Figures 4 and 5 show the generated line and phase to phase output voltage waveform respectively for a modulation index of $r = 1$ and their FFT analysis (simulation and experimental results), whereas Figs. 6 and 7 present the same waveforms and FFT analysis but for $r = 0.8$. Form these results it can be seen that all undesired harmonics (5th and 7th) are successfully eliminated.

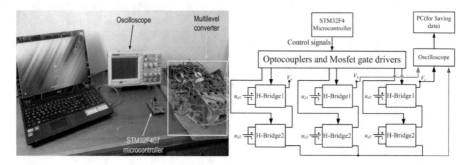

Fig. 3. Picture (left) and diagram (right) of the experimental setup.

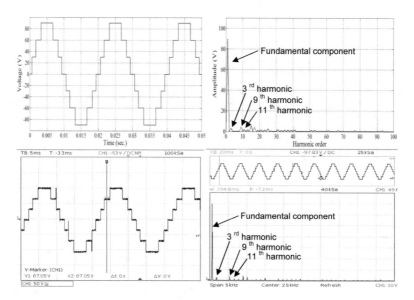

Fig. 4. Simulation (top) and experimental (bottom) results of generated phase voltage waveform and the corresponding FFT analysis for $r = 1$

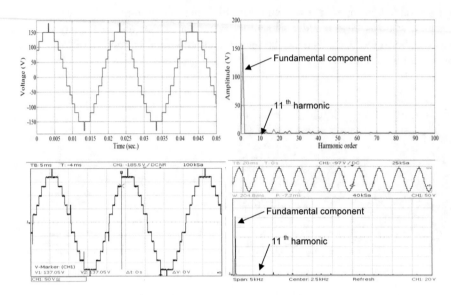

Fig. 5. Simulation (top) and experimental (bottom) results of generated phase to phase voltage waveform and the corresponding FFT analysis for $r = 1$.

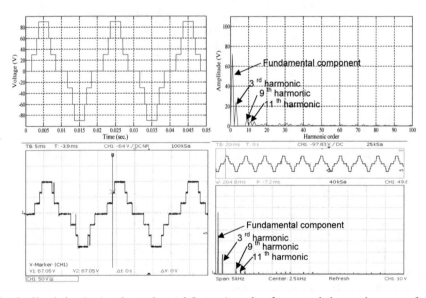

Fig. 6. Simulation (top) and experimental (bottom) results of generated phase voltage waveform and the corresponding FFT analysis for $r = 0.8$

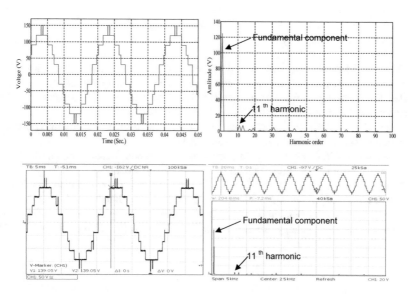

Fig. 7. Simulation (top) and experimental (bottom) results of generated phase to phase voltage waveform and the corresponding FFT analysis for $r = 0.8$.

4 Conclusions

This paper demonstrated the ability of the selective harmonic elimination strategy for seven level inverter with reduced number of switches of eliminating any undesired harmonics and maintain the fundamental component at a desired value, and also the possibility of using the harmony search algorithm to solve the optimal switching problem for multilevel inverters. The set of non-linear equations that describe the overall system are solved to obtain the optimal switching angles using the proposed optimization algorithm.

References

1. Jacob, J., Chitra, A., Thomas, R.V., Rakesh, E.: Hardware realization and analysis of a seven level reduced switch inverter with space vector modulation technique. In: Proceedings of the 2017 Second International Conference on Electrical, Computer and Communication Technologies (ICECCT), pp. 1–6 (2017)
2. Krismadinata, N.A., Rahim, H.W., Ping, J., Selvaraj, J.: Elimination of harmonics in photovoltaic seven-level inverter with Newton-Raphson optimization. Procedia Environ. Sci. **17**, 519–528 (2013)
3. Chabni, F., Taleb, R., Helaimi, M.: Output voltage waveform improvement of modified cascaded H-bridge multilevel inverter using selective harmonic elimination technique based on hybrid genetic algorithm. Rev. Roum. Sci. Tech.-Serie Electrotech. Energ. **62**(4), 405–410 (2017)

4. Deniz, E., Aydogmus, O., Aydogmus, Z.: Implementation of ANN-based selective harmonic elimination PWM using hybrid genetic algorithm-based optimization. Measurement **85**, 32–42 (2016)
5. Taleb, R., Meroufel, A., Massoum, A.: Control of a uniform step asymmetrical 13-level inverter using particle swarm optimization. Automatika **55**(1), 79–89 (2014)
6. Hiendro, A.: Multiple switching patterns for SHEPWM inverters using differential evolution algorithms. Int. J. Power Electron. Drive Syst. **1**(2), 94–103 (2011)
7. Geem, Z.W., Kim, J.H., Loganathan, G.V.: A new heuristic optimization algorithm: harmony search. Simulation **76**(2), 60–68 (2001)
8. Wang, X., Gao, X.Z., Zenger, K.: An Introduction to Harmony Search Optimization Method. Springer, Berlin (2015). https://doi.org/10.1007/978-3-319-08356-8
9. Geem, Z.W., Kim, J.H., Loganathan, G.V.: Harmony search optimization: application to pipe network design. Int. J. Model. Simul. **22**(2), 125–133 (2002)
10. Chabni, F., Taleb, R., Helaimi, M.: ANN-based SHEPWM using a harmony search on a new multilevel inverter topology. Turk. J. Electr. Eng. Comput. Sci. **25**(6), 4867–4879 (2017)

Performance of Shunt Active Power Filter Using STF with PQ Strategy in Comparison with SOGI Based SRF Strategy Under Distorted Grid Voltage Conditions

Khechiba Kamel[1(✉)], Zellouma Laid[2], Kouzou Abdallah[1],
and Khiter Anissa[1]

[1] Department of Electrical Engineering, Djelfa University, Djelfa, Algeria
khechibakamel@gmail.com, khiterhind@gmail.com,
kouzouabdallah@yahoo.fr
[2] Department of Electrical Engineering, El Oued University, El Oued, Algeria
Zelloumal3@yahoo.fr

Abstract. In the case of balanced and undistorted supply voltages, shunt APFs can achieve current harmonic cancellation and give unity power factors. However, this is not possible when grid voltage is non-sinusoidal and unbalanced. In this paper, we first show that the harmonic suppression performance of the well-known p-q and d-q theory deteriorates in non ideal grid voltage conditions. A technique for alleviating the detrimental effects of a distorted and unbalanced grid voltage is proposed that uses a self-tuning filter with p-q theory and we compare it with SOGI with d-q. The proposed control techniques gives an adequate compensating current reference even for non ideal voltage condition. The results of simulation study are presented to verify the effectiveness of the proposed control techniques in this study.

Keywords: Active power filter · Self-tuning filter
Second order generalized integrator · p-q theory
Non-ideal grid voltages · Synchronous reference frame

1 Introduction

In recent years, increases in the penetration of nonlinear loads have made the subject of power quality a major concern not only for utilities but consumers as well. These nonlinear loads draw non sinusoidal currents from the utility and cause a type of voltage and current distortion, namely harmonics. These harmonics cause various problems in power systems and in consumer products, such as equipment overheating, blown capacitors, transformer overheating, excessive neutral currents, low power factor, etc. Furthermore, high penetration of renewable energy such as inverter based systems brings power quality (PQ) challenges. This is as a result of the nonlinear interface of these renewable which pollute the power network with more harmonics causing the grid supply current to become distorted. The combined effect of PQ issues includes high power losses and reduced overall efficiency of utility and consumer

© Springer Nature Switzerland AG 2019
M. Hatti (Ed.): ICAIRES 2018, LNNS 62, pp. 421–428, 2019.
https://doi.org/10.1007/978-3-030-04789-4_45

sensitive loads. To address PQ problems, Shunt APFs have been employed largely to compensate current harmonics, load unbalancing and reactive power demand. The power converter of an active power filter is controlled to generate a compensation current that is equal to the harmonic and reactive currents. In order to determine the harmonic and reactive components of the load current, several techniques are introduced in the literature. Techniques for reference current generation may be put into two categories: time-domain and frequency-domain. Number of time-domain methods have been proposed, one of which was proposed by Akagi [1] called instantaneous active and reactive power theory (or p-q), and the conventional SRF method which is also known as d-q method, and it is based on a-b-c to d-q-0 transformation (park transformation), Similar to the p-q theory, using filters, the harmonics and fundamental components are separated easily and transferred back to the ab-c frame as reference signals for the filter. However, the p-q method only works correctly in the case when three phase grid voltages are balanced and undistorted. The distorted currents cause non-sinusoidal voltage drops and as a result the network voltages become distorted. The unbalanced voltages usually occur because of variations in the load – arising from differing phases of the load current due to for example different network impedances [2]. Also the SRF method is not depend on the voltage source and under ideal utility conditions, i.e., neither imbalance nor harmonic distortion, the PLL yields good results, but under voltage unbalance however, the bandwidth reduction is not an acceptable solution since the overall dynamic performance of the PLL system would become unacceptably deficient. In this paper, we propose the use of a self-tuning filter (STF) with the instantaneous reactive power theory in comparison with the synchronous reference frame using the SOGI in order to increase the harmonic suppression efficiency of active power filter in the case of non-ideal grid voltage condition.

A. Shunt Active Power Filter

The SAPF connects in parallel with the network and injects in real time the Harmonic components of currents absorbed by non-linear loads connected to network. Thus, the current supplied by the energy source becomes sinusoidal. The block diagram of a basic three phase active power filter (APF) connected to a general nonlinear load is shown in Fig. 1.

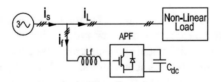

Fig. 1. Block diagram of the APF.

It is well known that the three-phase load current has a non unity power factor. Therefore, the current drawn by the possibly reactive load with harmonics and is given by

$$I_L(t) = I_1(t) + I_h(t) + I_q(t) \tag{1}$$

Which represent the load current, the fundamental current, the harmonic current and the reactive current. As is convention, APFs are operated as a current source that is parallel with the loads. The power converter of an APF is controlled to generate a compensation current, $If(t)$, which is equal to the harmonics and opposite phase, i.e.

$$I_f(t) = -\left(I_h(t) + I_q(t)\right) \tag{2}$$

by replacing Eq. (2) in (1), this yields a sinusoidal source current given by

$$I_s = (I_1 \sin \omega t) \tag{3}$$

2   Instantaneous Active and Reactive Power Pq Theory

The Clark transformation applied for the voltage and current variables is given by:

$$\begin{bmatrix} V_\alpha \\ V_\beta \\ V_0 \end{bmatrix} = \frac{\sqrt{2}}{\sqrt{3}} \begin{bmatrix} 1 & -1/2 & -1/2 \\ 0 & \sqrt{3}/2 & -\sqrt{3}/2 \\ 1/\sqrt{2} & 1/\sqrt{2} & 1/\sqrt{2} \end{bmatrix} \begin{bmatrix} v_a \\ v_b \\ v_c \end{bmatrix} \tag{4}$$

$$\begin{bmatrix} i_\alpha \\ i_\beta \\ i_0 \end{bmatrix} = \frac{\sqrt{2}}{\sqrt{3}} \begin{bmatrix} 1 & -1/2 & -1/2 \\ 0 & \sqrt{3}/2 & -\sqrt{3}/2 \\ 1/\sqrt{2} & 1/\sqrt{2} & 1/\sqrt{2} \end{bmatrix} \begin{bmatrix} i_a \\ i_b \\ i_c \end{bmatrix} \tag{5}$$

The inverse voltage and current transformations are respectively

$$\begin{bmatrix} V_a \\ V_b \\ V_c \end{bmatrix} = \frac{\sqrt{2}}{\sqrt{3}} \begin{bmatrix} 1/\sqrt{2} & 0 & 1 \\ 1/\sqrt{2} & \sqrt{3}/2 & -1/2 \\ 1/\sqrt{2} & -\sqrt{3}/2 & -1/2 \end{bmatrix} \begin{bmatrix} v_\alpha \\ v_\beta \\ v_0 \end{bmatrix} \tag{6}$$

$$\begin{bmatrix} i_a \\ i_b \\ i_c \end{bmatrix} = \frac{\sqrt{2}}{\sqrt{3}} \begin{bmatrix} 1 & -1/2 & -1/2 \\ 0 & \sqrt{3}/2 & -\sqrt{3}/2 \\ 1/\sqrt{2} & 1/\sqrt{2} & 1/\sqrt{2} \end{bmatrix} \begin{bmatrix} i_\alpha \\ i_\beta \\ i_0 \end{bmatrix} \tag{7}$$

Then, the active and reactive instantaneous powers 'P' and 'Q' are expressed in matrix form by:

$$\begin{bmatrix} p \\ q \\ 0 \end{bmatrix} = \begin{bmatrix} v_\alpha & v_\beta & 0 \\ -v_\beta & v_\alpha & 0 \\ 0 & 0 & v_0 \end{bmatrix} \begin{bmatrix} i_\alpha \\ i_\beta \\ i_0 \end{bmatrix} \tag{8}$$

The three phase reference current signal is obtained by:

1- After separating the continuous and alternating terms of active and reactive instantaneous power.

2- Applying the inverse Clark transform to the stationary reference current, therefore:

$$
\begin{bmatrix} i_{fa}^* \\ i_{fb}^* \\ i_{fc}^* \end{bmatrix} = \frac{\sqrt{2}}{\sqrt{3}} \begin{bmatrix} 1 & 0 \\ -1/2 & \sqrt{3}/2 \\ -1/2 & -\sqrt{3}/2 \end{bmatrix} \begin{bmatrix} i_{f\alpha} \\ i_{f\beta} \end{bmatrix}
\tag{9}
$$

However, in the case of distorted and unbalance grid voltage, this theory have drawback and adverse effect on the final parameters which will reduce the harmonic detection performance. We propose, a method in order to suppressing the effects of a non-ideal grid voltage by introducing the Self Tuning Filter.

3   Self-Tuning Filter

The self tuning filter is the most important part of this control which allows to make insensible to the disturbances and filtering correctly the currents in a ß axis. Hong-scok Song [3] The transfer function is defined as:

$$
H(s) = \frac{V_{xy}(s)}{U_{xy}(s)} = K \frac{s + j\omega}{s^2 + \omega^2}
\tag{10}
$$

In the stationary reference, the fundamental components are given by:

$$
\bar{v}_\alpha(s) = \frac{k}{s}[v_\alpha(s) - \bar{v}_\alpha(s)] - \frac{\omega}{s}\bar{v}_\beta(s)
\tag{11}
$$

$$
\bar{v}_\beta(s) = \frac{k}{s}[v_\beta(s) - \bar{v}_\beta(s)] - \frac{\omega}{s}\bar{v}_\alpha(s)
\tag{12}
$$

The Eqs. (11) and (12) represents the output of the STF (Figs. 2 and 3).

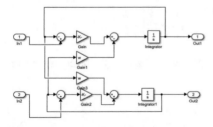

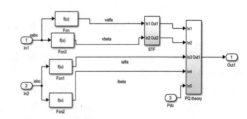

Fig. 2. Block diagram of the STF.   **Fig. 3.** The block scheme of the proposed control system using STF with the PQ method

4 Synchronous Reference Frame

As we mention in the previous section, the SRF method depends only on the load current and have similar identification of the three phase reference current to the PQ method by the use of the Clark and its inverse transform.

Under ideal utility conditions, i.e., neither unbalance nor harmonic distortion, the SRF-PLL yields good results [4]. But under voltage unbalance and distorted, the PLL system would become unacceptable. On the other hand, when the utility frequency is not constant, the positive-sequence detection system uses closed-loop adaptive methods in order to render it insensitive to input frequency variations. With the aim of simplifying, this work proposes the use of a second order generalized integrator (SOGI) [5].

5 Second Order Generalized Integrator

The overall proposed 2nd order LPF based control technique is shown in Fig. 4 and its characteristic transfer functions are given by [6]:

$$H(s) = \frac{k\omega s}{s^2 + k\omega s + \omega^2} \tag{13}$$

where ω and k set resonance frequency and damping factor of the SOGI respectively [7].

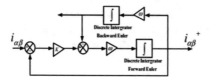

Fig. 4. Block diagram of the SOGI.

The extracted i_α^+ and i_β^+ are transformed to a rotating d-q reference frame using Park's transformation where the fundamental d-q load current component needed to generate the reference current signal is extracted. And to maintain synchronization with the source voltage, $sin\ \omega t$ and $cos\ \omega t$ are generated from a phase locked loop (PLL) such that:

$$\begin{bmatrix} i_d^+ \\ i_q^+ \end{bmatrix} = \begin{bmatrix} \sin \omega t & -\cos \omega t \\ \cos \omega t & \sin \omega t \end{bmatrix} \begin{bmatrix} i_\alpha^+ \\ i_\beta^+ \end{bmatrix} \tag{14}$$

6 Control Strategy

The control signals needed in semiconductors commutation are carried out from the technique of hysteresis band current control, which is the most suitable for all the applications of current controlled voltage source inverter in active power filters. This method has the advantages of good stability, fast response time and good precision. Figure 5 shows the principle of the hysteresis band current controller for three phase system. The hysteresis band current controller decides the switching pattern of APF. Each violation of this band gives an order of commutation. This control system is also characterized by a variable frequency of commutation. The hysteresis techniques have also a few undesirable features such as uneven switching frequency that causes acoustic noise and difficulty in designing input filter [8].

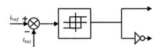

Fig. 5. Hysteresis current control principal.

7 Simulation Results

The control system and compensation by APF is simulated using MATLAB/Simulink and power system block set environment to verify the performance of the proposed techniques. RL type non-linear load is used to see dynamic performances of the APF. The system parameters used in these simulations are given in Table 1.

Table 1. Parameter of the analyzed system

Coupling inductance	3mH
Coupling resistance	0.01 Ω
DC link capacitance	1100 µF
Source inductance	2.3 mH
Source resistance	0.42 Ω
Load resistance	10 Ω
Load inductance	0.3 mH
Source voltage (r.m.s)	100 V
System frequency	50 Hz
Coupling inductance	3 mH

Given comparison in Table 2 is based on the conditions that have the same switching frequency, and the same load conditions. Simulation results show that the proposed method of the SRF with the SOGI give a better result in comparison with the PQ using the STF from the THD% point of view, and it can be used to filter the distorted a-β components in order to extract the sinusoidal and symmetrical voltage from the distorted and asymmetrical grid voltage (Fig. 6).

Table 2. Total harmonic distortion of system with and without filter

System	System without SAPF	System with SAPF using PQ and STF	System with SAPF using DQ and SOGI
THD%	16.26	2.43	2.31

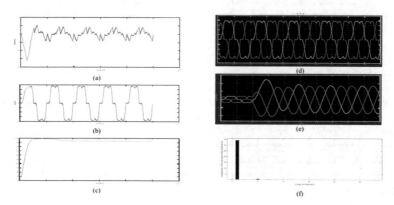

Fig. 6. Performance of the SAPF (a) reference current, (b) load current, (c) DC link side, (d) three phase source current without APF, (e) three phase source current with APF, (f) harmonic order of the proposed technique.

8 Conclusion

The case of distorted and unbalanced grid voltage condition has been considered in this paper. This study shows the performance of the d-q method using the second order generalized integrator which gives better results upon the p-q theory using the self tuning filter based active power filter (APF) degrades in the case of an unbalanced and distorted supply voltage condition. The use of a self-tuning filter (STF) and the SOGI are proposed in order to increase the harmonic suppression efficiency of APF. Simulation results show that the proposed method can improve the performance of active power filters under non-ideal grid voltage conditions.

References

1. Akagi, H., Kanazawa, Y., Nabae, A.: Generalized theory of the instantaneous reactive power in three-phase circuits. In: Proceedings of the International Power Electronics Conference (IPEC 83), pp. 1375–1386 (1983)
2. Biricik, S., Ozerdem, O.C., Redif, S., Kmail, M.O.I.: Performance improvement of active power filters based on P-Q and D-Q control methods under non-ideal supply voltage conditions. In: Proceedings of the 7th International Conference on Electrical and Electronics Engineering, pp. 312–316 (2011)
3. Abdusalama, M., Poureb, P., Karimia, S., Saadatea, S.: New digital reference current generation for shunt active power filter under distorted voltage conditions. Electr. Power Syst. Res. **79**(5), 759–763 (2009)
4. Bhattacharya, S., Divan, D.M., Banerjee, B.: Synchronous reference frame harmonic isolator using series active filter. In: Proceedings of 4th EPE, Florence, Italy, vol. 3, pp. 030–035 (1991)
5. Karimi-Ghartemani, M., Iravani, M.R.: A method for synchronization of power electronic converters in polluted and variable-frequency environments. IEEE Trans. Power Syst. **19**, 1263–1270 (2004)
6. Yuan, X., Merk, W., Stemmler, H., Allmeling, J.: Stationary-frame generalized integrators for current control of active power filters with zero steady-state error for current harmonics of concern under unbalanced and distorted operating conditions. IEEE Trans. Ind. Appl. **38**, 523–532 (2002)
7. Teodorescu, R., Blaabjerg, F., Borup, U., Liserre, M.: A new control structure for grid-connected LCL PV inverters with zero steady-state error and selective harmonic compensation. In: Proceedings of IEEE Applied Power Electronics Conference and Exposition (APEC 2004), vol. 1, pp. 580–586 (2004)
8. Beaulieu, S., Ouhrouche, M.: Real-time modelling and simulation of an active power filter. In: IASTED International Conference on Power and Energy Systems, PES 2007, Clearwater, Florida, USA (2007)

Validation of Three Level Solar Inverter Based on Tabu Search Algorithm

Houssam Eddine Benabderrahman[✉], Rachid Taleb,
M'hamed Helaimi, and Fayçal Chabni

Laboratoire Génie Electrique et Energies Renouvelables (LGEER), Electrical
Engineering Department, Hassiba Benbouali University, Chlef, Algeria
he.benabderrahman@gmail.com

Abstract. Providing electrical energy for oil and gas extraction sites can be a real challenge especially for a large oil and gas company like Sonatrach. Solar energy can be great solution for this issue. This study presents an efficient way to convert electrical energy from DC to AC using a modulation strategy called selective harmonic elimination (SHE); this method will be used to optimize the performance of a three level solar inverter for standalone power system. Harmonic pollution is a very common problem in the field of power electronics, this problem can cause multiple problems for power converters and electrical devices and also reduce their lifespan. The SHE modulation strategy allows the elimination of low order harmonics and also control the amplitude of the fundamental component of the output voltage spectrum. In this paper Tabu Search Algorithm (TSA) is used to determine optimum switching angles for a three level solar inverter. Simulation and experimental results are presented in this work.

Keywords: Solar inverter · Harmonic elimination · Tabu Search Algorithm
Optimization

1 Introduction

The global demand for energy is expected to increase significantly over the upcoming two hundred years, this increase will pose a major challenge for oil and gas companies such as ExxonMobil, Chevron, and Sonatrach. Oil production and refining process consume a lot of energy. Oil and gas fields are often located far away from cities and towns, so getting essential resources and necessary equipments to the sites present a big problem. Using solar energy in production sites will solve some of the logistical problems.

Energy efficiency is a very important issue in the field of power electronics, modern solar power systems and converters (DC/DC or DC/AC) destined for professional use must meet high efficiency and protection standards and also must generate high quality electrical power to ensure the proper functioning of the electrical devices on site.

The harmonic content in an AC voltage waveform generated by an inverter can affect significantly the performance of AC machines. For example harmonics can raise the temperature of an AC motor which decreases the lifetime of the insulation and

© Springer Nature Switzerland AG 2019
M. Hatti (Ed.): ICAIRES 2018, LNNS 62, pp. 429–436, 2019.
https://doi.org/10.1007/978-3-030-04789-4_46

consequently the lifetime of the motor itself. One way to fight this problem is by choosing the right modulation strategy.

Several modulation strategies have been proposed and studied for the control of multilevel inverters such as Sinusoidal Pulse width modulation (SPWM) [1] and space vector pulse width modulation (SVPWM) [2]. A more efficient method called selective harmonic elimination pulse width modulation (SHE-PWM) [3] is also used; the method offers a lot of advantages such as operating the inverters switching devices at a low frequency which extends the lifetime of the switching devices. The main disadvantage of this method is that a set of non-linear equations must be solved to obtain the optimal switching angles to apply this strategy.

The optimal firing (switching) angles are computed by solving a set of non linear equation that represents the desired waveform. Multiple algorithms have been used to solve the optimal switching problem for multilevel inverters such as Genetic Algorithm [4], Differential Evolution [5] and Particle swarm optimization [6] but these algorithms are hard to program and they can take a long time to solve the equations. Tabu Search Algorithm (TSA) [7–9] can be used to solve the optimal switching problem, it is really easy to program and it can solve the non linear equations in few seconds.

This study presents the use of a simple H-Bridge configuration controlled by TSA based selective harmonic elimination for solar application. The next section will present briefly the Selective harmonic elimination for multilevel inverters and the Tabu Search Algorithm. The last section presents the obtained simulation and experimental results.

2   Selective Harmonic Elimination for Solar Inverter

Standalone solar power system is an off grid electrical supply system. An as shown in Fig. 1 the system consists of an arrangement of several components including solar panels, batteries and charge controllers, and also inverters which are the parts responsible for changing the electric current from DC to AC.

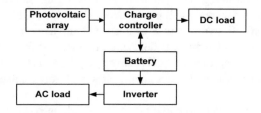

Fig. 1. Standalone solar power system

The configuration of the inverter chosen in this study is presented in Fig. 2 which consists of four switching elements assembled in an H-bridge configuration, this configuration can generate up to three voltage levels. The converter is powered by a direct current source.

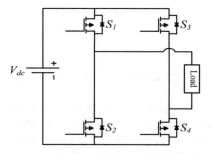

Fig. 2. The proposed three level solar inverter

The performance of an inverter using any modulation strategy is rated according to the harmonics in the generated voltage waveform. In order to control the fundamental voltage and eliminate low order harmonics the proposed inverter must generate a waveform similar to Fig. 3. The figure shows a three level voltage waveform with three switching angles θ_1, θ_2 and θ_3.

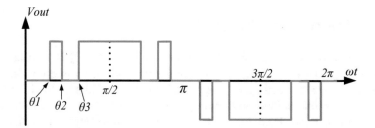

Fig. 3. Generalized three level waveform with multiple switching angles.

Fourier series expansion of the generalized three level waveform output waveform of the single-phase multilevel converter shown in Fig. 3 can be expressed as follows:

$$V(\omega t) = \sum_{n=1,3,5,\ldots}^{\infty} \left[\frac{4V_{dc}}{n\pi} \sum_{i=1}^{p} (-1)^{i+1} \cos(n\theta_i) \right] \sin(n\omega t) \quad (1)$$

where n is rank of harmonics, $n = 1, 3, 5,\ldots$, and $p = (N - 1)/2$ is the number of switching angles per quarter waveform., and θ_i is the ith switching angle, and N is the number of voltage levels per half waveform. The optimal switching angles θ_1, θ_2 and θ_3 can be determined by solving the following system of non-linear equations:

$$\begin{cases} \cos(\theta_1) - \cos(\theta_2) + \cos(\theta_3) = r\pi/4 \\ \cos(3\theta_1) - \cos(3\theta_2) + \cos(3\theta_3) = 0 \\ \cos(5\theta_1) - \cos(5\theta_2) + \cos(5\theta_3) = 0 \end{cases} \quad (2)$$

where $r = H_1/V_{dc}$ is the modulation index and H_1 is the amplitude of the fundamental component. The obtained solutions $(\theta_1, \theta_2$ and $\theta_3)$ must satisfy the following constraint:

$$0 < \theta_1 < \theta_2 < \theta_3 < \pi/2 \tag{3}$$

An objective function is necessary to perform the optimization operation, the function must be chosen in such way that allows the elimination of low order harmonics while maintaining the amplitude of the fundamental component at a desired value Therefore the objective function is defined as:

$$F(\theta_1, \theta_2, \theta_3) = \left(\sum_{n=1}^{3} (\cos(\theta_n) - r\pi/4) \right)^2 + \left(\sum_{n=3,5}^{3} (\cos(n\theta_n)) \right)^2 \tag{4}$$

The optimal switching angles are obtained by minimizing Eq. (4) subject to the constraint Eq. (3). The main problem is the non-linearity of the transcendental set of Eq. (2), the Tabu Search Algorithm (TSA) is used to overcome this problem.

3 Tabu Search Algorithm (TSA)

Tabu search which was first presented by Glover in 1986 [10] is a meta-heuristic optimization method for solving hybrid optimization problems based on local search algorithms to overcome their flaws. The overall structure of Tabu Search is as follows: To achieve the optimal solution in an optimization problem, Tabu Search starts to move from an initial solution. Then, the algorithm selects the best neighbor solution among neighbors of the current solution. If the solution is not on the Tabu list, the algorithm will move to the neighbor solution. Otherwise, the algorithm will check the aspiration criterion. Based on the aspiration criterion, if the neighbor solution is better than the best solution found so far, the algorithm will move towards that solution, even if it is in the Tabu list. After moving to neighbor solution, the Tabu list is updated; that is, the previous move to the neighbor solution is placed on the Tabu list to avoid return to that solution in a cycle. In fact, Tabu list is an instrument in the Tabu search algorithm by which the algorithm is prevented from falling into the local optimum. Then, a number of moves previously put in the Tabu list are removed from the list. The time when moves are placed in the Tabu list is determined by a parameter called as Tabu tenure. The move from the current solution to the neighbor solution continues until the stop criterion is met. Different stop criteria can be considered for the algorithm. For example, limited number of moves to the neighbor solution can be a stop criterion [11]. A flowchart of the TSA for SHE is shown in Fig. 4.

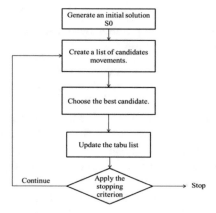

Fig. 4. Flowchart of Tabu Search Algorithm (TSA)

4 Simulation and Experimental Results

In order to prove the theoretical predictions and to test the effectiveness of the proposed algorithm, the control method and the mathematical model of the proposed inverter were developed and simulated using MATLAB scientific programming environment; the optimization program was executed on a computer with Intel(R) Core(TM) i3 CPU @ 2.13 GHz Processor and 4 GB of RAM, the optimization algorithm takes 1.002 s to complete the computation process.

An H-Bridge module was built to validate the results obtained from the simulation process; Irf640 MOSFETS were used as switching devices for the proposed inverter, 4N25 optocouplers were used to protect the microcontroller used in this experiment, Siglent SDS 1000 oscilloscope with FFT capability was used to preview the voltage waveforms and to perform Fast Fourier Transform (FFT) analysis. Figure 5 shows the experimental setup used in this study.

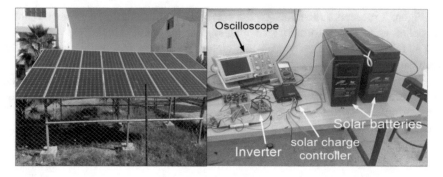

Fig. 5. Solar panels (left) and experimental setup used in this study (right)

To choose the set of switching angles with the lowest Total Harmonic Distortion (*THD*), the generated solutions from the Tabu Search Algorithm are examined for their corresponding *THD* using this equation:

$$THD(\%) = \frac{\sum_{n=3,5,7,\ldots}^{\infty} H_n^2}{H_1} \times 100 \qquad (5)$$

where H_n is the amplitude of a harmonic of rank n and H_1 is the amplitude of the fundamental component. The left side of Fig. 6 shows the generated switching angles for the three level inverter versus the modulation index r. The modulation index varies from 0 to 1.06 with a step size of 0.01. The right side of Fig. 6 presents the corresponding total harmonic distortion for each set of solutions and it can be clearly seen that the lowest harmonic content corresponds to $r = 1.05$ with a *THD* of 44.28%. So the values (in degrees) of the switching angles with lowest *THD* are: $\theta_1 = 20.08°$, $\theta_2 = 33.04°$ and $\theta_3 = 43.63°$.

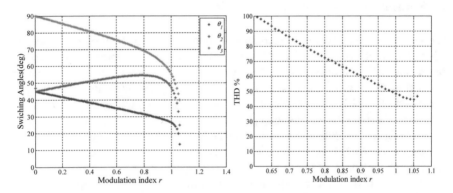

Fig. 6. Switching angles versus modulation index r (left), THD versus modulation index r (right)

Figure 7 presents simulation (on the left) and experimental (on the right) results of the output voltage waveform for switching angle values of $\theta_1 = 20.08°$, $\theta_2 = 33.04°$ and $\theta_3 = 43.63°$. It can be clearly seen that simulation and experimental waveforms are identical.

Figure 8 presents simulation (on the left) and experimental (on the right) results of FFT of the generated three level waveform. Simulation and experimental results are identical and as expected it can be seen from the results that the targeted harmonics (3rd and 5th) were successfully eliminated.

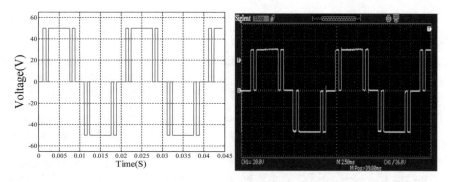

Fig. 7. Output voltage waveform for $r = 1.05$: simulation result (left) and experimental result (right)

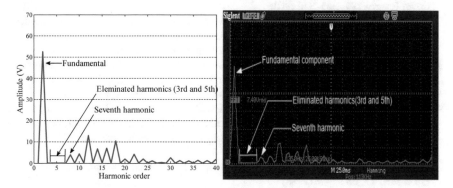

Fig. 8. FFT analysis of the generated voltage waveform: simulation result (left) and experimental result (right)

5 Conclusions

This paper demonstrated the ability of the selective harmonic elimination strategy for multilevel inverters of producing high quality voltage waveform with less harmonics and maintain the fundamental component at a desired value, and also the possibility of using the Tabu Search Algorithm (TSA) to solve the optimal switching problem for multilevel inverters. The SHE would be a very efficient method to control solar DC to AC converters.

References

1. Karami, B., Barzegarkhoo, R., Abrishamifar, A., Samizadeh, M.: A switched-capacitor multilevel inverter for high AC power systems with reduced ripple loss using SPWM technique. In: Power Electronics, Drives Systems and Technologies Conference, pp. 627–632 (2015)

2. Jana, K.C., Biswas, S.K.: Generalized switching scheme for a space vector pulse-width modulation-based N-level inverter with reduced switching frequency and harmonics. IET Power Electron. **8**(12), 2377–2385 (2015)
3. Sinha, Y., Nampally, A.: Modular multilevel converter modulation using fundamental switching selective harmonic elimination method. In: 2016 IEEE International Conference on Renewable Energy Research and Applications (ICRERA), Birmingham, pp. 736–741 (2016)
4. Deniz, E., Aydogmus, O., Aydogmus, Z.: Implementation of ANN-based selective harmonic elimination PWM using hybrid genetic algorithm-based optimization. Measurement **85**, 32–42 (2016)
5. Chabni, F., Taleb, R., Helaimi, M.: Differential evolution based SHEPWM for seven-level inverter with non-equal DC sources. Int. J. Adv. Comput. Sci. Appl. (IJACSA) **7**(9), 304–311 (2016)
6. Letha, S.S., Thakur, T., Jagdish, K.: Harmonic elimination of a photo-voltaic based cascaded H-bridge multilevel inverter using PSO (particle swarm optimization) for induction motor drive. Energy **107**, 335–346 (2016)
7. Xu, K., Fei, R., He, D.: A Tabu-search algorithm for scheduling jobs with precedence constraints on parallel machines. In: 13th IEEE Conference on Industrial Electronics and Applications (ICIEA), 31 May–2 June 2018, Wuhan, China (2018)
8. Abdullah, Z., Tsimenidis, C.C., Johnston, M.: Quantum-inspired Tabu search algorithm for antenna selection in massive MIMO systems. In: IEEE Wireless Communications and Networking Conference (WCNC), 15–18 April 2018, Barcelona, Spain (2018)
9. Meeruang, J., Dolwichai, T.: Optimum criterion of the vehicle navigation for saving the fuel consumption by Tabu search algorithm with non dominated technique. In: 8th International Conference on Computational Intelligence and Communication Networks (CICN), 23–25 December 2016, Tehri, India (2016)
10. Glover, F.: Future paths for integer programming and links to artificial intelligence. Comput. Oper. Res. **13**, 533–549 (1986)
11. Guan, C.H., Cao, Y., Shi, J.: Tabu search algorithm for solving the vehicle routing problem. In: IEEE Third International Symposium on Information Processing, pp. 74–77, 15–17 October (2010)

A Hybrid SPWM and OHSW for Nine-Level Inverter with Reduced Number of Circuit Devices

Houssam Eddine Benabderrahman$^{(\boxtimes)}$, Rachid Taleb,
M'hamed Helaimi, and Fayçal Chabni

Laboratoire Génie Electrique et Energies Renouvelables (LGEER), Electrical
Engineering Department, Hassiba Benbouali University, Chlef, Algeria
he.benabderrahman@gmail.com

Abstract. Multilevel inverters are well used in high power electronic applications because of their ability to generate a very good quality of waveforms, reducing switching frequency, and their low voltage stress across the power devices. This paper presents the Hybrid Pulse Width Modulation (HPWM) strategy base on Sinusoidal PWM (SPWM) and Optimized Harmonics Stepped Waveform (OHSW) of a nine-level asymmetrical inverter. The use of asymmetrical topologies allow to reduce the number of power devices. The HPWM approach is compared to the SPWM strategy. Simulation results demonstrate the better performances and technical advantages of the HPWM controller in feeding a High Power Induction Motor (HPIM).

Keywords: Asymmetrical inverter · Hybrid PWM · Sinusoidal PWM
Optimized harmonics stepped waveform · High power induction motor

1 Introduction

Inverters are widely used in modern power grids; a great focus is therefore made in different research fields in order to develop their performance. Three-level inverters are now conventional apparatus but other topologies have been attempted this last decade for different kinds of applications. Among them, Neutral Point Clamped (NPC) inverters, flying capacitors inverters also called imbricated cells, and series connected cells inverters called cascaded H-bridge inverters [1–3].

This paper is a study about a three-phase multilevel converter based on series connected single phase inverters (partial cells) in each phase. A multilevel converter with k partial inverters connected in serial is presented by Fig. 1. In this configuration, each cell of rank $j = 1\ldots k$ is supplied by a dc-voltage source u_{dj}. It has been shown that feeding partial cells with unequal dc-voltages (asymmetric feeding) increases the number of levels of the generated output voltage without any supplemental complexity to the existing topology [4, 5]. These inverters are referred to as "Cascaded H-bridge Asymmetrical Multilevel Inverters" or CHBAMI.

Some applications such as active power filtering need inverters with high performances [6]. These performances are obtained if there are still any harmonics at the

© Springer Nature Switzerland AG 2019
M. Hatti (Ed.): ICAIRES 2018, LNNS 62, pp. 437–445, 2019.
https://doi.org/10.1007/978-3-030-04789-4_47

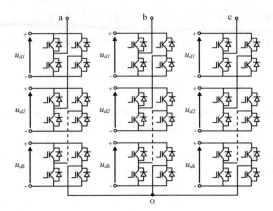

Fig. 1. Three-phase structure of a multilevel converter with k H-bridge inverters series connected per phase

output voltages and currents. Different Pulse-Width Modulation (PWM) control-techniques have been proposed in order to reduce the residual harmonics at the output and to increase the performances of the inverters [7, 8]. The most popular one is probably the Sinusoidal PWM technique (SPWM) which shifts the harmonics to high frequencies by using high-frequency carriers [9, 10].

To minimize the Total Harmonic Distortion (THD) of the output voltage of the CHBAMI, we have applied the Hybrid Pulse Width Modulation (HPWM) strategy combining SPWM and Optimized Harmonics Stepped Waveform (OHSW) methods [11, 12]. In this study we compare the SPWM strategy and HPWM strategy applied to the control of a Nine-level Uniform Step Cascaded H-bridge Asymmetrical Inverter (9-level USCHBAI). As well we compare the performances related to the association 9-level USCHBAI-HPIM for both strategies. Simulation results demonstrate the better performances and technical advantages of the HPWM controller in feeding a high power induction motor.

2 Uniform Step CHBAMI (USCHBAMI)

Multilevel inverters generate at the ac-terminal several voltage levels as close as possible to the input signal. The output voltage step is defined by the difference between two consecutive voltages. A multilevel converter has a uniform or regular voltage step, if the steps Δu between all voltage levels are equal. In this case the step is equal to the smallest dc-voltage, u_{d1} [13]. This can be expressed by:

$$u_{d1} = \Delta u = u_{s2} - u_{s1} = u_{s3} - u_{s2} = \ldots = u_{sN} - u_{s(N-1)} \tag{1}$$

If this is not the case, the converter is called a non uniform step CHBAMI or irregular CHBAMI. An USCHBAMI is based on dc-voltage sources to supply the partial cells (inverters) composing its topology which respects to the following conditions:

$$\begin{cases} u_{d1} \leq u_{d2} \leq \ldots \leq u_{dk} \\ u_{dj} \leq 1 + 2 \sum_{l=1}^{j-1} u_{dl} \end{cases} \tag{2}$$

where k represents the number of partial cells per phase and $j = 1\ldots k$.

The number of output voltage levels depends on the number of cells per phase and on the corresponding supplying dc-voltages. Equation 3 shows that in certain cases, there are many possibilities for setting the partial dc-voltages to obtain the same number of levels. These possible redundant solutions are an other degree of freedom for the designer.

$$N = 1 + 2\sigma_k \text{ where } \sigma_k = \sum_{j=1}^{k} u_{dj} \tag{3}$$

Table 1 gives some examples of the dc-voltages which can be set and the corresponding number of output voltage levels which can be obtained. In this example there are $k = 3$ series-connected single-phase inverters per phase.

Table 1. Examples of unequal dc-voltages in a 3 cells CHBAMI

u_{d1} (p.u.)	u_{d2} (p.u.)	u_{d3} (p.u.)	N
1	1	2	9
1	1	3	11
1	2	2	
1	1	4	13
1	2	3	

3 Multilevel Inverters Control Strategies

Among several modulation strategies, the multi-carrier sub-harmonic PWM technique has been receiving an increasing attention for symmetrical multilevel converters [14]. This modulation method can also be used to control asymmetrical multilevel power converters. Other kinds of modulation techniques can also be used in the case of CHBAMI. This section presents the both strategies SPWM and HPWM (SPWM-OHSW), these control strategies will be compared by computer simulations.

A. Sinusoidal Pulse-Width Modulation (SPWM)

The SPWM is also known as the multi-carrier PWM because it relies on a comparison between a sinusoidal reference waveform and vertically shifted carrier waveforms. $N - 1$ carriers are therefore required to generate N levels. The carriers are in continuous bands around the zero reference. They have the same amplitude A_c and the same frequency f_c. The sine reference waveform has a frequency f_r and an amplitude A_r. At each instant, the result of the comparison is 1 if the triangular carrier is greater than the reference signal and 0 otherwise. The output of the modulator is the sum of the different comparisons which represents the voltage level. The strategy is therefore characterized by the two following parameters, respectively called the modulation index and the modulation rate [10, 15]:

$$m = \frac{f_c}{f_r} \tag{4}$$

$$r = \frac{2}{N-1}\frac{A_r}{A_c} \tag{5}$$

The reference voltages are given as follows:

$$\begin{cases} u_{ri} = u_{rmax}\sin(2\pi f_r t - (j-1)2\pi/3) \\ (i,j) \in \{(a,1),(b,2),(c,3)\} \end{cases} \tag{6}$$

We propose to develop a nine-level uniform step cascaded H-bridge asymmetrical inverter composed of $k = 3$ partial inverters per phase with the following dc-voltage sources: $u_{d1} = 1p.u.$, $u_{d2} = 1p.u.$ and $u_{d3} = 2p.u.$. The output voltage u_{ab} and its frequency representation are presented by Fig. 2.

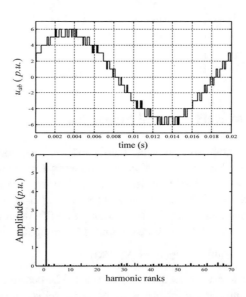

Fig. 2. Output voltage u_{ab} and its frequency representation of the 9-level USCHBAI controlled by the SPWM ($r = 0.8$, $m = 21$)

B. Hybrid Pulse Width Modulation (HPWM)

The hybrid PWM strategy associates the SPWM and Optimized Harmonics Stepped Waveform (OHSW) synthesis in higher power cell with high frequency PWM modulation for the lowest power. Figure 3 shows the principle scheme of HPWM.

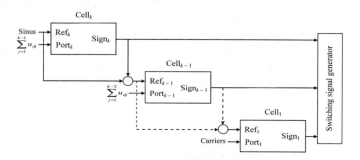

Fig. 3. Schematic diagram of the proposed method (HPWM)

In this strategy the reference signal of multilevel inverter is the same used with reference for the higher cell (cell three), that is compared with two levels dc voltage ($+\sigma_2$ and $-\sigma_2$). The difference between the reference signal and output voltage of cell three is the reference for the cell two, that is compared with ($+\sigma_1$ and $-\sigma_1$). The difference between the reference signal and the output voltage of cell two is the reference of cell one, which is compared with a high frequency triangle carrier signal. The output voltage u_{ab} and its frequency representation are presented by Fig. 4.

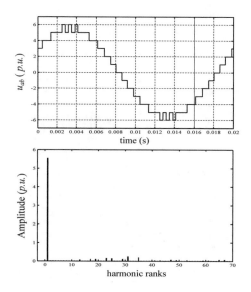

Fig. 4. Output voltage u_{ab} and its frequency representation of the 9-level USCHBAI controlled by the HPWM ($r = 0.8$, $m = 21$)

4 Application of High Power Induction Motor

In order to evaluate the performance of the proposed approach, a 9-level USCHBAI is used to supply a High Power Induction Motor (HPIM). This approach is compared to the SPWM strategy in controlling the 9-level USCHBAI. The objective is to use the proposed strategy in order to minimize the harmonics absorbed by the induction motor.

Figures 5 and 6 show the results of a high power induction motor with the following data: rated power $P_n = 20$ MW, rated voltage 5.5 kV, stator resistance $R_s = 0.397$ Ω, rotor resistance $R_r = 0.081$ Ω, stator inductance $L_s = 0.0089$H, rotor inductance $L_r = 0.0085$H, mutual inductance $L_m = 0.0082$H, number of pole pairs $P = 2$, rotor inertia $J = 1400$ kg m^2, viscous friction coefficient $K_f = 0.009$ Nm s rad^{-1}.

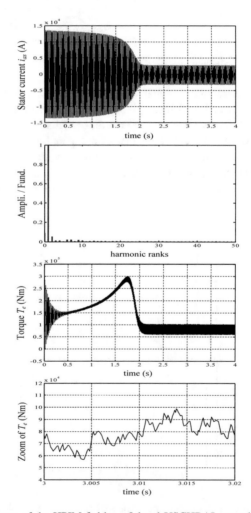

Fig. 5. Performance of the HPIM fed by a 9-level USCHBAI controlled by the SPWM

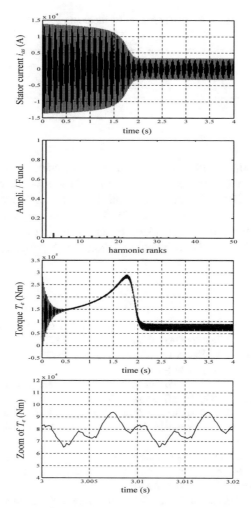

Fig. 6. Performance of the HPIM fed by a 9-level USCHBAI controlled by the HPWM (SPWM-)

Performances obtained with both methods are summarized in Table 2. The analysis of figures and Table 2, show that:

– For HPWM strategy: the THD measured on u_a is smaller than the one obtained with the SPWM method, the stator current closer to the sinusoid;

– For SPWM strategy: the electromagnetic torque continuously oscillates at a frequency f (50 Hz), because of the harmonics of rank tow which is present in the stator current. The torque oscillates at $2f$ (100 Hz) with the THIPWM approach.

Table 2. Performances of the control methods

Control method	u_{ab} THD (%)	i_{as} THD (%)	f_{Te} (Hz)	Δ_{Te} (kNm)
SPWM	13.41	6.38	50	37.71
Proposed method HPWM (SPWM-OHSW)	6.73	4.26	100	22.41

5 Conclusions

The improved performance of a drive system of an induction motor passes through the choice of a best strategy of the control inverter. In this work, we have shown by simulation that the HPWM strategy presents better performances than the SPWM strategy. Indeed, it ensures a highest quality torque and minimizes the current harmonics. Therefore the choice of this strategy in controlling a uniform step cascaded H-bridge asymmetrical multilevel inverter feeding a high power induction motor.

References

1. Rodriguez, J., Bernet, S., Steimer, P.K., Lizama, I.E.: A survey on neutral-point-clamped inverters. IEEE Trans. Ind. Electron. **57**(7), 2219–2230 (2010)
2. Huang, J., Corzine, K.A.: Extended operation of flying capacitor multilevel inverters. IEEE Trans. Power Electron. **21**(1), 140–147 (2006)
3. Babaei, E.: A cascade multilevel converter topology with reduced number of switches. IEEE Trans. Power Electron. **23**(6), 2657–2664 (2008)
4. Mariethoz, S.: Etude formelle pour la synthèse de convertisseurs multiniveaux asymétriques: topologies, modulation et commande (in french), Ph.D. thesis, EPF-Lausanne, Switzerland (2005)
5. Song-Manguelle, J.: Convertisseurs multiniveaux asymétriques alimentés par transformateurs multi-secondaires basse-fréquence: réactions au réseau d'alimentation (in french), Ph.D. thesis, EPF-Lausanne, Switzerland (2004)
6. Ghani, M.A.: A simple active power filter with 5-level NPC inverter. In: 4th International Conference on Engineering Technology and Technopreneuship, ICE2T 2014, Kuala Lumpur, Malaysia, 27–29 August 2014, pp. 330–334 (2014)
7. Bonan, S., Zhang, W., Gao, Q., Xu, D.: Research on the control strategy of modular multilevel converter. In: Third International Conference on Instrumentation, Measurement, Computer, Communication and Control, IMCCC 2013, Shenyang, 21–23 September 2013, pp. 1678–1683 (2013)
8. Zhao, T., Shen, W., Ji, N., Liu, H.: Study and implementation of SPWM microstepping controller for stepper motor. In: 13th IEEE Conference on Industrial Electronics and Applications (ICIEA), Wuhan, China, 31 May–2 June 2018 (2018)
9. Taleb, R., Meroufel, A., Massoum, A.: Control of a uniform step asymmetrical 13-level inverter using particle swarm optimization. Automatika J. Control Meas. Electron. Comput. Commun. **55**(1), 79–89 (2014)
10. Taleb, R., Meroufel, A.: Control of asymmetrical multilevel inverter using artificial neural network. Elektronika ir Elektrotechnika **96**(8), 93–98 (2009)

11. Hosseini Aghdam, M.G., Fathi, S.H., Gharehpetian, G.B.: Comparison of OMTHD and OHSW harmonic optimization techniques in multi-level voltage-source inverter with non-equal DC sources. In: 7th International Conference on Power Electronics, Daegu, South Korea, 22–26 October 2007 (2007)
12. Ghasem Hossein, M., Aghdam, I., Fathi, S.H., Ghasemi, A.: The analysis of conduction and switching losses in three-phase OHSW multilevel inverter using switching functions. In: International Conference on Power Electronics and Drives Systems, Kuala Lumpur, Malaysia, 28 November–1 December 2005 (2005)
13. Taleb, R., Derrouazin, A.: USAMI control with a higher order harmonics elimination strategy based on the resultant theory. In: International Conference on Technologies and Materials for Renewable Energy, Environment and Sustainability, TMREES 2014, Beirut, Lebanon, 10–13 April 2014 (2014)
14. Bouchafaa, F.: Etude et commande de différentes cascades à onduler à neuf niveaux à structure NPC: Application à la conduite d'une MSAP (in french), Ph.D. thesis, ENP, Algiers, Algeria (2008)
15. Taleb, R., Meroufel, A., Wira, P.: Control of a uniform step asymmetrical 9-level inverter based on artificial neural network strategy. Acta Polytech. Hung. 6(4), 137–156 (2009)

Power Quality Improvement in Power System with DPC Controlled PWM Rectifier

Tarik Mohammed Chikouche$^{(\boxtimes)}$, Kada Hartani,
and Mohamed Mankour

Electrotechnical Engineering Laboratory, Tahar Moulay University of Saida,
BP-138 En-nasr, Saida, Algeria
tchikouche@yahoo.fr, kada_hartani@yahoo.fr,
mankourmohamed312@yahoo.fr

Abstract. An improved switching table is proposed to solve the power ripple and achieve better performance of direct power control (DPC) for a three-phase PWM rectifier in this paper. This new control method is applied to overcome the instantaneous power ripple, to eliminate line current harmonics and therefore reduce the total harmonic distortion and to improve the power factor.

The new switching table is based on the analysis on the change of active and reactive power, to select the optimum switching state of the voltage rectifier. The effectiveness of the proposed control strategy is verified by simulation platform using Matlab/Simulink.

Keywords: Power quality · Instantaneous power · Direct power control
Switching table · Unity power factor

1 Introduction

In order to improve the characteristics of rectifying process in power electronics, a new generation of rectifiers called PWM rectifiers have replaced the conventional full bridge Diode/Thyristor rectifier duo to its advanced qualities, such as low THD of input ac current, the possibility of operation with a power factor unity, the control of the direction of the active power and DC-bus voltage control over a wide range [1–3]. Various control strategies proposed in [4–6] which is classified into two categories: (1) Voltage Oriented Control (VOC) similar to the vector control of electrical machines [7–9], and (2) Direct Power Control (DPC) similar to the direct torque control of electrical machines [10]. These strategies reach the same goals, such as the unity power factor and the sinusoidal input current waveform, but their principles are different. This method has some disadvantages such as coupling which occurs between the active and reactive components and the problem of coordinate transformation. However, the DPC controls the active and reactive power directly. Compared to VOC, the DPC can achieve very quick response with simple structure by selecting a voltage vector from predefined switching table. The latter is not accurate for it gives large power ripples [11, 12]. The main advantages of DPC are absence of coordinate transformation, no internal current control loop and no PWM modulator block. In the conventional DPC [13], the active and reactive power are estimated using grid voltage and current measurements based on instantaneous power

© Springer Nature Switzerland AG 2019
M. Hatti (Ed.): ICAIRES 2018, LNNS 62, pp. 446–456, 2019.
https://doi.org/10.1007/978-3-030-04789-4_48

theory [14]. The hysteresis band control technique is used to compare instantaneous errors of active and reactive power. The output of the two hysteresis controller and the position of the voltage vector constitute the inputs of the switching table which imposes the switching state of the PWM rectifier [15, 16].

The disadvantages of conventional DPC are high active power ripple and slow transient response to the step changes in power load. The switching table has a very important role in the performance of the direct power control [17]. However, the use of only one voltage vector in the conventional switching table during one control period leads to high power ripples. The conventional switching table illustrated in [18], demonstrates that it is not satisfactory in the controlling of the active and reactive power. The main aim of this paper is to propose a new switching method for DPC to improve DC-bus voltage regulation by directly controlling the This paper is arranged as follows: A model of a three-phase rectifier is presented in section two. A principle of the proposed DPC with a new switching table is carried out based on the analysis of the instantaneous active and reactive power, including steady state performance, dynamic response and robustness against external load disturbance. Simulations by using Matlab/Simulink are performed to study the characteristics and performance of the proposed method under steady state and transient conditions. To conclude, there is a thorough conclusion.

2 Model of Three-Phase PWM Rectifier

The topology of three-phase bidirectional voltage-source PWM rectifier (VSR) is shown in Fig. 1. The VSR is connected to the three phase ac source via smoothing L and internal resistance R. The inductance acts as a line filter for smoothing the line currents with the minimum ripples. Insulated Gate Bipolar Transistor (IGBTs) are used as the VSR power switches since IGBTs have features of high frequency switching applications. The dc-link capacitor C, is used for filtering the ac components so that dc voltage with minimum ripple can be achieved at the output of VSR. It is assumed that a pure resistive load R_L is connected at the dc-link capacitor C [19].

The model of the PWM rectifier can be expressed in (a, b, c) frame as:

$$[v_{abc}] = R[i_{abc}] + L\frac{d}{dt}[i_{abc}] + [v_{rabc}] \tag{1}$$

$$c\frac{dv_{dc}}{dt} = S_a i_a + S_b i_b + S_c i_c \tag{2}$$

The phase voltages at the poles of the converter are equal to:

$$\begin{bmatrix} v_{ra} \\ v_{rb} \\ v_{rc} \end{bmatrix} = \frac{v_{dc}}{3} \begin{bmatrix} 2 & -1 & -1 \\ -1 & 2 & -1 \\ -1 & -1 & 2 \end{bmatrix} \begin{bmatrix} S_a \\ S_b \\ S_c \end{bmatrix} \tag{3}$$

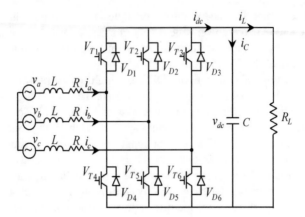

Fig. 1. Topology of three-phase PWM rectifier.

The instantaneous active power and reactive power at the grid side can be calculated from grid voltage and current as [20–22]:

$$\begin{cases} p = v_a i_a + v_b i_b + v_c i_c \\ q = \frac{1}{\sqrt{3}} \left[(v_b - v_c) i_a + (v_c - v_a) i_b + (v_a - v_b) i_c \right] \end{cases} \tag{4}$$

From the power model of PWM rectifier we can know that different switching states have different influences on the active and reactive power. It's possible to select optimal switching states to adjust active and reactive power.

3   Model of Switching Table for DPC

3.1   Principle of the DPC

The direct power control (DPC) technique is based on the direct control of active and reactive power of PWM rectifier. The instantaneous values of active p and reactive q power are estimated by (4). The active power reference is obtained from the voltage controller of the DC bus. However, the reactive power reference is set to zero to get unity power factor. As shown in Fig. 2, the output of the two hysteresis controllers constitute the inputs of the proposed switching table which selects the optimal switching states of PWM rectifier [10–14]. The digitized signals S_p, S_q which provided by a fix band hysteresis comparators can show whether should increase or reduce (decrease) the active or reactive power.

The power model of PWM rectifier is given as [21]:

$$\begin{cases} L\frac{dp}{dt} = -R_p - \omega L_q - \left(v_\alpha v_{r\alpha} + v_\beta v_{r\beta} \right) + \left(v_\alpha^2 + v_\beta^2 \right) \\ L\frac{dq}{dt} = -R_q + \omega L_p - \left(v_\alpha v_{r\alpha} - v_\beta v_{r\beta} \right) \end{cases} \tag{5}$$

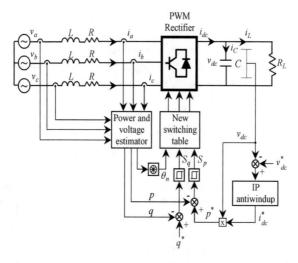

Fig. 2. Proposed DPC configuration of three-phase PWM rectifier.

From the power model of PWM rectifier, we can know that different switching states have different influences on the active and reactive power. It's possible to select the proper switching states to adjust the active and reactive power. The phase of the power-source voltage vector is converted to the digitized signal θ_n. For this purpose, the stationary coordinates are divided into 12 sectors, as shown in Fig. 3.

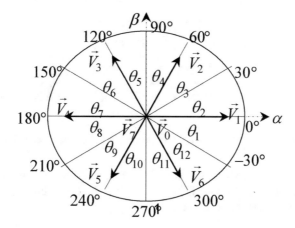

Fig. 3. Twelve sectors in stationary coordinates, to specify power-source voltage vector position and rectifier voltage vectors.

3.2 Vector Selection in the New Switching Table

The new switching table is formed from the output of the two hysteresis controllers (S_p, S_q) and the angular position θ_n of the voltage vector. $S_P = 1$ stands for the need to increase the active power, while $S_P = 0$ denotes the need to decrease the active power. So is the case of the S_q. According to the inputs S_P and S_q, together with the sector information, the proper rectifier input voltage vector can be chosen and the corresponding switching stable will be sent to trigger the IGBTs of the main circuit. Considering the value of R is small enough to be neglected, the instantaneous active and reactive power can be rewritten as:

$$\begin{cases} \frac{dp}{dt} = \frac{3}{2}\frac{V_M^2}{L} - \frac{V_M v_{dc}}{L}\cos\left[\omega t - \frac{\pi}{3}(k-1)\right] \\ \frac{dq}{dt} = -\frac{V_M v_{dc}}{L}\sin\left[\omega t - \frac{\pi}{3}(k-1)\right] + \omega p \end{cases} \quad (6)$$

Where $k = 1, 2, 3, 4, 5, 6$, corresponding to the no zero selected voltage vector number shown in Fig. 3. The variation of active power and reactive power versus grid voltage position for various rectifier voltage vectors are depicted in Fig. 4. In order to achieve better performance of the system, the switching table should be synthesized based on the variation of active and reactive power for various rectifier voltage vectors in each sector, as shown in Fig. 4. Take the first three sectors for example, the signs of slope in active and reactive power are illustrated in Table 1 [22].

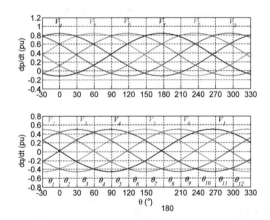

Fig. 4. Variation of active and reactive power for various rectifier voltage vectors.

Take the first three sectors for example, the signs of slope in active and reactive power are illustrated in Table 1 [22]. The new switching table for DPC of PWM rectifier can be summarized in Table 2.

Table 1. Table signs of slope in active and reactive power for first three sector.

dp/dt			dq/dt	
Sector	$\rangle 0\ (S_p = 1)$	$\langle 0\ (S_p = 0)$	$\rangle 0\ (S_q = 1)$	$\langle 0\ (S_q = 0)$
θ_1	V_2, V_3, V_4, V_5	V_1, V_6	V_1, V_2, V_3	V_4, V_5, V_6
θ_2	V_3, V_4, V_5, V_6	V_1, V_2	V_2, V_3, V_4	V_1, V_5, V_6
θ_3	V_3, V_4, V_5, V_6	V_1, V_2	V_2, V_3, V_4	V_1, V_5, V_6

Table 2. The switching table for DPC of PWM rectifier.

S_P	S_q	θ_1	θ_2	θ_3	θ_4	θ_5	θ_6	θ_7	θ_8	θ_9	θ_{10}	θ_{11}	θ_{12}
1	0	V_5	V_6	V_6	V_1	V_1	V_2	V_2	V_3	V_3	V_4	V_4	V_5
	1	V_2	V_2	V_3	V_3	V_4	V_4	V_5	V_5	V_6	V_6	V_1	V_1
0	0	V_6	V_1	V_1	V_2	V_2	V_3	V_3	V_4	V_4	V_5	V_5	V_6
	1	V_1	V_2	V_2	V_3	V_3	V_4	V_4	V_5	V_5	V_6	V_6	V_1

3.3 Control of DC-Link Voltage

The basic operation principle of VSR is to regulate the dc-link voltage v_{dc} at load, at a reference value v_{dc}^*, while maintaining a desired grid side power factor. The value of v_{dc}^* has to be high enough to keep the diodes of converter blocked and maintain the controller stability. Generally, the minimum dc-link voltage can be determined by the peak value of line-to-line grid, i.e. $V_{dc} \succ \sqrt{3}\sqrt{2}V_{(RMS)} = 2,45V_{(RMS)}$. The error between rectified voltage v_{dc} and reference v_{dc}^* is then fed to the anti-windup IP controller to obtain the current component command i_{dc}^* [19]. The product of rectifier voltage v_{dc} and the current reference obtained at the output of the anti-windup IP controller gives the active power reference.

4 Results and Discussion

To evaluate the performance of the proposed DPC with switching table, the simulation test is carried out on a two-level three-phase PWM rectifier. In this simulation test, we have introduced some changes the reference of the DC bus voltage (between $t = 0.30$ s and $t = 0.70$ s), and then introduced a perturbation characterized by a load resistance increasing between the instant $t = 0.45$ s and $t = 0.55$ s. The main parameters of the simulation circuit are given in Table 3.

Table 3. Rectifier parameters.

The input phase voltage :	$V = 125V$ / $f = 50Hz$
The input inductance :	$L = 37mH$
The input resistance :	$R = 0,3\Omega$
The output capacitor :	$C_{dc} = 1100\mu F$
The output voltage :	$V_{dc} = 350V$

Figure 5(a) shows simulation waveform that the load increasing from 500 Ω to 750 Ω at 0.4 s and at 0.6 s. We notice from Fig. 5(b) that the response of the DC-bus voltage v_{dc} follows perfectly its reference. There is a satisfactory steady state operation

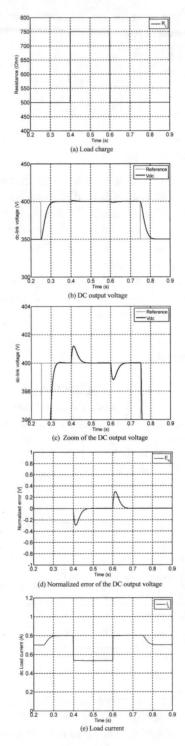

(a) Load charge

(b) DC output voltage

(c) Zoom of the DC output voltage

(d) Normalized error of the DC output voltage

(e) Load current

Fig. 5. Simulation result of the PWM rectifier with under load disturbance (50% variation of resistance at 0.45 s and 0.55 s).

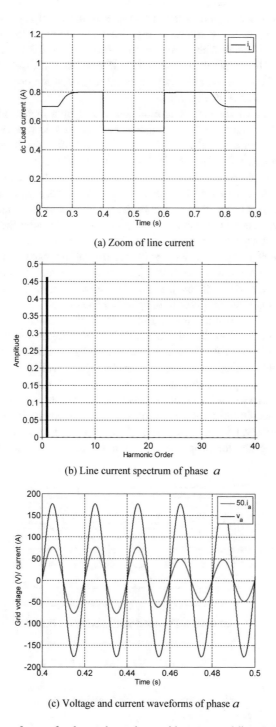

(a) Zoom of line current

(b) Line current spectrum of phase a

(c) Voltage and current waveforms of phase a

Fig. 6. The waveforms of voltage, three phase grid current and line current spectrum.

with no static error, which shows that the proposed analytical approach for the design of the IP regulator is fairly rigorous, Fig. 5(c). The application of the disturbance affects the DC bus voltage with a weak drop of the order of 0.3% for a brief period of 0.05 s, Fig. 5(c). This signifies that the voltage IP regulator acts well on the rejection of this disturbance. To show the efficiency of the IP regulator, the normalized error of the DC bus voltage is shown in Fig. 5(d).

The introduction of the perturbation characterized by a increase in the load resistance applied at the instant $t = 0.4$ s in steady state causes a decrease in the load current which responds instantaneously to this variation and after the instant $t = 0.6$ s, the current i_L is kept constant at its nominal value 0.8 A, Fig. 5(e).

The current response is practically instantaneous, as shown in Fig. 6(a), which represents the three currents at the input of the rectifier corresponding to the current operation. In transient mode, these currents show a transient with a rapid increase when the load is applied. Then they stabilize at amplitude of 1.25A after the instant $t = 0.6$ s. We notice that these grid currents are sinusoidal which gives a low rate of harmonic distortion. Figure 6(b) shows the harmonic spectrum of the response of the grid current i_a. It is noted that all the low render harmonics are well attenuated, which gives a rate of harmonic distortion (THD = 0.96%). Figure 6(c) shows that the grid current i_a is phase with the grid voltage, which gives a unity power factor.

The power response is illustrated in Fig. 7. The active power increases from 245 W to 406 W at $t = 0.25$ s, and then decreases to 270 W between $t = 0.3$ s and $t = 0.7$ s, and then increases to 406 W. After $t = 0.8$ s it stabilizes at the initial value (245 W). The proposed DPC with a new switching table adjusts well the active power in all sectors when the load power decreases. It is clearly seen that in Fig. 7, the reactive power is kept at zero to achieve a unity power factor. It can be seen that the proposed DPC achieves a decoupled control of active and reactive power. It can be seen that the proposed DPC achieves a decoupled control of active and reactive power. The simulation results prove that the proposed DPC is much better when the load changes.

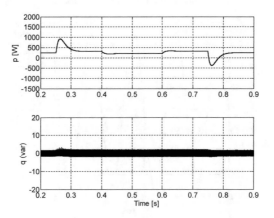

Fig. 7. Active power and reactive power.

5   Conclusion

The proposed method in this paper can select an appropriate switching state of the voltage rectifier by analysis of instantaneous active and reactive power. Consequently, the effectiveness of this proposed method is verified by simulation which can achieve unity power factor, and keep the instantaneous active and reactive power and dc-bus voltage at their desired values. The presented results show that the proposed approach gives better performances in steady state and dynamic response disturbance rejection.

References

1. Aissa, O., Moulahoum, S., Kabache, N., Houassine, H.: Improved power quality PWM rectifier based on fuzzy logic direct power controller. In: Proceedings of the 16th Annual International Conference on Harmonics and Quality of Power (ICHQP), pp. 219–223, June 2014
2. Hashmi, M.U.: Design and development of UPF rectifier in a microgrid environment. Masters Thesis, Department of Energy Science and Engineering, Indian Institute of Technology Bombay, India (2012)
3. Rodríguez, J.R., Dixon, J.W., Espinoza, J.R., Pontt, J., Lezana, P.: PWM regenerative rectifiers: state of the art. IEEE Trans. Ind. Electron. 52, 5–22 (2005)
4. Huang, J., Zhang, A., Zhang, H., Wang, J.: A novel fuzzy-based and voltage-oriented direct power control strategy for rectifier. In: 37th Annual Conference on IEEE Industrial Electronics Society, pp. 1115–1119 (2011)
5. Liutanakul, J.P., Pierfederici, S., Meibody-Tabar, F.: Application of SMC with I/O feedback linearization to the control of the cascade controlled-rectifier/inverter-motor drive system with small DC-link capacitor. IEEE Trans. Power Electron. 23, 2489–2499 (2008)
6. Norniella, J., Cano, J., Orcajo, G., Rojas, C., Pedrayes, J., Cabanas, M., et al.: Optimization of direct power control of three-phase active rectifiers by using multiple switching tables. In: International Conference of Renewable Energies and Power Quality (2010)
7. Ambade, A.P., Premakumar, S., Patel, R., Chaudhari, H., Tillu, A., Yerge, U., et al.: Direct power control PWM rectifier using switching table for series resonant converter capacitor charging pulsed power supply. In: IEEE International Conference on Recent Trends in Electronics, Information & Communication Technology (RTEICT), pp. 758–762 (2016)
8. Malinowski, M., Kazmierkowski, M.P., Trzynadlowski, A.M.: A comparative study of control techniques for PWM rectifiers in AC adjustable speed drives. IEEE Trans. Power Electron. 18, 1390–1396 (2003)
9. Razali, A.M., Rahman, M., Rahim, N.A.: Real-time implementation of DQ control for grid connected three phase voltage source converter. In: 40th Annual Conference of the IEEE Industrial Electronics Society, pp. 1733–1739 (2014)
10. Fekik, A., Denoun, H., Benamrouche, N., Benyahia, N., Badji, A., Zaouia, M.: Comparative analysis of direct power control and direct power control with space vector modulation of PWM rectifier. In: 4th International Conference on Control Engineering & Information Technology (CEIT), pp. 1–6 (2016)
11. Zhang, Y., Peng, Y., Yang, H.: Performance improvement of two-vectors-based model predictive control of PWM rectifier. IEEE Trans. Power Electron. 31, 6016–6030 (2016)
12. Zhang, Y., Qu, C., Li, Z., Zhang, Y.: Mechanism analysis and experimental study of table-based direct power control. In: Machines and Systems (ICEMS), pp. 2213–2218 (2013)

13. Ohnishi, T.: Three phase PWM converter/inverter by means of instantaneous active and reactive power control. In: Proceedings of International Conference on Control and Instrumentation, IECON 1991, pp. 819–824 (1991)
14. Razali, A.M., Rahman, M.: Performance analysis of three-phase PWM rectifier using direct power control. In: IEEE International Electric Machines & Drives Conference (IEMDC), pp. 1603–1608 (2011)
15. Bouafia, A., Krim, F., Gaubert, J.-P.: Direct power control of three-phase PWM rectifier based on fuzzy logic controller. In: IEEE International Symposium on Industrial Electronics, ISIE 2008, pp. 323–328 (2008)
16. Lamterkati, J., Khaffalah, M., Ouboubker, L.: Fuzzy logic based improved direct power control of three-phase PWM rectifier. In: International Conference on Electrical and Information Technologies (ICEIT), pp. 125–130 (2016)
17. Baktash, A., Vahedi, A., Masoum, M.: Improved switching table for direct power control of three-phase PWM rectifier. In: Power Engineering Conference, Australasian Universities, pp. 1–5 (2007)
18. Noguchi, T., Tomiki, H., Kondo, S., Takahashi, I.: Direct power control of PWM converter without power-source voltage sensors. IEEE Trans. Ind. Appl. **34**, 473–479 (1998)
19. Hartani, K., Miloud, Y.: Control strategy for three phase voltage source PWM rectifier based on the space vector modulation. Adv. Electr. Comput. Eng. **10**, 61–65 (2010)
20. Dalessandro, L., Round, S.D., Kolar, J.W.: Center-point voltage balancing of hysteresis current controlled three-level PWM rectifiers. IEEE Trans. Power Electron. **23**, 2477–2488 (2008)
21. Zhou, H., Zha, X., Jiang, Y., Hu, W.: A novel switching table for direct power control of three-phase PWM rectifier. In: Power Electronics and Application Conference and Exposition (PEAC), 2014 International, pp. 858–863 (2014)
22. Gong, B., Wang, K., Zhang, J., You, J., Luo, Y., Wenyi, Z.: Advanced switching table for direct power control of a three-phase PWM rectifier. In: IEEE Conference and Expo on Transportation Electrification Asia-Pacific (ITEC Asia-Pacific), pp. 1–5 (2014)

Speed Control for Multi-phase Induction Machine Fed by Multi-level Converters Using New Neuro-Fuzzy

E. Zaidi$^{(\boxtimes)}$, K. Marouani, and H. Bouadi

Ecole Militaire Polytechnique, LCM-UER-ELT, Bordj El-Bahri 16046, Algeria
zaidielyazid2@yahoo.fr, marouani_khoudir@yahoo.fr,
hakimbouadi@gmail.yahoo.fr

Abstract. This paper proposes a novel neuro-fuzzy control (NFC) for multi-phase induction machine (MPIM) fed by two multi-level converters (MLC) using venturing modulation algorithm. A four-layer artificial neural network (ANN) structure is utilized to train the parameters of the fuzzy logic controller (FLC) based on the minimization of the square of the error. In the proposed method indirect field oriented control (IFOC) is applied to the MPIM. The results are compared with the results obtained from a proportional–integral (PI) controller. Simulation results obtained are very satisfactory and showed that neuro-fuzzy control performance is enhanced using two multi-level converters (MLC) with introduces load disturbances.

Keywords: Multi-phase induction machine (MPIM)
Multi-level converters (MLC) · Neuro-fuzzy controller (NFC)

1 Introduction

The Multi-phase induction machine fed by voltage source multi-level converters has many advantages over conventional three-phase machines, such as reducing the harmonic currents of the rotor, reducing the current per phase without increasing the voltage per phase increasing reliability reducing the amplitude and reducing torque pulsation [1, 2]. The double star asynchronous machine (DSIM) is a typical example of the above-mentioned machines. It has two windings whose phases are spatially displaced by $\alpha = 30°$ electrical degrees with isolated neutrals [3].

The main disadvantages of conventional control algorithms such as proportional-integral-derivative (PID) and proportional-integral (PI) controllers are sensitivity to variations in system parameters and inadequate rejection of internal perturbations and load changes and the designs of these controllers depend on the exact machine model with precise parameters [4, 5]. On the other hand, intelligent controller designs do not need the exact mathematical model of the system. Therefore, an intelligent controller requires special attention for controlling the speed of high-performance induction machine drive systems.

© Springer Nature Switzerland AG 2019
M. Hatti (Ed.): ICAIRES 2018, LNNS 62, pp. 457–468, 2019.
https://doi.org/10.1007/978-3-030-04789-4_49

Numerous methods have been proposed to replace conventional controllers integral-proportional (IP) controllers proportional-integral (PI) and proportional-integral-derivative (PID), such as the fuzzy logic controller (FLC) [6–8] and artificial neural networks (ANN's) [9, 10]. A simple fuzzy controller implemented in the machine drive speed control has a narrow speed operation and needs much manual adjusting by trial and error if high performance is wanted [8]. On the other hand, it is extremely difficult to create a training data set for the artificial neural network (ANN) which can handle all modes of operation [11]. Neuro-fuzzy controller (NFC) developed in the early 90s by Jang [12], combines the concepts of fuzzy logic and neural networks to form a hybrid intelligent system that enhances the ability to automatically learn and adapt. An artificial neural network (ANN) is used to adjust the input and output parameters of membership functions in a Fuzzy logic controller (FLC). The back propagation-learning algorithm is used for training this network. Neuro-fuzzy controller, or simply NFC controller for the induction machine drive, which has the advantages of both FLC and ANN is proposed.

Concerning studies and control of multi-phase induction machines use a conventional voltage source multi-level converter (VSMLC) to supply machine. However, the induction machine drive fed by the multi-level converter is superior to the conventional pulse-width-modulation voltage-source multi-level converters (PWM–VSMLC) because the MLC offers many advantages [13, 14].

In this study, a robust control method based on neuro-fuzzy (NF) is proposed for speed control of a multi-phase induction machine (MPIM) supplied by two multi-level converters (MLC) in the associated to the indirect field oriented control (IFOC). The simulation results show that the proposed techniques can yield very satisfactory performances with introduces load disturbances and parameters variations. A complete simulation model for indirect field oriented control of MPIM supplied by two multi-level converters incorporating the proposed NFC is developed in MATLAB software.

This paper is organized as follows: In Sects. 2 and 3, the DSIM and IFOC is explained, in the next Sect. 4 the multi-level converter model is presented, the design of the NFC controller is shown in Sect. 5 results and discussion are submitted in Sect. 4, The conclusion is given in the last Sect. 7.

2 Modeling of the DSIM

A common type of multi-phase machine is the a dual stator induction machine (DSIM), where two sets of three-phase windings, spatially phase shifted by 30 electrical degrees, share a common stator magnetic core as shown in Fig. 1 [15, 16].

The stator six phases are divided into two wye-connected three phase sets labeled S_{as1} S_{bs1} S_{cs1} and S_{as2} S_{bs2} S_{cs2} whose magnetic axes are displaced by $\alpha = 30°$ electrical angles, the windings of each three-phase set are uniformly distributed and have axes that are displaced 120° apart, The three-phase rotor windings S_{ar}, S_{br} and S_{cr} are also sinusoidal distributed and have axes that are displaced apart by 120° [8–15].

The modeling and control of DSIM in the original reference frame would be very difficult. For this reason, it is necessary to obtain a simplified model to control this machine [6–16], The DSIM model is decomposed into two main sub-models noted

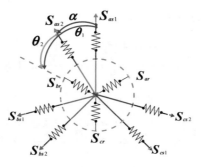

Fig. 1. DSIM windings.

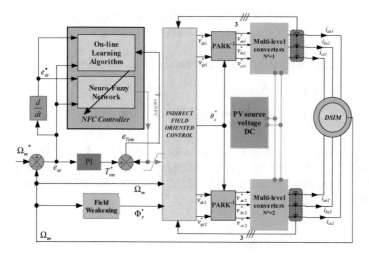

Fig. 2. Block diagram NFN with IFOC for speed control of DSIM.

(ds1-qs1) and (ds2-qs2) for the stator side and one sub model noted (dr-qr) for the rotor side. All sub-models are expressed in the synchronous reference frame [15, 16].

The stator voltage sub-model (ds1-qs1) is written as:

$$\begin{cases} v_{ds1} = R_{s1}i_{ds1} + \dfrac{d}{dt}\varphi_{ds1} - \omega_s\varphi_{qs1} \\ v_{qs1} = R_{s1}i_{qs1} + \dfrac{d}{dt}\varphi_{qs1} + \omega_s\varphi_{ds1} \end{cases} \tag{1}$$

The stator voltage sub-model (ds2-qs2) is written as:

$$\begin{cases} v_{ds2} = R_{s2}i_{ds2} + \dfrac{d}{dt}\varphi_{ds2} - \omega_s\varphi_{qs2} \\ v_{qs2} = R_{s2}i_{qs2} + \dfrac{d}{dt}\varphi_{qs2} + \omega_s\varphi_{ds2} \end{cases} \tag{2}$$

The rotor voltage model (dr-qr) is written as:

$$\begin{cases} 0 = R_r i_{dr} + \dfrac{d}{dt}\varphi_{dr} - (\omega_s - \omega_r)\varphi_{qr} \\ 0 = R_r i_{qr} + \dfrac{d}{dt}\varphi_{qr} + (\omega_s - \omega_r)\varphi_{dr} \end{cases} \tag{3}$$

The expressions for stator and rotor flux are:

$$\begin{cases} \varphi_{ds1} = L_{s1}i_{ds1} + L_m(i_{ds1} + i_{ds2} + i_{dr}) \\ \varphi_{qs1} = L_{s1}i_{qs1} + L_m(i_{qs1} + i_{qs2} + i_{dr}) \end{cases} \tag{4}$$

$$\begin{cases} \varphi_{ds2} = L_{s2}i_{ds2} + L_m(i_{ds2} + i_{ds2} + i_{dr}) \\ \varphi_{qs2} = L_{s2}i_{qs1} + L_m(i_{qs1} + i_{qs2} + i_{qr}) \end{cases} \tag{5}$$

$$\begin{cases} \varphi_{dr} = L_r i_{dr} + L_m(i_{ds1} + i_{ds2} + i_{dr}) \\ \varphi_{qr} = L_r i_{qr} + L_m(i_{qs1} + i_{qs2} + i_{qr}) \end{cases} \tag{6}$$

The electromagnetic torque and represented by the following equation:

$$\begin{cases} J\dfrac{d\Omega}{dt} = C_e - C_r - f\Omega \\ T_{em} = P\dfrac{L_m}{L_m + L_r}\left(\varphi_{rd}(I_{sq1} + I_{sq2}) + \varphi_{rq}(I_{sd1} + I_{sd2})\right) \end{cases} \tag{7}$$

Where ω_s, ω_s speed of synchronous reference frame and rotor electrical angular, L_{s1}, L_{s2} and L_r stator and rotor inductances, L_m resultant magnetizing inductance, P number of polepairs, f moment of inertia, C_r load torque, f total viscous friction coefficient.

3 DSIM Control Method

The IFOC theory applied to the DSIM aims at obtaining a decoupled control of the machine flux and torque. The d-axis is aligned with the rotor flux space vector [6–16]. Furthermore, in the case of rotor flux orientation, its components are controlled to ensure the following condition:

Firstly, The developed model can be simplified again in case of the application of the IFOC technique to the DSIM drive system, the control strategy is used to maintain the quadrature component of the flux equals to zero and the direct flux equals to the reference frame like shows (1) and (2) respectively: $\Phi_{rq} = 0, \Phi_{rq} = \varphi_r$.

$$T_{em} = P\frac{L_m}{(L_m + L_r)}\varphi_r(i_{qs1} + i_{qs2}) \tag{8}$$

The final formula expression of the direct currents id given by:

$$(i_{ds1} + i_{ds2}) = \varphi_r / L_m \tag{9}$$

$$(i_{qs1} + i_{qs2}) = \frac{L_m + L_r}{P\,L_m\varphi_r} T_{em} \qquad (10)$$

Finely, the slip angular frequency is:

$$\omega_{sl} = \frac{R_r L_m}{(L_m + L_r)\varphi_r}(i_{qs1} + i_{qs2}) \qquad (11)$$

Figure 1 presents a scheme showing the relationship between the inputs and outputs expressed by the above equations.

4 Multi-level Converter Model and Control

The different topologies for multi-level converter have been proposed, the most popular being the diode-clamped, flying capacitor and cascade H-bridge structures [17]. The cascaded H-bridge multi-level converters have more advantages than other topologies because it does not require any balancing capacitors and diodes. In this paper, a cascade inverter is studied for (RL) load and control schemes of DSIM.

The proposed multi-level three-phase cascaded H-bridge converter, for example, the topology of a five-level includes a standard three-leg converter (one leg for each phase) and H-bridge in series with each converter leg as shown in Fig. 3 [17, 18]. All signals for controlling the cascade H-bridge multi-level converter are created by a PWM signal modulated technique, for controlling the active devices, the most popular and easiest technique to implement uses several triangle carrier signals and one reference (Fig. 4).

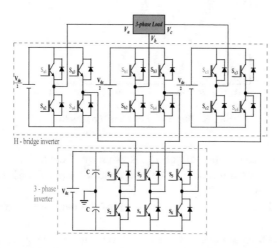

Fig. 3. Topology of the five-level three-phase cascaded H-bridge multi-level converter.

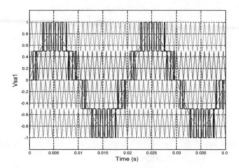

Fig. 4. Principe of multi carrier PWM strategy.

5 Design of the NFC Controller

The neuro-fuzzy controller (NFC) is composed of an on-line learning algorithm with a neuro-fuzzy network (show scheme in Fig. 2). The neuro-fuzzy network is trained using an on-line learning algorithm. Fuzzy inference system is generated with two inputs and one output, the two inputs correspond to speed error and its changing and take the name "e_ω" and "e_ω^*" respectively. The output corresponds to the torque command takes the name "T_{emNFC}" [19, 20]. In this four-layer ANN structure Fig. 5, the first layer represents for inputs, the second layer represents for Fuzzification, the third layer represents for fuzzy rule evaluation and the four layer for Defuzzification [20].

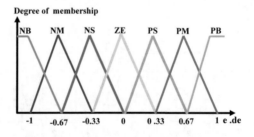

Fig. 5. Neuro-fuzzy network structure.

The desired electromagnetic torque T_{em}^* is given by PI controller, $e_{T_{em}}$ represents the error between the desired torque T_{em}^* and the control torque T_{emNFC} (its current value given by neuro-fuzzy controller).

The detailed discussions on different layers of the neuro-fuzzy network are given below.

First layer:Each input node in this layer corresponds to the specific input variable, the inputs of this layer are given by $net_1^I = e_\omega$ and $net_1^I = e_\omega^*$.

The outputs of this layer are given by:$y_1^I = f_1^I(net_1^I) = e_\omega$ and $y_2^I = f_2^I(net_2^I) = e_\omega^*$. The weights of this layer are unity and fixed.

Second layer: Each node performs a membership function that can be referred to as the fuzzification procedure, each input have seven Gaussian membership functions (MFs) as shown in Fig. 6.

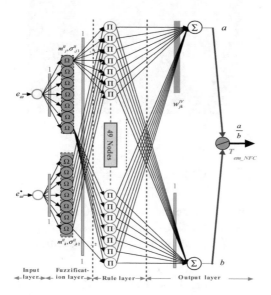

Fig. 6. Membership functions for inputs "e" and "de"

$$net_{1,j}^{II} = -\left(\frac{x_{1,j}^{II} - m_{1,j}^{II}}{\sigma_{1,j}^{II}}\right)^2 \tag{12}$$

$$net_{2,k}^{II} = -\left(\frac{x_{2,k}^{II} - m_{2,k}^{II}}{\sigma_{2,k}^{II}}\right)^2 \tag{13}$$

$$y_{1,j}^{II} = f_{1,j}^{II}\left(net_{1,j}^{II}\right) = e^{net_{1,j}^{II}} \tag{14}$$

where: $m_{1,j}^{II}$, $m_{2,k}^{II}$ represents the Gaussian MFs centers and $\sigma_{1,j}^{II}$, $\sigma_{2,k}^{II}$ determines the MFs widths.

$$y_{2,k}^{II} = f_{2,k}^{II}\left(net_{2,k}^{II}\right) = e^{net_{2,k}^{II}} \tag{15}$$

Third layer: Is called inference and decision layer; the output of every node is the product of all input signals. Based on 49 rule in rule-base of fuzzy inference system, the rule base of the neuro-fuzzy controller is given in Table 1, there are 49 nodes in this layer with function as:

Table 1. Rules base for NFC control.

de \ e	NB	NM	NS	ZE	PS	PM	PB
PB	ZE	NS	NM	NB	NB	NB	NB
PM	PS	ZE	NS	NM	NB	NB	NB
PS	PM	PS	ZE	NS	NM	NB	NB
ZE	PB	PM	PS	ZE	NS	NM	NB
NS	PB	PB	PM	PS	ZE	NS	NM
NM	PB	PB	PB	PM	PS	ZE	NS
NB	PB	PB	PB	PB	PM	PS	ZE

$$\text{net}_{j,k}^{III} = \left(x_{1,j}^{III} \times x_{2,k}^{III} \right), y_{j,k}^{III} = f_{j,k}^{III} \left(\text{net}_{j,k}^{III} \right) = \text{net}_{j,k}^{III} \tag{16}$$

The values of weights between second layer and third layer are unity.

Fourth layer: is called defuzzifier layer, the center of gravity method is used to determine the output of NFC, each node equation is specified as flowing:

$$a = \sum_j \sum_k \left(\omega_{jk}^{IV} y_{jk}^{III} \right), \quad b = \sum_j \sum_k \left(y_{jk}^{III} \right) \tag{17}$$

$$\text{net}_0^{IV} = \frac{a}{b}, \quad y_0^{IV} = f_0^{IV} = \frac{a}{b} \tag{18}$$

ω_{jk}^{IV} represent the values of the output membership functions used in the FLC as shown in Fig. 6. $\omega_0^{IV} y_0^{IV}$ is output of the Defuzzification layer. **a** and **b** are the numerator and the denominator of the function used in the center of area method, respectively. In the NFC, the goal of learning algorithm is adjustment the weights ω_{jk}^{IV}, the $m_{1,j}^{II}$, $m_{2,k}^{II}$ and $\sigma_{1,j}^{II}$, $\sigma_{2,k}^{II}$. For the learning algorithm we use the supervised gradient descent method. Therefore, the error E we take for describe the back propagation algorithm.

$$E(l) = \frac{1}{2} e_{T_{em}}^2 \tag{19}$$

$$e_{T_{em}} = d - y \tag{20}$$

where: d is the desired torque control T_{em}^* (The outputofPI controller) and y is the actual output (y is equal to the output of NFC (T_{em}_NFC).

6 Simulation Results and Discussion

The simulation of the proposed control NFC scheme has been implemented using Matlab/Simulink Many tests were performed to evaluate the performance of the NFL of an IFOC-DSIM, In order to compare the performances and robustness of the speed-control IFOC-DSIM, the same tests are made with the classical PI controller. The supply voltage is achieved form the voltage commands enforced by two multi-level converters with suitable PWM control method.

The simulations were performed such that a step reference speed control was applied to the drive of about 2500 rpm. Therefore, step changes of load torque and speed inversion were applied to evaluate the drive high performance and robustness. Table 2 shows the parameters simulation for the DSIM.

Figure 7 shows the stator current per phase response with the number of level variation of m = 2 to m = 7, when a step load change of 14 N m to −14 N.m at 1.0 s to 2.0 s is applied. According to the results found there that when the level of the converter voltage is m = 2 to m = 7 the output voltage approaches more and more perfect sinusoidal form. It has shown that the THD of voltages and currents decreases when the level inverter is increased.

The proposed controller offers a number of advantages with multi-level converter over with the conventional converter such as shown in Fig. 8. The Fig. 8 presents the performances of NFC and PI speed control when the step load is changed and reference

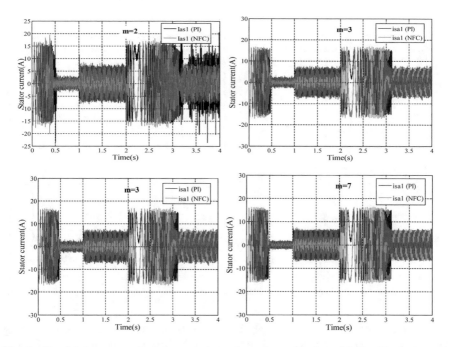

Fig. 7. Simulated responses to stator current per phase, the application of the load in the interval (Cr = [14 −14] N.m) at the instance in an interval (t = [1 2] s), with the number of level variation of m = 2 to m = 7.

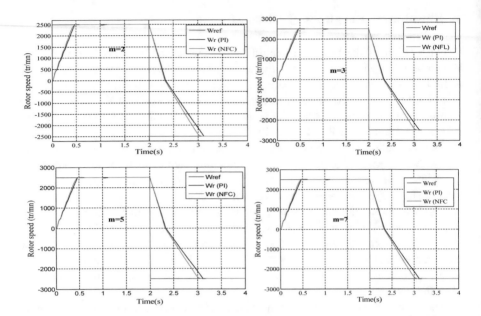

Fig. 8. Simulated responses to rotor speed,the application of the reference speed 2500 rpm to - 2500 rpm, by applying rated load torque in interval (Cr = [14 −14] N.m) at the instance in the interval (t = [1 2]s), with the number of level variation of m = 2 to m = 7.

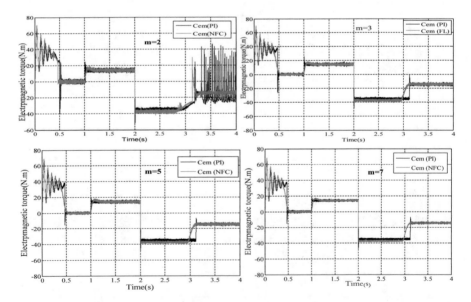

Fig. 9. Simulation results of the control, response of the electromagnetic torque by applying rated load torque the interval (Cr = [14 −14] N.m) at the instance in the interval (t = [1 2] s), with the number of level variation of m = 2 to m = 7.

speed inversion. Therefore, Fig. 8 we present the response of the DSIM with reference speed variations and with at t = 2 s and −2500 rd/s, there seems to be no speed change when a step load change of 14 N m to −14 N m at 1.0 s to 2.0 s is applied.

The Fig. 9 show that the electromagnetic torque ripple in steady state decreases progressively and gradually as the number of levels for the converter increases. According to the results found there that when the level of the inverter voltage is m = 2 to m = 7.

Both simulation results indicate that the proposed NFC controller gives better performances and robustness than the classical PI controller.

7 Conclusion

In this paper, a novel approach NFC is designed that capable for speed control of dual star induction machine (DSIM). The four layers of the fuzzy system control are implemented by using four-layer NN architecture, with two input including error speed and its derivative. The learning of NFC is based on the gradient decent descent method.

The simulation results the robustness tests show too that the NFC is more robust compared to a traditional controller PI. Therefore, the robustness tests show too that the NFC controller is more robust than the PI controller when load disturbances occurred, and when some machine parameters.

Appendix 1

See Table 2.

Table 2. Dual star induction machine parameters for simulation

Quantity	Symbol and magnitude
Rated power	Pn = 4.5 kW
Rated voltage	Vn = 220/380 V
Rated current	In = 6 A
Rated speed	Nn = 2753 rpm
Number of poles	2*p = 4
Rated frequency	f = 50 Hz
Stator resistance	$R_{S\ 1,2}$ = 3.72 Ω
Rotor resistance	Rr = 2.12 Ω
Stator inductance	$L_{S\ 1,2}$ = 0.022H
Rotor inductance	Lr = 0.006H
Mutual inductance	Mm = 0.3672H
Moment of inertia	J = 0.0662 kg m^2
Coefficient of viscous friction	f = 0.006 N ms/rad

References

1. Singh, G.K., Nam, K., Lim, S.K.: A simple indirect field-oriented control scheme for multiphase induction machine. IEEE Trans. Ind. Electron. **52**(4), 1117–1184 (2005)
2. Benyoussef, E., Meroufel, A., Barkat, S.: Neural network and fuzzy logic direct torque control of sensorless double star synchronous machine. Rev. Roum. Sci. Techn.– Électrotechn. Et Énerg **61**(3), 239–243 (2016)
3. Levi, E.: Recent developments in high performance variable-speed multi-phase induction motor drives.in: Sixth international symposium Nikola tesla, Belgrade, SASA, Serbia, pp. 18–20 (2006)
4. Yi, S.Y., Chung, M.J.: Robustness of fuzzy logic control for an uncertain dynamic system. IEEE Trans. Fuzzy Syst. **6**, 216–224 (1998)
5. Krishnan, R., Doran, F.: Study of parameters sensitivity in high performance inverter fed induction motor drive systems. IEEE Trans. Ind. **23**, 623–635 (1987)
6. Lekhchine, S., Bahib, T., Soufi, Y.: Indirect rotor field oriented control based on fuzzy logic controlled double star induction machine. Electr. Power Energy Syst. **57**, 206–211 (2014)
7. Sadouni, R., Meroufel, A.: Indirect rotor field-oriented control (IRFOC) of a dual star induction machine (DSIM) using a fuzzy controller. Acta Polytech. Hung. **9**(4), 177–192 (2012)
8. Merabet, E., Amimeur, H., Hamoudi, F., Abdessemed, R.: Self-tuning fuzzy logic controller for a dual star induction machine. J. Electr. Eng. Technol. **6**(1), 133–138 (2011)
9. Michael, U., Wishart, T., Harley, R.G.: Identification and control of induction machines usingartificial neural networks. IEEE Trans. Ind. Appl. **31**(3), 612–619 (1995)
10. Lekhchine, S., Bahi, T., Soufi, Y., Merabet, H.: Neural fuzzy speed control for six phase induction machines. In: International Conference of Control, Engineering & Information Technology (CEIT'13) Algeria, vol. 3, pp. 123–127 (2013)
11. Uddin, M.N., Abido, M.A., Rahman, M.A.: Development and implementation of a hybrid intelligent controller for interior permanent magnet synchronous motor drive. IEEE Trans. Ind. Appl. **40**(1), 68–76 (2004)
12. Jang, J.-S.R.: ANFIS: adaptive-network-based fuzzy inference systems. IEEE Trans. Syst. Man Cybern. **23**(3), 665–684 (1993)
13. Iffouzar, K., Benkhoris, M.F., Ghedamsi, K., Aouzellag, D.: Behavior analysis of a dual stars induction motor supplied by PWM multi-level inverters. Rev. Roum. Sci. Techn.– Électrotechn. Et Énerg **61**(2), 137–141 (2016)
14. Talaeizadeh, V., Kianinezhad, R., Seyfossadat, S.G., Shayanfar, H.A.: Direct torque control of six-phase induction motors using three-phasematrix converter. Energy Convers. Manag. **51**, 2482–2491 (2010)
15. Marouani, K., Nounou, K., Benbouzid, M., Tabbache, B.: Power factor correction of an electrical drive system based on multiphase machines. In: IEEE ICGE 2014, Mar 2014, Sfax, Tunisia, pp. 152–157 (2014)
16. Marouani, K., Baghli, L., Hadiouche, D., Kheloui, A., Rezzoug, A., Mai, A.: New PWM strategy based on 24-sector vector space decomposition for a six-phase VSI-fed dual stator induction motor. IEEE Trans. Ind. Electron. **55**(5), 1910–1920 (2008)
17. Corzine, K.A., Baker, J.R.: Reduced parts-count multi-level retifiers. IEEE Trans. Ind. Electron. **49**(3), 766–774 (2002)
18. Peng, F.Z.: A generalized multilevel inverter topology with self voltage balancing. IEEE Trans. Ind. Appl. **37**, 611–618 (2004)
19. Elmas, C., Hasan, O.U., Sayan, H.: A neuro-fuzzy controller for speed control of a permanent magnet synchronous motor drives. Expert Syst. Appl. **34**, 657–664 (2008)
20. Uddin, M.N., Wen, H.: Development of a self-tuned neuro-fuzzy controller for induction motor drives. IEEE Trans. Ind. Appl. **43**(4), 1108–1116 (2007)

Smart Building and Smart Cities

Modeling of Fuel Cell SOFC

M. Mankour$^{(\boxtimes)}$ and M. Sekour

Electrotechnical Engineering Laboratory, Department of Electrical Engineering,
Faculty of Technology, University Dr. Moulay Tahar Saida, Saida, Algeria
`Mankourmohamed312@yahoo.fr`

Abstract. Today, the development of devices of electrochemical conversion of energy to "high temperatures" Powerful and reliable fact appeal to different axs of research both technological and scientific. The expected progress on these devices requires conducting front of studies on the components as well as on their integration. The heart of these systems is the electrochemical cell to solid oxides whose structure is a multilayer complex behavior involving different areas of physics. The electrochemistry of solids, physico-chemistry of ceramics, the thermal or still the mechanics of materials are all areas to consider describing and analyzing properly the response of such systems. It is therefore essential to develop methods that are capable of understanding the behavior of electrochemical cells in their together taking into account the different couplings existing. In the field of fuel cells SOFC one of the important characteristics of the materials used as solid electrolytes is their ability to drive the ions at high temperature.

Keywords: Electrochemical solid oxides · Physico-chemistry · Fuel cells SOFC

1 Introduction

If the production, management and the control of energy are one of the major challenges for our societies in the coming decades, the storage of energy is a strategic issue. The storage is in particular the only way to offset in the time the production of energy demand. The system the most used today to store large quantities of energy during periods of overproduction is the hydraulic storage, which is to refit the water in the dams. If other means of storage on a large scale exist (compressed air, flywheels…), the most widely used remains unquestionably the electrochemical storage with the super capacitors and the batteries. This success is explained by the considerable advantage that they bring compared to other solutions: Mobility. The development of electrochemical generators performance is therefore of particular importance. The transition to all-electric vehicle with of autonomies in excess of 300 km will require for example the development of batteries with more energy as the Li-Si or metal-air (Zn-air or Li-air by example) [4].

When it moves to applications requiring the storage located (on-site) of very large quantities of energy, such as that of renewable energy for use in the electrical network, it may not appeal to the hydraulic storage. In these cases, the choice of modes of the storage is dictated more by the performance in energy per unit of mass (Wh/kg) or

© Springer Nature Switzerland AG 2019
M. Hatti (Ed.): ICAIRES 2018, LNNS 62, pp. 471–482, 2019.
https://doi.org/10.1007/978-3-030-04789-4_50

volume (Wh/L); the cost takes a more important place. The batteries with electrolyte circulation (Solid Oxide Fuel Cells – SOFC, in English) are particularly interesting in these applications because they store the energy in the electrolyte in the form of redox systems dissolved, electrolytes contained in tanks that can reach several hundreds of m3. The development of these systems remains however technologically complicated, which explains the difficulties of industrialization [8]. The hydrogen vector, with in particular the batteries to fuels and the electrolyzers (to produce hydrogen), also offers interesting prospects for these applications, large-scale storage where the conventional batteries also have an important role to play. The storage solution will therefore be multi-system and in this context the Electrical study of the association of several sources of energy - hybridization - takes all its meaning. The solid oxide fuel cells (Solid Oxide Fuel Cells - SOFC, in English) are electrochemical systems which allow the conversion of the chemical energy in electrical form. The principle is very simple since from a fuel such as hydrogen and an oxidant such as oxygen, water and electricity are products. Beyond the ecological interest that these systems have, since they do not release any greenhouse gases, they highlight the potential of the hydrogen as a vector of energy for the production of electricity but also the storage of energy in chemical form. The applications essentially referred to be the decentralized energy production with electrical powers provided that can go from a few kW to a few hundred kW on several hundreds of hours for the residential and urban areas.

2   Principle of Operation of a FC

A FC is a converter of chemical energy into electrical energy and heat. The reaction implementation is an electrochemical reaction between hydrogen H2 (fuel) and the oxygen O2 (oxidizer) with simultaneous production of water, electricity and heat according to the overall reaction of synthesis of the water:

H2 + 1/2 O2 H2 Unlike traditional batteries or batteries, energy is therefore not stored in the Finite Volume of the stack itself but in the tanks of gas that can supply power to the FC in a continuous manner. The flow of energy issued by the FC is the result of the movement of the fuel gas (H2) and the oxidizing gas (O2). The implementation of this reaction is carried out through two half-reactions The A to the Anode corresponding to the oxidation of hydrogen and the other to the cathode with the reduction of the oxygen producing water.

It is (see Fig. 1) an electrochemical redox and controlled of hydrogen and oxygen, with simultaneous production of electricity, water and heat, depending on the chemical reaction following overall, known:

$$H_2 + \frac{1}{2}O_2 \rightarrow H_2O \tag{1}$$

The electrodes

The electrochemical reaction takes place within a structure essentially composed of two electrodes [1] (the anode and the cathode) separated by a solid electrolyte, driver of ions O2- more specifically, the following reactions involved in the two electrodes:

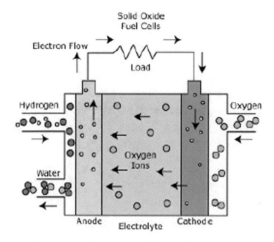

Fig. 1. Schematic of a Solid Oxide Fuel Cell

The anode

$$H_2 + O_2- \rightarrow H_2O + 2e- \qquad (2)$$

The material typically used at the anode is of the porous nickel or a mixture of nickel and of zirconium oxide doped with yttrium.

The cathode

$$^1/_2\, O_2 + 2e- \rightarrow O_2- \qquad (3)$$

The cathode materials Work in strong oxidizing conditions (air or oxygen + high temperature), which prohibits the employment of conventional materials and requires the use of noble materials and/or exotic (oxides semi-conductors, metal oxides drivers), more expensive therefore. The most widely used material to the cathode is a manganite of lanthanum doped strontium. The electrical resistances to the electrodes are the main sources of internal loss.

3 Electrochemical Model and Simulation

We present a static model based on equations of electrochemical of a cap of type SOFC. This model will allow us to find the characteristic experimental electric V (I) of the stack.

On the other hand, it will allow us to perform a parametric study in order to see the influence of the various parameters of operation (temperature, pressure) on the electrical characteristic.

The variation of free enthalpy of the chemical reaction is written [3]:

$$\Delta G = \Delta H - T\Delta s \tag{4}$$

With:

$$\Delta H = \Delta U + \Delta PV \tag{5}$$

And
$\Delta U = W + q$ (6)
Where:
ΔG: Variation of free enthalpy.
ΔH: Variation of enthalpy.
T: temperature.
ΔS: variation of entropy.
P: pressure.
Δv: Variation of the volume.
ΔU: Variation of internal energy.
W: work.
Q: heat.
(5) and (6) The Eq. (4) is written:

$$\Delta G = W + Q + P\Delta V - T\Delta s \tag{7}$$

The transformation is reversible and therefore:

$$Q = T\Delta s \tag{8}$$

And the Eq. (3.31) is written:

$$\Delta G = W + P\Delta V \tag{9}$$

In addition, w here includes the electric work (-nFE.) and the work of the forces of pressure
(-PΔV), therefore:

$$\Delta G = -nFE. - P\Delta V + P\Delta V \tag{10}$$

As a result, it was simply:

$$\Delta G = -nFE. \tag{11}$$

With:
E: electromotive force of the stack.
F: Constant of Faraday.
N: number of electrons transferred.

The free enthalpy of the chemical reaction (4) is:

$$\Delta G_r = G_{H_2O} - G_{H_2} - \frac{1}{2}G_{O_2} \tag{12}$$

The Eqs. (3.31) and (3.35) allow us to write:

$$\Delta G_r = G_r - NRT log \frac{P_{O_2}^{\frac{1}{2}} P_{H_2}}{P_{H_2O}} \tag{13}$$

With:

R: constant molar of gas.

$P_{O_2}, P_{H_2}, P_{H_2O}$ Are respectively the pressure of oxygen, hydrogen, and the pressure of water vapor.

By dividing the two members of the Eq. (13) by -nF, it comes:

$$E = E^0 + \frac{RT}{nF} log \frac{P_{O_2}^{\frac{1}{2}} P_{H_2}}{P_{H_2O}} \tag{14}$$

Real potential of the stack:

The fuel cell produces a potential V, lower than the ideal potential so that one writes:

$$V = E - \text{ losses.} \tag{15}$$

These losses are due to the loss of irreversible load also called, polarizations, which are polarization of activation, Ohmic polarization and the polarization of concentration.

A. Polarization of activation:

The polarization of activation is present when the rate of an electrochemical reaction on the surface of the electrode is controlled by the slowdown in the kinetics for this electrode. In other words, the polarization of activation is directly related to the rate of the electrochemical reaction. In the two cases, for a chemical or electrochemical reaction can start, the reagents must exceed a barrier of activation [4].

The losses of activation are expressed by the following equation [5]:

$$\eta_{act} = \frac{RT}{\alpha_c nF} log(\frac{i}{i_0}) \tag{16}$$

Where:

I: current density.

i_0: Exchange current density.

α_c: Coefficient of charge transfer.

B. Ohmic Polarization:

It is due to the resistance that meets the flow through the electrolyte and the electrical circuit, it also due to the electrical resistance in the plates of dissemination (Backing) [2].

The Ohmic losses are expressed by the following equation [4, 5].

$$\eta_{oh} = \alpha \exp\left[\beta\left(\frac{1}{T_0} - \frac{1}{T}\right)\right] \tag{17}$$

Where:

α and β: Coefficients of the Ohmic resistance.

T: the temperature of the cell.

C. Polarization of concentration

When at the anode there is a loss of potential due to the inability of the system to maintain, the initial concentration of reagents, then we have the formation of a concentration gradient.

Many factors can contribute to the polarization of concentration, the low diffusion of the gas through the porous electrodes and the dissolution of the reagents.

These losses are given by the following relationship [6–8].

$$\eta_{con} = \frac{RT}{nF} \ln\left(1 - \frac{i}{i_l}\right) \tag{18}$$

Where:

I: Density of currant of load.

i_l: Current Density Limit.

(12), (14), (16) and (18), the voltage output of the stack in taking into account the different losses is given by:

$$V_c = N_0\left(E^0 + \frac{RT}{nF}\log\frac{P_{O_2}^{1/2}P_{H_2}}{P_{H_2O}}\right) - \frac{RT}{nF}\text{Ln}\left(1 - \frac{i}{i_l}\right) - \frac{RT}{\alpha_c nF}\log\left(\frac{i}{i_0}\right)$$
$$- \alpha \exp\left[\beta\left(\frac{1}{T_0} - \frac{1}{T}\right)\right] \tag{19}$$

III.3.3. Standard Potential and Performance:

The variation of free energy increases when the temperature of the battery decreases, therefore the potential ideal of a stack is proportional to the change of the standard free energy [9].

At 25 °C and for the water produced in the liquid state, it was:

$\Delta GR = -237.19$ kJ/mol.

$\Delta Hr = -285.84$ kJ/mol.

The standard potential of the stack and:

$$E_0 = -\frac{\Delta G}{nF} = 1.229V$$

For the water produced in the vapor state:

$\Delta GR = -228.59$ kJ/mol.

$\Delta Hr = -241.83$ kJ/mol $\Rightarrow E0 = 1.18$ V.

But at 1000 °C (1273.15 °C), the free enthalpy, the enthalpy and the entropy of the reaction are calculated as follows [11]:

$$\Delta H_{1000\,°C} = \Delta H_{25\,°C} + \int_{25}^{1000} \Delta C_p dt \tag{20}$$

$$\Delta S_{1000\,°C} = \Delta S_{25\,°C} + \int_{25}^{1000} \frac{C_p}{T} dt \tag{21}$$

$$\Delta G_{1000\,°C} = \Delta H_{1000\,°C} - T\Delta S_{1000\,°C} \tag{22}$$

In this case, the empty voltage equal: V0 = 0.935 V.

II.3.4. Performance of the stack:

The chemical reaction is accompanied by a variation of entropy leaving to temperature and pressure data a free enthalpy, transformable in work, here in electrical energy, the rest being converted into heat in the course of the chemical reaction in the stack.

The performance thermodynamics, considered for a reversible transformation, corresponds to the report of the free enthalpy on the standard enthalpy of the reaction of formation of the water:

$$\Gamma_{max} = \frac{\Delta G^0}{\Delta H^0} \tag{23}$$

In the standard conditions, this performance is 83% for the formation of liquid water, to 95% for the formation of water under gaseous forms but at 1000 °C the theoretical performance is

$$\Gamma_{th} = 0.74\%$$

A. The Voltaic performance [9]:

Of the Eq. (3.42) One can define a voltaic performance to performance by voltage:

$$\Gamma_E = \frac{V_c(I)}{E^0} \tag{24}$$

B. The electrical performance [40]:

It is interesting to define the concept of electrical performance:

$$\Gamma_{électrique} = \Gamma_{max}\Gamma_E = \frac{\Delta G^0 V_c(I)}{\Delta H^0 E^0} \tag{25}$$

$$E_{max} = -\frac{\Delta G^0}{nF} \qquad \Gamma_{\text{électrique}} = \frac{V_c(I)}{E^0} \qquad (26)$$

By asking: It is found

It is then possible to directly measure this electrical performance experimentally by a voltage measurement, in other words draw the characteristic V(I) returns to plot the electrical performance to T Constant, this performance is strongly dependent on the temperature and it is therefore necessary to recalculate Emax for each temperature.

C. Performance of Operation [7, 10]:

It defines

$$\Gamma_{\text{Fonctionnemnt}} = \frac{V_c(I)}{\text{Evide}} = \frac{V_c(I)}{V_c(0)} \qquad (27)$$

This performance is more simple to experimentally measure, but it FOUT also take into account the performance faradique, which is reported between the value of the current flow through the stack for a voltage V and that of the theoretical current corresponding to the total transformation of reagents:

$$\Gamma_f = \frac{I}{I_m} \qquad (28)$$

In general the performance faradique is close to 1.

Finally the overall performance of a stack is the product of the three previously yields defined:

$$\Gamma_{\text{pile}} = \Gamma_{max}\Gamma_E\Gamma_f \qquad (29)$$

4   The Results of the Simulation

The values of the parameters used for the simulation are grouped in the following table (Table 1).

Table 1. Simulation parameters of the SOFC

The variables	The values
E_0	1.18 V
R	8.314 (i/mole°K)
F	96500 (c/mol)
A	0.2
B	−2870
I_t	0.8 A/cm2

5 Characteristic of Voltage and Current of the Stack

The characteristic (voltage/current) of a cell obtained is represented on the Fig. 2.

Normally the thermodynamic potential equal theoretical 1.18 V, but the open circuit voltage (I = 0) equal 0.94 V (Fig. 2).

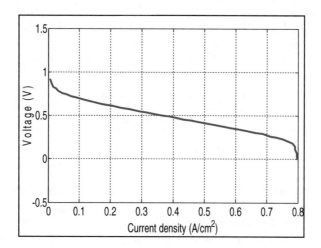

Fig. 2. characteristic (voltage/current) of a cell

This first reduction is linked to the irreversibility the electrochemical reactions, including the reduction of the oxygen at the cathode. In addition, for the low current densities, [12, 13] (1.10–3 to 0.1A/cm^2), against reactions on the electrodes, whose importance is linked to their kinetics, generate power surges of activation (Fig. 3).

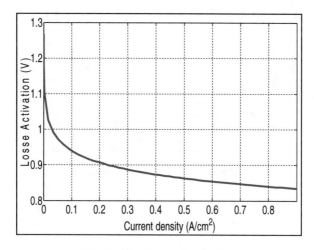

Fig. 3. The losses of activation

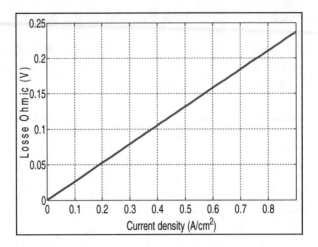

Fig. 4. The Ohmic losses

Then, for the linear portion of the curve, this are the losses related to electronic resistors and internal ionic (Fig. 4), which decrease the voltage between electrodes.

Finally, for the high current densities, c is the kinetics of dissemination of gas through the electrodes which becomes the limiting factor. This phenomena is the loss of concentration (Fig. 5).

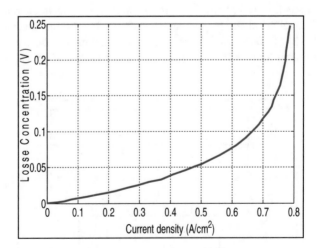

Fig. 5. The loss of concentration

6 The Power

The Fig. 6 represents the curve of the power density as a function of the current density. It is almost parabolic, she believes and then reaches a maximum and end decreases rapidly. This behavior is due to the effect of the polarizations which are directly connected to the current density.

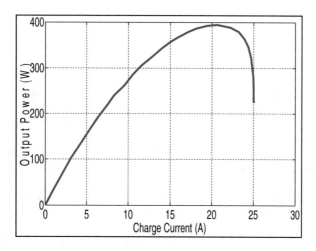

Fig. 6. The power density

7 Electrical Performance

The Fig. 7 shows the variation of the electrical performance as a function of the current density. We note that the electrical performance to Open Circuit equal to 0.825, and then it decreased until the 0.15, therefore this performance is directly proportional to the load [14, 15].

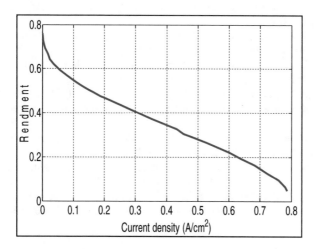

Fig. 7. Electrical performance

8 Conclusion

The modelling and the simulation in static regime have allowed us to find the characteristics of the SOFC consistent with that found in the literature.

The study according to the various parameters of operation also shows that the influence of these parameters joined those obtained experimentally.

References

1. Steffen, Christopher J., Jr., Freeh, Joshua E., Larosiliere, Louis M., Off- Design Performance Analysis of a Solid Oxide Fuel Cell/Gas Turbine Hybrid for Auxiliary Aerospace Power, Third International Conference of Fuel Cell Science and Technology, May 23-25, 2005, FUELCELL2005- 74099
2. Srikar, V.T., Turner, K.T., Ie, T.Y.A., Spearing, S.M.: Structural design considerations for micro machined solid-oxide fuel cells. J. Power Sources (2004)
3. Yuan, J., Rokni, M., Sunden, B.: Three-dimensional computational analysis of gas and heat transport phenomena in ducts under for anode-supported solid oxide fuel cells. Int. J. Heat Mass Transf. (2003)
4. Suzuki, M., Fukagata, K., Shikazono, N., Kasagi, N.: Numerical analysis of temperature and potential distributions in Planar-Type SOFC. In: Thermal and Fluids Engineering Conference. Jeju- Korea (2005)
5. Damm, D.L., Fedorov, A.: Radiation heat transfer in SOFC materials and components. J. Power Sources (2005)
6. Daun, K.J., Beale, S.B., Liu, F., Smallwood, G.J.: Radiation heat transfer in planar SOFC electrolytes. J. Power Sources (2005)
7. Sedghisigrarchi, K., Feliachi, A.: Dynamic and Transient analysis of power distribution systems with fuel cells-Part I: fuel-cell dynamic model. IEEE Trans. Energy Convers. 19(2), 423–428 (2004)
8. He, W.: Dynamic Simulation of Fuel-Cell Molten-Carbonate Systems. Delft Univ. Press, Delft (2000)
9. Rémi, S.: Contribution to the systemic study of energy devices electrochemical components. Formalism Graph Bond applied to fuel cells, batteries Lithium-Ion, Solar Vehicle, doctoral thesis. 2 April 2004
10. Schott, P.: 'Document CEA internal/Alstom: Sizing of a fuel cell system CFMEP 400 kW for a transport application, NT DTEN N°2002/90 (2002)
11. Lachaize, J.: Study of strategies and structures of control for the control of energy systems to Fuel Cell (CAP) intended for the Traction. September 2004
12. Hatziadoniu, C.J., Lobo, A.A., Pourboghrat, F., Daneshdoost, M.: A simplified dynamic model of grid-connected fuel cell generators. IEEE Trans. Power Deliv. 17(2), 467–473 (2002)
13. Freeh, J.E.: A presentation on solid oxide fuel cell/gas turbine hybrid systems for auxiliary power, center for advanced power system, April 19, 2005
14. Samuesen, S.: Turbo fuel cell report: fuel cell/gas turbine hybrid systems, National Fuel Cell Research Center, University of California, Irvine, CA (2004)
15. Theachichi, A.: Modeling and stability of a hybrid regulator of current: Application to converters for fuel cell. Doctoral thesis. University of Franche Comte (2005)

A Comparison Between Parallel Plates and Packed Bed in Electrocaloric Refrigerator Based on Hydrogen Liquefier

Kehileche Brahim[1]([⊠]), Chiba Younes[2], Henini Noureddine[1], Tlemçani Abdelhalim[1], and Mimene Bakhti[1]

[1] Faculty of Technology, Department of Electrical Engineering,
University of Medea, Medea, Algeria
kehileche@yahoo.fr
[2] Faculty of Technology, Department of Mechanical Engineering,
University of Medea, Medea, Algeria
Chiba.younes@univ-medea.dz

Abstract. Electrocaloric refrigerator based on Hydrogen liquefier is a new environmentally friendly cooling technology with a potential for high energy efficiency. The technology is based on the electrocaloric effect; the electrocaloric effect is a phenomenon in which a material shows a reversible temperature change under an applied electric field.

In this work, we studied the effect of parameters (thermal performance) in elctrocaloric refrigerator based on hydrogen liquefier: (1) a packed bed and (2) a parallel plates. The temperature distribution (solid - fluid) is determined by the standard heat transfer equation implemented in COMSOL multiphysics, and they indicate under which operating conditions packed bed configuration is to be preferred to parallel plates and vice versa.

Keywords: Hydrogen liquefier · Electrocaloric effect · COMSOL multiphysics

1 Introduction

This work deals with the study of the hydrogen liquefier operating through an active electrocaloric refrigerator cycle. For this purpose, the two-dimensional numerical model (packed bed and parallel plates) has been developed for predicting the thermal efficiency of such a liquefier. A Refrigerant is a gas/liquid that is employed in the air conditioning systems and refrigerators. Without refrigerant, Air Conditioners, Refrigerators or any other freezing technology will not be possible. Liquid hydrogen (LH2) is the liquid state of the element hydrogen. Hydrogen is found naturally in the molecular H2 form. To exist as a liquid, H2 must be cooled below hydrogen's critical point of 33 K. However, for hydrogen to be in a fully liquid state without boiling at atmospheric pressure, it needs to be cooled to 20.28 K (252.87 C). At the present time [8–10], a number of hydrogen refrigerators and liquefiers are being used to maintain experimental apparatus at low temperatures. For example, several 21–27 K hydrogen refrigerators are used for continuous refrigeration of liquid hydrogen bubble chambers.

© Springer Nature Switzerland AG 2019
M. Hatti (Ed.): ICAIRES 2018, LNNS 62, pp. 483–490, 2019.
https://doi.org/10.1007/978-3-030-04789-4_51

Refrigeration at this temperature level is also being planned for experimental apparatus associated with nuclear reactors and electromagnets. Certain applications may require temperature below those obtainable with hydrogen [8–13].

2   Description of an Electrocaloric Refrigeration Device

Design and operation of the device of elctrocaloric is presented in Fig. 1. The main steps of the elctrocaloric cycle are:

- The elctrocaloric material is polarized adiabatically by application of an electrical field.
- By circulation of carrier fluid in the regenerator bed for exchanging the heat.
- Adiabatic depolarization of the elctrocaloric material under zero electrical fields.
- Cold recovery by moving of carrier fluid in the regenerator bed for exchanging the heat.

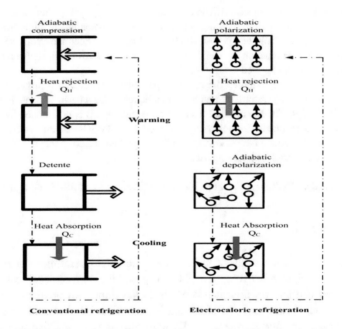

Fig. 1. Schematic diagram of an electrocaloric refrigerator device.

3 Modelling Geometry

By neglecting boundary effects in the transversal direction (the z-direction in Fig. 2), the Active electrocaloric regenerator can be confined to two dimensions in COMSOL multiphysics. Figure 2 shows a schematic of the full 2-D geometry (packed bed and parallel plates) considered in the development of the mathematical model (Table 1).

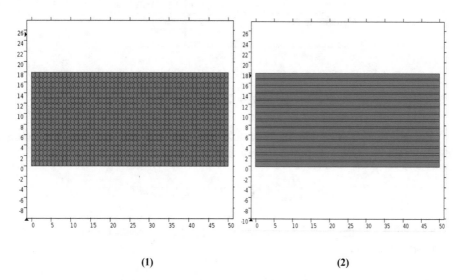

(1)   (2)

Fig. 2. Close up of the regenerator geometry: (1) a packed bed and (2) a parallel plates

4 Modelling Geometry

The governing equations of an active elctrocaloric refrigerator mathematical model consist of a set of coupled partial differential equations, which were solved with the commercial software COMSOL Multiphysics [9, 12, 13].

The velocity distribution in the fluid is determined by solving the momentum (Eq. 1) and continuity equations (Eq. 2) as implemented in COMSOL for an incompressible fluid with constant (temperature independent) properties [11–13]

$$\rho_f\left(\frac{dU}{dt} + (U.\nabla)U\right) - \mu_f\nabla^2 U + \nabla p = 0 \tag{1}$$

$$\nabla.U = 0 \tag{2}$$

For the solid domains (Electrocaloric materials), the temperature distributions determined by the standard heat transfer equation.

Table 1. Geometric parameters of the regenerator

Geometry (mm) parallel plates	L	l	e_s	e_f
	50	18	1	0.3
Geometry (mm) packed bed	L	l	R	
	50	18	0.5	
Polarization (s)	t_{Pol}	t_{Depol}	Δt	
	0.1	0.1	0.1	
Velocity (m/s)	u			
	0.2			

$$\rho_{p,s}\frac{\partial T_s}{\partial t} - k_s\nabla^2 T_s = 0 \tag{3}$$

The temperature distribution in the fluid is determined by the heat transfer equation implemented in COMSOL for an incompressible fluid with convective terms.

$$\rho_f c_{p,f}\left(\frac{\partial T_f}{\partial t} + (U.\nabla)T_f\right) - k_f\nabla^2 T_f = 0 \tag{4}$$

The solids and the fluid are assumed in perfect thermal contact with the following boundary condition.

$$\left(k_f\frac{\partial T_f}{\partial y}\right)_{y=E_{f1}} = \left(k_s\frac{\partial T_s}{\partial y}\right)_{y=E_{f1}} \tag{5}$$

Once steady cyclic state is obtained, the resulting cooling capacity and coefficient of performance can be calculated as follows [14]:

$$\dot{Q}_C = \dot{m}_{nf}c_{nf}(T_C - T_L) \tag{6}$$

And coefficient of performance:

$$COP = \frac{\dot{Q}_C}{\dot{Q}_H - \dot{Q}_C + \dot{W}_p} \tag{7}$$

5 Results and Discussions

Figures 3 and 4; shows the temperature distribution in the active an electrocaloric refrigerator (the hot blow/the cold blow) respectively, and the temperature profiles in the x-direction at various times during the steady state electrocaloric refrigerator cycle with the present operating parameters. The temperature profiles are determined at the middle of the regenerator plate and at the middle of the fluid channel (Liquid hydrogen) of the active an electrocaloric refrigerator geometry shown in Fig. 3.

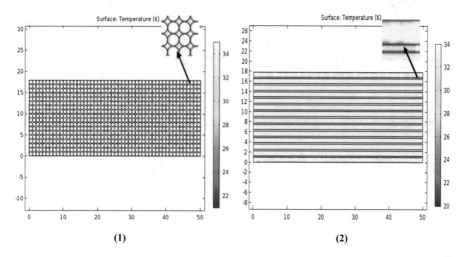

(1)   (2)

Fig. 3. Temperature distribution during the hot blow: (1) a packed bed and (2) a parallel plates

After the polarization and depolarization, the temperature span was 34 K (from 1 K to 35 K: a packed bed) and 32 K (from 2 K to 34 K: a parallel plate). Since electrocaloric refrigerator based on hydrogen liquefier generated a growing interest of industrial, policy makers and researchers. Experimental researches as well as theoretical research are oriented today in several domains such as, research on new materials presenting a high level of electrocaloric effect,

Figure 5 show evolution of coefficient of performance as function of temperature span. As a matter of fact the results provide an indication about the operating conditions under which packed bed configuration (COP = 5.1) has to be preferred to parallel plates (COP = 3)

In Fig. 6, it has shown the Tspan detected for both the regenerator geometries under each fluid flow velocity investigated. Such parameter has been obtained by evaluating the difference between T_c and T_h.

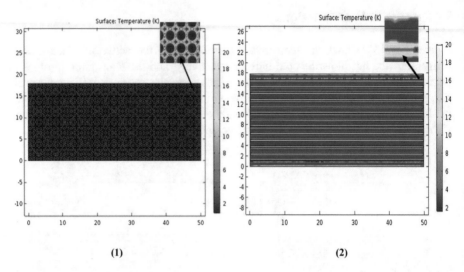

Fig. 4. Temperature distribution during the cold blow: (1) a packed bed and (2) a parallel plates

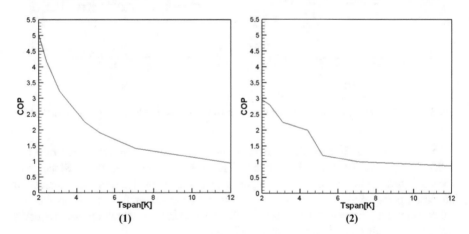

Fig. 5. Evolution of coefficient of performance as function of temperature span: (1) a packed bed and (2) a parallel plates

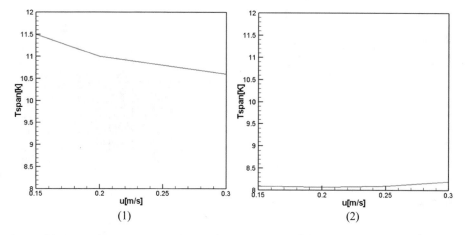

Fig. 6. Evolution of temperature span as function of fluid flow velocity: (1) a packed bed and (2) a parallel plates

6   Conclusion

This work described the development of a 2-D mathematical model of an active electrocaloric refrigerator based on hydrogen liquefier with a regenerator made of: (1) a packed bed and (2) a parallel plates. The electrocaloric refrigerator model was developed using the commercial software COMSOL Multiphysics.

Packed bed structure is the most rational solution for electrocaloric refrigerator based on hydrogen liquefier. It provides specific heat raising and operational point temperature controlling.

Packed bed structures fabrication, investigation, thermal and physical characterisation should be considered as a perspective trend in electrocaloric-based cooling structure research.

Finally, the cost of producing liquid from gaseous hydrogen is an economic barrier to the early adoption of liquid hydrogen as an energy carrier. Decreasing the liquefaction costs is partially dependent on increasing the efficiency of the liquefaction process above the current 30 to 35%.

Nomenclature

Symbols	Meaning	Unit
COP	coefficient of performance	
Cp	specific heat	$J\ kg^{-1}{}^{\circ}C^{-1}$
E	electric field	$V\ m^{-1}$
P	polarization	$C\ m^{-2}$
T	temperature	$^{\circ}C$
k	thermal conductivity	$Wm^{-2}K^{-1}$
L	length	Mm
l	width	Mm

(*continued*)

(continued)

Symbols	Meaning	Unit
e	thickness	Mm
p	pressure	Pa
$\dot{m}_f$	mass-flow rate	kg s^{-1}
$\dot{Q}$	heat rate	W
μ_f	viscosity	kgm^{-1}s^{-1}
ρ	density	kgm^{-3}

References

1. Chiba, Y., Smaili, A., Mahmed, C., Balli, M., Sari, O.: Thermal investigations of an experimental active magnetic regenerative refrigerator operating near room temperature. Int. J. Refrig. **37**(1), 36–42 (2014). https://doi.org/10.1016/j.ijrefrig.2013.09.038
2. Aprea, C., Greco, A.: A comparison between electrocaloric and magnetocaloric materials for solid state refrigeration. Int. J. Heat Technol. **35**(1), 225–234 (2017). https://doi.org/10.18280/ijht.350130
3. Pirc, R., Roi, B., Kutnjak, Z., Blinc, R., Li, X., Zhang, Q.: Electrocaloric effect and dipolar entropy change in ferroelectric polymers. Ferroelectrics **426**(1), 38–44 (2012). https://doi.org/10.1080/00150193.2012.671101
4. Ozbolt, M., Kitanovski, A., Tusek, J., Poredo, A.: Electrocaloric refrigeration: thermodynamics, state of the art and future perspectives. Int. J. Refrig. **40**, 174–188 (2014). https://doi.org/10.1016/j.ijrefrig.2013.11.007
5. Correia, T., Zhang, Q.: Electrocaloric Materials New Generation of Coolers. Springer, Berlin (2014)
6. Scot, J.F.: Electrocaloric Materials. Cambridge University, Cavendish Laboratory (2011)
7. Aprea, C., Greco, A., Maiorino, A., Masselli, C.: Electrocaloric refrigeration: an innovative, emerging, eco-friendly refrigeration technique. J. Phys. **796**(1), 012019 (2017). https://doi.org/10.1088/1742-6596/796/1/012019
8. Kamiya, K., Takahashi, H., Numazawa, T.: Hydrogen liquefaction by magnetic refrigeration. In: International. Cryocooler Conference, Inc., Boulder, CO (2007)
9. Chelton, D.B., Dean, J.W., Strobridge, T.R.: Helium Refrigeration and Liquefaction Using a Liquid Hydrogen Refrigerator for Precooling. Office of technical services, USA (1960)
10. Kirstein, K., Henri, J.: Numerical modeling and analysis of the active magnetic regenerator. Technical University of Denmark (2010)
11. Frank, T., Nini, P., Anders, S.: Numerical modeling and analysis of a room temperature magnetic refrigeration system. Technical University of Denmark (2008)
12. Lionte, S., Vasile, C., Siroux, M.: Approche multiphysique et multi-échelle d'un régénérateur magnéto thermique actif. Institut National des Sciences Appliques INSA (2015). https://doi.org/10.13140/2.1.4864.0807
13. Chiba, Y., Smaili, A., Sari, O.: Enhancements of thermal performances of an active magnetic refrigeration device based on nanofluids. Mechanika **23**(1), 31–38 (2017). https://doi.org/10.5755/j01.mech.23.1.13452

Artificial Neural Networks Modeling
of a Shallow Solar Pond

Abdelkrim Terfai[(✉)], Younes Chiba, and Mohamed Najib Bouaziz

LBMPT, Mechanical Engineering Department, Faculty of Technology,
University Yahia Fares, Medea, Algeria
terfaiabdelkrim@gmail.com, chibayounes@gmail.com,
mn_bouaziz@email.com

Abstract. The aim of this work is to use multi-layered feed-forward back-propagation artificial neural networks and multiple linear regressions models to predict the efficiency of the shallow solar pond. For this purpose, the experimental data collection including wind speed, solar radiation, ambient air temperature, inlet temperature of fluid and mass flow rate of the heat transfer fluid was used in order to predict pond water temperature, outlet temperature of the fluid, rate of heat the heat transfer fluid and instantaneous collection efficiency of a shallow solar pond. In addition, the obtained results are presented and discussed.

Keywords: Renewable energy · Solar energy · Shallow solar pond
Artificial neural networks · Numerical simulation

1 Introduction

Renewable energy sources are a good alternative to reducing multiple fossil fuel problems, such as the cost of extraction, pollution, etc., so scientific research has begun to develop them in the last few decades [1]. One of the most important renewable energy sources is solar energy because it is a clean energy that is abundant in all over the earth and not compromise or add to the global warming. One way to collect and store the energy is to use shallow solar ponds [2, 3]. The shallow solar pond is a solar collector that can be utilized to collecting and storing solar energy. The name suggests that the profundity of water in the SSP is relatively little, commonly 4–15 cm, Solar energy is changed over to thermal energy by warming the water amid the day [4]. Solar pond is utilized for different warm applications greenhouse warming, process warm in dairy plants, desalination, Mechanical process warming and power generation [5, 6]. Neural networks have wide pertinence to scientific issues. Indeed, they have been generally utilized in a wide scope of applications. These applications include pattern recognition, function approximation optimization, simulation, and estimation, automatic, among numerous other application regions. Besides, research has produced expansive number of system ideal models. These days, ANNs have been trained to solve complex problems that are difficult by conventional approaches [7, 8]

The objective of this study is to establish and train four models of neural networks with good performance to predict pond water temperature, outlet temperature of the

M. Hatti (Ed.): ICAIRES 2018, LNNS 62, pp. 491–496, 2019.
https://doi.org/10.1007/978-3-030-04789-4_52

fluid, rate of heat collected by the heat transfer fluid and instantaneous collection efficiency of a shallow solar pond.

2   Neural Networks Description

A neural network is an enormously parallel-disseminated processor made up of straightforward preparing units that has a characteristic penchant for putting away experiential information and making it accessible for utilize. It looks like the cerebrum in two respects:

- The network from its environment through a learning process acquires knowledge.
- Interneuron connection strengths, known as synaptic weights, are used to store the acquired knowledge [9].

The principle of artificial neural networks has been shown in the Fig. 1.

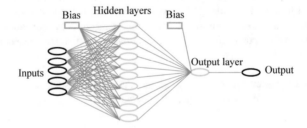

Fig. 1. A diagram showing the neural network used.

- The role of the input variable represents the information that is fed into the artificial neural network.
- The role of each hidden variable is determined by the activities of the input variables and the weights on the connections between the input and the hidden variables.
- The output variable depends on the activity of the hidden variable and the weights between the hidden and output variable.

Neural networks are viewed as general interpolator, and they work best if the system you are utilizing them to show has a high resistance for error. However, they work extremely well for:

- The connection is hard to depicting the issue by utilizing ordinary approaches.
- Variables number or diversity of the points is great.
- The connection between factors is vaguely comprehended.

3   Experimental Setup

The experiment was carried out by means of a measuring system mounted in the Laboratory of renewable energies in arid zones, University of Ouargla. The pond is a rectangular container with a black background and filled with colorful water, the heat is extracted by Water circulation in a PVC coil arranged in the waterbed. In a previous work [10], the thermal performance of the pond with double glass covers was found to be better than that with a single glass cover. Therefore, in the present thermal analysis, it is assumed that the pond has double glass covers.

The dimensions of the shallow solar pond are illustrated in the Fig. 2.

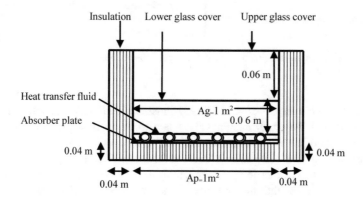

Fig. 2. A diagram showing the dimensions of the shallow solar pond used

The experiment was carried out on 2 May 2005. With the following scenario:

- The start of the solar pond from 08:00 to 17:00.
- Measured quantities: solar radiation, temperature of the heat transfer fluid and of the basin fluid, the wind speed and the useful power.

The error values between the experimental data and the predicted data Artificial Neural Networks can be expressed by Eq. (1)

$$\text{Error}(\%) = \left| \frac{\text{Data}_{\text{Experimental}} - \text{Data}_{\text{Predicted}}}{\text{Data}_{\text{Experimental}}} \right| \times 100 \tag{1}$$

4   Results and Discussions

The general detail of an architecture used for predict pond water temperature, outlet temperature of the fluid, rate of heat collected by the Heat Transfer Fluid of a shallow solar pond models is presented in Figs. 3 and 4.

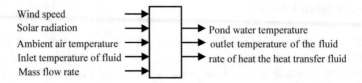

Wind speed
Solar radiation
Ambient air temperature
Inlet temperature of fluid
Mass flow rate

Pond water temperature
outlet temperature of the fluid
rate of heat the heat transfer fluid

Fig. 3. Neural networks architecture used for predict Pond water temperature, outlet temperature of the fluid and rate of heat the heat transfer fluid for a shallow solar pond

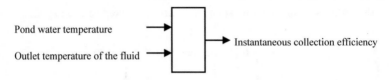

Pond water temperature

Outlet temperature of the fluid

Instantaneous collection efficiency

Fig. 4. Neural networks architecture used for predict Instantaneous collection efficiency for a shallow solar pond.

A correlation coefficient of R = 0.99981 illustrated in Fig. 5 was obtained between the predicted and the experimental values according to data presented in Table 1, which acts to study the effect solar radiation on the pond water temperature of a shallow solar pond during time. The total average error value obtains experimental and predicted results of pond water temperature equals to 0.2949%. The value of this error is very low and indicates that the capacity of the proposed model is very good for the prediction of pond water temperature of a shallow solar pond.

A coefficient of R = 0.99999 illustrated in Fig. 6 was obtained between the predicted and the experimental values according to data presented in Table 1, which acts to study the effect local time on the Outlet temperature of the fluid of a shallow solar

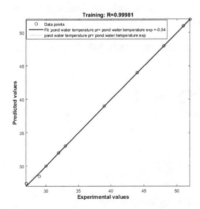

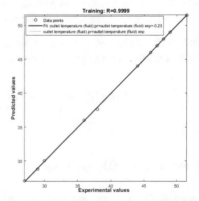

Fig. 5. Regression analysis plots for the optimum model between Output and target of pond water temperature of a shallow solar pond.

Fig. 6. Regression analysis plots for the optimum model between Output and target of Outlet temperature of the fluid of a shallow solar pond

Table 1. Training parameters values used in Neural Networks model for the shallow solar pond.

Neural Networks parameters	Values and nomination in MATLAB
Network type	Feed-forward backprop
Number of neurons	10
Train function	TRAINLM
Transfer function	TANSIG
Performance function	MSE
Error after learning	0.001
Train epochs	1000
Adaption learning function	LEARNGDM

pond. The total average error value obtained between experimental and predicted results of temperature span equals to 0.3792%. The value of this error is very low and indicates that the capacity of the proposed model is very higher to predict of Outlet temperature of the fluid of a shallow solar pond.

A correlation coefficient of R = 0.99858 illustrated in Fig. 7 was obtained between the predicted and the experimental values according to data presented in Table 1, which acts to study the effect local time on the rate of heat collected by the Heat Transfer Fluid. The total average error value obtained between experimental and predicted results of temperature span equals to 3.7851%. The value of this error is very low and indicates that the capacity of the proposed model is very higher to predict of rate of heat collected by the Heat Transfer Fluid of a shallow solar pond.

A correlation coefficient of R = 0.99993 illustrated in Fig. 8 was obtained between the predicted and the experimental values according to data presented in Table 1, which acts to study the effect local time on the instantaneous collection efficiency of a shallow solar pond. The total average error value obtained between experimental and

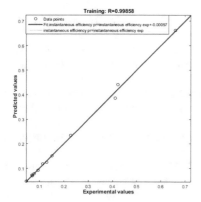

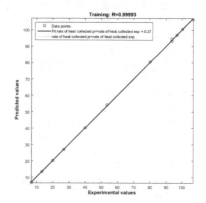

Fig. 7. Regression analysis plot for the optimum model between Output and target of rate of heat collected by the Heat Transfer Fluid of a shallow solar pond

Fig. 8. Regression analysis plot for the optimum model between Output and target of instantaneous collection efficiency of a Shallow solar pond

predicted results of temperature span equals to 5.8845%. The value of this error is low and indicates that the capacity of the proposed model is higher to predict of instantaneous collection efficiency of a shallow solar pond.

5 Conclusion

The study was made in order to develop artificial neural networks models used for predicting performances of a shallow solar pond. The study carried out in this work showed the feasibility of using a simple neural network and to predict pond water temperature, outlet temperature of the fluid, rate of heat collected by the Heat Transfer Fluid and instantaneous collection efficiency of a shallow solar pond. The following conclusion can be noted is the good performance of artificial neural networks model for the prediction of temperature span and coefficient of performance with correlation coefficient about 0.99981 and 0.99999 and 0.99858 and 0.99993 respectively corresponding to regression experimental values.

References

1. Saidur, R., BoroumandJazi, G., Mekhlif, S., Jameel, M.: Exergy analysis of solar energy applications. Renew. Sustain. Energy Rev. **16**(1), 350–356 (2012)
2. Solangi, K.H., Islam, M.R., Saidur, R., Rahim, N.A., Fayaz, H.: A review on global solar energy policy. Renew. Sustain. Energy Rev. **15**(4), 2149–2163 (2011)
3. Karakilcik, M., Dincer, I.: Exergetic performance analysis of a solar pond. Int. J. Therm. Sci. **47**(1), 93–102 (2008)
4. El-Sebaii, A.A., Aboul-Enein, S., Ramadan, M.R.I., Khallaf, A.M.: Thermal performance of shallow solar pond under open and closed cycle modes of heat extraction. Sol. Energy **95**, 30–41 (2013)
5. Velmurugan, V., Srithar, K.: Prospects and scopes of solar pond: a detailed review. Renew. Sustain. Energy Rev. **12**(8), 2253–2263 (2008)
6. El-Sebaii, A.A., Ramadan, M.R.I., Aboul-Enein, S., Khallaf, A.M.: History of the solar ponds: a review study. Renew. Sustain. Energy Rev. **15**(6), 3319–3325 (2011)
7. Esen, H., Ozgen, F., Esen, M., Sengur, A.: Artificial neural network and wavelet neural network approaches for modelling of a solar air heater. Expert Syst. Appl. **36**(8), 11240–11248 (2009)
8. Mohanraj, M., Jayaraj, S., Muraleedharan, C.: Applications of artificial neural networks for refrigeration, air-conditioning and heat pump systems—a review. Renew. Sustain. Energy Rev. **16**(2), 1340–1358 (2012)
9. Haykin, S.: Neural Networks and Learning Machines, 3rd edn. Prentice Hall International, Hamilton, ON (2009)
10. Aboul-Enein, S., El-Sebaii, A.A., Ramadan, M.R.I., Khallaf, A.M.: Parametric study of a shallow solar-pond under the batch mode of heat extraction. Appl. Energy **78**(2), 159–177 (2004)

Dynamic Economic Dispatch Solution with Firefly Algorithm Considering Ramp Rate Limit's and Line Transmission Losses

H. Mostefa[1], B. Mahdad[1], K. Srairi[1], and N. Mancer[2(✉)]

[1] Department of Electrical Engineering, University Biskra,
Biskra 07000, Algeria
[2] Department of Electrical Engineering, Constantine 1 University,
Constantine 25017, Algeria
hamed_elect2008@yahoo.fr, namancer@yahoo.fr

Abstract. The practical dynamic economic dispatch (DED), with consideration of valve-point effects and ramp rate limit's, considered as a non-linear constrained optimization problem. In this paper, a new strategy optimization based Firefly algorithm is proposed to solve this problem. To illustrate of the efficiency of the proposed algorithm, a comparative study between other algorithms is applied. The obtained results show clearly the potential of the investigation strategy based the proposed algorithm with a qualitative calculated solution and converging characteristics.

Keywords: Dynamic economic dispatch · Valve-point effects
Ramp rate limit · Algorithm · Particle swarm optimization
Biogeography based optimization · Transmission line losses

1 Introduction

Dynamic economic dispatch (DED) consider as one of important problem of power generation, its consist to determine the schedule of power unit generation to predicted load demands to reduce the fuel cost under a several operational constraints. Considering the non smooth cost function (Shichang et al. 2018), the minimum and maximum operational units constraint's, ramp rate limit's (Kumar et al. 2016 and Li et al. 2014), and line losses the (DED) shows a non-linear and complicated optimization problem. Recently a lot of stochastic methods are applied and reported in literature to tackle the (DED) problem in a more efficient and quality convergence, (Attia and El-Fergany 2011) proposed the Genetic Algorithm to solve the (SED) with smooth and non smooth cost function considering transmission losses. (Bhattacharya 2010) applied the Biogeography algorithm to solve the Economic Dispatch. (Wu et al. 2017) proposed Two-phase mixed integer programming to solve the non convex economic dispatch with consideration of spinning reserve. (Ghasemi 2013) proposed Honey bee algorithm to find the feasible optimal solution of the environmental economic power dispatch (EED) problem with considering operational constraints of the generators. (Jadoun et al. 2015) use the Modulated particle swarm optimization to solve the economic

© Springer Nature Switzerland AG 2019
M. Hatti (Ed.): ICAIRES 2018, LNNS 62, pp. 497–505, 2019.
https://doi.org/10.1007/978-3-030-04789-4_53

emission dispatch. author's in (Mahdad and Srairi 2012) present a Hierarchical Adaptive PSO to resolve multi-Objective OPF Considering Emissions Based Shunt FACTS, Differential Evolution based Dynamic Decomposed Strategy is used by Belkacem Mahdad and al to resolve the large practical economic dispatch (Mahdad and Srairi 2014). (Jayabarathi et al. 2016) applied the grey wolf optimizer to resolve the ED problem. [11]. (Meng et al. 2015) applied the Crisscross optimization algorithm to treat the large scale dynamic economic dispatch.

In this paper, a novel Firefly algorithm is proposed to deal with the Dynamic economic dispatch problem considering, valve point effect, ramp rate limit's and transmission losses calculate by Beta coefficient formula, the effectiveness of the proposed approach is compared with the results found by BBO and PSOTVAC in one hand, and the other results presented in literature in the other hand. The proposed algorithm's are demonstrated on a practical electrical network 10 unit test system.

2   Problem Formulation

The objective function of (DED) problem is to minimize the total production cost over the operation period which can be written by the non linear characteristic represented by the following equation (Mahdad and Srairi 2014):

$$\min TC = \sum_{t=1}^{T} \sum_{i=1}^{ng} C_{it}(P_{it}) \tag{1}$$

Where C_{it} is the unit i production cost at time t, ng is the number of generation units and P_{it} is the power output of it unit at time t. T is the total number of hours in the operation period. The cost function is nonlinear characteristic which can be represented by the following formula:

$$F(p_{gi}) = \sum_{i=1}^{ng} a_i + b_i p_{it} + c_i p_{it}^2 + \left| e_i \sin(f_i(P_{it}^{min} - P_{it})) \right| \tag{2}$$

Where a_i, b_i, c_i, e_i, f_i are the Cost generators coefficients.

The objective function of the (DED) problem should be minimized subject to following equality and inequality constraints (Mahdad et al. 2009):

2.1   The Equality Constraint

$$\sum_{i=1}^{ng} P_{i,t} = P_d(t) + P_{loss} \tag{3}$$

2.2   Inequality Constraints (Mahdad et al. 2009)

$$P_i^{min} \leq P_i \leq P_i^{max} \quad i = 1....ng \tag{4}$$

Where P_i^{max}, P_i^{min} are the maximum and the minimum of unit's production.

2.3   Ramp Rate Limit's

The Ramp Rate limit's (Mahdi et al. 2018) are represented by the following equation

$$P_{it} - P_{i(t-1)} \leq UR_i \tag{5}$$

$$P_{i(t-1)} - P_{it} \leq DR_i \tag{6}$$

3   Algorithm

This algorithm is based on the principle of attraction between fireflies and the simulative swarm comportment of the in nature, what gives many similarities with other meta-heuristics based on group collective intelligence such as PSO algorithm. An algorithm is governed by the three following rules:

1. All the fireflies are unisex; this means that their attraction does not occur to their sex.
2. The attraction is proportional according to their luminosity; hence, for two fireflies, the less luminous reverses towards the highly luminous. In case, when there is only one particular luminous, the latter will reverse randomly.
3. 's luminosity is determined in function by an objective function (an optimized one).

3.1   Attractiveness

The basic mathematical formulation of the attractiveness function between fireflies is expressed by the following equation:

$$\beta(r) = B_0 \exp(\gamma r^m), \quad with\, m \geq 1 \tag{7}$$

Where r is the distance between any two fireflies, B_0 is the initial attractiveness at $= 0$, and γ is an absorption coefficient which controls the decrease of the light intensity.

3.2   Distance

In the search space, the distance between two fireflies i and j at positions x_i and x_j can be defined by the following relation:

$$r_{ij} = ||x_i - x_j|| = \sqrt{\sum_{k=1}^{d} (x_{i,k} - x_{j,k})^2} \tag{8}$$

Where $x_{i,k}$ is the ith component of the spatial coordinate x_i, x_j and d is the number of dimension.

3.3   Movement

The movement of a i which is attracted by a more attractive j is given by the following equation:

$$x_i^{t+1} = x_i^t + B_0 \exp(-\gamma r_{ij}^2) * (x_i - x_j) + \infty(rand - 0.5) \tag{9}$$

Where the first term is the current position of a, the second term is used for considering a's attractiveness to light intensity seen by adjacent fireflies, and the third term is for the random movement of a in case there are not any brighter ones. I_i and I_j are two variables which reflect the light intensity that is associated with a specified fitness function of particles to be evaluated (Mahdad and Srairi 2015). The flowchart of the optimization algorithm is represented as bellow:

Algorithm: Algorithm

Initialize population of m fireflies, $x_i,\ i=1,2,3,...m$.
Compute Light intensity $f(x_i)$, for all $i=1,2,3,...m$.
While   (stopping criteria is not met)   **do**
    for   i=1 to m
        for   j=1 to m
            if $f(x_i) > f(x_j)$ **then**
                Move i towards j using
            end if
        end for

    end for
Update Light intensity $f(x_i)$ for all $.i=1,2,3,...m$.
Rank the fireflies and find the current best
end while

---

4   Simulation Results

Table 1 shown the result of FA algorithm for the 10 test system, system data, ramp rate limit's and beta-coefficients are taken from (Arul et al. 2013), (Mohammadi-ivatloo et al. 2012) respectively.

Table 1. Best solution of FA algorithm

H	Pg1	Pg2	Pg3	Pg4	Pg5	Pg6	Pg7	Pg8	Pg9	Pg10	Loss (MW)	Cost ($)
1	150.0000	222.2779	171.3511	60.0000	73.0000	122.4377	129.5898	47.0000	20.0000	55	14.6398	28575
2	150.0000	222.2693	197.1005	60.0000	122.8894	122.5142	129.6052	47.0000	20.0000	55	16.3860	30188
3	226.6569	222.2462	234.1733	60.0000	122.8708	159.9955	129.5890	47.0000	20.0000	55	19.5491	33819
4	303.2547	222.2697	309.4328	60.0000	122.9049	160.0000	129.6031	47.0000	20.0000	55	23.4568	36866
5	379.8746	222.2623	309.2867	60.0545	122.9099	160.0000	129.5837	47.0000	20.0000	55	25.9956	38542
6	456.5046	302.2475	340.0000	60.0000	122.8744	129.0570	129.6161	47.0085	20.0000	55	34.3270	42165
7	456.4962	309.5180	297.4507	91.4800	172.7394	160.0000	129.5839	47.0000	20.0000	55	37.2989	43839
8	456.5392	389.4972	300.7239	60.0032	172.7245	160.0000	129.5979	47.0000	49.9657	55	45.0528	45667
9	456.5041	460.0000	316.4921	109.9861	222.6014	160.0000	129.6012	47.0076	20.0121	55	53.2360	49515
10	456.5177	460.0000	340.0000	159.9745	240.3889	160.0000	129.6085	76.9985	50.0115	55	56.4828	53886
11	456.5388	460.0000	340.0000	195.4525	243.0000	160.0000	129.6156	85.3195	80.0000	55	58.9424	55818
12	456.4895	460.0000	340.0000	241.2428	242.0184	159.9889	129.6288	115.3145	80.0000	55	59.7017	57658
13	456.4986	396.8010	324.8272	241.2243	222.6035	122.5130	129.5892	119.9813	52.0602	55	49.1218	52719
14	456.5067	396.8096	301.1725	191.3157	172.7326	122.5293	129.6347	119.9996	22.0900	55	43.7845	49092
15	456.3807	316.8119	296.4855	170.8892	122.8657	122.4090	129.5828	119.9898	20.0000	55	34.4033	45703
16	379.9245	236.8120	319.3170	120.9517	73.000	122.5040	129.6118	119.9868	20.0004	55	23.1379	40704
17	303.2290	222.2814	284.9302	120.2963	122.8711	122.4162	129.5940	119.9932	20.0000	55	20.6258	38678
18	303.2457	222.2623	299.8295	170.2897	172.7644	160.0000	129.6045	120.0000	20.0105	55	25.0043	42313
19	379.9092	302.2602	325.3638	180.8425	172.7043	122.4734	129.5584	120.0000	20.0000	55	32.1381	45738
20	456.4873	382.2547	340.0000	230.8105	222.5971	133.2020	129.6040	120.0000	49.9539	55	47.9218	53316
21	379.8932	396.8053	302.7534	180.8574	222.6135	160.0000	129.5914	120.0000	20.0000	55	43.5753	49035
22	303.1977	396.7966	243.0434	130.9248	172.7554	122.4279	129.5887	90.0095	20.0000	55	35.7757	42531
23	226.6210	316.8037	185.1964	92.0218	122.9442	122.4865	129.5757	85.2955	20.0000	55	23.9831	35622
24	150.0000	309.5460	199.5909	60.0000	73.000	122.5701	129.5879	85.3102	20.0000	55	20.6122	31956
											845.1526	1043945

5   Result and Discussion

In this paper, a comparative study is elaborated to solve the dynamic economic dispatch considering several practical constraints. Three algorithms are investigated, FA, BBO, PSOTVAC, A test system consists of 10 units is considered. The optimized active power of thermal units during 24 h is achieved considering valve point effect, ramp rate limits and transmission losses. For fair comparison between different methods, the population size for all methods is taken 50. Table 1 show the details of the optimized active power of 10 thermal units during 24 H of the proposed algorithm. The FA achieves the best solution **1043945** $ at 150 iterations, the corresponding execution time is 33.69 min, the convergence characteristic is shown in Fig. 3, The transmission losses calculated by beta coefficient formula corresponding is **845.1526 MW,** the BBO Fig. 1 achieves the best total cost **1058700$** which is higher than FA, also this algorithm requires large number of iteration (200), at a relatively reduced execution time

23.93 min compared to FA. Table 2 depicts details about the performances of several algorithms in solving DED in terms of the best, the mean and the maximum value. Figure 2 show the convergence behavior for PSOTVAC when achieve to a total cost **1049200$** at a competitive time (1.8445 min). it is also important that the best solution achieved by FA without violation of constraints, however the degree of permissible violation constraints for the proposed algorithm are near 10-2.

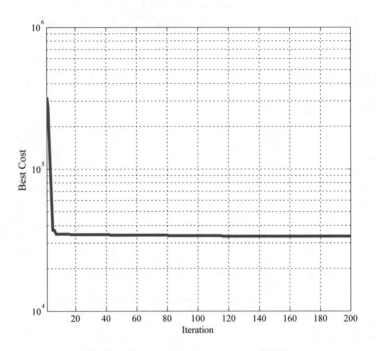

Fig. 1. Characteristic Convergence of BBO

Table 2. Best solution of FA, BBO, PSOTVAC, FA-PSOTVAC, BBO-PSOTVAC.

Method	Pop size	Max iteration	Best solution	Worst solution	Mean value	Losse (MW)	Min-max of Balance Demande Violation	Time (min)
FA	50	150	**1043945**	7744900	3350500	**845.1526**	0.0168–0.0611	33.69
BBO	50	200	1058700	1144100	42709	800.30	0.3386–0.3378	23.93
PSOTVAC	50	1000	1049200	28619000	13785000	914.3054	0.0044–0.0256	1.8445
CSO (Meng et al. 2015)	30	1000	1038320	1042518	1039374	802.62	–	1.481
GA (Hemamalini and Simon 2011a, 2011b)	–	–	1052251	1062511	1058041	–	–	3.444
IPSO (Yuan et al. 2009)	–	–	1046275	–	1048154	–	–	0.180
AIS (Hemamalini and Simon 2011a, 2011b)	–	–	1045715	1048431	1047050	835.62	–	30.973
ECE (Selvakumar 2011)	–	–	1043989	–	1044470	–	–	0.644

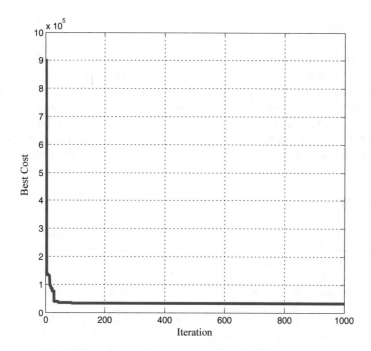

Fig. 2. Characteristic Convergence of PSOTVAC

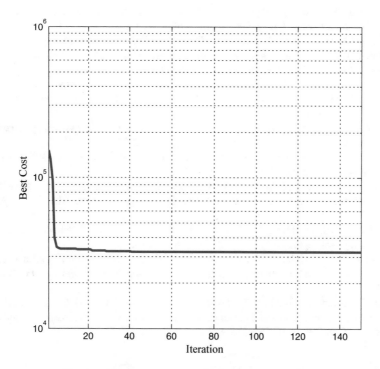

Fig. 3. Characteristic Convergence of FA Algorithm.

6 Conclusion

In this study, three algorithms the FA, PSOTVAC, BBO, have been adapted and applied to solve the DED considering three practical constraints simultaneously such as the valve point effect, ramp rate limits, and transmission losses. The performances of FA algorithm in terms of solution quality and number of generation required have been improved and validated on practical test system 10 units to solve the DED considering simultaneously three constraints. The total cost achieved using FA is competitive in term of quality and reduction of the number of convergence characteristics.

References

Shichang, C., Wang, W., Lin, X., Liu, X.-K.: Distributed auction optimization algorithm for the non convex economic dispatch problem based on the gossip communication mechanism. Int. J. Electr. Power Energy Syst. **95**, 417–426 (2018)

Kumar, N., Panigrahi, B.K., Singh, B.: A solution to the ramp rate and prohibited operating zone constrained unit commitment by GHS-JGT evolutionary algorithm. Int. J. Electr. Power Energy Syst. **81**, 193–203 (2016)

Li, Y.Z., Li, M.S., Wu, Q.H.: Energy saving dispatch with complex constraints: prohibited zones, valve point effect and carbon tax. Int. J. Electr. Power Energy Syst. **63**, 657–666 (2014)

El-Fergany, A.A.: Solution of economic load dispatch problem with smooth and non-smooth fuel cost functions including line losses using genetic algorithm. Int. J. Comput. Electr. Eng. **3**(5), 706 (2011)

Bhattacharya, A.: Biogeography-based optimization for different economic load dispatch problems. IEEE Trans. Power Syst. **25**(2), 1064–1077 (2010)

Wu, Z.L., Ding, J.Y., Wu, Q.H., Jing, Z.X., Zhou, X.X.: Reserve constrained dynamic economic dispatch with valve-point effect: a two-stage mixed integer linear programming approach. CSEE J. Power Energy Syst. **3**(2), 203–211 (2017)

Ghasemi, A.: A fuzzified multi objective interactive honey bee mating optimization for environmental/economic power dispatch with valve point effect. Electr. Power Energy Syst. **49**, 308–321 (2013)

Jadoun, V.K., Gupta, N., Niazi, K.R., Swarnkar, A.: Modulated particle swarm optimization for economic emission dispatch. Electr. Power Energy Syst. **73**, 80–88 (2015)

Mahdad, B., Srairi, K.: Hierarchical Adaptive PSO for Multi-Objective OPF Considering Emissions Based Shunt FACTS. IEEE (2012)

Mahdad, B., Srairi, K.: Multi objective large power system planning under sever loading condition using learning DE-APSO-PS strategy. Energy Convers. Manag. **87**, 338–350 (2014)

Jayabarathi, T., Raghunathan, T., Adarsh, B.R., Suganthan, P.N.: Economic dispatch using hybrid grey wolf optimizer. Energy **111**, 630–641 (2016)

Meng, A., Hanwu, H., Yin, H., Peng, X., Guo, Z.: Crisscross optimization algorithm for large-scale dynamic economic dispatch problem with valve-point effects. Energy **93**, 2175–2190 (2015)

Mahdad, B., Srairi, K., Bouktir, T.: Dynamic strategy based parallel GA coordinated with FACTS devices to enhance the power system security. In: Power & Energy Society General Meeting, 2009. PES '09. IEEE (2009)

Mahdi, F.P., Vasant, P., Kallimani, V., Watada, J., Fai, P.Y.S., Abdullah-Al-Wadud, M.: A holistic review on optimization strategies for combined economic emission dispatch problem. Renew. Sustain. Energy Rev. **81**(2), 3006–3020 (2018)

Mahdad, B., Srairi, K.: Security optimal power flow considering loading margin stability using hybrid FFA–PS assisted with bra instorming rules. Appl. Soft Comput. **35**, 291–309 (2015)

Arul, R., Ravi, G., Velusami, S.: Chaotic self-adaptive differential harmony search algorithm based dynamic economic dispatch. Electr. Power Energy Syst. **50**, 85–96 (2013)

Mohammadi-ivatloo, B., Rabiee, A., Ehsan, M.: Time varying acceleration coefficients IPSO for solving dynamic economic dispatch with noncom-smooth cost function. Energy Convers. Manag. **56**, 175–183 (2012)

Hemamalini, S., Simon, S.P.: Dynamic economic dispatch using artificial bee colony algorithm for units with valve-point effect. Eur Trans Electr Power **21**(1), 70–81 (2011a)

Yuan, X., Su, A., Yuan, Y., Nie, H., Wang, L.: An improved PSO for dynamic load dispatch of generators with valve-point effects. Energy **34**(1), 67–74 (2009)

Hemamalini, S., Simon, S.P.: Dynamic economic dispatch using artificial immune system for units with valve-point effect. Int. J. Electr. Power Energy Syst. **33**(4), 868–874 (2011b)

Selvakumar, I.A.: Enhanced cross-entropy method for dynamic economic dispatch with valve-point effects. Electr Power Energy Syst **33**(3), 783–790 (2011)

Voltage Stability Improvement of Practical Power System Based STATCOM Using PSO_TVAC

N. Mancer[1($\boxtimes$)], B. Mahdad[2], K. Srairi[2], and H. Mostefa[2]

[1] Department of Electrical Engineering, Constantine 1 University,
25017 Constantine 1, Algeria
namancer@yahoo.fr
[2] Department of Electrical Engineering, University Biskra,
07000 Biskra, Algeria

Abstract. Voltage stability improvement is an important issue in power system planning and operation. Voltage stability of a system depends on the network topology and settings of reactive compensation devices. To ensure reliable operation of modern power system characterized by large integration of inter-mittent renewable sources and multi types of FACTS devices, the voltage stability index becomes a challenge for expert and industrials. In This preliminary study a particle swarm optimization with time varying acceleration (PSO-TVAC) algorithm is proposed for monitoring and improving voltage stability. The proposed technique is based on the minimization of the maximum of L-indices of load buses considering multi STATCOM device under normal case and contingency situation. The proposed algorithm has been tested on IEEE 57-bus test systems and successful results have been obtained.

Keywords: Voltage stability · Reactive power planning · STATCOM PSO-TVAC

1 Introduction

Due to economic reasons arising out of deregulation and open market of electricity, modern day power systems are being operated closer to their stability limits. The voltage stability (Kessel and Glavitsch 1986) problem has become a major concern in power system, especially for a system is one of the challenging problems faced by the utilities. Many different criteria to establish which solutions correspond to a stable equilibrium point have been developed. The classic voltage stability criterion known as dQ/dV criterion is base on the capability of the system to supply the reactive power required by loads, dividing the reactive power demand from the real power demand. Another criterion is the dE/dV criterion in which the equivalent emf E can be expressed as a function of the load voltage and also the dQG/dQL based on the relation between the reactive power generation QG(V) and the load reactive demand QL(V) (Jayasankara et al. 2010). Yang et al. (2012) have proposed optimal setting of reactive compensation devices with an improved voltage stability index for voltage stability enhancement. In Mehta et al. (2018) authors have proposed Optimal selection of

© Springer Nature Switzerland AG 2019
M. Hatti (Ed.): ICAIRES 2018, LNNS 62, pp. 506–515, 2019.
https://doi.org/10.1007/978-3-030-04789-4_54

distributed generating units and its placement for voltage stability enhancement and energy loss minimization, they have used the L-index proposed in Mehta et al. (2018) for voltage stability assessment. The authors in Rabiee et al. (2014), considered the desired LM and satisfaction of this LM in Corrective voltage control scheme considering demand response and stochastic wind power while loadability margin (LM) of power system maximized. The authors in Mohseni-Bonab et al. (2016) focused on the effect of Voltage stability constrained multi-objective optimal reactive power dispatch under load and wind power uncertainties. In Ref Ratra et al. (2018), voltage stability assessment in power systems using line voltage stability index have been proposed. In Nikkhah and Rabiee (2018) authors have been proposed a study for optimal wind Power Generation Investment, Considering Voltage Stability of Power Systems. While the different methods listed above give a general picture of the proximity of the system to voltage collapse, the index proposed in reference Kessel and Glavitsch (1986) Kessel gives a scalar number to each load bus called the L-index.this index value ranges from 0 (no load of system) to 1 (voltage collapse). The bus with the highest L-index value will be the most vulnerable bus in the system and hence this method helps in identifying the weak areas in the system which need critical reactive power support. The advantage of this method lies in the simplicity of the numerical calculations and expressiveness of the results (Ratra et al. 2018). The modern power systems are facing increased power flow due to increasing demand and are difficult to control. The rapid development of fast acting and self commutated power electronics converters, well known as FACTS controllers, introduced in 1988 by Hingorani and Gyugyi (2000) are useful in taking fast control actions to ensure security of power systems. FACTS devices are capable of controlling the voltage angle, voltage magnitude at selected buses and/or line impedance of transmission lines. The work in this paper describes the steady state model of facts device and its integration in an existing ORPF under normal and contingency condition. The STATCOM which is one of the most effective FACTS devices, is considered in the study (Zhang et al. 2004). The proposed algorithm for optimal reactive power flow control achieves the goal by setting suitable values for generator terminal voltages, transformer tap settings and parameter setting of the STATCOM. This work proposes a coordinated control of all parameters of reactive power control and the system to enhance the performances of practical electrical power system at critical situation.

2 Voltage Stability Index (L_Index)

Voltage stability analysis involves both static and dynamic factors. As dynamic computations are time consuming, the static approach is generally preferred for stability assessment and control. Static voltage stability analysis involves determination of an index called voltage stability index (Kessel and Glavitsch 1986). This index is an approximate measure of closeness of the system to voltage collapse. There are various methods of determining the voltage stability index. One such method is L-index proposed in Mandal and Roy (2013). It is based on load flow analysis. Its value ranges from 0 (no load condition) to 1 (voltage collapse). The bus with the highest L-index value will be the most vulnerable bus in the system. The L-index calculation for a power system is briefly discussed below. Consider a N-bus system in which there are N_g generators. The relationship between voltage and current can be expressed by the following expression:

$$\begin{bmatrix} \overline{I}_G \\ \overline{I}_L \end{bmatrix} = \begin{bmatrix} \overline{Y}_{GG} & \overline{Y}_{GL} \\ \overline{Y}_{LG} & \overline{Y}_{LL} \end{bmatrix} \begin{bmatrix} \overline{V}_G \\ \overline{V}_L \end{bmatrix} \tag{1}$$

Where I_G, I_L and V_G, V_L represent currents and voltage at the generator buses and load buses, Rearranging the above equation we get:

$$\begin{bmatrix} \overline{V}_L \\ \overline{I}_G \end{bmatrix} = \begin{bmatrix} \overline{Z}_{LL} & \overline{F}_{LG} \\ \overline{K}_{GL} & \overline{Y}_{GG} \end{bmatrix} \begin{bmatrix} \overline{I}_L \\ \overline{V}_G \end{bmatrix} \tag{2}$$

Here:

$$\overline{F}_{LG} = -\begin{bmatrix} \overline{Y}_{LL} \end{bmatrix}^{-1} \begin{bmatrix} \overline{Y}_{LG} \end{bmatrix} \tag{3}$$

The L-index of the jth node is given by the expression:

$$L_j = \left| 1 - \sum_{i=1}^{Ng} \overline{F}_{ji} \frac{\overline{V}_i}{\overline{V}_j} \angle (\theta_{ij} + \delta_i - \delta_j) \right| \tag{4}$$

Where: V_i, V_j are the voltage magnitude of i_{it} and j_{it} generator, θ_{ij} is phase angle of the term F_{ij}, δ_i, δ_j are the voltage phase angle of i_{it} and j_{it} generator unit. The values of F_{ij} are obtained from the matrix F_{LG}. The L-indices for a given load condition are computed for all the load buses and the maximum of the L-indices ($L^{\max}$) gives the proximity of the system to voltage collapse. The indicator $L^{\max}$ is a quantitative measure for the estimation of the distance of the actual state of the system to the stability limit.

3 Problem Formulation

The main objective of this work is to determine the optimal parameter setting of the multi-STATCOM, generator voltages and transformer tap; in the optimal reactive power flux we have optimize one objective function only with represent the voltage stability index so the problem can be formulated as:

$$Minimize \quad (L\_max) + \sum_{i \in npv} R \left(Q_{gi} - Q_{gi}^{\lim} \right)^2 \tag{5}$$

This optimization while satisfying several equality and inequality constraints and to eliminate or minimize the reactive power generation violations under the critical single contingencies.

3.1   Equality Constraints

The equality constraints are the load flow equation:

$$P_{Gi} - P_{Di} = V_i \sum_{j=1}^{N_B} V_j \left(G_{ij} \cos \theta_{ij} + B_{ij} \sin \theta_{ij} \right) \quad for \quad i = 1, \ldots, N_B - 1 \quad (6)$$

$$Q_{Gi} - Q_{Di} = V_i \sum_{j=1}^{N_B} V_j \left(G_{ij} \sin \theta_{ij} - B_{ij} \cos \theta_{ij} \right) \quad for \quad i = 1, \ldots, N_{PQ} - 1 \quad (7)$$

N_B, N_{PQ}: is the set of numbers of total buses excluding slack bus and numbers of PQ buses.

3.2   Inequality Constraints

Generator constraints: generator voltages V_G and reactive power outputs are restricted by the limits as the below relation:

$$V_{G,i}^{\min} \leq V_{G,i} \leq V_{G,i}^{\max}, \quad i = 1, \ldots, N_{PQ} \quad (8)$$

$$Q_{G,i}^{\min} \leq Q_{G,i} \leq Q_{G,i}^{\max}, \quad i = 1, \ldots, N_{PV} \quad (9)$$

$$V_{L,i}^{\min} \leq V_{L,i} \leq V_{L,i}^{\max}, \quad i = 1, \ldots, N_{PQ} \quad (10)$$

$$-50 \leq Q_{STATCOM} \leq 50 \quad (11)$$

Where N_{PV}, N_{PQ} are the set numbers of generator and load bus respectively.

4   Modeling of STATCOM with Power Flow

STATCOM is a second generation of FACTS device used for shunt reactive power compensation. According to the IEEE, STATCOM system is a static synchronous generator operated as a static compensator connected in parallel whose output current (inductive or capacitive) can be controlled independently of the AC system voltage (Mahdad 2011). The bus at which the STATCOM is connected is represented as a PV bus, this dispositive can be generated or absorbed reactive power would reach to the maximum limit. Figure 1a shows the basic configuration of STATCOM (Mancer et al. 2012).

4.1   Power Flow Equation with STATCOM

An alternative way to model the STATCOM in a Newton-Raphson power flow algorithm is described in Mahdad (2010). The power transmission line between two bus system can be represented by:

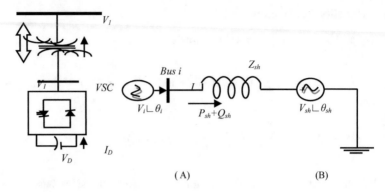

Fig. 1. (A) Basic configuration of STATCOM. (B) STATCOM equivalent circuit.

The active and reactive power transmitted

$$P = \frac{V_{i*}V_{sh}}{X}\sin(\delta_i - \delta_{sh})$$
$$Q = \frac{V_i^2}{X} - \frac{V_{i*}V_{sh}}{X}\cos(\delta_i - \delta_{sh})$$

$$(12)$$

Where V_i, V_{sh} are the voltage at the nodes, $(\delta_i - \delta_{sh})$ the angle between the voltage and X the line impedance. After performing some complex operations, the following active and reactive power equations are obtained as follows:

$$P_{sh} = V_i^2 g_{sh} - V_i V_{sh}(g_{sh}\cos(\theta_i - \theta_{sh}) + b_{sh}\cos(\theta_i - \theta_{sh})) \quad (14)$$

$$Q_{sh} = -V_i^2 b_{sh} - V_i V_{sh}(g_{sh}\sin(\theta_i - \theta_{sh}) - b_{sh}\sin(\theta_i - \theta_{sh})) \quad (15)$$

Where: $g_{sh} + j b_{sh} = Z_{sh}$ and

g_{sh}: Equivalent conductance of the STATCOM.

b_{sh}: Equivalent susceptance of the STATCOM.

Z_{sh}: Equivalent impedance of the STATCOM.

4.1.1 PSO Strategy

PSO is relatively a modern heuristic search method motivated from the simulation of the behavior of social systems such as fish schooling and birds flocking (Mancer et al. 2012). The classical particle swarm optimization (PSO) first introduced by Kennedy and Eberhart (1995), The motivation behind this concept is to well balance the exploration and exploitation capability for attaining better convergence to the optimal solution. The modified velocity and position of each particle can be calculated using the current velocity and the distance from Pbest to Gbest as shown in the following formulas general:

$$\begin{cases} V(t+1) = K(w * V(t) + C_1 rand_1 * (Pbest_i - X(t)) + C_2 rand_2 * (Gbest_i - X(t))) \\ X(t+1) = X(t) + V(t+1) \end{cases}$$

$$(16)$$

where $V(t)$ is the current velocity, $V(t+1)$ is the velocity (modified velocity) $rand_1$ and $rand_2$ are the random numbers between 0 and 1, $Pbest_i$ is the best value found by particle i, $Gbest_i$ is the best particle found in the group, $X(t)$ is the current position $X(t+1)$ the current position (modified searching point), Here w is the inertia weight parameter, K is constriction factor, C_1; C_2 are cognitive and social coefficients,. A large inertia weight helps in good global search while a smaller value facilitates local exploration.

4.1.2   PSO Based Time Varying Acceleration Coefficients (PSO-TVAC)

The idea behind time varying acceleration (TVAC) is to enhance the global search in the early part of the optimization and to encourage the particles to converge towards the global optima at the end of the search (Mancer et al. 2015). This is achieved by changing the acceleration coefficients C_1 and C_2 with time in such a manner that the cognitive component is reduced while the social component is increased as the search proceeds. The acceleration coefficients are expressed as:

$$\begin{cases} C_1 = (C_{1f} - C_{1i}) \frac{iter}{iter_{max}} + C_{1i} \\ C_2 = (C_{2f} - C_{2i}) \frac{iter}{iter_{max}} + C_{2i} \end{cases}$$

$$(17)$$

Were C_{1f}, C_{1i}, C_{2f} and C_{2i} are social acceleration factors and initial and final values of cognitive respectively. The concept of time varying inertial weight was introduced in Mancer et al. (2015) is suggested to decrease linearly from 0.9 to 0.4 ruing the run. The inertial weights formulated as in (18)

$$w = (w_{max} - W_{min}) * \frac{(iter_{max} - iter)}{iter_{max}} + W_{min}$$

$$(18)$$

Where iter is the current iteration number while *itermax* is the maximum number of iterations.

4.1.3   Simulation Results

In this study, The program was validated and tested for the networks IEEE 57 nodes. This test system consists of 57 nodes, 63 branches, 15 transformers and seven controlled nodes. For all tests, the maximum limits of capacitive and inductive compensation to install at node i are equal to 50 Mvar, and the minimum and maximum limits on tap transformers are 0.90 p.u and 1.10 p.u respectively. Different study cases are considered:

- Case 1. Power flow solution under base loading condition (nominal point).
- Case 2. Power flow solution under uniform load variation of +20 per cent condition.

- Case 3. Optimization without STATCOM with uniform load variation of +20 per cent from base case.
- Case 4. Optimization with STATCOM with uniform load variation of +20 per cent from base case.

In this section the objective is to the verify the feasibility and performance of the PSO_TVAC based FACTS using multi STATCOM Controllers by adjusting dynamically their parameters setting. The objective function considered is voltage stability index L-index (Modarresi et al. 2016).

Table 1. Optimized control variables: cases 1–4.

	Case 1	Case 2	Case 3	Case 4
V_{G1}	1.0400	1.0400	1.0513	1.0881
V_{G2}	1.0100	1.0100	1.0447	1.0076
V_{G3}	0.9850	0.9850	1.0161	1.0341
V_{G6}	0.9800	0.9800	1.0115	1.0063
V_{G8}	1.0050	1.0050	1.0881	1.0322
V_{G9}	0.9800	0.9800	0.9803	0.9953
V_{G12}	1.0150	1.0150	1.0504	1.0147
T_{4-18}	0.9700	0.9700	0.9595	0.9615
T_{4-18}	0.9780	0.9780	0.9336	1.0009
T_{21-20}	1.0430	1.0430	1.0421	0.9926
T_{24-26}	1.0430	1.0430	1.0963	0.9020
T_{7-29}	0.9670	0.9670	1.0025	0.9797
T_{34-32}	0.9750	0.9750	0.9514	0.9731
T_{11-41}	0.9550	0.9550	0.9611	0.9662
T_{15-45}	0.9550	0.9550	0.9345	0.9369
T_{14-46}	0.9000	0.9000	0.9033	1.0021
T_{10-51}	0.9300	0.9300	1.0384	0.9672
T_{13-49}	0.8950	0.8950	0.9134	0.9946
T_{11-43}	0.9580	0.9580	0.9650	0.9635
T_{40-56}	0.9580	0.9580	0.9642	0.9431
T_{39-57}	0.9800	0.9800	0.9386	0.9807
T_{9-55}	0.9400	0.9400	0.9373	0.9945
$V_{G25\text{-}STATCOM}$	0.0000	0.0000	0.0000	0.9988
$V_{G31\text{-}STATCOM}$	0.0000	0.0000	0.0000	0.9408
$V_{G33\text{-}STATCOM}$	0.0000	0.0000	0.0000	1.0359
$Min(V)$	0.9001	0.8248	0.8851	0.9256
$Max (V)$	1.0571	1.0400	1.0681	1.0881
$Ploss$	28.4550	61.6580	60.6240	57.42
L_index	0.3217	0.4478	0.4022	0.2903

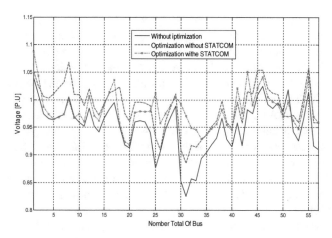

Fig. 2. Voltage profiles at all EEE 57-Bus considering multi STATCOM

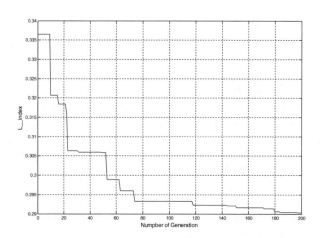

Fig. 3. Convergence characteristic of voltage stability index based PSO-TVAC.

Table 2. Results of optimized reactive power of generating units and STACOM: cases 1–4.

	QG $_{MIN}$	QG $_{MAX}$	Case 2	Case 3	Case 4
QG-1	2.0000	3.0000	1.6441	1.1566	2.2997
QG-2	0.1700	0.5000	0.3609	**0.5021**	0.1015
QG-3	0.1000	0.6000	0.4132	0.3084	0.3964
QG-6	0.0800	0.2500	0.1459	0.0782	0.1391
QG-8	1.4000	2.0000	1.1140	1.7916	0.9781
QG-9	0.0300	0.0900	**0.4249**	**0.1515**	0.0799
QG-12	1.5000	1.5500	1.3921	1.4802	1.0672
QG-25 STATCOM	−0.5000	0.5000	0	0	0.2211
QG-30 STATCOM	−0.5000	0.5000	0	0	0.0946
QG-33 STATCOM	−0.5000	0.5000	0	0	0.2219

The best results of the objective function optimized is given in Table 1. Figure 2 shows the voltage profiles at all buses at three cases: base case, Case 2 (general case), case 3 (Optimization without STATCOM controllers), and case 4 (Optimization considering multi STATCOM controllers). Figure 3 shows the convergence characteristics of the stability index 'L-index' (voltage stability index) based on the proposed algorithm variant PSO-TVAC, The Table 1 shown clearly that the optimal solution achieved with multi STATCOM is 0.2903 p.u better than that calculated by the Optimization without installation STATCOM controllers and worth 0.4022 pu, this significant difference explains the role of the incorporation of this device (SATCOM). Secondly, it can be concluded that the location of this STATCOM has the effect of generating a low stability index with consideration of all constraints. Table 2 presents the optimized reactive powers at the levels of the production units and those injected by the STATCOM for all the cases addressed, it is clear that the reactive powers produced of all the units are within their admissible limits (without violation of the constraints).

5 Conclusion

This paper proposes a particle swarm optimization with time varying acceleration (PSO-TVAC) algorithm considering dynamic shunt controllers based FACTS technology for monitoring and improving voltage stability under sever loading conditions. In this study the voltage stability index called L-index is proposed to evaluate the performances of practical electrical power system at critical situation. The proposed variant based modified PSO is applied to the standard IEEE 57-Bus. Results obtained in term of voltage stability index, and power transmission losses are improved compared to the based case (without optimization) and to the case without considering integration of STATCOM controllers. In our future works, the present paper can be extended to adopt new combined algorithms such as Firefly Algorithm (FA) and biogeography optimization (BBO) to address the problem of congestion management in power system with other type of FACTS devices such UPFC and SSSC. Also, it is interesting to include the uncertainties and intermittent aspect associated with various renewable sources in the optimal reactive power dispatch (ORPD) problems.

References

Kessel, P., Glavitsch, H.: Estimation the voltage stability of a power system. IEEE Trans. Power Deliv. 1(3), 346–354 (1986)

Jayasankara, V., Kamaraj, N., Vanaja, N.: Estimation of voltage stability index for power system employing artificial neural network technique and TCSC placement. Neurocomputing 73, 3005–3011 (2010)

Yang, C.-F., Lai, G.G., Lee, C.-H., Su, C.-T., Chang, G.W.: Optimal setting of reactive compensation devices with an improved voltage stability index for voltage stability enhancement. Electr. Power Energy Syst. 37, 50–57 (2012)

Mehta, P., Bhatt, P., Pandya, V.: Optimal selection of distributed generating units and its placement for voltage stability enhancement and energy loss minimization. Ain Shams Engineering Journal 9, 187–201 (2018)

Rabiee, A., Soroudi, A., Mohammadi-Ivatloo, B., Parniani, M.: Corrective voltage control scheme considering demand response and stochastic wind power. Power Syst IEEE Trans. **29**, 2965–2973 (2014)

Mohseni-Bonab, S.M., Rabiee, A., Mohammadi-Ivatloo, B.: Voltage stability constrained multi-objective optimal reactive power dispatch under load and wind power uncertainties: a stochastic approach. Renew. Energy **85**, 598–609 (2016)

Ratra, S., Tiwari, R., Niazi, K.R.: Voltage stability assessment in power systems using line voltage stability index. Comput. Electr. Eng. Available online 5 January 2018 In Press

Nikkhah, S., Rabiee, A.: Optimal wind power generation investment, considering voltage stability of power systems. Renew. Energy **115**, 308–325 (2018)

Hingorani, N.G., Gyugyi, L.: Understanding FACTS—Concepts and Technology of Flexible AC Transmission Systems. IEEE Press, New York (2000)

Zhang, X.P., Handschin, E., Yao, M.: Multi-control functional static synchronous compensator (STATCOM) in power system steady-state operations. Electr. Power Syst. Res. **72**, 269–278 (2004)

Mandal, B., Roy, P.K.: Optimal reactive power dispatch using quasi-oppositional teaching learning based optimization. Electr. Power Energy Syst. **53**, 123–134 (2013)

Mahdad, B.: Contribution to the improvement of power quality using multi hybrid model based Wind-Shunt FACTS. In: 10th EEEIC International Conference on Environment and Electrical Engineering, Italy (2011)

Mancer, N., Mahdad, B., Srairi, K.: Multi objective optimal reactive power flow based STATCOM using three variant of PSO. Int. J. Energy Eng. **2**(2), 1–7 (2012)

Mahdad, B.: Optimal power flow with consideration of FACTS devices using genetic algorithm: application to the algerian network. Doctorate Thesis, Biskra University Algeria (2010)

Kennedy, J., Eberhart, R.: Particle swarm optimization. In: Proceedings of the IEEE Conference on Neural Networks (ICNN'95), vol. IV. Perth, Australia, pp. 1942–1948 (1995)

Mancer, N., Mahdad, B., Srairi, K.: Optimal coordination of directional overcurrent relays using PSO-TVAC considering series compensation. Adv. Electr. Electron. Eng. **13**(2), 96–106 (2015)

Modarresi, J., Gholipour, E., Khodabakhshian, A.: A comprehensive review of the voltage stability indices. Renew. Sustain. Energy Rev. **63**, 1–12 (2016)

1-D Fluid Modeling of Methane Dissociation in Radiofrequency Capacitively Coupled Plasma

Abdelatif Gadoum$^{(\boxtimes)}$, Djilali Benyoucef,
and Mohamed Habib Allah Lahoual

Laboratoire Génie Electrique et Energies Renouvelables,
Chlef University, Chlef, Algeria
gdabdelatif@gmail.com

Abstract. The use of the hydrogen gas as a fuel for the energy produced by fuel cells is increasingly being studied in several areas. Among techniques of hydrogen production is the dissociation of methane in cold plasma. This work represents the modelling of methane dissociation in radio frequency capacitively coupled plasma by using fluid model, and by considering 21 species (i.e. in total; neutrals, radicals, ions, and electrons) and more than 30 reactions (electronic impact with CH_4, neutral-neutral, neutral-ions and surface reactions).

Keywords: Radiofrequency capacitively coupled plasma · Fluid model
Methane · Hydrogen

1 Introduction

Radiofrequency plasma is known by richness of these reactive elements and recent applications in various fields such as medicine, semiconductors processing, biology etc. [1]. Recently, large efforts have been made to understand the physical and chemical behavior occurring the radiofrequency plasma, for this purpose, modeling is very useful tool for understanding and predicting discharge plasma parameters, many modeling works as a very effective tool for the proper understanding of this phenomenon [2]. Methane is the simplest poly-atomic hydrocarbon, and is present in the atmosphere of most planets and the interstellar surface [3]. Low-pressure plasma created by radiofrequency discharge in methane is used in many fields of technology [4], this molecular gas is considered to be a good test gas [5], and several studies that were experimentally interested for example [6], or by numerical (analytical) study by mixing it with other gases such as H_2, N_2, etc. [7], or pure [8].

The main objective of this work is modeling of capacitively coupled plasma (CCP) in methane. In the simulation of a molecular gas, the electron molecule process takes into account different number of ionic species [9]. In our simulation of the CH_4 discharge we include a complete set of electron-impact processes (elastic momentum transfer, vibration and dissociative ionization and neutral to neutrals dissociation). In comparison to 2D fluid model, 1D fluid model has a great advantage in obtaining the

© Springer Nature Switzerland AG 2019
M. Hatti (Ed.): ICAIRES 2018, LNNS 62, pp. 516–522, 2019.
https://doi.org/10.1007/978-3-030-04789-4_55

characteristics of the plasma and the density of the species, in addition the reduction of the calculation time [10]. The fluid model used in this work is based on the coupling of the moments of the Boltzmann equation with the Poisson's equation [11], this model is sufficient to predict the dominant reactions and the most existing species in methane. The description and the equations of the model are presented in Sect. 2, and the simulation results are discussed in Sect. 3.

2 Description of the Model

This section is divided into two parts: the first is reserved for the chemical model and in the second for the mathematical physics model.

2.1 Chemical Model

The model contains, in aggregate 42 reactions; the considered species in the model are shown in Table 1, impact electronic reactions in Table 2, reactions and rate coefficient in Table 3.

Table 1. Species taken into account in the methane plasma model

Electron	Ions	Radicals & Neutrals
e	CH_4^+, CH_5^+, CH_3^+, CH_2^+, CH^+, C^+, $C_2H_4^+$, $C_2H_5^+$, H_2^+, H^+	CH_4, CH_3, CH_2, CH, C, H_2, H, C_2H_4, C_2H_5, C_2H_6

2.2 Mathematical Physics Model

The fluid model is contains the first three moments of Boltzmann equation (mass conservation equation, momentum conservation equation for electrons, and energy conservation), the first two moments of Boltzmann equation for ions (mass conservation equation, and momentum conservation equation), and only the first equation for the neutrals (mass conservation equation), where these equations are coupled with the Poisson's equation [15]. The discharge is powered by alternating voltage generator (frequency RF = 13.56 MHz), where the grounded electrode is taken at the position $x = 0$, while the powered electrode at the position $x = d$, where d is taken 6 cm. The discharge conditions are: 0.1 Torr of gas pressure in 300 K of temperature, and 200 V of radiofrequency. The ion flux density at the electrode levels is due solely to field drainage and therefore the scattering components are null at all times:

$$\frac{dn_p}{dx}\bigg|_{x=0,d} = 0 \qquad (1)$$

Table 2. Electron impact reactions and threshold energy

No	Reactions	Threshold (eV)
01	$e + CH_4 = e + CH_3 + H$	8.8
02	$e + CH_4 = e + CH_2 + H2$	9.4
03	$e + CH_4 = e + CH + H_2 + H$	12.5
04	$e + CH_4 = e + C + 2H_2$	14
05	$e + CH_4 = 2 + CH_4^+$	12.6
06	$e + CH_4 = 2e + H + CH_3^+$	14.3
07	$e + CH_4 = 2e + CH_2^+ + H_2$	16.2
08	$e + CH_4 = 2e + CH^+ + H_2 + H$	22.2
09	$e + CH_4 = 2e + C^+ + 2H_2$	22.0
10	$e + CH_4 = 2e + H_2^+ + CH_2$	22.3
11	$e + CH_4 = 2e + H^+ + CH_3$	21.1

Table 3. Reactions with rate coefficient $K = A.T_g^n.\exp(- E/RT_g)$ where T_g is the gas temperature [12–14].

No	Reaction	Rate coefficient		
		A[m3/s.mol]	E[J/mol]	n
01	$CH_4 + CH_2 => CH_3 + CH_3$	0.0713E−16	41,988	0
02	$CH_4 + CH => C_2H_4 + H$	153E−16	–	− 0.9
03	$CH_4 + H => CH_3 + H_2$	2.2E−26	33,632	3
04	$CH_3 + CH_3 => C_2H_6$	4.66E−16	–	− 0.37
05	$CH_3 + CH_3 => C_2H_4 + H_2$	170E−16	1,133,030	0
06	$CH_3 + CH_3 => C_2H_5 + H$	0.5E−16	56,540	0
07	$CH_3 + H_2 => CH_4 + H$	1.1E−26	39,410	2.74
08	$CH_3 + H => CH_2 + H_2$	1E−16	63,190	0
09	$CH_2 + H_2 => CH_3 + H$	0.19E−16	53,212	0.17
10	$CH_2 + H => CH + H2$	2.2E−16	–	–
11	$CH + H_2 => CH_2 + H$	5.46E−16	16,155	–
12	$CH + H => C + H_2$	1.31E−16	665	–
13	$C + H_2 => CH + H$	6.64E−16	97,278	–
14	$CH_4^+ + CH_4 = CH_5^+ + CH_3$	11.5E−16	–	–
15	$CH_4^+ + H_2 => CH_5^+ + H$	1.086E−16	− 300	− 0.14
16	$CH_4^+ + H => CH_3^+ + H_2$	0.1E−16	–	–
17	$CH_5^+ + H => CH_4^+ + H_2$	1.5E−16	–	–
18	$CH_3^+ + CH_4 => C_2H_5^+ + H_2$	9.6E−16	–	–
19	$CH_3^+ + H => CH_2^+ + H_2$	7E−16	87,800	–
20	$CH_2^+ + CH_4 => C_2H_5^+ + H$	2.88E−16	–	–
21	$CH_2^+ + CH_4 => C_2H_4^+ + H_2$	5E−16	–	–

The density of the electrons due to the speed of recombination of the latter with the surface of the anode, is zero and therefore:

$$n_e|_{x=d} = 0 \tag{2}$$

The overall diffusion coefficient D_j of neutral species j in the gas can be calculated by the Blank's law:

$$\frac{P_{tot}}{D_j} = \sum_i \frac{P_j}{D_{ij}} \tag{3}$$

where D_{ij} is the binary diffusion coefficient of species j in every background gas i

$$D_{ij} = \frac{3}{16} \frac{k_B T_{gas}}{P_{tot}} \frac{(2\pi k_B T_{gas}/m_{ij})^{1/2}}{\pi \sigma_{ij}^2 \Omega_D(\psi)} \tag{4}$$

where k_B is the Boltzmann constant, T_{gas} is the gas temperature in Kelvin, P_{tot} is the total gas pressure in Pascal, m_{ij} is reduced molecular mass in amu, σ_{ij} is the binary collision diameter in $\overset{\circ}{A}$,

$$\sigma_{ij} = \frac{\sigma_i + \sigma_j}{2} \tag{5}$$

$\Omega_D(\psi)$ is the integral diffusion collision given by :

$$\Omega_D = \frac{1.06036}{\psi^B} \frac{0.19300}{e^{D\psi}} \frac{1.03587}{e^{F\psi}} \frac{1.76464}{e^{H\psi}} \tag{6}$$

with $\psi = T_{gas}/\varepsilon_{ij}$, $\varepsilon_{ij} = (\varepsilon_i + \varepsilon_j)^{1/2}$ and constant, $B = 0.15610$, $D = 0.47635$, $F = 1.52996$, and $H = 3.89411$.

For ions both mobility and diffusion coefficients are considered. The ion mobility of an ionic species j in the background gas i can be calculated fro, the law electric field Langevin mobility expression:

$$\mu_{ij} = 0.514 m_{ij}^{1/2} \frac{T_{gas}}{P_{tot}} \alpha_i^{-1/2} \tag{7}$$

where α_i in $A^{\circ 3}$ is polarizability of the background gas. The overall ion mobility μ_j of ion j in the gas mixture can again be obtained by Blank's law; see above.

The ion diffusion coefficient can be derived Einstein's relation:

$$\alpha_i^{-1/2} = \frac{k_B T_{ion}}{e} \mu_j \tag{8}$$

where T_{ion} represents the ion temperature which is assumed to be equal to the gas temperature.

3 Results and Discution

Modeling is performed for a pure CH_4 discharge. The neutral gas distribution is assumed uniform during the simulations. The results shown in this section are about neutrals and ionic densities as a function of distance between electrodes. For the neutral species shown in the Fig. 1 the most abundant element is CH_3, this result is in agreement with the results of Tachibana et al. [16] and Chien-Wei Chang et al. [17], and this due to the fact that the large rate of CH_3 production by electron impact with CH_4 as shown in the Table 2 (reactions (1) and (11)—as primary production). In addition the reaction radical-molecule of CH_2 with CH_4 (reaction (1) in Table 3) contribute mainly to this result. The other reactions contribute to the production of CH_3 are (3) and (5) in Table 2. The second most abundant element is H, this due to the fact that is among the primaries product of electron collisions with CH_4, the reactions (1), (3), (6), and (8) in Table 2, and the reactions between CH_4 and CH (the reaction (2) Table 3) and others reactions such as (2), (7), (9), ... in the Table 3. The third species is H_2. In same raison (primary product by (2), (3), (4), (7), (8), (9) and (10) in the Table 2 and others reactions between CH_4 and H... see Table 2. The CH_2 species is also mainly produced; because it is produced by electron impact with CH_4 (reactions (2) in Table 2). We can see that H_2 is most abundant than CH_3 and H at the sheaths, because of the H_2 is mainly formed through the wall reactions (many charged species are taken to be returned as H_2 molecule). C_2H_4 is produced mainly one essential reaction—the reaction of the radical-molecule reaction of CH with CH_4 (reaction (2) in Table 3). CH is also primary product by collision of electron with CH_4, see Table 2, nevertheless, his destruction by the radical-molecule reaction is quite faster, therefore his number density become small in the plasma; a considerable loss of CH is in the reaction with CH_4 (reaction (2) in Table 3) and contribute to the formation of C_2H_4 and H. The production of others species (C_2H_5, C_2H_3, C_2 and C) is not much as previous elements because of their small production rates.

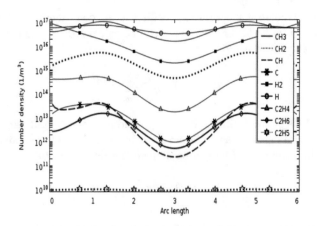

Fig. 1. Neutral species densities

Figure 2 shows that CH_5^+ and $C_2H_5^+$ followed by CH_4^+ and CH_3^+ are the most abundant ions, which is in agreement with the study of Rhallabi and Cathrine [18]. The figure shows also that he species CH_4^+ and CH_3^+ decrease in center of discharge, where the formation of ions CH_5^+ and $C_2H_5^+$ increase; this due to the reactions of CH_4^+ and CH_3^+ with CH_4 with high reaction rates, which contribute to increase the densities of CH_5^+ and $C_2H_5^+$; see reactions (14) and (18) in Table 3. The same behavior is observed for CH_2^+; decreasing in the center and contribute to the formation of $C_2H_4^+$, see the reaction (21) in the Table 3.

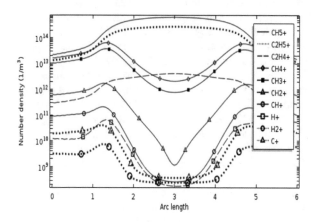

Fig. 2. Ions species densities

The charged species like CH_2^+, CH^+, H_2^+, H^+ and C^+ are produced by collision of electron CH_4 with large quantities, but the destruction rate by ions–molecule reactions is large, in fact their number densities in the plasma is very small.

4 Conclusion

The study achieved out during this work concerns the numerical modeling of a capacitively coupled cold electric discharge plasma. The adopted model considers the plasma as a continuous fluid medium. In this model, ions and electrons are described by the continuity and momentum transfer equations for ions and electrons and the conservation energy equation for electrons, these equations are coupled with Poisson's equation.

The work carried out has highlighted a number of phenomena taking place in the methane plasma especially the densities of existing species. The most abundant neutrals species in the methane plasma are CH_3, H, H_2, CH_2 and C_2H_4, and for ions: CH_5^+, $C_2H_5^+$, CH_4^+ and CH_3^+. These results are in agreement with the previous works [16, 17] for the neutral species and with the work of Rhallabi and Catherine [18] for the ions. The small differences found in magnitudes, resulting from a choice differs from the boundary conditions and transport parameters (mobility coefficient, diffusion, ionization, etc.). These need further study to correlate more adequately with the different physical parameters of the discharge such as gas pressure, applied voltage, secondary emission coefficient, etc.

References

1. Benyoucef, D., Yousfi, M.: Particle modelling of magnetically confined oxygen plasma in low pressure radio frequency discharge. Phys. Plasmas **22**(1), 013510 (2015)
2. Benyoucef, D.: Modélisation particulaire et multidimensionnelle des décharges hors équilibre à basse pression excitées par champs électromagnétiques. Diss. Université de Toulouse, Université Toulouse III-Paul Sabatier, 2011
3. Liu, X., Shemansky, D.E.: Analysis of electron impact ionization properties of methane. J. Geophys. Res. Space Phys. **111**(A4), 29–33 (2006)
4. Mao, M., Bogaerts, A.: Investigating the plasma chemistry for the synthesis of carbon nanotubes/nanofibres in an inductively coupled plasma enhanced CVD system: the effect of different gas mixtures. J. Phys. D Appl. Phys. **43**(20), 205201 (2010)
5. Davies, D.K., Kline, L.E., Bies, W.E.: Measurements of swarm parameters and derived electron collision cross sections in methane. J. Appl. Phys. **65**(9), 3311–3323 (1989)
6. Wei, B., et al.: The relative cross section and kinetic energy distribution of dissociation processes of methane by electron impact. J. Phys. B: At. Mol. Opt. Phys. **46**(21), 215205 (2013)
7. Snoeckx, R., et al.: Influence of N_2 concentration in a CH_4/N_2 dielectric barrier discharge used for CH_4 conversion into H_2. Int. J. Hydrog. Energy **38**(36), 16098–16120 (2013)
8. Ziółkowski, M., et al.: Modeling the electron-impact dissociation of methane. J. Chem. Phys. **137**(22), 22A510 (2012)
9. Donkó, Z., et al.: Fundamental investigations of capacitive radio frequency plasmas: simulations and experiments. Plasma Phys. Control. Fusion **54**(12), 124003 (2012)
10. Herrebout, D., Bogaerts, A., Yan, M., Gijbels, R., Goedheer, W., Dekempeneer, E.: One-dimensional fluid model for an rf methane plasma of interest in deposition of diamond-like carbon layers. J. Appl. Phys. **90**(2), 570–579 (2001)
11. Herrebout, D., et al.: One-dimensional fluid model for an rf methane plasma of interest in deposition of diamond-like carbon layers. J. Appl. Phys. **90**(2), 570–579 (2001)
12. http://kida.obs.u-bordeaux1.fr. Accessed 22 Jan 2018
13. http://kinetics.nist.gov/kinetics/. Accessed 22 Jan 2018
14. http://udfa.ajmarkwick.net/index.php. Accessed 22 Jan 2018
15. Boeuf, J.P., Pitchford, L.C.: Two-dimensional model of a capacitively coupled RF discharge and comparisons with experiments in the gaseous electronics conference reference reactor. Phys. Rev. E **51**(2), 1376 (1995)
16. Tachibana, K., et al.: Diagnostics and modelling of a methane plasma used in the chemical vapour deposition of amorphous carbon films. J. Phys. D Appl. Phys. **17**(8), 1727 (1984)
17. Chang, C.-W., Davoudabadi, M., Mashayek, F.: One-dimensional fluid model of methane plasma for diamond-like coating. IEEE Trans. Plasma Sci. **38**(7), 1603–1614 (2010)
18. Rhallabi, A., Catherine, Y.: Computer simulation of a carbon-deposition plasma in CH/sub 4. IEEE Trans. Plasma Sci. **19**(2), 270–277 (1991)

The Performance Analysis of a Solar Chimney Power Plant and Production of Electric Power

Nouar Hadda[1]([⊠]), Tahri Toufik[2], and Chiba Younes[3]

[1] Department of Electrotechnology and Renewable Energies,
University of Hassibba Ben Bouali, Chlef, Algeria
haddahanine@yahoo.com
[2] Department of Mechanical Engineering,
University of Hassibba Ben Bouali, Chlef, Algeria
ntahritoufik@yahoo.fr
[3] Department of Mechanical Engineering, University Yahia Fares,
Medea, Algeria
Chiba.younes@univ-medea.dz

Abstract. The solar power plant is a thermal system that generates electricity using both the buoyancy effect of hot air generated within the greenhouse by solar radiation and chimney effect. The solar chimney is used in hot areas such as Chlef with high intensity of solar radiation. This work, the solar chimney plant (collector, chimney and turbine) is modeled theoretically, and the global solar radiation, the air temperature was measured during the period of 01/01/2015 to 1/06/2016. The model is employed in simulation to predict the electrical power produced by the solar chimney plant using the meteorological data of Chlef region. the effect of the geometrical parameters on the performances of the solar chimney plant was studied. The influence of chimney height on electrical power produced is simulated in order to have the optimum SCPP to respond to the electricity demand in Chlef region. The performance characteristics of solar chimneys indicate that the plant size, the air temperature and solar radiation flux are important parameters for performance enhancement.

Keywords: The solar chimney power plant · Temperature ambient
Power electrical

1 Introduction

The fuels that we use, and which are fully linked to our lives, may one day lead us to a frightening and painful land, because the sources of energy we consume have precipitated the problems of the Earth, and the problem has escalated by increasing global pollution, which has led to the search for renewable and environmentally friendly sources. Among the types of energy renewable the solar energy, the most important uses of solar energy converted into electric using solar chimney power plant.

The solar power plant is a thermal system that generates electricity using both the buoyancy effect of hot air generated within the greenhouse by solar radiation and chimney effect. The solar chimney is used in hot areas such as Chlef with high intensity of solar radiation. the air of system is heated by solar radiation under the influence of

© Springer Nature Switzerland AG 2019
M. Hatti (Ed.): ICAIRES 2018, LNNS 62, pp. 523–531, 2019.
https://doi.org/10.1007/978-3-030-04789-4_56

the greenhouse, and thus a difference in temperature is created between ambient temperature and temperature within the chimney. As hot air is lighter than cold air, it rises up the chimney tower. At last, the air flow out the tower through the chimney exit. The turbine (or turbines) set at the path of airflow converts the kinetic energy of air into electricity [1].

In this present work, a theoretical model was developed to simulate the physical process of solar chimney in order to supply electricity for inhabitants in Chlef region, Algeria. The input parameters are solar radiation and ambient temperature, while the output is the electrical power. The meteorological data are taken in the site of Ouled Fares at Chlef, every quarter hour during 2015. The simulation was down for two days from the year (winter and summer).

2   Mathematical Model

The solar chimney is composed of three main parts. The first one is the collector (Fig. 1). The second part is the chimney. The third part is the turbine and generator.

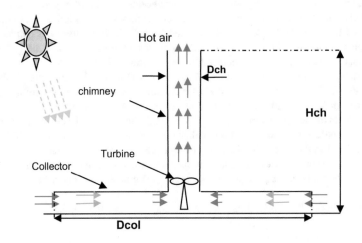

Fig. 1. Schematic of solar chimney power plant

The total efficiency is calculated by multiplying the efficiency of the collector, efficiency of the chimney and efficiency of turbo-generator.

The overall efficiency turbine pressure drop ratio for an unloaded SCPP.

$$\eta_{SC} = \eta_{col} \cdot \eta_{tg} \cdot \eta_{ch} \cdot f_t \tag{1}$$

Where η_{SC}, η_{col}, η_{ch} and η_{tg} are the overall efficiency, collector efficiency, chimney efficiency and turbo-generator efficiency respectively. f_t is the turbine pressure drop ratio. Total power produced by the plant is given by Eq. 2:

$$W_e = Q_s.\eta_{SC} \tag{2}$$

Where W_e is the total electrical power and Q_s is the total solar energy input in the collector which is given by Eq. 3:

$$Q_s = I_s.A_{col} \tag{3}$$

Where, I_s is the solar radiation and A_{col} is the collector area.

2.1   The Solar Collector

In the collector, the air inside the system is heated by the sun due to the greenhouse effect, because the roof is open around its periphery and rises to the chimney. Collector efficiency can be calculated by dividing the amount of heat obtained from air to the available heat from the sun irradiance [2], which is defined as:

$$\eta_{coll} = \frac{Q_u}{Q_s} = \frac{m^{\cdot}.c_p.\Delta T}{I_s.A_{coll}} \tag{4}$$

Where m is the air mass flow rate inside the collector.

$$Q_u = A_{col}[\alpha.I_s - \beta.\Delta T_0] \tag{5}$$

Where ΔT_0 is the difference in temperature between collector surface and ambient temperature and ΔT is the difference in temperature between average hot air in the collector and ambient air. α is effective absorption coefficient of collector with a typical value of '0.75–08' and β is an adjusted heat transfer coefficient that allows for radiation and convection losses.

By substituting Eq. 5 in Eq. 3, the collector efficiency is calculated by Eq. 6:

$$\eta_{coll} = \alpha - \frac{\beta.\Delta T_0}{I_s} \tag{6}$$

The calculation of average temperature of hot air and absorber surface is fussy. The assumption for the above equation is that the absorber surface temperature is equal to the average air flow temperature as follows (Eq. 7).

$$\Delta T = \Delta T_0 \tag{7}$$

where $\Delta T = T_f - T_a$ and $\Delta T_0 = T_0 - T_a$ with T_f, T_a and T_0 are the average hot air temperature in the collector, ambient air temperature and collector outlet temperature respectively.

$$m = -\frac{\eta_{col}}{\alpha} \tag{8}$$

The collector efficiency of the Eq. 8 can be calculated and an appropriate method was adopted to determine the relationship between m and the two environmental factors according to Eq. 9:

$$m = -18.8354 \times I_s^{0.0504} \times T_a^{-0.6143} \tag{9}$$

2.2 The Solar Chimney

Hot air is moving towards the centre of the roof and up the chimney, hot air flow obtained from the collector is converted into kinetic and potential energy, which causes convection currents in the chimney and drops pressure in the wind turbine [2]. Chimney efficiency is given by Eq. 10:

$$\eta_{ch} = \frac{W_{ch}}{Q_u} \tag{10}$$

Where W_{ch} is the power contained in the air flow at bottom of chimney.

$$W_{ch} = \Delta p_{ch}.V_{ch}.A_{ch} \tag{11}$$

Where Δp_{ch} is the pressure difference produced between chimney base and the ambient.

According to Schlaich [3], the pressure difference in the chimney and the equation of maximum velocity is given by Eqs. 12 and 13:

$$\Delta p_{ch} = g.\rho_{ch}.k.H_{ch}.(T - T_a)/T_a \tag{12}$$

$$V_{ch} = \sqrt{2.g.H_{ch}.\Delta T_0/T_a} \tag{13}$$

V_{ch} is updraft velocity in the entrance of chimney respectively.

The efficiency of the chimney is related to its rise according to Eq. 14, the value of chimney height is proportional to the chimney efficiency [2].

$$\eta_{ch} = \frac{g.H_{ch}}{c_p.T_a} \tag{14}$$

2.3 The Turbine

Hot air in the chimney creates an updraft; the updraft drives the turbine installed at the chimney base to generate electrical power. The maximum mechanical power produced by the turbine is given by Eq. 15:

$$W_t = \frac{2}{3}W_{ch} \tag{15}$$

The electrical power produced by the generator is calculated as follows Eq. 16:

$$\eta_g = \frac{W_e}{W_t} \tag{16}$$

The turbo-generator efficiency is given by Eq. 17:

$$\eta_{tg} = \frac{W_e}{W_{ch}} \tag{17}$$

Where $\eta_{tg} = \eta_t.\eta_g$ combined turbine and generator efficiency. η_{tg} Can be treated as a constant in this work with a value of 0.80.

According to Guo et al. [4], the optimal turbine pressure drop ratio for the Spanish prototype can be written as follows Eq. 18:

$$f_{opt} = \frac{2 - m}{3} \tag{18}$$

3   Results and Discussion

In the present paper, simulations that incorporate solar radiation for Chlef region are conducted using the geometrical parameters of Prototype at the Center for Energy Research and Technology (CRTEn), Borj Cedria, in northern Tunisia to explore the electrical power produced by SCPP. The main dimensions of the Tunisian prototype selected as the physical model are listed in Table 1.

Table 1. Geometrical dimensions of the pilot plant in CRTEn of Borj Cedria, northern Tunisia.

Parameter value	Value
Mean collector radius	4 m
Chimney height	4 m
Chimney radius	0.3 m

The geographical coordinates of the data collection site at Chelf were 36°23′N Latitude, 23°1′W Longitude and 143 m altitudes above mean sea level. The environmental parameters such as global solar radiation and ambient air temperature are collected from 1 January 2015 till now. Figure 2a shows the variation of solar radiation together with ambient air temperature of 8 January 2015. It is found that, the maximum global radiation is observed with 531 W/m^2 at 12:30 h, whereas the ambient air temperature is found maximum with 17.1 °C at 14:45 h. The variation of solar radiation together with ambient air temperature of 1 June 2016 is shown in Fig. 2b. It is found that, the maximum ambient air temperature is 29.3 °C at 15:00 h although the maximum global radiation is observed by 946 W/m^2 at 12:45 h.

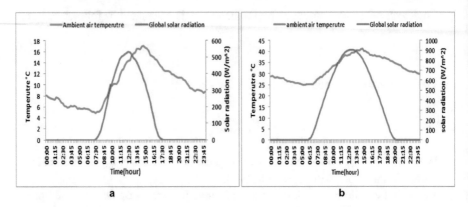

Fig. 2. Solar radiation and air temperature measured at Chlef: a for January 2015 and b for June 2016

Figure 3 illustrates that the total energy is very important in summer compared to winter and the maximum value is 46000 W at 13:00 h. It can be noted that the two plots of total solar energy follow the same trend of solar radiation.

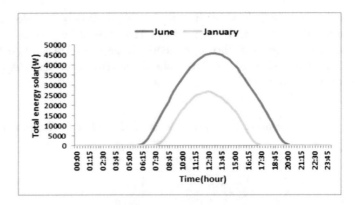

Fig. 3. Total solar energy received by the collector of solar chimney

The updraft velocity in the entrance of chimney according to the hours of the day of January 2015 and June 2016 are represented in Fig. 4. We find that the two plots of velocity follow the same trend of solar radiation and the maximum value is 1.27 m/s at 13:00 h in summer and 0.97 m/s at 12:30 h in winter. It can be seen that the updraft velocity values were observed only during the interval from 05:45 to 20:15 for June 2016, from 07:30 to 17:30 for January 2015 and it was almost null during the night time.

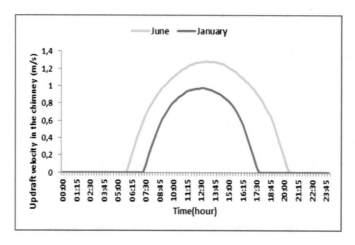

Fig. 4. Updraft velocity in the entrance of chimney

Figure 5 show the electric power produced by the SCPP at Chlef is 2.7 W at 12:30 h in winter, and 4.4 W at 12:45 h in summer. Also, we can find that the power output follow the same trend of solar radiation.

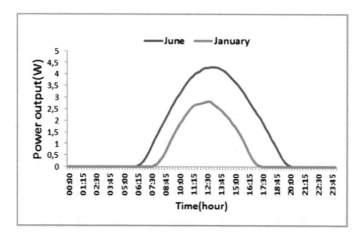

Fig. 5. Power output produced by SCPP at Chlef

Figure 6 illustrates the influence of the chimney tower height on electric power output for a day of June month at Chlef with 4 m diameter collector.

We used the Eq. (14) of the model and found the effect of the chimney height. It is found that the power output is very depends on chimney height. The higher the chimney height the higher the power output.

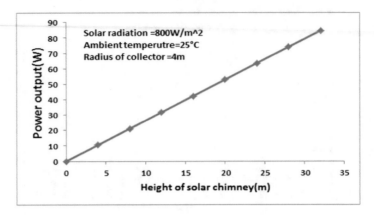

Fig. 6. Influence of chimney height in power output

Figure 7 shows the effect of the collector radius on electric power output for a day of June at Chlef with chimney height value of 4 m.

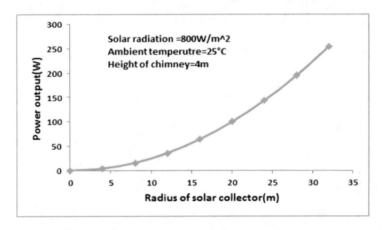

Fig. 7. Influence of the collector radius in power output

The more the collector radius is important the more the power output is important.

4   Conclusion

The work presented in this study is developed to the performance analysis of a solar chimney power plant at Chlef, Algeria. For this purpose, a theoretical model was developed to describe the process of the solar chimney. The results indicated that the performance characteristics of solar chimney indicate that the plant size and solar radiation are important parameters for the power performance enhancement, when we increase the hight of tower and the collector diameter.

References

1. Schlaich, J., Bergermann, R., Schiel, W., Weinrebe, G.: Design of commercial solar updraft tower systems—utilization of solar induced convective flows for power generation. J. Sol. Energy Eng. **127**, 117–124 (2005)
2. Dehghani, S., Mohammadi, A.H.: Optimum dimension of geometric parameters of solar chimney power plants—a multi-objective optimization approach. Sol. Energy **105**, 603–612 (2014)
3. Schlaich, J.: The Solar Chimney—Electricity from the Sun. Axel Menges, Stuttgart (1995). https://scihub22266oqcxt.onion.link/
4. Guo, P., Li, J., Wang, Y., Wang, Y.: Evaluation of the optimal turbine pressure drop ratio for a solar chimney power plant. Energy Convers. Manag. **108**, 14–22 (2016)

Wind Energy and Induction Machine

Comparative Analysis and Simulation of Highest Efficiency Multi-junction Solar Cells for Space Applications

A. Hadj Dida[1,2(✉)], M. Bourahla[1], and M. Bekhti[2]

[1] Departement d'Electrotechnique, Faculté de Génie Electrique,
Laboratoire d'Electronique de Puissance Appliquée,
Université des Sciences et de la Technologie d'Oran Mohamed Boudiaf,
USTO-MB, BP 1505, El M'naouer, 31000 Oran, Algeria
abdelkader.hadjdida@univ-usto.dz
[2] Departement de Recherche et Instrumentation Spatiale,
Centre de Développement des Satellites, Agence Spatiale Algérienne,
Po Box 4065 Ibn Rochd USTO, Bir El Djir, Oran, Algeria

Abstract. Nowadays, solar energy is promising the primary source of energy for space missions that have a great potential to generate power for an extremely low operating cost when compared to already existing power generation technologies. The development of space systems is affected to the study of space itself, the science of materials and especially the field of energy. Solar arrays are the only non-nuclear means that enable space vehicles and satellites in orbit to be fed continuously. Increasing the efficiency of solar cells is a major goal and the prominent factor in space photovoltaic system research. The most efficient technology for generation of electricity from solar irradiation is multi-junction solar cells. The materials used in these structures are a prime factor, controlling device efficiency. Current triple junction solar cells reach 30% and the next generation will have a high efficiency to peak at 40%. The aim of this work is to simulate, investigate and correlate their performance in terms of efficiency, fill factor, and other electrical parameters. Then, we had made an efficiency comparison between a various solar cells measured under one-sun AM0 spectral conditions to determine the best choice of these which will bring a good performance to be used in solar array design of space photovoltaic systems.

Keywords: Solar energy · Multi-junction solar cells
Characterization performance · Space photovoltaic systems

1 Introduction

The development of space systems is affected especially to the field of energy. Currently, solar energy is promising the primary source of energy for space missions. Indeed, solar panels are the only non-nuclear means that enable space systems to be fed continuously. The efficiency of these solar cells increases. Twenty years ago, silicon solar cells allowed 12%. With gallium arsenide single junction, appeared a decade ago, we rose to 20%. Now, the space solar cells are multi-junctions. Current triple junction

© Springer Nature Switzerland AG 2019
M. Hatti (Ed.): ICAIRES 2018, LNNS 62, pp. 535–543, 2019.
https://doi.org/10.1007/978-3-030-04789-4_57

solar cells reach 30% [1]. Due to their high conversion efficiency, this type of solar cell is mainly used in solar panels for space applications. In this research work, we had simulated, investigated and correlated the performance of multi-junctions in terms of efficiency, fill factor, and other electrical parameters. Subsequently, the influence of irradiation and temperature on the electrical parameters of solar cells must be measured under one–sun AM0 spectral conditions (0.1353w/cm², 28 °C) to guarantee a certain level of reliability and performance of these. For this application, we had made an efficiency comparison between a various solar cells to determine the best choice which will bring a good performance to be used in the design of solar array of our space photovoltaic systems.

2 Modeling of Highest Efficiency Multi-junctions Solar Cells

The electrical behavior is one the most important aspects that characterize a solar cell device and it's important to know what the fundamental equations are and how they are linked to the physics parameters. Moreover, the equation which characterizes the electrical behavior gives the possibility to build an equivalent circuit to easily make simulations that permit to extract some important I-V and P-V curves [2]. Due to increasing requirement for power, mass and area traditional silicon solar cells will be more and more replaced by high efficiency multi-junction solar cells on space solar generators. Triple junction solar cells consist of sub cells connected in series in a way similar to the conventional one, to get high efficiency through current-matching among the generated current from each sub cells band gap energy should be strictly combined. Moreover, layers are connected together with tunnel junction and are also provided with window layers. The current generated is highly dependent on amount of light exposure. Series resistance models the tunnel resistance between each cell and the shunt resistances represent the cell material resistance. GaAs based solar cells have the

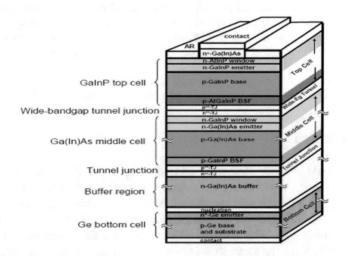

Fig. 1. Triple junction solar cell equivalent structure

potential to reach efficiencies greater than 30%. Figure 1 below demonstrates the equivalent structure of "InGaP/GaAs/Ge" triple junction solar cells and the equivalent electrical circuit model of these cells are shown in Fig. 2 [3, 7].

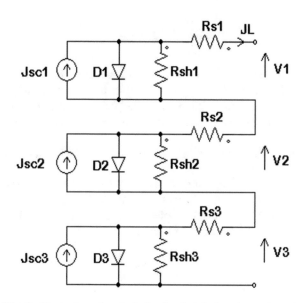

Fig. 2. Equivalent electrical circuit of triple junction solar cells

From the equivalent electrical circuit, the total current density J_L of triple junction solar cell is given by the following equation:

$$J_L = J_{sc,i} - J_{o,i}\left(\frac{q\left(V_i + AR_{s,i}J_L\right)}{n_iK_BT} - 1\right) - \frac{V_i + AR_{s,i}J_L}{AR_{sh,i}} \tag{3}$$

Where: i, J_{sc}, J_o, J_L, q, V, R_{sh}, R_s, K_B, n, A, T represent the number of sub cells (1 = high, 2 = medium and 3 = low), photon current density in A/cm², reverse saturation current density in A/cm², operating current density of cells in A/cm², electrical charge constant (1.602 x 10⁻¹⁹ C), operating voltage per cell in V, shunt resistance in Ω, series resistance in Ω, Boltzmann constant (1.38064852 x 10⁻²³ m²kg s⁻²k⁻¹), diode ideality factor (typically constant between 1 and 2), cell area in cm² and operating temperature in K respectively [4].

It's assumed that temperature of the cell is uniform. The diode's dark saturation current is strongly dependent on temperature and is calculated by the following equation:

$$J_{o,i} = k_i \times T^{\frac{3+\gamma_i}{2}} \times e^{\frac{-q \times E_{gi}}{n_i \times K_B \times T}} \tag{4}$$

Where: γ_i is a constant (typically between 0 and 2) and E_g is the band gap energy in eV. Band gap energy differs from one cell to another due to the difference in material used. The relation between band gap energy and temperature is described by the equation below:

$$E_{gi} = E_g(0) - \frac{\alpha_i \times T^2}{T + \sigma_i} \tag{5}$$

Where: $E_g(0)$ is the band gap energy at 0 °C, α and σ are constants dependent on the materials used. The band gap energy given in Eq. (6) is strongly affected by the mixture alloys of the Indium Gallium Arsenide "InGaAs" and Indium Gallium Phosphate "InGaP". However, Germanium "Ge" is referred as pure material.

$$E_{gi}(A_{1-x}B_x) = (1 - x)E_g(A) + xE_g(A) + xE_g(B) - x(1 - x)P \tag{6}$$

Where: $A_{1-x} B_x$ is the composition of the alloy material and P is an alloy dependent parameter that takes into account deviations from the linear approximation in eV. The voltage at terminals of triple junction solar cell is given in Eq. (7) by the sum of the voltages of three sub-cells connected in series.

$$V = \sum_{i=1}^{3} V_i \tag{7}$$

Where:

$$V_i = \frac{n_i K_B T}{q} \ln\left(\frac{J_{sc,i} - J_L}{J_{o,i}} + 1\right) - J_L AR_{s,i} \tag{8}$$

Open circuit voltage per cell is achieved by canceling J and it's described by the equation below:

$$V_i = \frac{n_i K_B T}{q} \ln\left(\frac{J_{sc,i}}{J_{o,i}} + 1\right) \tag{9}$$

The open circuit voltage is inversely proportional to temperature and can be also demonstrated using I-V curve characteristics representation. The maximum power point "MPP" is obtained by canceling the term dP/dJ_L where: $P = J_L VA$.

3   Simulations Results Comparison of Various Solar Cells

The electrical characteristic is very important to well understand the behavior of solar cells in term of device efficiency. To do that, we had introduced programs under PSpice software to determine the electrical characteristics I-V and P-V curves of the sub-cells solar of "InGaP, GaAs, Ge, Si" and that of the triple junction solar cells

"InGaP/GaAs/Ge" respectively. I-V and P-V curves were the most efficient techniques for extracting short circuit current density J_{sci}, open circuit voltage V_{oci}, maximum power P_m, filling factor FF and the power conversion efficiency of cell η. The influences of irradiation and temperature on the electrical parameters of cells were investigated. A comparison between solar cells efficiency was demonstrated to determine the best choice of solar cell that will be used in space solar panel. In order to simulate and calculate the filling factor FF and the power conversion efficiency η of solar cells, the radiation intensity or solar constant of air mass zero (AM0) was considered equal to 1.353 kw/m^2 and a given temperature equal to 28 °C for the area of 1 cm^2. Figures 3, 4, 5, 6 and 7 shows the simulation results of the electrical characteristics I-V and P-V curves of the "InGaP", "GaAs", "Ge", "Si" and "InGaP/GaAs/Ge" solar cells respectively. Simulated and calculated results values of the electrical parameters extracting from the I-V and P-V curves of each solar cells are shown in Table 1. Where fill factor is the ratio between the maximum power and the multiplication of open circuit voltage and short circuit current. Fill Factor's, maximum power are given by [5, 6]:

$$P_m = V_m * J_m \tag{10}$$

$$FF = \frac{P_m}{V_{oc} * J_{sc}} \tag{11}$$

Table 1. Simulated and calculated results values of electrical parameters of solar cells

Electrical parameters of solar cells	InGaP	GaAs	Ge	Si	InGaP/GaAs/Ge
Voc (V)	1.43	1.04	0.22	0.63	2.69
Jsc (mA/cm^2)	16.50	28.80	47.65	38.40	16.52
Vm (V)	1.34	0.94	0.16	0.53	2.55
Jm (mA/cm^2)	16.00	27.96	42.12	37.25	16.09
Pm (mW/cm^2)	21.44	26.47	6.93	19.99	41.12
FF (%)	90.74	88.39	65,65	82.65	92.53

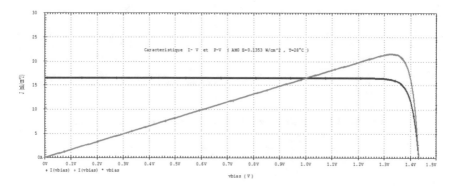

Fig. 3. Electrical characteristic curve I-V and P-V of InGaP solar cell

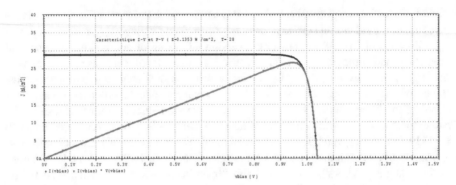

Fig. 4. Electrical characteristic curve I-V and P-V of GaAs solar cell

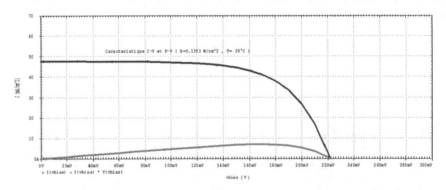

Fig. 5. Electrical characteristic curve I-V and P-V of Ge solar cell

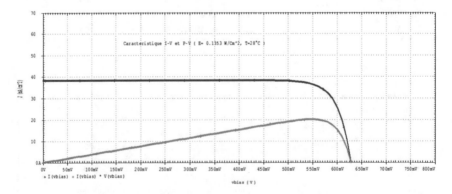

Fig. 6. Electrical characteristic curve I-V and P-V of Si solar cell

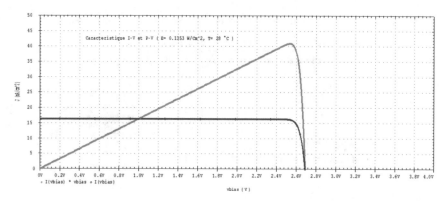

Fig. 7. Electrical characteristic curve I-V and P-V of InGaP/GaAs/Ge Triple Junction Solar cells

3.1 Simulation Results Discussion

The above figures show the simulation results of the electrical characteristics I–V and P–V of each solar sub-cell "GaInP, GaAs, Ge, Si" and that of the triple-junction solar cell "InGaP/GaAs/Ge". From Fig. 5; it can be seen that solar sub-cell of "Ge" has a higher current density than the other solar sub-cells due to its low gap which allows all absorbed photons that have an energy equal to or greater than its gap to produce a high current density at this level. As the gap increases, the voltage increases, so that the photons absorbed by the solar sub-cell have insufficient energy to be all absorbed from where the density of the current decreases. Figure 7 show that the triple junction solar cell "InGaP/GaAs/Ge" has a very small short-circuit current density (J_{sc}) and an important open circuit voltage (V_{co}) cause of the series connection of solar sub-cells.

3.2 Typical Comparison Performance of Solar Cells

A typical comparison of performance between the various types of solar cells at given temperature equal to 28 °C and constant radiation intensity equal to 0.1353 w/cm^2 is shown in Table 2. It is noted that the efficiency of the triple junction solar cell is better compared to other solar cells. The power conversion efficiency of solar cells equation is given by [9]:

$$\eta = \frac{P_m}{P_{incidente}} \tag{12}$$

Table 2. Typical performance comparison between a various types of solar cells

Comparative table of solar cells efficiency		
Power conversion efficiency (%)	InGaP/GaAs/Ge	30.40
	GaInP	15.84
	GaAs	19.56
	Ge	5.12
	Si	14.78
Temperature (°C)	28	
Radiation intensity (W/cm^2)	0.1353	

4   Conclusion

The aim of this research work was to simulate, investigate and correlate the performance of multi-junction solar cells in terms of efficiency, fill factor and other electrical parameters that used in the design of space power systems. To do this task, we started by a theoretical study to understand the multi-junction solar cells structure and equivalent electrical circuit works. Secondly, we had modeled and simulated a various solar cells under Pspice software program. From the comparison and simulation results obtained of the electrical characteristics I-V and P-V of each solar cell, we conclude that the triple junction solar cells deliver a high efficiency compared to other solar cells, they are also expensive. Due to their high cost, MJSC are the main component for producing the electricity from solar energy, they are primarily used in the design of systems in outer space and as concentrator cells where a large amount of sunlight is reflected onto the cell [8]. The performance of these cells must be also high in space environment in order to have a minimum mass at equal power. Solar cells being mounted in satellite and space systems will have to withstand the vibrations of the launch and then support huge deviations of the temperature as well as devastating radiations. The highest efficiencies for photovoltaic devices to day are achieved with triple junction solar cells based on "InGaP/GaAs/Ge". Depending on the applications in space or terrestrial concentrator applications, the efficiency is determined either under AM0 or AM1.5d spectral conditions. Solar cells had a high performance when temperature is low and radiation is high.

References

1. Nishioka, K., Takamoto, T., Agui, T., Kaneiwa, M., Uraoka, Y., Fuyuki, T.: Evaluation of InGaP/InGaAs/Ge triple junction solar cell and optimization of solar cell's structure focusing on series resistance for high-efficiency concentrator photovoltaic systems. Solar Energy Mater. Solar Cells **90**, 1308–1321 (2006)
2. Patel, J., Sharma, G.: Modeling and simulation of solar photovoltaic module using matlab/simulink. IJRET Int. J. Res. Eng. Technol. **2** (2013). ISSN 2319-1163

3. Rashel, M.R., Albino, A., Tlemcani, M., Gonçalves, T,C.F., Rifath, J.: MATLAB simulink modeling of photovoltaic cells for understanding shadow effect. In: IEEE International Conference on Renewable Energy Research and Applications (ICRERA 2016), Birmingham, UK, 20–23 November 2016

4. Nishioka, K., Sueto, T., Uchina, M., Ota, Y.: Detailed analysis of temperature characteristics of an InGaP/InGaAs/Ge triple-junction solar cell. J. Electron. Mater. **39**, 704–708 (2010)

5. Otakwa, R.M., Simiyu, J., Waita, S.M., Mwabora, J.M.: Application of dye-sensitized solar cell technology in the tropics: effects of air mass on device performance. Int. J. Renew. Energy Res. IJRER **2**(3), 369–375 (2012)

6. Fatemi, N., Pollard, H., Hou, H., Sharps, P.: Solar array trades between very high-efficiency multi-junction and Si space solar cells. In: Conference Record of the 28th IEEE Photovoltaic Specialists Conference (PVSC 2000), pp. 1083–1086. IEEE, New York (2000)

7. Liu, Jialin, Hou, R.: Solar cell simulation model for photovoltaic power generation system. Int. J. Renew. Energy Res. IJRER **4**(1), 49–53 (2014)

8. Das, N., Al Ghadeer, A., Islam, S.: Modelling and analysis of multi-junction solar cells to improve the conversion efficiency of photovoltaic systems. In: 2014 Australasian Universities Power Engineering Conference (AUPEC) (2014)

9. Stan, M.A., Aiken, D.J., Sharps, P.R., Fatemi, N.S., Spadafora, F.A., Hills, J., Yoo, H., Clevenger, B.: 27.5% efficiency InGaP/InGaAs/Ge advanced triple junction (ATJ) space solar cells for high volume manufacturing. In: Conference Record of the Twenty-Ninth IEEE Photovoltaic Specialists Conference, New Orleans, LA, USA, 22 April 2003

Study and Simulation of the Power Flow Distribution of an Optical Channel Drop Filter in Structure Based on Photonic Crystal Ring Resonator for Different Organic Liquids

Mehdi Ghoumazi[1,2(✉)], Abdesselam Hocini[1],
and Messaoud Hameurlain[2]

[1] Département d'électronique, Université Mohamed Boudiaf, M'Sila, Algeria
g_mehd@yahoo.fr, hocini74@yahoo.fr
[2] Unité de Recherche en Optique et Photonique (UROP), Centre de
Développement de Technologies Avancées (CDTA), Sétif, Algeria
hameurleinm@cdta.dz

Abstract. The following work represents a propagation and power flow investigation for different organic liquids of an optical channel drop filter on a 2D photonic crystal flower ring resonator. The structure is composed of dielectric rods immersed in air and based on photonic crystal ring resonator. The ring resonator is formed as a flower shape and is sandwiched by two wave guides. For analyzing this structure, plane –wave expansion (PWE) approach and finite element method is applied. The numerical results shows the propagation and the power flow variations for different organic liquids used. This variation is due to their refractive index which varies from one material to another. In this work, we fixed the radius 'r' and the lattice constant 'a' by interesting to the refractive index which is an important parameters of each materials used.

Keywords: Photonic crystals · Ring resonators · Channel drop filter
Organic liquids

1 Introduction

Photonic crystals are nano-sized periodic structures and devices [1] having refractive index that is modulated with the wavelength-scale periodicity in one to three dimensions [2] this periodicity results in a wavelength region in which propagation of optical waves is prohibited. This wavelength region is called the photonic band gap (PBG) [3] due to the remarkable importance of the PBG, many applications of PCs depend on their PBG's proprieties [4, 5]. By creating appropriate defects, different optical devices based on PCs can be developed, they include, optical switches [6], optical waveguides [7], optical converters [8], optical power dividers [9], optical filters [10], optical sensors [11]. We remove some rods or holes inside the Pc structure in order to have a ring shape.

In recent years, several researches were based on CDF ring resonator of square and triangular lattices as: David and Abrishamian who work on a multichannel –drop filter

© Springer Nature Switzerland AG 2019
M. Hatti (Ed.): ICAIRES 2018, LNNS 62, pp. 544–551, 2019.
https://doi.org/10.1007/978-3-030-04789-4_58

with PhcRR by using two different refractive indexes in the 2D–PC with square lattice [12]. Other authors like Mehdizadeh et al. studied the effect of several parameters such as refractive index of dielectric rods and so on…which are important parameters for tuning the filter [3].

In our work will study a design of channel drop filter (CDF) based on new configuration of 2D photonic crystal flower ring resonator (PhCFRR) by using finite element method (FEM).

2 Design of Optical Filter

The finite-element (FEM) numerical method and plane wave expansion method (PWE) have been used to obtain the photonic band gap (PBG), propagation and the power flow for different organic liquids in the structure designed by COMSOL software. In this paper, we have used a ring resonator as a flower shape to design optical filter. In the proposed optical filter as shown in Fig. 2, the ring resonator is sandwiched by two parallel waveguides; one is created by removing a complete row of dielectric rods in the Γ-M direction to create the bus waveguide and the second by removing some rods in the Γ-M direction to create the output waveguide.

Our proposed structure is a two dimensional photonic crystal composed of 35 × 28 square lattice of dielectric rods immersed in air with lattice constant a = 0.623 μm. The refractive index of the rods is taken to be 3.46 and for the air background is 1. The radius of rods of perfect PC (with no defects) is r = 0.19a.

In this paper, we studied the photonic band structure PBGs, propagation and power flow calculations of 2D photonic by using finite element method (FEM). By solving Maxwell's Equations, we studied electromagnetic wave propagation in a photonic crystal structure where following form for magnetic field is given by [4]:

$$\nabla \times \left(\frac{1}{\varepsilon(\vec{r})} \nabla \times H(\vec{r}) \right) = \left(\frac{\omega}{c} \right)^2 H(\vec{r})$$

ω is the angular frequency, c is the speed of light in vacuum, and ε(r) is the relative permittivity of material.

The FEM calculation is applied for the dispersion relation of square lattice pattern for TE and TM polarizations. Light propagation is considered in the (xy) plane of the square lattice structure. The PBGs of the Phc with aforementioned values is depicted in Fig. 1. In the next section our simulations are adapted to the proposed structure of a channel drop filter based on 2D photonic crystal ring resonator.

Figure 1, displays three PBGs, one for TE mode (green zone) and the two for TM mode (yellow zone). The TE PBGs range 'a/λ' is [0.85, 0.86] and the TM PBGs have two ranges, the first 'a/λ' is [0.32, 0.44] and the second is 'a/λ' is [0.77, 0.78]. Only the first PBG in TM mode is large enough for covering the sufficient wavelengths for optical communication applications. In order to have maximum compatibility with optical communication ranges, we took a = 623 nm where the study will be in 1415 nm < λ < 1946 nm range in TM mode.

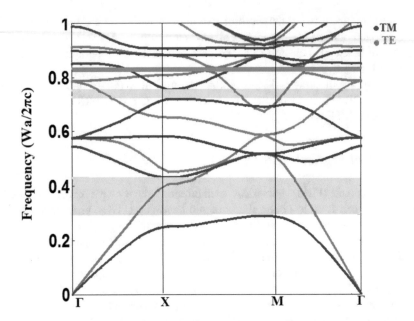

Fig. 1. The band structure of the proposed structure

As explained above, to realize the proposed filter in a fundamental platform, we removed a complete row of dielectric rods in the Γ-M direction, and some rods in the G-M direction we created the bus waveguide and the output waveguide respectively. Next step is creating resonant ring between bus waveguide and the output waveguide, first we removed from a 15 × 15 array of dielectric rods at the appropriate place some rods for creating a flower form. This flower shape is created by 13-fold quasi crystal which is quasi-periodic structure and composed of one central air pore as core rod. The radius of the flower form is the same as the radius of all other rods in the initial Pc structure. The final schematic diagram of the proposed filter is depicted in Fig. 2.

The proposed filter has 4 ports: The input port is marked as port (1), the output waveguide is marked as port (2) and called as the forward transmission terminal. The port (3) and (4) of waveguide is denoted as forward dropping (see Fig. 2). Optical waves enter the structure from port (1) to exit from port (2), but at the desired wavelength the optical wavelengths drop to drop waveguide through the resonant ring and travel toward port (4). Flower shape has the same radius and refractive index as the initial structure. Furthermore, as depicted in Fig. 2, The refractive index of the frame surrounding the flower resonator has been changed for different organic liquids such as water, Ethanol, Carbon tetrachloride, Benzene and Carbon Disulfide. The refractive index rods changed are labelled with red circle as shown in Fig. 3.

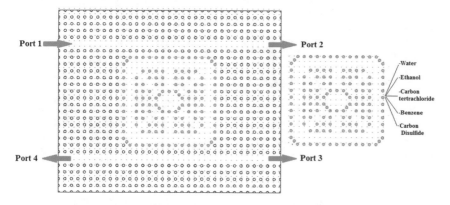

Fig. 2. The final sketch of our structure

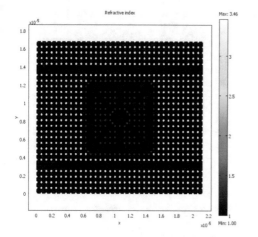

Fig. 3. The distribution of the refractive index in the proposed structure

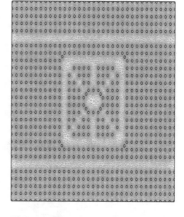

Fig. 4. The proposed structure with fine mesh

3 Simulation and Results

Our numerical results are obtained from the finite element method using the COMSOL software [13]. In this work, the study of the propagation and the power flow as a function of the refractive index for different organic liquids such as Water, Ethanol, Carbon tetrachloride, Benzene, Carbon disulfide in a structure of a two-dimensional square lattice based on photonic crystals. The plane of propagation of the electromagnetic wave is (xz). This wave is the telecommunications wavelength at 1.55 µm.

Figure 6, the power flow for the two indices of refraction, water (n = 1.333), and Ethanol (n = 1.361) will pass through maximums. For water, we observe that the

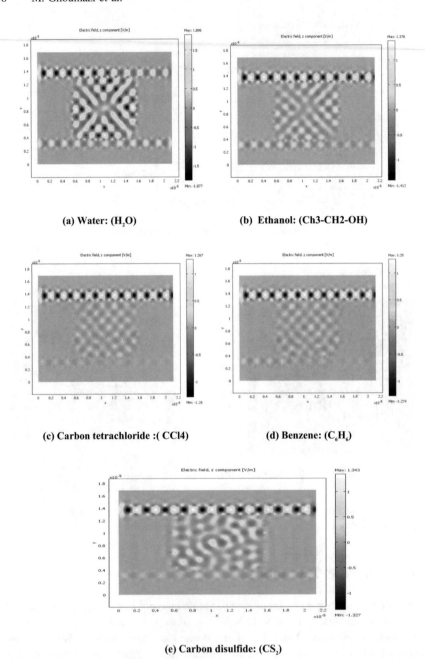

(a) Water: (H₂O)

(b) Ethanol: (Ch3-CH2-OH)

(c) Carbon tetrachloride :(CCl4)

(d) Benzene: (C₆H₆)

(e) Carbon disulfide: (CS₂)

Fig. 5. The propagation of the field distribution for different refractive index of (a) Water (H$_2$O), (b)Ethanol (Ch$_3$-Ch$_2$-OH) (c) Carbon tetrachloride(CCl$_4$), (d) Benzene (C$_6$H$_6$) and (e) carbon disulfide (CS$_2$) respectively at λ = 1550 nm

biggest transfer of the power flow from port 1 to port 4 reaches a maximum of 13.095 nW/m^2 versus a cross section line at 2.5 μm (Figs. 5a, 6), as well as for the case of Ethanol which reaches a medium power flow of 3.4958 nW/m^2 for the same a cross section line (see Figs. 6, 5b) and for Carbon Tetrachloride (n = 1.461), Carbon disulfide (n = 1.628) and benzene (n = 1.501) they completes a weak power flow maximum of 0.2885 nW/m^2, 0.4458 nW/m^2 and 0.6450 nW/m^2 respectively, also for the cross section line (see Figs. 5c–e, 6).

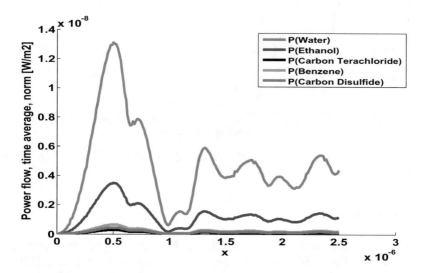

Fig. 6. The power flow of: (a) Water (H$_2$O), (b) Ethanol (Ch$_3$-Ch$_2$-OH) (c) Carbon tetrachloride (CCl$_4$), (d) Benzene (C$_6$H$_6$) and (e) carbon disulfide (CS$_2$) versus cross section line

So, Fig. 7 summarizes our work, we note that the two organic liquids Water and Ethanol are much more important in terms of transfer of power flow to port 4 than benzene, Carbon Disulfide and Carbon Tetrachloride where the resonance is maximum for all organic liquids. Knowing that, these maximums intersect in the same cross section line at 2.5 μm.

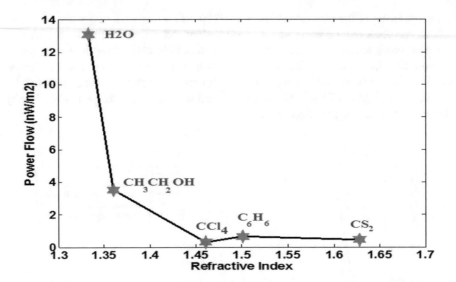

Fig. 7. The power flow versus refractive index of: (a) Water, (b) Ethanol (c) Carbon tetrachloride, (d) Benzene and (e) carbon disulfide

4   Conclusion

In this paper, we proposed a channel drop filter (CDF) based on photonic crystal flower ring resonator (PhCFRR). We used plane wave expansion method (PWE) and COMSOL software to simulate the propagation and the power flow of different organic liquids. From the simulation we observed a different behavior of the signal which depends on refractive index of each material. This refractive index is a good parameter that can be used for detection in biosensor for several research areas such as health, environment...etc.

Acknowledgements. The present work was supported by the Ministry of Higher Education and Scientific Research of Algeria

References

1. Abbaspour, A., Alipour Banaei, H., Andalib, A.: The new method for optical channel drop filter with high quality factor based on triangular photonic crystal design. J. Articficial Intell. Electr. Eng. **2**(6), 1 (2013)
2. Bendjelloul, R., Bouchemat, T., Bouchemat, M.: An optical channel drop filter based on 2D photonic crystal ring resonator. J. Electromagn. Waves Appl. 2402–2410 (2016). https://doi.org/10.1080/09205071.2016.1253508
3. Mehdizadeh, F., Alipour-Banaei, H., Serajmohammadi, S.: Channel-drop filter based on a photonic crystal ring resonator. J. Opt. **15**, 075401 (2013)
4. Joannopoulos, J.D., Johnson, S.G., Winn, J.N., et al.: Photonic Crystals: Molding the Flow of Light. Princeton University Press, Princeton (2008)

5. Skoda, K.: Optical Proprieties of Photonic Crystals. Springer, Berlin (2001)
6. Serajmohammadi, H.Alipour-Banaei, Mehdizadeh, F.: All optical decoder switch based on photonic crystal ring resonators. Opt. Quantum Electron. **47**, 1109–1115 (2015)
7. Janrao, N., Zafar, R., Janyani, V.: Improved design of photonic crystal waveguides with elliptical holes for enhanced slow light performance. Opt. Eng. **51**, 064001-1-064001-7 (2012)
8. Mehdizadeh, F., Soroosh, M., Alipour-Banaei, H., Farshidi, E.: A novel proposal for all optical analog-to-digital converter based on photonic crystal structures. IEEE Photon. J. **9**, 1–11 (2017)
9. Mesri, N., Alipour-Banaei, H.: An optical power divider based on twodimensional photonic crystal structure. J. Opt. Commun. **38**, 129–132 (2017)
10. Seifouri, M., Fallahi, V., Olyaee, S.: Ultra-high-Q optical filter based on photonic crystal ring resonator. Photon Netw. Commun. **2**, 1–6 (2017)
11. Olyaee, S., Mohebzadeh-Bahabady, A.: Designing a novel photonic crystal nano-ring resonator for biosensor application. Opt. Quant. Electron. **47**, 1881–1888 (2015)
12. Taalbi, A., Bassou, G.: Mahmoud M Y Optik at pres. https://doi.org/10.1016/j.ijleo.2012.01.045 (2012)
13. Comsol. [Online]. www.comsol.com

Insight of Electronic and Thermoelectric Properties of CdSiAs$_2$ Ternary Chalcopyrite from First Principles Calculations

Nacera Si Ziani[1]([✉]), Hamida Bouhani-Benziane[1], Melouka Baira[1], Abdelkader Belfedal[2], and Mohamed Sahnoun[1]

[1] Laboratoire de Physique Quantique de la Matière et Modélisation Mathematique (LPQ3M), Faculté des Sciences Exactes, Université Mustapha Stambouli, Mascara, Algeria
sizianinacera@gmail.com
[2] Laboratoire de Chimie-Physique de Macromolécule et Interface Biologique, Faculté des Sciences et de la vie, Université Mustapha Stambouli, Mascara, Algeria

Abstract. Electronic and thermoelectric properties of ternary chalcopyrite type CdSiAs$_2$ were studied using the first principles density functional calculations performed in the full potential linear augmented plane wave (FP-LAPW) method as implemented in the WIEN2k code. The thermoelectric properties are calculated by solving the Boltzmann transport equation within the constant relaxation time approximation. The calculated band gap using the Tran-Blaha modified Becke- Johnson potential (TB-mBJ) of CdSiAs$_2$ compound is in good agreement with the available experimental data. Thermoelectric properties like thermopower, electrical conductivity scaled by relaxation time and electronic thermal conductivity scaled by relaxation time are calculated as a function of temperature.

Keywords: Electronic properties · Thermoelectric properties
FP-LAPW method · TB-MBJ

1 Introduction

Chalcopyrite-type materials are well-known semi-conductors with energy band gap ranging from 1 to 3 eV [1, 2]. The semiconducting nature of these materials attracts the present researcher, as these materials find promising applications in solar cells, detectors, nonlinear optic devices and many more. The important application of the chalcopyrite-type compounds includes the energy conversion and serve as thermoelectric and photovoltaic energy converters. Ternary chalcopyrite semiconductors and pnictide compounds of the form ABC$_2$ have attracted researchers because of their special physical properties like high melting point, high refractive index, high nonlinear optical susceptibility, and many more [3, 4]. Because of the above mentioned diverse properties, both the chalcopyrites and pnictides are identified as promising materials for different device applications like electronic devices [5], nonlinear optical devices [6, 7], photovoltaic cell [8] etc. These materials crystallize in the tetragonal structure with

© Springer Nature Switzerland AG 2019
M. Hatti (Ed.): ICAIRES 2018, LNNS 62, pp. 552–559, 2019.
https://doi.org/10.1007/978-3-030-04789-4_59

space group I_42d and are considered as the ternary analogue of zinc blende structure [9, 10]. In this work; we employ first-principles calculations to study the structural, electronic and thermoelectric properties of ternary chalcopyrite CdSiAs$_2$.

2 Computational Details

Present calculations are based on the first principles density functional theory. We have used full potential linearized augmented plane wave (FP-LAPW) method implemented in WIEN2k package [11, 12]. The calculations using standard Generalized Gradient Approximation (GGA) schemes for the exchange-correlation potential underestimate the band gaps of semiconductors, and generally we have used the modified GGA known as the Tran-Blaha modified Becke-Johnson potential (TBmBJ) [13, 14] which is good in reproducing the experimental band gaps.

3 Results and Discussion

3.1 Structural Properties

The investigated compound CdSiAs$_2$ crystallize in tetragonal chalcopyrite type structure with space group I4_2d. Chalcopyrite crystal structure can be considered as the ternary analogue of zinc blende structure with slight tetragonal distortion. The chalcopyrite structure is closely related to the zinc blende structure. Generally, the chalcopyrite lattice is characterized by three structural parameters: the lattice constants "a" and "c", as well as the dimensionless anion displacement parameter u (Table 1).

Table 1. Calculated equilibrium lattice constant a (in Å), c/a ratio, distortion parameter (u).

Compound	a	c/a	U
CdSiAs$_2$	5.778	1.790	0.232
	5.885 [15–17]	1.849 [15–17]	

3.2 Electronic Properties

It is a well known fact that LDA and GGA exchange-correlation potentials underestimate the band gap of semiconductors [18]. Therefore in order to obtain the band structure accurately, we have used mBJ potential [19]. In the absence of spin-orbit interactions our calculated energy band structure of CdSiAs$_2$ along some high symmetry directions of Brillouin zone is presented in Fig. 1.

Figure 1 shows the calculated band structure along some high symmetry directions at the equilibrium lattice parameter for CdSiAs$_2$ using the TB-mBJ. Both the top of the valence band and the bottom of the conduction band are found to be at Γ point in the Brillouin zone (BZ), indicating that the fundamental band gap is direct (Γ-Γ). The computed band gap value for the CdSiAs$_2$ is given in Table 2 along with experimental data for comparison.

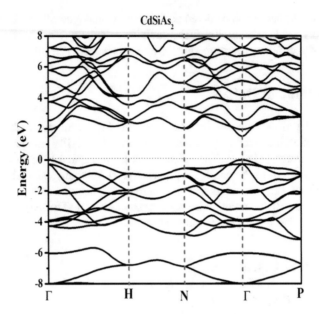

Fig. 1. Band structure for CdSiAs$_2$

Table 2. Band gap values of CdSiAs$_2$

Compound	Exchange	Band gap (eV)
CdSiAs$_2$	Tb.mbJ	1.54
	exp	1.55 [15–17]

The total (TDOS) and partial (PDOS) densities of states for CdSiAs$_2$ chalcopyrite compound are plotted in Fig. 2. In CdSiAs$_2$ chalcopyrite material, the Cd-s electrons contribute in the energy range from −5.0 to −3 eV while the upper valence band mainly constructed from As p states. In addition, the lowest conduction band (CB) is primarily derived from the s and p states of Si atoms.

3.3   Thermoelectric Properties

The temperature dependent thermoelectric properties of CdSiAs$_2$ is obtained by solving the Boltzmann transport equation as implemented in the BOLTZTRAP [21] code. The calculated thermoelectric properties are the Seebeck coefficient, the electrical conductivity divided by the scattering time (i.e. (σ/τ)), the thermal conductivity divided by the scattering time (i.e. (κ/τ)), and the merit's figure as a function of temperature in the temperature range from 100 K to 1200 K.

We have plotted the Temperature dependence of Seebeck coefficient and is presented in Fig. 3. For CdSiAs$_2$ compound. It initially increases and reaches its maximum of about 320 μV K^{-1} at 650 K, then its decreases down when the temperature increase, the seebeck coefficient, S for our compound is positive, indicating that the

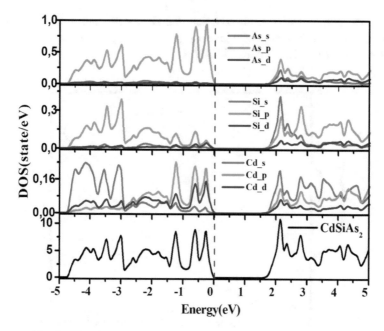

Fig. 2. The electronic total and partial density of states for CdSiAs₂

CdSiAs₂ is p-type semiconductors, where holes are predominant charge carriers. It is quite evident that our material have maximum thermopower values at low temperature due to the low carrier concentration contribution, as often the for low band gap semiconductor and due to the heavy conduction band.

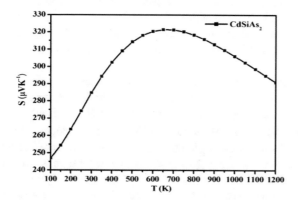

Fig. 3. The variation of the Seebeck coefficient as a function of Temperature for the CdSiAs₂ compound

Figure 4 presents the selected results of electrical conductivity measurements scaled by relaxation time, σ/τ for CdSiAs₂ compound as a function of temperature from 100 to 1200 K, One can observe that, the electrical conductivity spectra of the examined compound increases with increasing, and reaches its maximum values noted to be 7×10^{18} at the temperature of 1200 K.

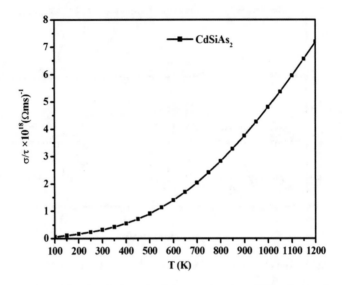

Fig. 4. The variation of the electrical conductivity as a function of Temperature for the CdSiAs$_2$ compound

The electronic thermal conductivity per relaxation time of the CdSiAs$_2$ compound each as a function of temperature is shown in Fig. 5. At the temperature of 100 K, the CdSiAs$_2$ compound have a thermal conductivity value of 0 W/mKs, after this value, One can observe that, over the entire temperature range, κ/τ increases with increasing temperature and reaches its maximum value at 1200 K noted to be 8.5×10^{14} W m^{-1} K^{-1} s^{-1}. This shows that the CdSiAs$_2$ compound is a better thermal conductor at all temperatures.

The dimensionless thermoelectric figure of merit as a function of temperature is plotted in Fig. 6, this parameter is calculated based on the measured values of the seebeck coefficient, electrical and electronic thermal conductivities from the equation $ZT = S^2\sigma T/\kappa$, of the CdSiAs$_2$ ternary chalcopyrite compound. Our ZT values will first increase with the increasing temperature and reach its maximum value of about 0.857 at the temperature of 200 K. After this temperature value its starts to decrease with increasing temperature. From our results of the merit's figure, it is very clear that the CdSiAs$_2$ is a good thermoelectric compound for low-temperature applications.

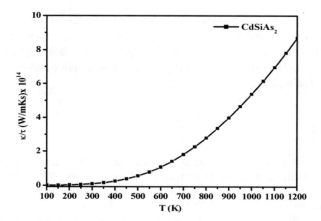

Fig. 5. The variation of the electronic thermal conductivity as a function of Temperature for the CdSiAs₂ compound

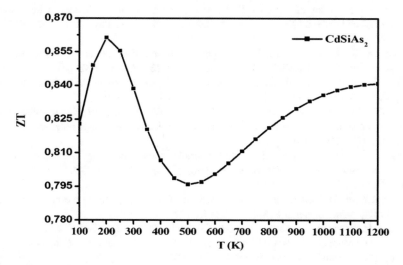

Fig. 6. The variation of the merit's figure as a function of Temperature for the CdSiAs₂ compound

4 Conclusion

In this work, we study the structural and electronic properties of pure CdSiAs₂, by means of the density functional theory (DFT) and the linearized augmented plane wave method (LAPW). The calculated structural parameters obtained are in good agreement with the experimental reports and other theoretical results. The calculation results of the electronic properties show that CdSiAs₂ has a semi conductor character, which is of interest for thermoelectric properties.

The electronic properties obtained from Wien 2K calculations and Boltzmann transport theory implemented in BoltzTrap code were employed to calculate the thermoelectric properties of $CdSiAs_2$ as function of temperature raging from 100 to 1200 K. Our results reveal that $CdSiAs_2$ has a largest figure of merit (ZTmax) with value of ≈ 0.9 at temperature at low temperature.

References

1. Yu, L., Zunger, A.: Identification of potential photovoltaic absorbers based on first-principles spectroscopic screening of materials. Phys. Rev. Lett. **108**, 068701 (2012)
2. Xiao, H., et al.: Accurate band gaps for semiconductors from density functional theory. J. Phys. Chem. Lett. **2**, 212–217 (2011)
3. Shay, J.L., Wernick, J.H.: Ternary Chalcopyrite Semiconductors Growth, Electronic Properties and Applications. Pergamon Press, Oxford (1975)
4. Kumar, V., Tripathy, S.K.: First-principle calculations of the electronic, optical and elastic properties of ZnSiP2 semiconductor. J. Alloys Compd. **582**, 101–107 (2014)
5. Turowski, M., Margaritondo, G., Kelly, M.K., Tomlinson, R.D.: Photoemission studies of CuInSe 2 and CuGaSe 2 and of their interfaces with Si and Ge. Phys. Rev. B **31**, 1022–1027 (1985)
6. Medvedkin, G.A., Voevodin, V.G.: Magnetic and optical phenomena in nonlinear optical crystals ZnGeP2 and CdGeP2. J. Opt. Soc. Am. B **22**, 1884–1898 (2005)
7. Das, S.: Pump tuned wide tunable noncritically phase-matched ZnGeP2 narrow line width optical parametric oscillator. Infrared Phys. Technol. **69**, 13 (2015)
8. Xu, Y., Ao, Z.M., Zou, D.F., Nie, G.Z., Sheng, W., Yuan, D.W.: Strain effects on the electronic structure of ZnSnP2 via modified Becke–Johnson exchange potential. Phys. Lett. A **379**, 427–430 (2015)
9. Yao, B., et al.: High-power Cr 2+: ZnS saturable absorber passively Q-switched Ho: YAG ceramic laser and its application to pumping of a mid-IR OPO. Opt. Lett. **40**, 348–351 (2015)
10. Zhang, Z., et al.: Femtosecond-laser pumped CdSiP 2 optical parametric oscillator producing 100 MHz pulses centered at 6.2 μm. Opt Lett **38**, 5110 (2013)
11. Blaha, P., Schwarz, K., Madsen, G.K.H., Kvasnicka, D., Luitz, J.: WIEN2k. In: Schwarz, K. (ed.) An Augmented Plane Wave þ Local Orbitals Program for Calculating Crystal Properties. Techn Universitat Wien, Austria (2001)
12. Blaha, P., Schwarz, K., Sorantin, P.I., Tricky, S.B.: Thermoelectric properties of binary LnN (Ln = La and Lu): first principles study. Comput. Phys. Commun. **59**, 399 (1990)
13. Becke, A.D., Johnson, E.R.: A simple effective potential for exchange. J. Chem. Phys. **124**, 221101 (2006)
14. Tran, F., Blaha, P.: Accurate band gaps of semiconductors and insulators with a semilocal exchange-correlation potential. Phys. Rev. Lett. **102**, 226401 (2009)
15. Zunger, A.: Order-disorder transformation in ternary tetrahedral semiconductors. Appl. Phys. Lett. **50**(3), 164 (1987)
16. Jaffe, J.E., Zunget, A.: Second-harmonic generation and birefringence of some ternary pnictide semiconductors. Phys. Rev. B **29**(4), 1882 (1984)
17. Rashkeev, S.N., Limpijumnong, S., Lambrecht, W.R.L.: Second-harmonic generation and birefringence of some ternary pnictide semiconductors. Phys. Rev. B **59**(4), 2737 (1999)
18. Shi et al., L.: J. Alloys Compd. **611**, 210–218 (2014)

19. Shaposhnikov, V.L., Krivosheeva, A.V., Borisenko, V.E.: Ab initio modeling of the structural, electronic, and optical properties of A II B IV C 2 V semiconductors. Phys. Rev. B **85**, 205201 (2012)
20. Vaipolin, A.A.: Fiz. Tverd. Tela 15 (1973) 1430 [Sov. Phys. Solid State 15 (1973) 965]
21. Reshak, A.H., Khenata, R., Kityk, I.V., Plucinski, K.J., Auluck, S.: X-ray photoelectron spectrum and electronic properties of a noncentrosymmetric chalcopyrite compound HgGa2S4: LDA, GGA, and EV-GGA. J. Phys. Chem. B **113**, 5803 (2009)
22. Tran, F., Blaha, P.: Accurate band gaps of semiconductors and insulators with a semilocal exchange-correlation potential. Phys. Rev. Lett. **102**, 226401 (2009)
23. Jaffe, J.E., Zunger, A.: Theory of the band gap anomaly in ABC2 chalcopyrite semiconductors. Phys. Rev. B **29**, 1882 (1984)

A Multi Stage Study of Thermoelectric Device Based on Semi-conductor Materials

Younes Chiba[1(✉)], Yacine Marif[2], Abdelali Boukaoud[3],
Kehileche Brahim[1], Abdelhalim Tlemcani[1], and Noureddine Henini[1]

[1] Faculty of Technology, University of Medea, Medea, Algeria
Chiba.younes@univ-medea.de, kehileche@yahoo.fr
[2] Faculté des Mathématiques et Sciences de la Matière, LENREZA,
Université de Ouargla, Ouargla, Algeria
yacine.marif@yahoo.fr
[3] Faculty of Sciences, University of Medea, Medea, Algeria

Abstract. The present work dedicated to the study and analysis of a multi stage study of thermoelectric cooler based on hydrogen liquefier. For this purpose, the Seebeck and Peltier effect are used for simulate thermoelectric cooler. The energy equation has been used under the following conditions for prediction the efficiency of hydrogen liquefier. The properties of semi-conductor materials corresponding of cryogenics process were defined. A multi stage process has been proposed for decrease cold temperature of cold exchanger. The obtained results, including power and coefficient of performance are presented and discussed.

Keywords: Effect of seebeck · Peltier effect · Thermoelectric
Refrigerator · Cryogenics

1 Introduction

The clean energy has been considered an alternative technology for several industrial processes; electricity production, water supply, heating and cooling production (Chiba 2017; Chiba et al. 2014). In this research work, we are interested to study the cooling production by using a thermoelectric process.

Thermoelectric devices are alternative technology which directly converts electrical energy to cooling energy by using Peltier effect and Seebeck effect by applying a heat charge for electrical energy production (Goupil et al. 2011; Semenyuk 2014). Thermoelectric refrigeration was considered a clean cooling energy compared with other process recently studied (Pooja 2016).Thermoelectric cooler based on semi-conductor materials procured by electrical current, this phenomenon produces different heat exchanger; hot blow and cold blow between two junctions of different semi-conductor materials (Li et al. 2011). Thermoelectric refrigerators devices based on the Peltier effect, discovered in 1834 (Pooja 2016). In this subject, the researchers have been focused for cooling production (Meng et al. 2011) and electrical energy production (Badillo-Ruiz et al. 2018). The thermoelectric interest as follows area such as numerical modeling and design of system and experimental and characterization of material used

© Springer Nature Switzerland AG 2019
M. Hatti (Ed.): ICAIRES 2018, LNNS 62, pp. 560–565, 2019.
https://doi.org/10.1007/978-3-030-04789-4_60

for cooling, electrical applications or other application (Waldrop and Morelli 2014). This paper deals to study a multi stage of thermoelectric cooler based on hydrogen liquefier. The theory and numerical modeling of hydrogen liquefier by using the energy equation are analyzed and studied. The operation parameters of thermo electrical cooler were defined. The main results including coefficient of performance, efficiency has been discussed and analyzed.

2 Thermoelectric Materials

A multi semi-conductor material is used in thermoelectric devices, according their domains of applications. The performance of all materials in thermoelectric was characterized by a factor of merit ZT. Figure 1 illustrates in detail factor of merit ZT for some various materials (Pooja 2016).

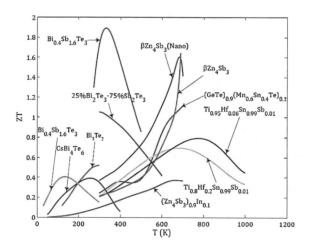

Fig. 1. Factor of merit ZT for different thermoelectric material (Pooja 2016)

3 Description of Thermoelectric Devices

The Peltier effect phenomenon was used for constructing the thermo cooler device under following assumption

- A multi stage of the thermoelectric module has been used.
- The heat transfer between two stages has been considered uniform.
- All losses are neglected.
- The heat conduction phenomenon was considered unidirectional.
- Hot blow rejection and cold blow absorption exchanger were characterized by hot exchanger Q_H and cold exchanger Q_C respectively.

Figure 2 shows in detail the description of a multi stage of thermo cooler device.

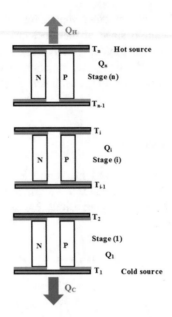

Fig. 2. Schematic diagram of a multi stage of hydrogen liquefier

4   Governing Equations

The relationship using to calculate heat cold blow, heat hot blow, voltage, power and coefficient of performance can be defined as;
   For single stage;

$$\Delta T_1 = T_H - T_C \tag{1}$$

$$Q_C = \alpha_1 T_c I - 0.5 R_1 I^2 - K_1 \Delta T_1 \tag{2}$$

$$V = \alpha_1 \Delta T_1 + R_1 I \tag{3}$$

$$P = VI \tag{4}$$

$$Q_H = P + Q_C \tag{5}$$

$$COP = Q_C / P \tag{6}$$

   For two-stage;

$$\Delta T_2 = T_2 - T_C \tag{7}$$

$$Q_C = \alpha_2 T_c I - 0.5 R_2 I^2 - K_2 \Delta T_2 \tag{8}$$

$$V = \alpha_2 \Delta T_2 + R_2 I + \alpha_1 (T_H - T_2) + R_1 I \tag{9}$$

$$T_2 = (0.5(R_2 + R_1)I^2 + K_1 T_H + K_2 T_c)/(\alpha_1 - \alpha_2)I + K_1 + K_2 \tag{10}$$

The maximum efficiency of thermoelectric cooler can be defined respectively under expression:

$$COP_C = T_C/(T_H - T_C) \tag{11}$$

$$COP_{\max} = COP_C \left(\sqrt{1 + ZT} - \frac{T_C}{T_H} \right) / \left(\sqrt{1 + ZT} + 1 \right) \tag{12}$$

Where,

T_H, T_C: is temperature of hot and cold source respectively, Q_H, Q_C: is heat rejection and cold absorption respectively, V: is voltage, I: is electrical current, α: is Seebeck coefficient, P: is electrical power, Z: is factor of merit, COP: is coefficient of performance and R: is resistance.

5 Results and Discussion

The numerical simulation of thermoelectric cooler has been carried out under parameters presented in Fig. 3. The Fig. 3 illustrates the evolution of Seebeck coefficient and thermal conductivity as a function of temperature range.

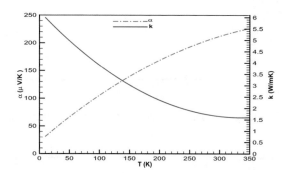

Fig. 3. Change of Seebeck coefficient and thermal conductivity as a function of temperature range.

Figure 4 show Change of maximum *COP* as function of cold temperature of hydrogen liquefier for different values of merit factor under hot temperature 310 K of exchanger hot blow. The big difference has been noted between Carnot coefficient and real coefficient.

Figure 5 illustrate the change of cooling power as a function of electrical current, the cooling power for hydrogen liquefier increase with different values of electrical

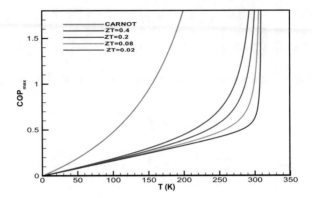

Fig. 4. Change of maximum COP as function of cold temperature of hydrogen liquefier.

current until it reaches its maximum value and then begins to decrease Also, the Fig. 6 shows the evolution of efficiency as a function of temperature of hydrogen liquefier. The efficiency for hydrogen liquefier increase with different of cold temperature until it reaches its maximum value and then begins to decrease.

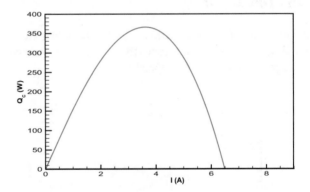

Fig. 5. Change of cooling power as a function of electrical current

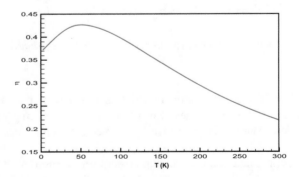

Fig. 6. Evolution of efficiency as a function of temperature of hydrogen liquefier

6   Conclusion

The main objective of this study is to predict the performance hydrogen liquefier by using thermoelectric systems and the purpose the methodology of liquefaction. Based on the results, the following conclusions are made;

- A multi-stage system is the best solution for hydrogen liquefaction.
- The cost of thermo cooler does not expensive compared with traditional system
- The thermo cooler system works for 12 volts. This system is ecologic and friend to our environment.
- The produced cooling power of a multi-stage system will still be higher.

References

Chiba, Y., Smaili, A., Mahmed, C., Balli, M., Sari, O.: Thermal investigations of an experimental active magnetic regenerative refrigerator operating near room temperature. Int. J. Refrig. **37**, 36–42 (2014). https://doi.org/10.1016/j.ijrefrig.2013.09.038

Chiba, Y.: Experimental and numerical study on the behavior of a multilayer for active magnetic refrigerator based on La-Fe-Co-Si. IEEE Trans. Magn. **53**, 1–4 (2017). https://doi.org/10.1109/TMAG.2017.2703100

Goupil, C., Seifert, W., Zabrocki, K., Muller, E., Snyder, G.J.: Thermodynamics of thermoelectric phenomena and applications. Entropy **13**, 1481–1517 (2011). https://doi.org/10.3390/e13081481

Semenyuk, V.: A comparison of performance characteristics of multistage thermoelectric coolers based on different ceramic substrates. J. Electron. Mater. **43**, 1539–1547 (2014). https://doi.org/10.1007/s11664-013-2777-7

Pooja, I.M.: Design, modeling and simulation of a thermoelectric cooling system (tec). Master's thesis, Western Michigan University (2016)

Li, L., Chen, Z., Zhou, M., Huang, R.: Developments in semiconductor thermoelectric materials. Front. Energy **5**, 125–136 (2011). https://doi.org/10.1007/s11708-011-0150-1

Meng, F., Chen, L., Sun, F.: Performance prediction and irreversibility analysis of a thermoelectric refrigerator with finned heat exchanger. Acta Phys. Pol. A **120**, 397–406 (2011)

Badillo-Ruiz, C.A., Olivares-Robles, M.A., Ruiz-Ortega, P.E.: Performance of segmented thermoelectric cooler micro-elements with different geometric shapes and temperature-dependent properties. Entropy **118**, 1–17 (2018). https://doi.org/10.3390/e20020118

Waldrop, S., Morellim, D.: Low-temperature thermoelectric properties of PtSb2_xTex for cryogenic peltier cooling applications. J. Electron. Mater. (2014). https://doi.org/10.1007/s11664-014-3480-z

Author Index

© Springer Nature Switzerland AG 2019
M. Hatti (Ed.): ICAIRES 2018, LNNS 62, pp. 567–569, 2019.
https://doi.org/10.1007/978-3-030-04789-4

Printed in the United States
By Bookmasters